Springer Tracts in Advanced Robotics

Volume 62

Editors: Bruno Siciliano · Oussama Khatib · Frans Groen

Andrew Howard, Karl Iagnemma,
and Alonzo Kelly (Eds.)

Field and Service Robotics

Results of the 7th International Conference

 Springer

Professor Bruno Siciliano, Dipartimento di Informatica e Sistemistica, Università di Napoli Federico II, Via Claudio 21, 80125 Napoli, Italy, E-mail: siciliano@unina.it

Professor Oussama Khatib, Artificial Intelligence Laboratory, Department of Computer Science, Stanford University, Stanford, CA 94305-9010, USA, E-mail: khatib@cs.stanford.edu

Professor Frans Groen, Department of Computer Science, Universiteit van Amsterdam, Kruislaan 403, 1098 SJ Amsterdam, The Netherlands, E-mail: groen@science.uva.nl

Editors

Andrew Howard
CalTech Jet Propulsion Laboratory
Mail Stop 198-235
4800 Oak Grove Drive
Pasadena, CA 91109
USA
E-mail: andrew.b.howard@jpl.nasa.gov

Alonzo Kelly
National Robotics Engineering Center
Robotics Institute
Carnegie Mellon University
10 Fortieth Street
Pittsburgh, PA 15201
USA
E-mail: alonzo@cmu.edu

Karl Iagnemma
MIT
77 Massachusetts Ave
Room 3-435a
Cambridge, MA 02139
USA
E-mail: kdi@mit.edu

ISBN 978-3-642-13407-4   e-ISBN 978-3-642-13408-1

DOI 10.1007/978-3-642-13408-1

Springer Tracts in Advanced Robotics   ISSN 1610-7438

Library of Congress Control Number: 2010928124

Typeset & Cover Design: Scientific Publishing Services Pvt. Ltd., Chennai, India.

Printed in acid-free paper

5 4 3 2 1 0

springer.com

STAR (Springer Tracts in Advanced Robotics) has been promoted under the auspices of EURON (European Robotics Research Network)

Foreword

Robotics is undergoing a major transformation in scope and dimension. From a largely dominant industrial focus, robotics is rapidly expanding into human environments and vigorously engaged in its new challenges. Interacting with, assisting, serving, and exploring with humans, the emerging robots will increasingly touch people and their lives.

Beyond its impact on physical robots, the body of knowledge robotics has produced is revealing a much wider range of applications reaching across diverse research areas and scientific disciplines, such as: biomechanics, haptics, neurosciences, virtual simulation, animation, surgery, and sensor networks among others. In return, the challenges of the new emerging areas are proving an abundant source of stimulation and insights for the field of robotics. It is indeed at the intersection of disciplines that the most striking advances happen.

The *Springer Tracts in Advanced Robotics (STAR)* is devoted to bringing to the research community the latest advances in the robotics field on the basis of their significance and quality. Through a wide and timely dissemination of critical research developments in robotics, our objective with this series is to promote more exchanges and collaborations among the researchers in the community and contribute to further advancements in this rapidly growing field.

The Seventh edition of *Field and Service Robotics* edited by Andrew Howard, Karl Iagnemma and Alonzo Kelly offers in its eleven-chapter volume a collection of a broad range of topics spanning: design, perception and control; tracking and sensing; localization and mapping; multi-robot cooperation and human-robot interaction; mining, maritime and planetary robotics. The contents of the forty-five contributions represent a cross-section of the current state of robotics research from one particular aspect: field and service applications, and how they reflect on the theoretical basis of subsequent developments. Pursuing technologies aimed at realizing robots operating in complex and dynamic environments, as well as robots working closely with humans, is the big challenge running throughout this focused collection.

Rich by topics and authoritative contributors, FSR culminates with this unique reference on the current developments and new directions in field and service robotics. A fine addition to the series!

Naples, Italy   Bruno Siciliano
March 2010   STAR Editor

Preface

Field and Service Robotics (FSR) is one of the (presently) five major conferences founded by the International Federation of Robotics Research (IFRR). As such, FSR is the leading single track conference dedicated to research related to development of robots that do real work, whether that work is hard labor or the performance of useful services. Field robots are often purpose-built machines that are highly adapted to their jobs, and hence their surroundings; they exhibit high mobility and they typically interact forcefully with their environments. By contrast, service robots are more adapted to assisting humans and they interact with their surroundings with a somewhat lighter touch.

The FSR conference is held every two years. Dating from 1997 it has followed a regular three continent rotation. It has been held in Canberra, Australia (1997), Pittsburgh, USA (1999), Helsinki, Finland (2001), Mount Fuji, Japan (2003), Port Douglas, Australia (2005), Chamonix, France (2007) and most recently in Cambridge, USA (2009).

This year we had 80 submissions of which 45 were selected for oral presentations.

The conference chairs were:

Alonzo Kelly (CMU), Karl Iagnemma (MIT) and Andrew Howard (Caltech-JPL)

The conference is overseen by members of the international organizing committee, who also serve on the program committee:

Hajime Asama	U Tokyo, Japan
Raja Chatila	LAAS/CNRS, France
Henrik Christensen	Georgia Tech, USA
Peter Corke	CSIRO, Australia
Aarne Halme	Helsinki U of Tech, Finland
John Hollerbach	U of Utah, USA
Andrew Howard	JPL / Cal Tech, USA
Karl Iagnemma	MIT, USA
Alonzo Kelly	CMU, USA
John Leonard	MIT, USA
Christian Laugier	INRIA, France
Eduardo Nebot	U Sydney, Australia
Erwin Prassler	U Applied Science Bonn, Germany
Jonathan Roberts	CSIRO, Australia
Daniela Rus	MIT, USA
Sanjiv Singh	CMU, USA
Roland Siegwart	ETH Zurich, Switzerland

Salah Sukkarieh   U Sydney, Australia
Chuck Thorpe   CMU, Qatar
Sebastian Thrun   Stanford, USA
David Wettergreen   CMU, USA
Kazuya Yoshida   Tohoku U, Japan
Alex Zelinsky   CSIRO Australia

The following researchers also served on the program committee for FSR09.

Timothy Barfoot   University of Toronto
Martin Beuller   iRobot Corp.
Wolfram Burgard   Albert-Ludwigs-Universität Freiburg
Toshio Fukuda   Nagoya University
Satoshi Kagami   AIST Digtal Human Research Center
Simon Lacroix   LAAS-CNRS
David P Miller   University of Oklahoma
Paul Newman   Oxford University
Liam Pedersen   NASA Ames
Miguel Angel Salichs   Carlos III University of Madrid.
Gaurav S Sukhatme   University of Southern California
Satoshi Tadokoro   Tohoku University
Takashi Tsubouchi   University of Tsukuba
Arto Visala   Helsinki University of Technology
Uwe Zimmer   Australian National University

The conference was sponsored by the US Army Research Office, iRobot Corporation, the US Army TARDEC, and US Army Corps of Engineers ERDC.

Mihail Pivtoraiko, Colin Green, and Chris Ward gave generously of their time to help arrange many aspects of the social and technical program and publicity.

Contents

Part III: Tracking and Servoing

Part IV: Localization

Part X: Maritime Robotics

Part XI: Planetary Robotics

Part I
Mechanism Design

Terrain Modeling and Following Using a Compliant Manipulator for Humanitarian Demining Applications

Marc Freese, Surya P.N. Singh, William Singhose,
Edwardo F. Fukushima, and Shigeo Hirose

Abstract. Operations with flexible, compliant manipulators over large workspaces relative to the manipulator are complicated by noise, vibration, and measurement bias. These difficulties are compounded in unstructured environments, such as those encountered in humanitarian demining. By taking advantage of the static structure of the terrain and the manipulator's fundamental mechanical characteristics, a series of adaptive corrections and filters refine noisy topographical measurements. These filters along with a shaped actuation scheme can generate smooth and well-controlled trajectories that allow for terrain surface following. Experimental testing was performed on a field robot with a compliant, 3 *m* long hybrid manipulator and a stereo vision system. The proposed method provides a vertical tracking precision of ±5 *mm* on a variety of ground clearings, with tip scanning speeds of up to 0.5 *m/s*. As such, it can agilely move the attached sensor(s) through precise scanning trajectories that are very close to the ground. This method improves overall detection and generation of precise maps of suspected mine locations.

1 Introduction

While robust manipulation in difficult field conditions is still in its infancy, environmental modeling using computer vision has progressed with several applications, including autonomous Martian mapping [10]. With regards to manipulation,

Marc Freese, Edwardo F. Fukushima, and Shigeo Hirose
Department of Mechanical and Aerospace Eng., Tokyo Institute of Technology
e-mail:{marc@sms,fukusima@mes,hirose@mes}.titech.ac.jp

Surya P.N. Singh
Australian Centre for Field Robotics, University of Sydney
e-mail: spns@usyd.edu.au

William Singhose
School of Mechanical Eng., Georgia Institute of Technology
e-mail: singhose@gatech.edu

A. Howard et al. (Eds.): Field and Service Robotics 7, STAR 62, pp. 3–12.

Fig. 1 Photograph of Gryphon in a test humanitarian demining field.

variation and noise are routinely minimized by stiffening the structure or constraining the operation [1, 2]. In humanitarian demining, the increased sensitivity needed by the metal detector precludes stiffening through the addition of proximal metal content and requires minimal mean and variance in the air gap to the ground. In demining, a myriad of sensing technologies have been proposed [7]. However, relatively little attention has been directed towards the field manipulation requirements for automating the dangerous and tedious ground scanning task.

An autonomous mobile robot named Gryphon [5] has been developed to assist the mine detection process. As shown in Figure 1, the robot is based on an all terrain vehicle (ATV) [4] to which a custom hybrid robotic manipulator [6] is added. This lightweight and counter-balanced 3-DOF arm is made from glass-fiber reinforced plastic (GFRP) which is compatible with sensitive metal detection (MD) operation that requires minimal metal near the sensor [3]. As shown in Figure 2, this design has a tip flexure of up to 5 *cm* (for a 2 *m* link) due to inertial forces only. While the structure could have been stiffened, the compliance also provides some safety in the event of a collision. Secondary motion from the ATV suspension is reduced, but not eliminated, through counterbalancing the manipulator. Hence, there is uncertainty in sensor location with respect to the ground.

As illustrated by the control architecture in Figure 3, Gryphon operates by driving close to a region of interest, then while the ATV is stationary, generates a **stereo map**. As detailed in Section 2, these are iteratively refined to construct a geometric **terrain model**. By iteratively operating using a local model, absolute rectification is not required because later processing stages account for aberrations through command shaping [5, 11]. This approach adds robustness without the need to identify the origin of imprecision. However, its use of linear models means its highest accuracy is near regions used to perform the system-level calibration where errors are small, hence defining the *zone of effectiveness*. From this terrain model a desired **path** for the manipulator endpoint is generated and corrected for the height errors and the travel of a detector body (as opposed to endeffector frame), as detailed in Section 3. The final path is close to the ground

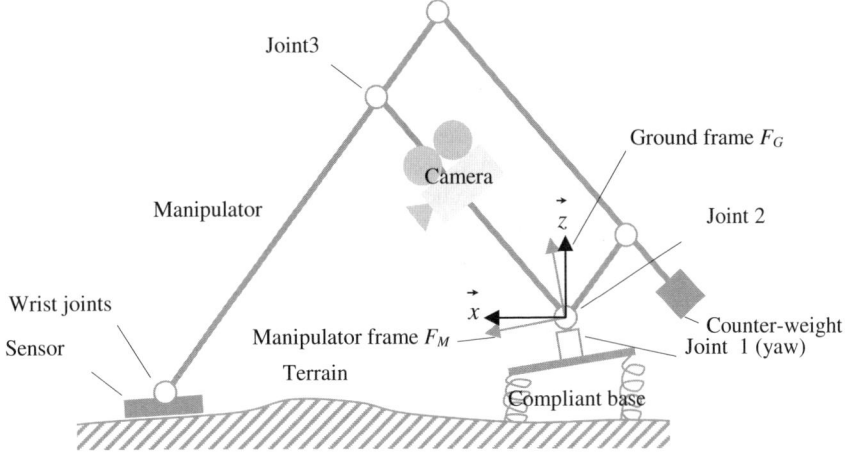

Fig. 2 Schematic of Gryphon and its principal frames of reference.

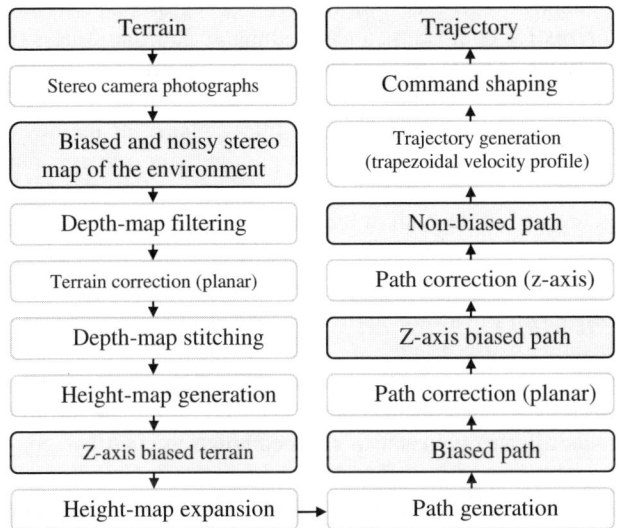

Fig. 3 Terrain scanning motion generation architecture. (Darker boxes indicate process outputs). while maintaining the best possible orientations for the detector. Experimental tests of the manipulator and the control architecture are presented in Section 4.

2 Terrain Modeling

In order to plan operations that map the location of suspected mines found by the sensor, it is first necessary to form a geometric model of the terrain.

The terrain modeling begins by sensing the environment using stereo vision. A *BumbleBee* (model BB-HICOL-60, Point Grey Research) stereo vision camera (location shown in Figure 2), acquires several depth maps from different manipulator configurations to cover the region of interest. Compared to laser mapping, this generates maps quickly, but more noise, especially in regions lacking texture. Map accuracy depends on object range and camera aberrations. Acquired raw depth maps (or disparity maps) are checked for consistency by subdividing them into *coherent* patches. Patches that fail to comply with certain criteria (e.g. do not represent possible terrain locations) are then discarded.

2.1 Conditional Planar Filtering

Due to limitations on camera resolution and calibration, a raw model generated from the camera data is significantly degraded by noise. Median filtering techniques are primarily effective against the shot noise, but do not remove high-frequency components. Simply smoothing (i.e., spatial low-pass filtering) is insufficient as this results in a degradation of features, especially at obstacles boundaries, which could lead to a collision between the end-effector and the terrain.

Thus, an adaptive filter based on the average region planarity is used to adjust filter kernel sizes for both Gaussian smoothing and median filters (i.e., the *conditional planar filter [CPF]*) [6]. The planarity of the selected region is determined by calculating the mean deviation between each point in the region and the region's corresponding (least-squares) best-fitting plane. If the deviation is small, the region is assumed to be planar, hence giving a conservative, but rapidly computed terrain classification. Based on this, the strength of the Gaussian and median kernels are varied depending on the deviation magnitude. Applying this to the perceived data yields a less noisy, but still potentially offset, or biased, map.

2.2 Height Map Generation

To simplify and speed-up terrain data processing, the depth maps are transformed to a common height map function. This map represents the terrain as a series of height (or z-coordinate) values at point locations specified by a uniform mesh in the ground plane of the ground frame. This is done, for instance, via Delauney triangulation methods with increased precision obtained through spatial weighted averages of the sensed data. The obtained height map offers the advantage of two dimensionally indexed queries on the terrain model and facilitates, by linear interpolation, a mechanism to fill holes and patches that may have arisen due to occlusions or lack of texture.

2.3 Height Map Expansion

At this stage of the process, the height map is a good approximation of the underlying topographical information; however, it is often desired to perform the scanning at a constant distance from the ground (i.e., a scanning gap). For that purpose

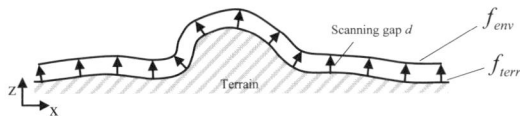

Fig. 4 Terrain surface expansion. f_{env} is an expansion of the terrain map given by f_{terr}.

we expand the height map via an envelope expansion. That is, we solve for a new surface, f_{env}, whose distance from f_{terr} is given by the scanning gap. This is illustrated in Figure 4. Note that the simple approach of shifting the height map vertically would be a bad approximation for highly curved or inclined terrain.

Considering the continuous case, the desired surface is easily obtained with the following parametric equations:

$$x_{env} = x_{terr} - \lambda \cdot \partial f_{terr}(x_{terr}, y_{terr})/\partial x \tag{1}$$

$$y_{env} = y_{terr} - \lambda \cdot \partial f_{terr}(x_{terr}, y_{terr})/\partial y \tag{2}$$

$$z_{env} = f_{terr}(x_{terr}, y_{terr}) + \lambda = f_{env}(x_{env}, y_{env}) \tag{3}$$

$$\lambda = d \cdot \left(\left(\partial f_{terr}(x_{terr}, y_{terr})/\partial x \right)^2 + \left(\partial f_{terr}(x_{terr}, y_{terr})/\partial y \right)^2 + 1 \right)^{\frac{-1}{2}} \tag{4}$$

3 Path Generation

The path generation for the manipulator takes as inputs the expanded terrain model from the previous section and the manipulator configuration, taking into account joint limitations of the wrist mechanism.

3.1 Scanning Scheme

Two different linear incrementing scanning schemes are available: in joint space (giving a circular profile since the base joint is revolute) or in workspace (giving a rectangular profile). The trajectories are smoothly combined at their extremities to reduce unnecessary slowdowns during direction changes. Performing this in the joint-space of the robot simplifies joint coordination by reducing simultaneous motions and velocity variation, and thus is dynamically more stable. This reduces power consumption and arm vibration and improves detection performance by reducing sensor location uncertainty.

3.2 Terrain Sampling

Either of the above described scanning schemes produce a planar path, where only the x-y coordinate components are defined. The z coordinate is obtained by sampling the corresponding position on the expanded terrain. Orientation of the detector at this position is calculated using the normal vector to the expanded terrain using the following relationships:

$$z_{path} = f_{env}(x_{path}, y_{path}) \tag{5}$$

$$\vec{n}_{path} = \left[-\gamma \cdot \partial f_{env}(x_{path}, y_{path}) / \partial x \quad -\gamma \cdot \partial f_{env}(x_{path}, y_{path}) / \partial y \quad \gamma \right]^{T} \tag{6}$$

$$\gamma = \left(\left(\partial f_{env}(x_{path}, y_{path}) / \partial x \right)^{2} + \left(\partial f_{env}(x_{path}, y_{path}) / \partial y \right)^{2} + 1 \right)^{-\frac{1}{2}} \tag{7}$$

3.3 Advanced Terrain Following (ATF)

The planned path has focused on guarantying that the center of the sensor follows the expanded terrain. With the consequence that the rest of the sensor body is still free to enter the scanning gap or even to collide with the terrain at positions where the curvature is concave, thus, it is important to consider the sensor as an extended body attached to the manipulator end point. This is performed by modeling the sensor via a series of control points along its lower surface; where, the number and position of points has been chosen to best approximate the contour and surface.

For each sensor position on the trajectory, all control points are tested for possible collisions with the expanded terrain. For each point under the grid, a small correction rotation (e.g., ½°) of the sensor is performed in the order of vertical deviation as illustrated in Figure 5a. This is iteratively repeated until an equilibrium orientation is reached (cf. Figure 5d). The sensor orientation is then corrected to respect configuration-space constraints by limiting wrist joint values to within allowable ranges and the sensor's z-coordinate is modified so as to have no control point in the scanning gap (cf. Figure 5e). Such an approach can lead to large gaps between the detector center and the terrain at concave positions, but it guarantees a constant minimum distance between the sensor as a whole and the terrain.

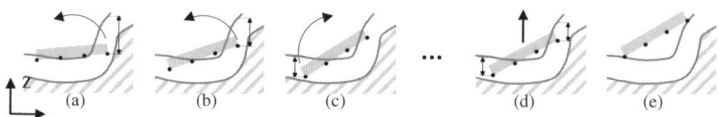

Fig. 5 Path correction scheme with the dots indicating control points.

A reactionary approach, by comparison, would simply lift the sensor without changing its orientation and thus is not sufficient as it would lead to non-ideal configurations with large and changing air gaps and potentially excessive motion.

3.4 Partial Path Correction

While perception inaccuracies were corrected for the horizontal plane in the *Terrain Modeling* Section, the model only partially accounted for mechanical inaccuracies. The type of mechanical inaccuracies corrected for are typically those arising from mechanical compliances (in the links and base) and calibration errors. Within the limitations of the *zone of effectiveness*, the path's x and y components are modified appropriately. This second part of the system-level calibration consists of applying a radial offset and scaling:

$$\rho' = \rho \cdot scaling_\rho + offset_\rho \tag{8}$$

and an angular offset:

$$\varphi' = \varphi + offset_\varphi \tag{9}$$

The above correction parameters are obtained by measuring the discrepancy between the real and computer-model manipulator tip position. Orientation of the detector is kept unchanged. The path becomes:

$$\begin{bmatrix} x_{path} & y_{path} & z_{path} \end{bmatrix}^T = \begin{bmatrix} \rho' \cdot \cos\varphi' & \rho' \cdot \sin\varphi' & f_{env}(\rho \cdot \cos\varphi, \rho \cdot \sin\varphi) \end{bmatrix}^T \tag{10}$$

3.5 Final Path Correction

The last modification of the path, and the final step of the system-level calibration, is the correction along the z-axis. Each x-y position is assigned an individual vertical correction factor that is obtained by linear interpolation between values of an Overall Calibration Matrix (OCM):

$$z'_{path} = z_{path} + f_{ZCorrection}(x_{path}, y_{path}) \tag{11}$$

The OCM is obtained by mapping the terrain, generating a non-expanded height map, and then manually driving the sensor to touch the terrain at workspace corner-points. At each spot, the necessary correction factor is given by the deviation of the computer model of the sensor from the terrain contact. Spots are then used to generate a Delaunay triangulated surface whose height at a given position gives the amount of correction needed. The OCM is directly extracted from that surface at regular grid intervals along the x-y plane in the manipulator frame.

4 Experiments and Results

Several experiments were conducted with Gryphon to assess performance of the terrain following method described in the previous sections. Precise quantification was performed by replacing the detector payload (i.e., metal detectors and ground penetrating radar) with a laser rangefinder (SICK DME 2000) to track the scanning gap at the region of maximum detector sensitivity. The effects of command shaping, the implementation of which on Gryphon is detailed in [5], were analyzed by attaching a 3-axis accelerometer to the tip of the manipulator.

4.1 Filter Performance

Other operating conditions, where possible, were adjusted to match those of a typical payload (e.g., weight was added to the laser rangefinder). A sandy terrain

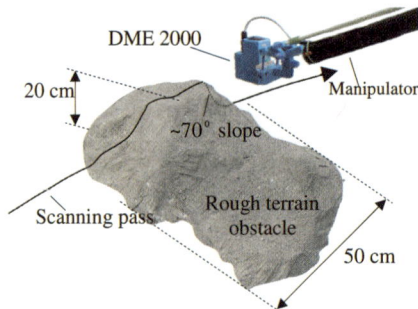

Fig. 6 Scanning pass used during experiments.

under natural light conditions (i.e., a fair weather day) was used with a circular scanning scheme with a scanning rate of 100 *mm/s* to evaluate the filter performance, and the effects of the advanced terrain following.

This set of tests used a 10 cm height map expansion and no ATF. The map filter was varied between three algorithms: (a) *minimal filtering* (a control case based on a raw depth map with median filtering), (b) *Gaussian Smoothing,* and (c) *Conditional Planar Filter (CPF).*

Filter performance was tested on two different terrain profiles: (i) relatively flat terrain and (ii) the same terrain with a rather challenging hump (or obstacle). In both cases, the scanning procedure comprised several passes.

The obstacle, shown in Figure 6, represents extreme slopes and contours for expected demining conditions. The pass selected for comparison was the most challenging and includes a 70° curvature.

The advantage of the CPF over Gaussian smoothing is visible in Figure 7. The Gaussian degrades terrain features, which results in a loss/gain of obstacle height.

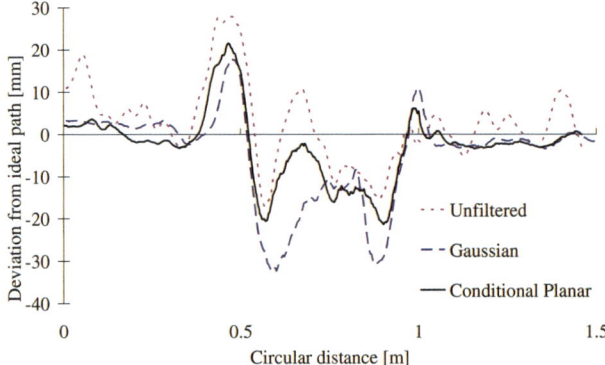

Fig. 7 Scanning deviations over a 200-mm feature by various filtering methods.

4.2 Effects of Command Shaping

The experiments also validate the use of robust control to address sensing and planning aberrations. In the case of Gryphon, an unshaped trajectory is first calculated by generating a trapezoidal velocity profile in joint space. The trajectory is then modified by a Zero Vibration and Derivative (ZVD) shaper (cf. ref. [11]). The sensor loading is large enough to that arm compliance is visible. In particular, measured residual vibration at the tip of the arm is approximately ±20 *mm*. This makes it difficult for a feedback control system relying on the joint encoders to adequately control the endpoint vibration. With command shaping, the residual vibration is approximately ±2 *mm*.

Smooth trajectories are of importance for best metal detector sensitivity, especially for straight-line rectangular motion as this requires coordinated motion of the arm joints. As shown in Figure 8, command shaping is significantly smoother.

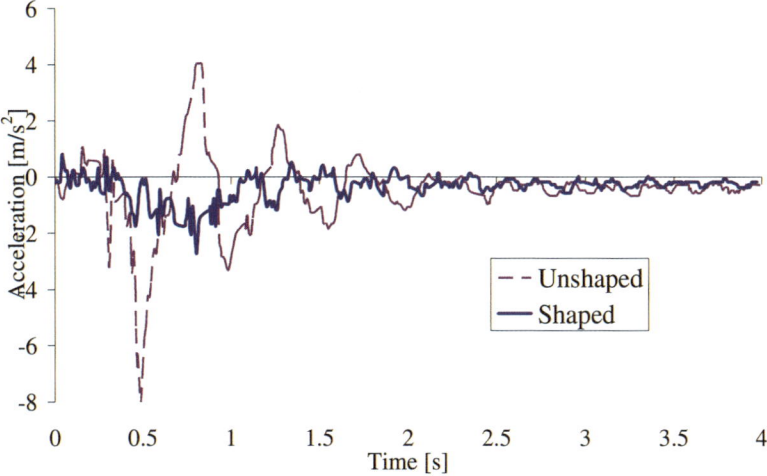

Fig. 8 Endpoint acceleration at the end-effector for a typical scanning pass for a straight line scanning pass of 2.4 *m* with a maximum speed and acceleration of 0.8 *m/s* and 7.5 *m/s²*.

4.3 In-Field Testing

Gryphon is an integrated, weather-proof system built to be robust against dust, humidity and rain, and resistant to extended temperature ranges. It has been field tested for 95 days on flat ground, bumpy terrain, and slopes. This includes operations on various test mine fields, including tests conducted under the supervision of the Japan Science and Technology Agency (JST) in Japan in early 2005 [8] and early 2006 in Croatia [9]. Two Gryphon robots (each with a different detector configuration) also took part in extensive trials in Cambodia (in cooperation with by the Cambodian Mine Action Center). Operated by Cambodian deminers, the Gryphon machines performed tests for more than 150 hours of semi-autonomous operation.

5 Conclusion

Field operations of long-reach manipulators are complicated by noise, measurement bias and vibration that lead to end-effector positional uncertainty. This drastically reduces the number of field applications that can be performed accurately and safely. The architecture described in this paper combines a novel filtering method, a decoupled system-level calibration procedure and a vibration reduction technique to yield an effective framework for obstacle identification, trajectory planning and generation. Experimental testing on Gryphon shows considerable deviation reduction when applying the framework. Combined with an advanced terrain following technique, it effectively avoids collision with the terrain for a successful scanning operation. The methods demonstrated improved automated mine detection performance and tracking on the Gryphon robot.

Acknowledgments. We warmly appreciate the cooperation with Cambodian deminers and Cambodian Mine Action Center staff. This work is supported by the Japan Science and Technology Agency and the Grant-in Aid for the 21st Century COE Program by the Ministry of Education, Culture, Sports, Science and Technology. The second author also acknowledges the support of the Rio Tinto Centre for Mine Automation and the ARC Centres of Excellence Programme funded by the Australia Research Council and New South Wales State Government.

References

1. Book, W.J.: Modeling, Design, and Control of Flexible Manipulator Arms: A Tutorial Review. In: Proc. of the 29th Conf. on Decision and Control, Honolulu (December 1990)
2. Craig, R.R., Bampton, M.: Coupling of Substructures for Dynamic Analysis. Journal of AIAA 6(7), 1313–1319 (1968)
3. CEIA Corporation. CEIA Metal Detector MIL-D1, Operator Manual
4. Debenest, P., Fukushima, E.F., Tojo, Y., Hirose, S.: A New Approach to Humanitarian Demining, Part 1: Mobile Platform for Operation on Unstructured Terrain. Journal of Autonomous Robots 18, 303–321 (2005)
5. Freese, M., et al.: Robotics Assisted Demining with Gryphon. Advanced Robotics 21(15), 1763–1786 (2007)
6. Freese, M., Singh, S., Fukushima, E., Hirose, S.: Bias-Tolerant Terrain Following Method for a Field Deployed Manipulator. In: Proc. Int. Conf. Robotics & Automation, pp. 175–180 (2006)
7. Ghaffari, M., Manthena, D., Ghaffari, A., Hall, E.L.: Mines and human casualties, a robotics approach toward mine clearing. In: SPIE Intelligent Robots and Computer Vision XXI: Algorithms, Techniques, and Active Vision, vol. 5608 (2004)
8. Ishikawa, J., Kiyota, M., Furuta, K.: Evaluation of Test Results of GPR-based Antipersonnel Landmine Detection Systems Mounted on Robotic Vehicles. In: Proc. of the IARP Int. Workshop on Robotics and Mechanical Assistance in Humanitarian Demining, Tokyo (June 2005)
9. Ishikawa, J., Kiyota, M., Pavkovic, N., Furuta, K.: Test and Evaluation of Japanese GPR-EMI Dual Sensor Systems at Benkovac Test Site in Croatia. Technical Report JST-TECH-MINE06-002, Japan Science and Technology Agency (2006)
10. Olson, C.F., Matthies, L.H., Wright, J.R., Li, R., Di, K.: Visual Terrain Mapping for Mars Exploration. In: Proc. IEEE Aerospace Conf., March 2004, vol. 2, pp. 762–771 (2004)
11. Singer, N.C., Seering, W.P.: Preshaping command inputs to reduce system vibration. J. of Dyn. Syst., Meas. Contr. 112, 76–82 (1990)

Towards Autonomous Wheelchair Systems in Urban Environments

Chao Gao, Michael Sands, and John R. Spletzer

Abstract. In this paper, we explore the use of synthesized landmark maps for absolute localization of a smart wheelchair system outdoors. In this paradigm, three-dimensional map data are acquired by an automobile equipped with high precision inertial/GPS systems, in conjunction with light detection and ranging (LIDAR) systems, whose range measurements are subsequently registered to a global coordinate frame. The resulting map data are then synthesized *a priori* to identify robust, salient features for use as landmarks in localization. By leveraging such maps with landmark meta-data, robots possessing far lower cost sensor suites gain many of the benefits obtained from the higher fidelity sensors, but without the cost. We show that by using such a map-based localization approach, a smart wheelchair system outfitted only with a 2-D LIDAR and encoders was able to maintain accurate, global pose estimates outdoors over almost 1 km paths.

1 Introduction

We are interested in developing smart wheelchair systems capable of autonomous navigation in unstructured, outdoor environments. Our primary work to date in this area has been with the Automated Transport and Retrieval System (ATRS) [1]. ATRS enables independent mobility for drivers in wheelchairs by automating the stowing and retrieval of the driver's wheelchair system. While ATRS has been commercialized, and its smart-chair system does in fact navigate autonomously, its autonomy is limited to an area immediately adjacent to the host vehicle. We would like to build on these results to support a greater range of smart-chair applications. Key to this objective is a robust means for outdoor localization.

Localization is a fundamental enabling technology for mobile robotics, and as a result a very active research area. Although the problem in structured, indoor environments might be considered solved, robust localization outdoors is still an open

Chao Gao, Michael Sands, and John R. Spletzer
Lehigh University, Bethlehem, PA, USA
e-mail: {chg205,mjs207,josa}@lehigh.edu

A. Howard et al. (Eds.): Field and Service Robotics 7, STAR 62, pp. 13–23.

research problem. While the community has made significant strides recently in terms of vehicle autonomy outdoors [2], much of this has been achieved through sensor suites using tightly coupled inertial/GPS navigation systems costing up to $100K or more. Such a solution is impractical in terms of both size and cost for applications such as ours. In the absence of reliable GPS measurements, fall-back strategies are similar to those used indoors and involve extracting strong features from RADAR, LIDAR, vision sensors, *etc.*, and tracking their relative position and uncertainty estimates over time [3, 4, 5]. However, such approaches are more fragile when used outdoors due to the absence of continuous structure and the much larger problem scale.

Urban environments represent an interesting and important middle-ground as over 80% of the U.S. population resides in cities and suburbs [6]. The availability of GPS measurements in these areas for pose estimation can typically be assumed, but multi-path errors from buildings, trees, *etc.*, can significantly compromise its accuracy. Fortunately, these same structures can be used as landmark features to yield accurate relative position estimates. In this paper, we investigate a paradigm where a smart wheelchair system relying upon lower cost sensors localizes with the assistance of three-dimensional maps generated by a vehicle equipped with a high fidelity sensor suite. These maps are synthesized *a priori* to identify robust, salient features which can be used as landmarks for robot localization. By leveraging these maps, the wheelchair gains many benefits obtained with the higher-fidelity sensors but without the cost.

2 Related Work

Our work relates to research efforts in three-dimensional mapping as well as robot localization and mapping. Generating three-dimensional maps of urban areas has been investigated by several groups, so there is significant previous work that can be leveraged [7, 8, 9, 10, 11, 12]. Unlike most of this work, our focus is generating and processing three-dimensional maps with respect to a global frame, which will be reusable and readily extended by any user. With the explosion of data services, we expect the availability of such maps to be commonplace in the future [13]. Our motivation is that by leveraging such maps, lower fidelity sensors could be employed. This has been demonstrated routinely indoors (*e.g.*, MCL with sonar vs. SLAM with LIDAR), and we believe the analogy will hold outdoors. Many features in urban environments are viable candidates for landmarks. For example, corner features are often used in EKF localization and mapping approaches as their position can be reduced to a single point [14]. Building facades and walls might also be used [8]. Indeed even signage can be detected and recognized [15]. While we are ultimately interested in integrating aspects of each of these features within our synthesized map representation, in this paper we limit our focus to pole features as landmarks. In this context, pole features would correspond to street lamps, trees, parking meters, street signs, *etc.* Such features are prevalent in most urban areas.

The use of poles features as landmarks has been investigated by other researchers. This includes the work of [16, 17], among others. The primary focus of these efforts was SLAM with a ground vehicle (*i.e.*, an automobile) where "cylinder" features were segmented using vision and/or LIDAR systems, and tracked over time. This technique enabled mobile localization and mapping in outdoor, unstructured environments over relatively long distances (*e.g.*, 100s of meters). We propose to build upon these efforts by first building large-scale three-dimensional maps, synthesizing these maps to identify strong landmark features, introducing a refinement stage to improve map consistency, and then leveraging these maps with an ultimate goal of improving localization performance outdoors.

3 Map Generation

Data Acquisition. Our vehicle for data acquisition was "Little Ben," which previously had served as the Ben Franklin Racing Team's entry in the DARPA Urban Challenge [18]. Vehicle pose was provided by an Oxford Technical Solutions RT-3050, which uses a Kalman filter based algorithm to fuse inertial measurements, GPS updates with differential corrections, and odometry information from the host vehicle. It provides 6-DoF pose updates at 100 Hz with a stated accuracy of 0.5 meters circular error probable (CEP). Range and bearing measurements from a pair of roof mounted, vertically scanning Sick LMS291-S14 LIDAR systems were then registered to the current vehicle pose to generate high-resolution range maps. The two LIDARs are highlighted (circled red) in Fig. 1 (Left). We used two LIDARs to improve map reconstruction by reducing scene occlusion and for redundant measurements to reduce noise effects. During data acquisition, Ben was driven at 8-12 km/hr. This corresponded to a LIDAR scan spacing of $\approx$ 4-6 cm, which allowed even thin pole features (*e.g.*, street signs, parking meters) to be captured reliably.

LIDAR Calibration. Ultimately, we need to register the acquired range scans to a common world frame W. This registration requires knowledge of the extrinsic parameters (rotation R and translation T) of both the vehicle frame V, and the front

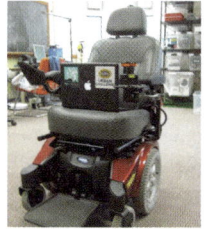

Fig. 1 Development Platforms. (Left) Vehicle used for data acquisition. Ben integrates an OXTS RT-3050 and a pair of vertically scanning Sick LMS291-S14 LIDARs (circled red). (Center-Right) Our smart-chair platform integrates LIDAR, vision, GPS, and odometry sensors. Its computer interface and on-board power distribution enable a range of sensors and accessories to be quickly integrated for prototyping purposes.

and back LIDAR frames (F,B) versus time with respect to W. The vehicle parameters $R_V^W(t), T_V^W(t)$ are estimated directly by the OXTS RT-3050 at 100 Hz. As the LIDARs are related to the vehicle frame by a rigid transformation, we need only recover the extrinsic parameters of the LIDARs with respect to the vehicle frame. For this work, we developed a novel approach to simplify the calibration process. Noting that points in the world frame and LIDAR frames were related by rigid transformations, we could recover the LIDAR extrinsic parameters with a sufficient number of point correspondences between the front and back LIDARs. To facilitate correspondence tracking over time, 6 poles on a $\approx 50 \times 100$ meter calibration loop were used as landmarks. On each pole, two retro-reflective targets were placed 1 meter apart for a total of 12 target points $X_T = [x_1,y_1,z_1,\ldots,x_{12},y_{12},z_{12}]^T$ in W. These targets were automatically segmented from the environment by thresholding the LIDAR remission measurements. A sample landmark pole is shown in Fig. 2 (left).

Point correspondences were obtained by driving multiple cycles around our calibration loop. Our calibration process then consisted of two stages. The first was to remove deterministic error between successive calibration loops caused by GPS jumps. This step was accomplished by using the first loop to generate reference landmark positions $X_1 = [x_{1,1},y_{1,1},z_{1,1},\ldots,x_{12,1},y_{12,1},z_{12,1}]^T$, where $x_{i,j}$ denotes the estimated x-position of the ith target in W during the jth calibration loop. The deterministic shift $S_j = [s_{xj},s_{yj},s_{zj}]^T$ of the jth loop was estimated by

$$S_j^* = \underset{S}{\text{argmin}}||X_1 - X_j - S_{j\times 12}||^2 \tag{1}$$

where $S_{j\times 12} \in \mathbb{R}^{36}$ is the estimated deterministic error S_j replicated for each target point. This minimization problem was solved using a least-squares approach. S_j was then treated as a bias, and the value of X_j for each loop was adjusted accordingly.

The second stage was to remove the influences of random error in the calibration process. In doing so, we needed to estimate the extrinsic parameters for each LIDAR $(R_B^V,T_B^V,R_F^V,T_F^V)$, as well as the positions of our 12 targets (X_T) in W. We note that for the same world point $\mathbf{x}_W$,

$$R_{VF}^W(R_F^V\mathbf{x}_F + T_F^V) + T_{VF}^W = R_{VB}^W(R_B^V\mathbf{x}_B + T_B^V) + T_{VB}^W \tag{2}$$

where the VF and VB subscripts are used to denote the vehicle transformation to the world frame corresponding to the different vehicle poses when $\mathbf{x}_W$ was observed by the front and back LIDARs, respectively. Thus, we can solve for both the LIDAR extrinsics as well as the target positions with a minimum of 16 point correspondences between the front and back LIDARs. Since the vehicle pose varies with each calibration loop, 12 unique correspondences can be obtained from each loop cycle. As a result, a large number of correspondences can be acquired very quickly. We solved for (2) using a non-linear minimization solver. However, one final enhancement was added first to remove measurement outliers. For this, we employed a "constrained" RANSAC approach [19], where we instantiated each model hypothesis with a small number (2-4) of correspondences at random from *each target*. This enhancement ensured that the error residuals were balanced across the entire calibration loop.

Fig. 2 (Left) One of six landmark poles used during the LIDAR calibration process. The retro-reflective targets could be automatically segmented to track correspondences across multiple calibration loops. Reconstruction results before and after our calibration phase. Improvements in data fusion with the front (red) and back (blue) LIDARs from the calibration phase are clearly visible.

Fig. 2 shows representative results of merging front and back LIDAR data using the measured extrinsics (center) and those estimated from the calibration process (right). Improvements in data fusion and scene reconstruction from the calibration process are clearly visible. The mean absolute error (MAE) for the error residuals between the two LIDAR reprojections was 12.67 cm. As the performance of the pose system is 50 cm CEP, these results were considered satisfactory.

Landmark Synthesis. Segmenting pole features was accomplished using a two-step clustering approach: (1) recursively cluster points within each scan, and (2) merge clusters in successive scans where appropriate. In both steps, a Euclidean distance threshold was used as the clustering criterion. For a cluster to be accepted as a pole feature, validation gates were placed on cluster size. Furthermore, only strongly vertical clusters were accepted by examining the covariance matrix C of the associated feature points' positions. Specifically, the eigenvector associated with the largest eigenvalue λ_{max} of C should be close to $[0,0,1]^T$. Only after clearing these validation gates was the cluster accepted as a landmark in the synthesized map.

4 Wheelchair Localization

The smart-chair used in this work is based upon an Invacare M91 Pronto power wheelchair with Mk5 electronics. It integrates high resolution optical encoders, a Hokuyo UTM-30LX LIDAR system, a 1024x768 Point Grey digital video camera, and a Garmin 18 WAAS enabled GPS system. For this work, the UTM-30LX was the wheelchair's sole exteroceptive sensor. When compared to the ubiquitous Sick LMS2xx LIDARs, it is extremely compact. In our current integration, the LIDAR and camera system are mounted on the opposite arm as the manual joystick controller as shown in Fig. 1 (center). The configuration is comparable in size to the joystick controller box.

Landmark Detection. In our localization paradigm, the wheelchair employs LIDAR and odometry sensors in conjunction with the synthesized landmark map. Implicit in this approach is the assumption that the landmarks can be reliably segmented. However, unlike the landmark synthesis phase, the wheelchair LIDAR must rely entirely upon two-dimensional scan data. To compensate for this, our landmark detection strategy used two approaches dependent upon pole feature geometry. The first step in either approach was clustering registered point returns from the wheelchair LIDAR scan in Euclidean space. Cluster diameters were then used to discriminate between larger diameter pole features (trees, telephone poles, street lamps, *etc.*) and narrower ones (parking meters, traffic sign posts, *etc.*).

Larger diameter clusters with 5 or more points were fit as circle features. Several additional validation gates followed based upon circle geometry and residual fitting error before a feature was accepted as a landmark candidate. We found empirically that larger diameter landmarks could be reliably detected at ranges $\approx 75\times$ the feature radius, meaning that a landmark with a radius of 10 cm would typically be detected at a range of about 7.5 meters. Circle fitting was not appropriate for smaller diameter clusters. As such, these features were tracked over time. If they were persistent and no other clusters were detected within a given distance threshold, they were accepted as landmark candidates. Using such an approach, smaller diameter features could reliably be detected at ranges ≤ 5 meters. Candidate landmarks were passed to the data association module for additional processing.

Data Association. There are inherent limitations in using two-dimensional LIDAR measurements to segment three-dimensional landmarks. As a result, the landmark detection process may erroneously validate other environmental features as pole features. The impact of these false positives on localization performance was mitigated through a data association phase.

Several sources of uncertainty exist in the synthesized landmark locations within our map. These sources include uncertainty introduced by the pose system, errors in LIDAR extrinsic calibration, LIDAR range errors, and noise associated with the landmark synthesis process itself to name but a few. Uncertainty in landmark position was modeled by associating a covariance matrix Σ_l with the position of each landmark while Σ_w and Σ_o denoted covariance matrices associated with the uncertainty in wheelchair pose and LIDAR range and bearing observations, respectively. With these so defined, we used the Mahalanobis distance D between the predicted and observed sensor measurements as our quality metric for data association, defined as

$$D = \sqrt{\mathbf{z}^T \left(\Sigma_o + H_w \Sigma_w H_w^T + H_l \Sigma_l H_l^T \right)^{-1} \mathbf{z}} \qquad (3)$$

where, H_w and H_l are the Jacobians of the observation model with respect to wheelchair pose and landmark location, respectively,

$$H_w = \begin{bmatrix} \frac{x_w - x_l}{z_r} & \frac{y_w - y_l}{z_r} & 0 \\ \frac{y_l - y_w}{z_r^2} & \frac{x_w - x_l}{z_r^2} & -1 \end{bmatrix}, \quad H_l = \begin{bmatrix} -\frac{x_w - x_l}{z_r} & -\frac{y_w - y_l}{z_r} \\ -\frac{y_l - y_w}{z_r^2} & -\frac{x_w - x_l}{z_r^2} \end{bmatrix} \qquad (4)$$

and $\mathbf{z} = \mathbf{z}_l - \mathbf{z}_o$ is the difference between the predicted and actual range and bearing measurements for the wheelchair LIDAR. A threshold on D served to filter out potential false positives observed during the landmark detection phase. For the case of closely located landmarks where multiple possible detection-landmark associations might be possible, the association minimizing the total assignment cost was used.

Localization Approach. Extended Kalman Filters (EKFs) have been one of the most popular techniques for state estimation in mobile robotics [3, 20, 14], and we took a similar approach for estimating the wheelchair pose $\mathbf{x} = [x_w, y_w, \theta_w]^T$. In the prediction step, linear and angular velocities (v, ω) were estimated from the encoders using a differential drive model for the wheelchair. For the correction step, the observation functions were based upon LIDAR estimates for the range z_r and bearing z_α to the segmented landmark at position (x_l, y_l)

$$z_r = \sqrt{(x_w - x_l)^2 + (y_w - y_l)^2}, \quad z_\alpha = \arctan\left(\frac{y_w - y_l}{x_w - x_l}\right) - \theta_w \qquad (5)$$

and the Kalman gain was calculated as $K = PH_w^T(H_w PH_w^T + H_l \Sigma_l H_l^T + \Sigma_o)^{-1}$ where H_w and H_l were as defined in (4). The process then followed a traditional EKF implementation with updates of 2-5 Hz dependent upon vehicle velocity.

Landmark Position Refinement. A shortcoming with relying heavily upon GPS for map generation is that changes in satellite geometry/visibility can lead to "jumps" in vehicle pose. These discontinuities affect map consistency. One approach to address this would be to integrate additional sensing onto the data acquisition platform and run SLAM in parallel with the data acquisition phase [12]. We propose an alternate refinement stage where SLAM is actually run by the map client – in our case the wheelchair – during an initial route traversal akin to a learning phase. This is something we envision would be done by the wheelchair user's care provider prior to enabling completely autonomous operations. An advantage of this approach is that the landmark refinement would be tuned to the actual sensor geometries employed by the client vehicle. For our implementation, we extended our EKF localization using a SLAM approach as in [3, 20] to further refine the landmark positions. The landmark locations were then updated with the SLAM-refined landmark positions and covariance Σ_L estimates. While this did not improve the global map accuracy, it significantly improved the map consistency.

5   Experimental Results

To investigate the viability of the proposed approach, our first experiments were conducted in the parking lots around Lehigh's Stabler Arena. Admittedly, this area was *not* representative of urban environments. However, it served as a low-traffic proving ground with sufficient pole features to first validate the concept. Fig. 3 (left) shows the raw registered range data acquired by driving Ben through the area. These data were then synthesized as outlined above, and the resulting map with embedded landmarks is shown at Fig. 3 (right). Validating the fidelity of this reconstruction is difficult due to the lack of absolute ground truth. However, we measured the distance between 25 pairs of landmarks using a Bosch DLE50 laser distance measure and compared these to the distances of corresponding synthesized landmark pairs.

Fig. 3 Registered raw (left) and synthesized map data (right). The relative distance differences between synthesized landmark pairs and their real-world counterparts was about 7 cm.

The mean absolute difference between the sets was 7.2 cm. We also reviewed the reliability of the landmark synthesis approach. All 71 pole features present in the area surveyed were positively detected and synthesized into the map.

Using the original synthesized map and landmark positions, the wheelchair was manually driven over a route network 960 meters in length at a fast walking speed (1.6 m/s). This first loop constituted the landmark refinement stage discussed in Section 4, and the wheelchair localized using an EKF SLAM approach with the segmented landmark positions. Using SLAM, the wheelchair was able to accurately maintain its pose over the entire 960 meter loop. We then repeated this same experiment 3 separate times using map-based localization with the updated landmark position and uncertainty estimates. All other parameters for data association and localization remained fixed. Representative results are at Fig. 4 (left). The landmark positions are denoted by red circles. The wheelchair path as estimated by the map-based localization approach is denoted by the blue line. The path as estimated by the wheelchair's own GPS is also shown for comparison purposes (green line). The initial pose estimate of the GPS was also used to initialize the pose for map localization. Using the SLAM-refined landmark positions, all 3 trials were successfully completed. To characterize the localization accuracy, the wheelchair was driven over 6 ground-truth points (shown as "+") whose positions relative to landmarks were measured by hand. The average position errors was 20 cm, with $3 < 1\sigma$, $5 < 2\sigma$, and all $6 < 3\sigma$ based upon the covariance estimates for Σ_W and Σ_l.

To motivate the need for the landmark refinement phase, we also ran these same trials using map-based localization (not SLAM) with the original landmark positions. Each of these trials ended in failure. This typically occurred at a portion of

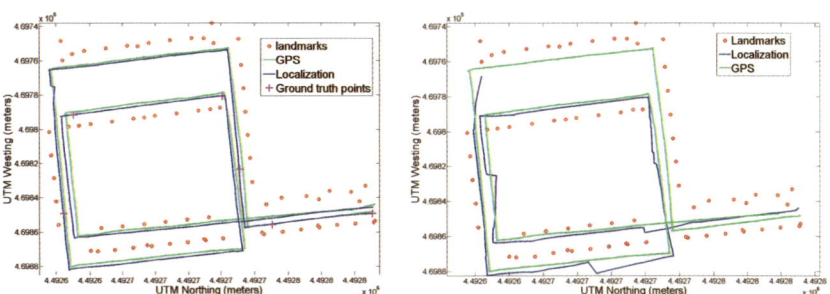

Fig. 4 Localization results using refined (left) and original landmark position estimates (right). Improving the consistency of landmark positions dramatically improved localization performance.

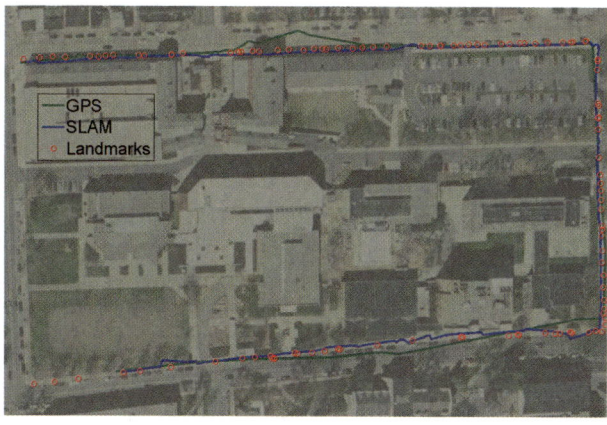

Fig. 5 Map-based localization of the wheelchair (blue line) vs. GPS position estimates (green line).

the course where the inter-landmark spacing required the wheelchair to rely upon dead reckoning for over 20 meters of travel. Significant error and uncertainty in wheelchair pose accumulated during this time, resulting in an incorrect feature association. This is shown in Fig. 4 (right). However, the open-loop travel was not the sole culprit. From a subsequent analysis, we determined that a fairly significant GPS shift occurred during the data acquisition phase for building the map. As a result, a fraction of the map exhibited a shift >1 meter with respect to the maps initial coordinate frame. This shift significantly contributed to the data association failure, and the correct robot pose could not be recovered by the EKF. As the SLAM algorithm updated the landmark positions on the fly, it was robust to this shift error. The subsequent landmark refinement stage mitigates the impact of GPS jump.

Our final experiment involved a similar test in South Bethlehem, PA adjacent to Lehigh's campus. This was a representative urban environment, with a significantly higher density of landmarks than seen during the Stabler testing. During this test, the wheelchair was manually driven approximately 720 meters along the sidewalk at a velocity of ≈ 1 m/s. Results from this trial using SLAM are shown at Fig. 5, where the landmarks positions (red circles), the SLAM estimated path (blue line) and GPS path (green line) are superimposed on a satellite image. While completing the loop was prevented by ongoing building construction, the end position was consistent with localization estimates. Again, results compared favorably to the wheelchair's WAAS-corrected GPS estimate. We have not been able to repeat this trial using localization with the SLAM-enhanced map due to seasonal weather conditions, but expect to in the near future.

6 Discussion

In this paper, we investigated the acquisition, synthesis and application of three-dimensional maps by a smart wheelchair for map-based localization. Since the maps were registered to a global frame, they provide a means for absolute position estimation in urban areas even in the absence of GPS. In our experiments, our wheelchair system was able to maintain accurate pose estimates after traveling hundreds of meters using such an approach. While we are satisfied with the results to date, we do realize this is just a first step. Pole features were an obvious first choice for landmarks, and we are now beginning to synthesize additional features into the map (*e.g.*, building corners). We are also interested in vision based signage detection, as these can provide nearly-unique IDs for inferring global position. We also assume the ability to automatically segment pedestrians from the environment. Our current implementation fuses results from vision and LIDAR systems. The camera uses the Haar-like feature based classifier for face detection from OpenCV [21], while the LIDAR segments candidate clusters based upon geometry constraints. Individually, both systems have high rates of false positives. However, this can be reduced dramatically by only accepting tracks when both sensors report a detection. A downside is that significant false-negatives remain. We are continuing to refine this approach.

References

1. Gao, C., Hoffman, I., Miller, T., Panzarella, T., Spletzer, J.: Autonomous Docking of a Smart Wheelchair for the Automated Transport and Retrieval System (ATRS). Journal of Field Robotics 25(4-5), 203–222 (2008)
2. Buehler, M., Iagnemma, K., Singh, S.: Special issues on the 2007 darpa urban challenge. The Journal of Field Robotics 25(8-10) (August-October 2008)
3. Dissanayake, M., Newman, P., Clark, S., Durrant-Whyte, H., Csorba, M.: A solution to the simultaneous localization and map building (SLAM) problem. IEEE Trans. on Robotics and Automation 17(3), 229–241 (2001)
4. Newman, P., Leonard, J.: Pure range-only subsea slam. In: IEEE International Conference on Robotics and Automation, Taiwan, China (September 2003)
5. Masson, F., Nieto, J., Guivant, J., Nebot, E.: Robust autonomous navigation and world representation in outdoor environments. Mobile Robots Perception and Navigation, 299–320 (2007)
6. United Nations, Population Division: World Urbanization Prospects (February 2008)
7. Zhao, H., Shibasaki, R.: Reconstructing textured cad model of urban environment using vehicle borne laser range scanners and line cameras. In: Second International Workshop on Computer Vision System (ICVS), Vancouver, Canada, pp. 284–295 (2001)
8. Frueh, C., Zakhor, A.: Reconstructing 3d city models by merging ground-based and airborne views. In: García, N., Salgado, L., Martínez, J.M. (eds.) VLBV 2003. LNCS, vol. 2849, pp. 306–313. Springer, Heidelberg (2003)
9. Georgiev, A., Allen, P.: Localization Methods for a Mobile Robot in Urban Environments. IEEE Trans. on Robotics 20(5), 851–864 (2004)
10. Howard, A., Wolf, D.F., Sukhatme, G.S.: Towards 3d mapping in large urban environments. In: IEEE/RSJ Int. Conf. on Intelligent Robots and Systems, Sendai, Japan, September 2004, pp. 419–424 (2004)
11. Jensen, B., Weingarten, J., Kolski, S., Siegwart, R.: Laser range imaging using mobile robots: From pose estimation to 3d-models. In: 1st Range Imaging Research Day, Zürich, Switzerland, September 2005, pp. 129–144 (2005)

12. Borrmann, D., Elseberg, J., Lingemann, K., Nüchter, A., Hertzberg, J.: Globally consistent 3d mapping with scan matching. Journal of Robotics and Autonomous Systems (JRAS) 56(2), 130–142 (2008)
13. Educating Silicon, Google Street View - Soon in 3D? (April 2008),
    http://www.educatingsilicon.com/
14. Altermatt, M., Martinelli, A., Tomatis, N., Siegwart, R.: SLAM with Corner Features Based on a Relative Map. In: IEEE/RSJ IROS (October 2004)
15. Kingston, T., Laflamme, C.: Automated road sign detection and recognition. Journal of the International Municipal Signal Association, 46–49 (January/February 2007)
16. Ramos, F., Nieto, J., Durrant-Whyte, H.: Recognising and Modelling Landmarks to Close Loops in Outdoor SLAM. In: IEEE ICRA (March 2007)
17. Nieto, J., Bailey, T., Nebot, E.: Recursive scan-matching SLAM. Robotics and Autonomous Systems 55, 39–49 (2007)
18. Bohren, J., Derenick, J., Foote, T., Keller, J., Kushleyev, A., Lee, D., Satterfield, B., Spletzer, J., Stewart, A., Vernaza, P.: Little Ben: The Ben Franklin Racing Team's Entry in the 2007 DARPA Urban Challenge. Journal of Field Robotics (2008) (accepted for publication)
19. Fischler, M., Bolles, R.: Random sample consensus: A paradigm for model fitting with applications to image analysis and automated cartography. Communications of the ACM (1981)
20. Thrun, S.: Robotic mapping: A survey. Tech. Rep. CMU-CS-02-111, Carnegie Mellon University (2002)
21. Intel Corporation, OpenCV, http://sourceforge.net/projects/opencv/

Tethered Detachable Hook for the Spiderman Locomotion (Design of the Hook and Its Launching Winch)

Nobukazu Asano, Hideichi Nakamoto, Tetsuo Hagiwara, and Shigeo Hirose

Abstract. This paper introduces a new concept of "tethered detachable hook (TDH)" and its launching winch. TDH system is the device which will be mounted on a mobile robot and enhances its traversability over extremely hostile terrain by launching detachable hook to nearby objects, producing large traction force by the tether and detaching/recovering the hook to the launcher again. In this paper the authors first of all introduce several prototype models of the TDH. We then discuss the design of latest model which features pneumatic detaching mechanism, the pneumatic launcher and the reel mechanism having three motion states; active rotation, free rotation and braking. Finally, the result of several experiments of constructed TDH model will be explained.

1 Introduction

Many types of mobile robots have been developed so far to move on off-the-road terrains, such as modified wheel, track, legs, and snake-like configuration. Even jumping can be considered as one of the means for high mobility [1]. However, if long and steep slope or ditch much wider than the size of the robots is on the way, terrain adaptability of these conventional methods is not enough. In this paper, we propose a new type of locomotion method which assists the mobility of these mobile robots. It consists of "tethered detachable hook" and its launcher and winch system, which assists the mobile robot as shown in Fig.1(a). Here the mobile robot is going to climb the steep slope and the "tethered detachable hook" is launched to

Nobukazu Asano and Shigeo Hirose
Tokyo Institute of Technology, 2-12-1, Ookayama, Meguro-ku, Tokyo, 152-8552, Japan
e-mail: nasano@robotics.mes.titech.ac.jp, hirose@mes.titech.ac.jp

Hideichi Nakamoto
Toshiba Corporation, 1, KomukaiToshiba-cho, Saiwai-ku, Kawasaki-shi, 212-8582, Japan

Tetsuo Hagiwara
KinderHeim Corporation, 335, kaminakazato-cho, Isogo-ku, Yokohama-shi, 235-0045, Japan

A. Howard et al. (Eds.): Field and Service Robotics 7, STAR 62, pp. 25–34.
springerlink.com © Springer-Verlag Berlin Heidelberg 2010

the branch of a tree. When the hook is connected to the branch, the winch winds the tether and produces large traction force to assist the robot to go over the steep slope. After the motion is over, the hook is detached from the branch by changing the shape of the hook to smooth and linear shape, rewinding the winch, and finally restoring the hook in the launcher again to prepare for the next launching task. If the mobile robot has more than two tethered detachable hooks and their launcher-winch system, it can even lift itself from the ground and move from branch to branch as shown in Fig.1(b). This is like the motion of long-armed ape in forest or the spiderman flying from building to building.

Until now several tethered robots were already proposed, such as TITAN VII [2] or DANTE II [3]. They are supported by tethers which are anchored beforehand at the top of the slopes. Cliffbot [4] is supported by an anchor robot which stays at the top of the cliff and connected by the tether. Casting manipulator [5] has the tether with a gripper at the end and casted to catch an object. Although the objective of this casting manipulator was for the manipulation of a remotely located object, the concept can easily be extended to the supporting system for mobile robot. The automated tether management system [6] is most closely related to our concept. It used a tether with a gripper which can be remotely operated to lock or detach it and help the flying motion of a space robot. But as it is designed for the activity in micro-gravity environment, the system can not directly apply for the application of field robotics which we are targeting in this paper.

This paper is organized as follows. Section 2 describes the design of former models of TDH. Section 3 presents design of the latest model of TDH and its launching winch. Section 4 reports experimental results of the constructed TDH and its launching winch. Section 5 concludes the results and proposes future works.

2   Former Models of Tethered Detachable Hook

2.1   Model I

As the first model of the "tethered detachable hook", the authors developed the Model I as shown in Fig.2(a). Although we call it as "tethered detachable hook", we did not selected the hook but gripper. As shown in Fig.2(a) and (b), the gripper is designed to grip an object when the tip end of the rod contact the object and hold

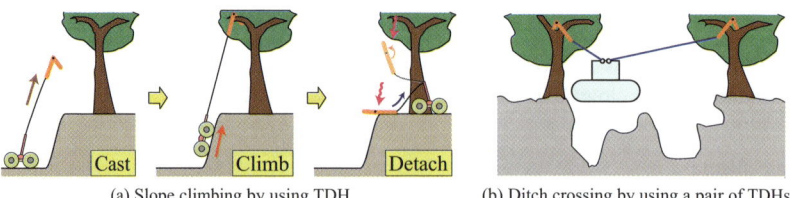

(a) Slope climbing by using TDH   (b) Ditch crossing by using a pair of TDHs

Fig. 1 Concept of tethered locomotion by using tethered detachable hook(TDH)

its gripping state by the ratchet mechanism. Ability to hold the object tight was the main reason why we selected gripper configuration for this first model. Release of the gripping motion is designed to be done by a mechanical memory system. The mechanical memory system is already used in the ballpoint pen with different colors. It consists of a cylindrical cam with zigzag grooves (Fig.2(c)) in which the pin fixed to the external cylinder is inserted. When the tension of the tether connected to the cylindrical cam changes and drives the cam to make reciprocating motion, the cam is driven by the pin and starts to rotate in one direction. In the three zigzag grooves, one of the grooves is made longer, so the ratchet release rod is inserted in the ratchet trigger to release the ratchet and open the finger every three times of the pulling motion of the tether.

For this Model I we also made simple launcher and made the experiment to cast it to the branch. Once it is gripped the branch, it showed strong connection and release motion was also very smooth. But the problem of this first model was its difficulty of aiming at the target object (branch). The gripper should be aimed at the object precisely in position and also in orientation; otherwise the gripper could not hold the object successfully. This is the big problem if we hope to make automatic launching system. Another problem of this first model was the shape of the gripper in open state. It is not streamlined and there is always the danger to be stacked in narrow gap.

2.2 Model II

To solve the difficulty of precise aiming of the target, we selected hook for the following models. The Model II of the tethered detachable hook is shown in Fig.3. As shown in Fig.3, the Model II also adopted the mechanical memory system, and it could be released by pulling the tether for few times. Repeated traction of the tether rotate the cylindrical cam and it drives the lock lever, and release the stopper to open the claws.

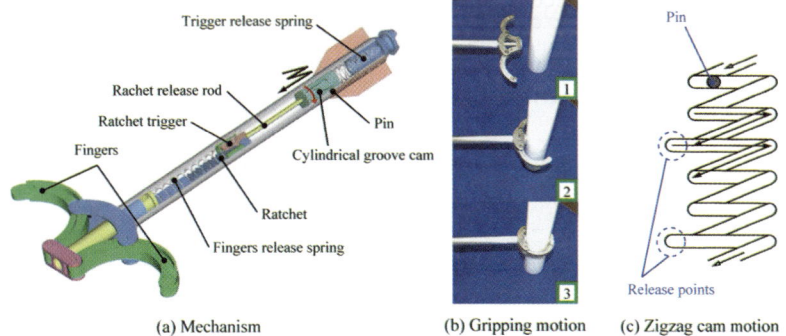

(a) Mechanism   (b) Gripping motion   (c) Zigzag cam motion

Fig. 2 Mechanism of the TDH model I

The Model II has four claws attached radially at the end of the hook so that the hook can be anchored to the target object much easier than the gripper. It can be hooked the target object only by throwing it over the target. The Model II was much easier to connect to the target objects than former model, however it still remained several problems. One of them is its weight. As there are four claws, it is heavy and powerful launcher is needed. The second problem is the shape of the hook in release state. Although the shape is more streamlined than that of Model I, the shape is still in wedge like and there remained the possibility to be stack in narrow gap while it is recovering to the launcher. The third problem is the possibility of mal-operation of the mechanical memory system, for the tether will always be affected by accidental pulling and releasing motion and it may be released by chance.

2.3   Model III

To solve the problems mentioned above, we developed Model III. The model III has three important modifications as follows;

1. reduce the number of claws from four to one
2. change the shape of the hook as a simple rod in the release state
3. introduce active detaching mechanism

Modification 1 is done to reduce the weight of the hook. We selected four claws configuration for the Model II to secure reliable anchor action in any posture of the hook. However, we found that even one claw hook can exhibit similar action only by adding enough length and weight to the claw. Effect of this configuration is observed in the experiment of Fig.10. In this experiment, when the hook pulled slowly over a branch (in this case a pipe), the hook will rotate around its stem and let the claw lower on the branch as the claw is heavy, and thus the claw grips the branch.

Modification 2 is done to minimize the stack action while the hook is in retracting state. As shown in Fig.4, the hook is designed to change from L-shaped state to linear-shaped state. Difficulty of realizing this shape was in the joint design of the claw. As large torque is applied at the joint to support large traction force of the tether, the joint mechanism has to produce large torque and the joint tends to be bulky and heavy. To solve this problem, we introduced tether supported joint

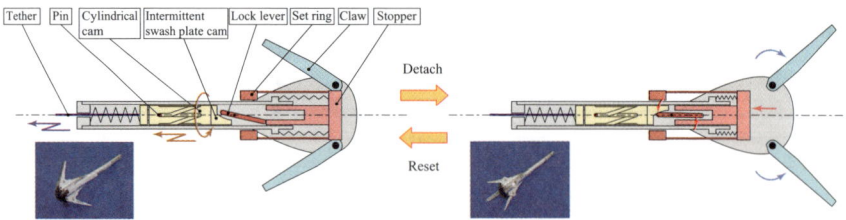

Fig. 3 Mechanism of the TDH model II

mechanism. In this mechanism, the end of the tether is connected to the claw and joint torque is directly supported by the traction of tether as shown in Fig.4. The tether is designed to go out of the joint to produce large torque in this hooked state. In the linear-shaped state, to the contrary, the tether retracts inside the joint and the joint is made slender.

Modification 3 is done to eliminate the expected malfunction of the release motion depending upon the accidental pulling action of the tether. The mechanical memory system was ideal because normal rope can be used as the tether and lock-and-release mechanism of the hook in Model II, or gripper in Model I could be made comparatively simple. To make comparative system with high reliability, we considered electric and pneumatic types of trigger driving mechanisms.

Design of the electric detachable mechanism is shown in Fig.5. A stopper is connected to the tether and the stopper is locked by a trigger that is fixed by the rod (Phase 1). In this state, the hook holds L shaped configuration and act as an anchor. Release motion of the hook is done by rotating the screw by the small motor and slides the rod fixing the rotation of the trigger. This motion frees the stopper and the tether automatically slides to open the claw in a release state by the spring attached around the joint (Phase 2). Required electric current is very small and it can be supplied by small diameter electric wire inside the tether. As the tether has to support large traction force, we used a Kevlar® fiber-reinforced wire together with the fine electric wires. The authors have already used it as the "Hyper Tether" system [7] and Anchor climber [8].

A pneumatic detachable mechanism is shown in Fig.6. The tether consists of air tube and wire, and the end of the tube is connected to a small air bag which is located inside a stopper. The stopper can slide inside a pipe fixed to the hook and the stopper

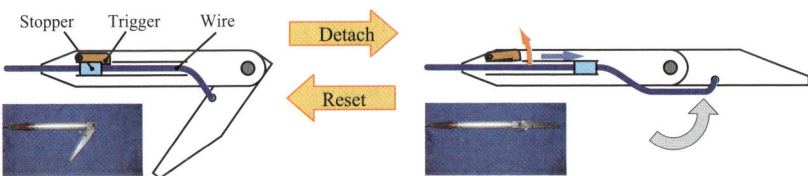

Fig. 4 Basic structure of the TDH model III

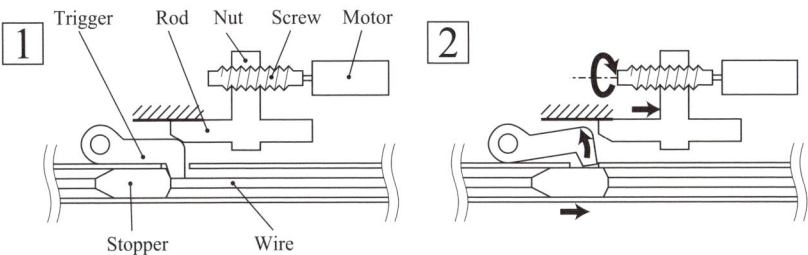

Fig. 5 Electric detach type of the TDH model III

Table 1 Specifications of the TDH model II & III

Type	Mass	Total length	Length of claw
Model II	840 [g]	435 [mm]	85 [mm]
Model III (Electric)	395 [g]	325 [mm]	150 [mm]
Model III (Pneumatic)	390 [g]	345 [mm]	150 [mm]

is locked by the trigger as shown in Fig.6 (Phase 1). The wire inside the air tube is connected to the stopper and the end of the wire is connected to the claw. Detaching motion of the hook can be done by supplying pressurized air in the air tube and expand the airbag. Expanded airbag pushes out the projection point of the bag and drive the trigger out of the hole and release the stopper (Phase 2). It enables the stopper and the wire connected to slide freely and open the claw. Although the gap between air tube and wire (polyethilene line) is small, we found that the pressurized air could easily be transmitted along the air tube longer than 10[m] with ease.

We successfully made both types of detachable mechanisms and verified their motions. Specifications of these types are shown in Table 1 with those of Model II. Between these types, we selected the pneumatic type, because tether of the pneumatic type can be lighter and driving mechanism of the hook can be lighter and rugged enough to be protected against the shock.

3 Design of Launching Winch for the Tethered Detachable Hook

A casting device developed for TDH Model III is shown in Fig.7. It consists of a launcher for the hook and a winch to wind the tether.

First we describe a launcher. Among the spring type and pneumatic cylinder type, we found that the pneumatic one is better because it can generate powerful and high speed launching motion with lightweight mechanism. We already adopted pneumatic hook detachable system and selection of the pneumatic system for the launcher will have other effect to make the total system simple. One of the most

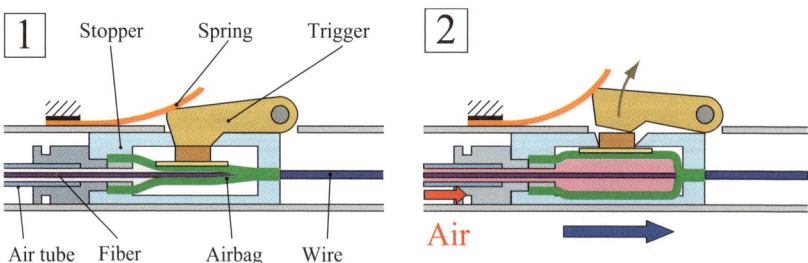

Fig. 6 Pneumatic detach type of the TDH model III

important parts of the launching system is in its trigger mechanism, and designed mechanism is shown in Fig.8. At first, a piston is locked by a ball type trigger, and is pressed by high-pressured air from "Port A" as shown in Fig.8(a). At this time, high-pressured air from "Port B", that is for the control rod, is also supplied. Launch motion can be done by decompress the air for "Port B" and drive the control rod out of the balls and let the air pressure from "Port A" drives the piston go right direction and launch the hook(Fig.8(b)). Compared with the trigger mechanism using normal valve, introduced mechanism can makes the trigger motion smoothly and as the pressurized air gives pressure to the hook from the beginning, it can increase the initial speed of the hook and enable it to cast in longer distance.

Next we explain a winch for the TDH. It is designed to have three modes; drive mode, free mode and brake mode.

The "drive mode" is used when it is used as winch, and large traction force should be generated to support a robot. The "free mode" is used when the hook is going to be launched. As the spool have to rotate in high speed, the actuator to produce large traction force in drive mode should be mechanically disconnected. The "brake mode" is needed for two reasons, one of them is to support the suspended robot without energy loss and the other is to adjust the rotational speed of the winch when it is in "free mode" and launching the hook. As the hook is launched by pneumatic pressure, winch in free mode tends to keep rotating while the hook is flying. However the speed of the hook decelerates while flying and the tether tends to excessively goes out of the reel and entangle around the reel. This phenomenon

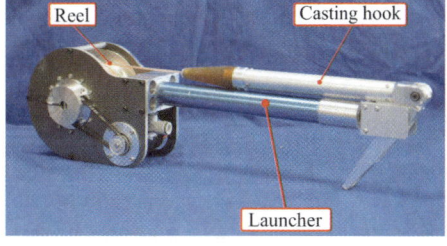

Fig. 7 Overview of a launching winch with pneumatic TDH model III

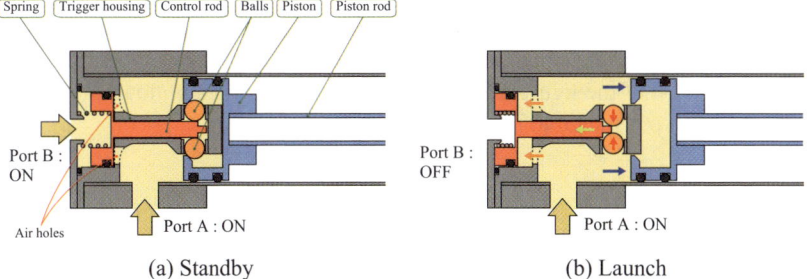

(a) Standby   (b) Launch

Fig. 8 Pneumatic trigger mechanism of the launcher

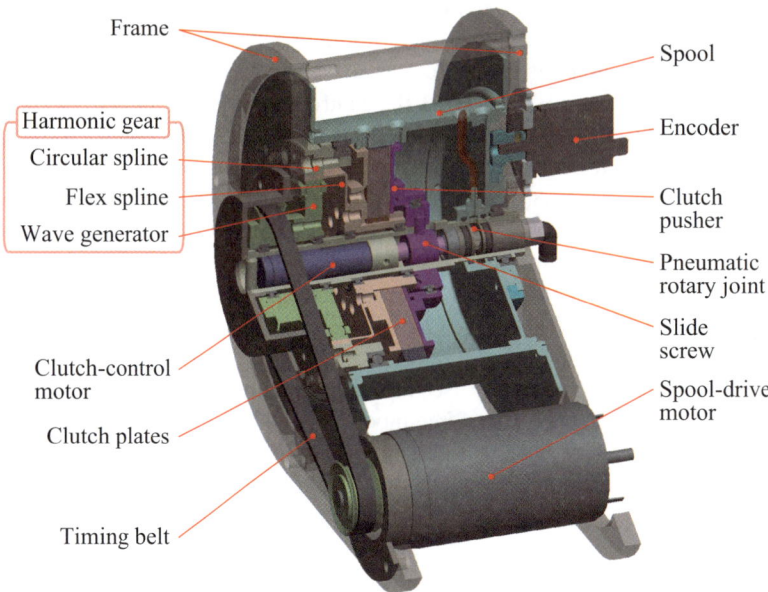

Frame

Spool

Encoder

Harmonic gear
Circular spline
Flex spline
Wave generator

Clutch
pusher

Pneumatic
rotary joint

Slide
screw

Clutch-control
motor

Spool-drive
motor

Clutch plates

Timing belt

Fig. 9 Mechanism of the winch

is called as "backrush" among anglers for their fishing reel motion. To prevent the backrush we need proper braking of the winch in the free mode.

To switch these 3 modes, we adopted a multi-plate clutch mechanism which is installed inside the spool as shown in Fig.9. Multiple input and output clutch plates are piled up to increase the braking torque. Maximum torque of this clutch mechanism T_c can be estimated as follows;

$$T_c = N_p \mu F_p r_e \tag{1}$$

where, N_p is the number of friction surfaces between clutch plates, μ is the coefficient of static friction, F_p is a pushing force for clutch plates generated by the motor thrusting force and r_e is an effective radius of friction surfaces.

As the winch rotate infinitely and pressurize air have to be supplied to the air tube to connect the hook, a rotary pneumatic joint is introduced. Air is supplied from a joint in right section of the figure, and it pass through holes on a hollow shaft. Two movable O-rings are installed to prevent the leak of the air.

4 Experiment

The authors confirmed motion performance of tethered detachable hook of Model III. Basic motion to hook the object was examined as shown in Fig.10. As is discussed before, the claw was automatically lowered and gripped the branch when the hook was pulled and located above the object. From this experiment, we confirmed the validity of introducing the one claw configuration for the TDH.

Next, the authors made a simple experiment of casting the Model III TDH to the real branch of the tree as shown in Fig.11. From this experiment, the anchoring function of the hook and its detaching motion was successfully demonstrated. Besides, smooth collection was achieved because of its straight shape after the detaching motion.

With the casting device mentioned above, the authors also made the experiments to verify the effectiveness of the prevention of "backrush" by the braking. Fig.12(a) shows the comparison of the tether on the after launching the hook. Left is the result without braking and right is the result with proper braking. From the comparison of these results, we know the importance of the braking of the winch in free mode. Fig.12(b) shows measurement results of outer circumferential velocity of the winch. They were measured by an encoder connected to the winch. From this figure too, we can know that proper braking enable to increase the launching speed. In Fig.12(b) we can observe the speed change of the reel at the time near the 1.2[sec]. It is caused by the falls of the hook on the ground.

Fig. 10 Sequential motion to show the self adjustment of the claw direction to the target branch

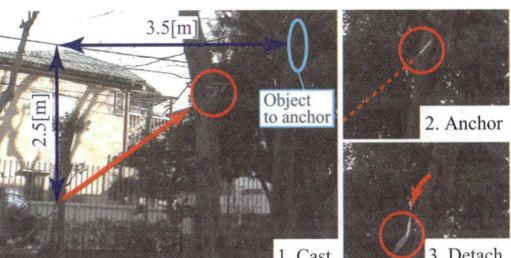

Fig. 11 Real launching experiment of the TDH model III to the branch of a tree

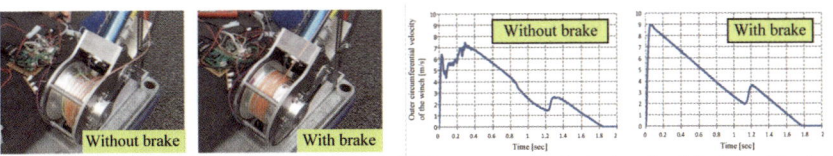

(a) Comparison of the tether with and without braking control   (b) Measurement results of outer circumferential velocity of the winch

Fig. 12 Comparison experiment with and without braking control

5 Conclusions and Future Works

This paper introduces a new concept of "tethered detachable hook (TDH)" and its launching winch for use as the locomotion assisting device for mobile robots. This paper firstly discusses about several prototype models of the TDH and elaborate latest model, such as pneumatic lock and release mechanism of the hook, pneumatic launcher and the reel mechanism which exhibits three motion states; active rotation, free rotation, and braking of the rotation. Performance of developed TDH and its launching winch are successfully demonstrated by the constructed mechanical model. Study of the proposing gtethered detachable hook (TDH)h is still at the starting point, and there remained many interesting research subjects to be studied on the hook, launcher, and winch mechanisms and their control. We are hoping to study further on these points and realize the mobile robots having TDH and move around mountainous area or disaster site by successively casting the tethers around the environment just like Spiderman does among buildings in near future.

References

1. Tanaka, T., Hirose, S.: Development of leg-wheel hybrid quadruped "AirHopper" design of powerful light-weight leg with wheel. In: IROS 2008, pp. 3890–3895 (2008)
2. Hodoshima, R., Doi, T., Fukuda, Y., Hirose, S., Okamoto, T., Mori, J.: TITAN VII: quadruped walking and manipulating robot on a steepslope. In: ICRA1997, pp. 494–500 (1997)
3. Krishna, M., Bares, J., Mutschler, E.: Tethering System Design for Dante II. In: ICRA 1997, pp. 1100–1105 (1997)
4. JPL Robotics, http://www-robotics.jpl.nasa.gov/
5. Arisumi, H., Komoriya, K.: Catching Motion of Casting Manipulation. In: IROS 2000, pp. 2351–2357 (2000)
6. Minor, M.A., Hirschi, C.R., Ambrose, R.O.: An Automated Tether Management System for Microgravity Extravehicular Activities. In: ICRA 2002, pp. 2289–2295 (2002)
7. Fukushima, E.F., Kitamura, N., Hirose, S.: A New Flexible Component for Field Robotics System. In: ICRA 2000, pp. 2583–2588 (2000)
8. Kitai, S., Tsuru, K., Hirose, S.: The proposal of Swarm Type Wall Climbing Robot System "Anchor Climber": The Design and Examination of Adhering mobile unit. In: IROS 2005, pp. 475–480 (2005)

New Measurement Concept for Forest Harvester Head

Mikko Miettinen, Jakke Kulovesi, Jouko Kalmari, and Arto Visala

Abstract. A new measurement concept for cut-to-length forest harvesters is presented in this paper. The cut-to-length method means that the trees are felled, delimbed and cut-to-length by the single-grip harvester before logs are transported to the roadside. The concept includes measurements done to standing trees before felling to calculate optimal length of logs. The modern forest harvesters use mechanical measurements for diameter and length.

In this paper, we will discuss different non-contact methods of measuring a tree stem before felling and during the cut-to-length process. Standing tree stems are measured with a 3D scanner and a computer vision systems. Trunk processing is measured with a computer vision system. Based on these new measurements, tree cutting pattern could be optimized and harvester automation increased, resulting in higher resource utilization.

1   Introduction

A long-term vision of the work presented here is that forest harvesters could be automatized to improve the overall efficiency and quality using advanced measurement technology. The Metrix project is a broad scale effort to realize this vision. The project studies new measurement technologies for forest harvester heads. The research goals are to measure and estimate the tree trunk dimensions and other quality variables with non-contact methods. The estimation is done in real time so that the optimal cutting pattern can be calculated and more detailed trunk information can be sent to processing mills. This research focuses on applying signal processing methods to machine vision, laser measurement and other optical measurement technologies in demanding forest environment.

Mikko Miettinen, Jakke Kulovesi, Jouko Kalmari, and Arto Visala

Department of Automation and Systems Technology, Helsinki University of Technology, PO Box 55000, 02015 TKK, Finland

e-mail: `firstname.lastname@tkk.fi`

A. Howard et al. (Eds.): Field and Service Robotics 7, STAR 62, pp. 35–44.

A new measurement concept for cut-to-length forest harvesters is presented in this paper. 3D scanner and machine vision based measurements are combined for measuring standing tree stems before felling. To help machine vision on the approach stage, structured light is also studied as a part of the project [6]. Felled tree trunks are fed through the head and cut to logs at desired lengths. Diameter and length of the processed logs are measured using machine vision system.

The research is a continuation to the work done in the Forestrix project [5, 8, 9]. The Forestrix project studied forest and tree trunk measurement technologies, signal processing methods and algorithms for semiautomatic control of forest harvesters. The main focus was to produce and update an accurate 2D tree map in real time, with diameter at breast height (DBH) information. In forest thinning operations, the tree map can support the harvester operator to select the right trees and to achieve optimal stand density. Semiautomatic harvester operation with tree map information was tested on a simulator. The collected data improves the verifiability of forest operations and the data can be used for planning future forest tasks.

This paper consists of the following sections: First, modern harvesters, tree parameters and measurement platforms are discussed. Second, the results including tree stem laser measurement, motion vision based structure estimation of the standing tree stem and machine vision based trunk measurement system for tree processing are presented. Finally, some conclusions about the applicability of the tested measurement concept are given.

1.1 Modern Forest Harvesters

Modern forest harvesters are already very efficient machines. Harvester heads have several functions including the felling, delimbing, diameter and length measuring, cutting to length, color marking and stump treatment. John Deere 745 harvester heads (Fig. 5 and 1) have been used in real forest tests during this project. Measurement and data gathering are very important in modern tree harvesting. The forest owner and the harvester contractors are paid according to the harvester measurements. Forest companies use harvester information to plan the subsequent forest operations.

In modern forest harvesters, the diameter sensors are connected to the feeding roller arms or the delimbing knives. The sensors are usually potentiometers. Diameter measurement depends on how wide the feeding arms open. The length of the tree is measured usually with a 2 channel incremental encoder. During processing, the tree trunk is pressed firmly against a measurement roller disc. see Fig. 1). Weather conditions influence the measurement accuracy, e.g., measurement roller spikes penetrate deeper into unfrozen wood than into frozen wood. Calibration is done regularly to guarantee the measurement quality.

1.2 Tree Parameters

Measurements of standing tree stems should include parameters like height of crown base, taper, sweep and lean. Calculating these parameters with ground-based laser

Fig. 1 Timberjack 1070 harvester with John Deere 745 head on test site on the left and standing tree measurement sensors attached to a 745 head on the right.

scanners for forest inventories have been studied [4, 10, 12]. But the possibility of measuring tree parameters with a moving forest harvester before felling is a novel one. Presently, the measurements performed in moder harvester heads are obtained too late for true cut-to-length pattern optimization. Thus, having measurements of the tree stem before felling gives valuable information that is not obtainable with traditional measuring implementations.

1.3 Measurement System

The measurement system consists of different sensors used for measuring standing tree stems and cut-to-length parameters. The system has been used on different platforms to collect measurement data. Different variations of the measurement systems have been developed during the project. This section describes the latest system.

The measurement system used in all terrain vehicle (ATV) and harvester head tests shown in Fig. 2 and in 1. The front box on ATV in Fig. 2 acts as a stand for the sensors while the box in the back contains a 24 volt battery for system power. The scanners and the measurement PC operate directly from the battery. The system sensors in Fig. 2 consist of 2D and 3D laser range finders, GPS receiver and MEMS inertial measurement unit (IMU). IMU is used to provide pose information of the platform and is essential for combining various measurements together. ATV measurements have been done with different development versions of the forest 3D scanner system. A stereo camera pair (Fig. 1 and 2) is used to measure standing tree structure using visual motion [7] and to tree trunk cut-to-length processing (5).

Forest harvester tests are done with the same measurement system as the ATV tests. Harvester head sensor arrangement for standing tree measurement is shown in Fig. 1. Harvester head cut-to-length processing measurement system is shown in Fig. 5. Processing measurements are done with a stereo camera pair. Measurement system sensors are mounted on a specially designed mounting attached to John Deere 745 harvester head. See Fig. 5 and 1.

Fig. 2 ATV platform with 3D scanner measurement equipment on the left and stereo camera pair on the right.

2 Standing Tree Stem Laser Measurement

Robust robot navigation in unstructured and outdoor environments is an unsolved problem. The absence of simple features leads to the need for more complex perception and modeling. This leads to a big variety of navigation algorithms and map representations, depending on the kind of environment, the degree of structuring and the target application. Many different outdoor Simultaneous Localization and Mapping (SLAM) algorithms have been studied in recent years [2, 13] and [1].

In this case, a scan correlation based method is used for short term sensor-based dead reckoning. There are numerous different scan correlation methods available to be used to sensor-based dead reckoning. The Iterative Closest Point (ICP) and Sum of Gaussian (SoG) methods are among the most popular. Different scan correlation methods are presented e.g. by Bailey [1].

Scan correlation is not enough to combine rotating 3D scanner measurements into meaningful tree measurement data. A SLAM based approach to harvester head localization is used to calculate head movement and combine measurements together. The method combines 2D laser localization with IMU measured pose information and height from the 3D scanner system to calculate 6 degrees of freedom (DOF) movement of the head and measurement data of the tree stem. The research with laser scanners, scan correlation and SLAM presented here is continuation to the work done in the Forestrix project [5, 8, 9]. Different filtering methods have been tested to provide the best possible 6DOF movement but the work is still in progress.

Tree stem 3D scanner measurement is studied to get parameters like height of crown base, taper, sweep, trunk dimensions, branches and lean. 3D point data measurement count depends on for how long the 6DOF movement estimation is accurate. If the movement can be accurately estimated for the whole movement to the tree, the point cloud collected is more precise and parameter calculations are easier. If the movement accuracy is low, the measurement error in the point cloud is too great for precise parameter calculation. From the measurements taken in movement

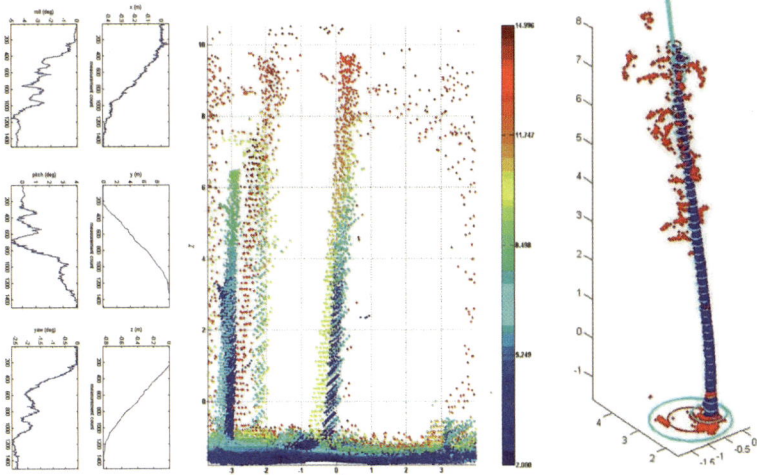

Fig. 3 Estimated 6DoF movement for one approach measurement in well defined pine forest (ATV) (left). 3D laser scan from movement (center). Darker measurements when the laser is close to the tree. Calculated tree parameters (right).

shown in Fig. 3 we can see that in this long approach (approximately 7 meter), accuracy of the movement has not been the best possible to combine all measurement points into one point cloud. The points measured from close do not match the points measured from a greater distance. In this case, it is better to use shorter 3D measurements and calculate parameter values from each short measurement together to get better results. 6DOF movement calculated for the same approach is shown in Fig. 3. The height estimate has the highest variance of the 6Dof because the rough forest ground and under foliage provide a poor height estimate. The ground height estimation from 3D scanner system is better if the ground is flat and smooth.

The ground level is searched from the measurement point cloud using RANSAC algorithm [3]. The ground level detection helps us to find the location the tree stump. The trunk diameters are calculated in different height segments of the trunk depending on the point count. The diameter is calculated using cylinder fitting ([4, 10] and [12]) or simple circle fitting. Taper information is used to estimate the true stem and extract the branches in the next trunk piece. Crown base height can be estimated from the trunk diameter calculations. The extracted branches are used to determine the height of the dry branches and the crown height. The sweep and lean can be estimated from the tree stem diameter and center points. The branches are excluded from the diameter calculations and are shown in different color in Fig. 3. The forest 3D scanning system with 90 degree field of view is designed for measuring only tree stems up to crown height. Goal is to measure tree stem diameters with error less than 1 cm. The research done so far, indicates that this can be achieved.

3 Motion Vision Based Structure Estimation in Natural Forest Environment

Motion vision can be used to determine world structure from a video sequence. A general research problem in this context is to acquire relevant information through relative motion of cameras and the environment. Using cameras to sensing the near surroundings is beneficial due to mass production of camera components (low cost) and wide availability (good support). In addition, cameras are generally applicable to various environmental conditions assuming that related research problems are solved. Forest is a good example of a challenging environment where e.g. occlusion, varying light conditions and natural uncontrolled conditions provide difficulties for computer vision tasks and algorithms.

The precise objective of this work was to measure trees from a distance using a motion vision approach. In addition, the whole visible environment structure is reconstructed. Monochrome digital video cameras were attached on an ATV and a video sequence was recorded while the ATV approaches a tree. Data from a single camera was used but a calibrated stereo camera system was present to gather richer data for future use. Based on the measurements, tree cutting could be optimized and harvester automation increased, resulting in higher resource utilization efficiency. An example of a structure estimate for an instantaneous moment in forest is shown in Fig. 4.

A broad range of motion vision methods were explored. Based on the findings, the final solution consists of three sub-components: block matching, motion estimation and triangulation. A consequent pair of images from a video sequence is used as the source data. The block matching algorithm uses a hierarchical image pyramid approach with three phases at each level of hierarchy. First, integer pixel precision solution for a dense optical flow field is computed with a limited range full search. Second, linear sub-pixel interpolation is used to fine-tune the integer precision results. Third, the dense optical flow field is filtered adaptively to reduce noise and to compensate for occlusion errors. Estimating motion from the optical flow is done by using a selected subset of points for which an optical flow fit error is minimized using numerical optimization to solve for both the motion parameters and depth for the selected points simultaneously. Finally, solving for the dense structure from the motion and optical flow can be formulated as a linear triangulation problem and is thus easy to calculate. Both block matching and triangulation can be computed in parallel. Thus, parallel computation power of multiple processor cores or modern graphics cards can be used in the future to meet the relative high computation cost of the algorithms.

The results obtained show robustness with respect to environmental challenges and the main objective of tree segmentation for measurement is achieved. In addition, overall depth map construction quality is sufficient for a more broad range of potential applications. In summary, the results prove that motion vision methods can be applied in uncontrolled forest environment conditions. More detailed explanation on motion vision based structure estimation in natural forest environment can be found in [7].

Fig. 4 A structure estimate for an instantaneous moment in a forest. Color brightness indicates the depth; the brighter, the further away.

4 Machine Vision Based Trunk Measurement System

The objective was to develop a system capable of measuring the length and thickness of a processed trunk. Current solutions for measuring the length utilize a cog wheel that follows the surface of the wood. Accuracy of present systems depends on the qualities of the wood, e.g., how soft the bark is and how many branches there are. The measurement error is usually around 1-2 cm. The thickness of the trunk is measured mechanically with two arms.

Stereo camera pair was used to track both the pose of the harvester head and the movement of the trunk. The system uses two black and white Foculus FO134S cameras having a resolution of 640x480. In the test, the capture rate used was 60 frames per second. The cameras were fixed to the upper part of the harvester head. Calibration of camera and stereo parameters was done in advance. As lighting we

Fig. 5 Harvester head with stereo camera pairs and lights attached (above). Left and right stereo images showing selected features, trunk estimate and the tracked chessboard pattern.

used originally halogen lamps and in later test LED lights synchronized with the cameras. Stereo camera pair in harvester head is shown in Fig. 5.

The basic problem to be solved was that there are usually three different motions in respect to the cameras that have to be distinguished; harvester heads motion, trunk's motions and background's motion. All of these motions have to be presented using both rotation and translation.

The basic algorithm begins by selecting some good Harris corners from the trunk. Then the features were stereo matched and tracked in consecutive frames so that 3D motion for corresponding points could be calculated. To ease the selection and matching of features, a cylindrical estimate of the trunk was used (Fig. 5). It was found that the number of false matches between left and right images was reduced when a priori estimate of the z-coordinates was used.

Lower part of the harvester head rotates around a single axis and for that reason it is possible to reduce the six parameter pose estimation problem of the harvester head to a one parameter estimation problem. The pose of the harvester head can be determined by tracking selected features (e.g. [11]) or by estimating the pose of the chessboard pattern fixed in the harvester head (Fig. 5.

Two different approaches were used to determine the length of the wood. The first method tracks harvester head and trunk separately. A single point in the end of the trunk is selected and its location is estimated from the translation and rotation of the trunk. The length of the wood is then determined by the distance of the original point and the estimated end point.

The second approach is based on the assumption that the trunk's rotation is nearly identical to harvester head's rotation. When the rotation and translation of the harvester head is estimated, it is possible to remove it from the trunk's motion vectors. When harvester head's motion is compensated, the translational movement of the trunk can be extracted and the length of the trunk determined.

Seven trees have been cut to length, estimated and compared to handmade measurements. All the measured trunks were about three meters long. The absolute differences between estimates and the real lengths with the first approach was in five cases no more than 1 mm, with largest estimate error being 12 mm. The second method gave slightly worse results, but still the error was in most cases 6 mm or less.

The data measured has been relatively easy and the more challenging data has not yet been fully analyzed. When the trunk moves faster and harvester head swings more, estimation of motion becomes harder. Algorithms for estimating the 3D structure and thickness of the trunks are still under development.

5   Conclusions and Future Work

A new concept to cut-to-length forest harvester remote sensing system was presented in this paper. The concept includes new non-contact measurements done before felling and during cut-to-length processing. The research goals are to measure and estimate the tree stem dimensions and other quality variables affecting the processing of the trunk. This research focuses on applying signal processing methods to machine vision, laser measurement and other optical measurement technologies in demanding forest environment. Based on the research, tree cutting pattern could be optimized and harvester automation increased, resulting in higher resource utilization.

The standing tree stem laser measurement, the machine vision based trunk measurement and the motion vision based structure estimation systems were presented. The current measurement and estimation methods implemented show promise that the measurement goals set for the project will be met in well defined forests. However, dense and cluttered forest environment is tough for precision measurements and will need more research.

The work presented in this paper is a part of an ongoing research project. Harvester head tests are ongoing and algorithm development is unfinished. Precision of the measured parameters is extremely important and research to better compare forest machine, hand, laser and machine vision measured tree stem parameters is underway.

Acknowledgements. The authors gratefully acknowledge the contribution of the Finnish Funding Agency for Technology and Innovation (Tekes), participating companies and reviewers' comments.

References

1. Bailey, T.: Mobile Robot Localisation and Mapping in Extensive Outdoor Environments. PhD thesis, University of Sydney, Australian Centre for Field Robotics (2002)
2. Brenneke, C., Wulf, O., Wagner, B.: Using 3D Laser Range Data for SLAM in Outdoor Environments. In: Proceedings of the 2003 IEEE/RSJ Intl. Conference on intelligent Robots and Systems, Las Vegas, Nevada, USA (2003)
3. Fischler, M.A., Bolles, R.C.: Random sample consensus: A paradigm for model fitting with applications to image analysis and automated cartography. Communications of the ACM 24, 381–395 (1981)
4. Hopkinson, C., Chasmer, L., Young-Pow, C., Treitz, P.: Assessing forest metrics with a ground-based scanning lidar. Canadian Journal of Forest Research (2004)
5. Jutila, J., Kannas, K., Visala, A.: Tree measurement in forest by 2D laser scanning. In: Proc. International Symposium on Computational Intelligence in Robotics and Automation, Jacksonville, USA, pp. 491–496 (2007)
6. Kauhanen, H.: Close Range Photogrammetry - Structured Light Approach for Machine Vision Aided Harvesting. In: Proc. ISPRS XXI Congress, Beijing, China, pp. 75–80 (2008)
7. Kulovesi, J.: Structure from visual motion in forest environment. Master's Thesis, Helsinki University of Technology, Department of Automation and Systems Technology, Finland (2008)
8. Miettinen, M., Ohman, M., Visala, A., Forsman, P.: Simultaneous localization and mapping for forest harvesters. In: Proc. IEEE International Conference on Robotics and Automation, Rome, Italy, pp. 517–522 (2007)
9. Ohman, M., Miettinen, M., Kannas, K., Jutila, J., Visala, A., Forsman, P.: Tree measurement and simultaneous localization and mapping system for forest harvesters. In: The 6th International Conference on Field and Service Robotics, Chamonix, France, p. 1 (2007)
10. Pfeifer, N., Gorte, B., Winterhalder, D.: Automatic Reconstruction of single trees from terrestrial laser scanner data. In: IAPRS, Istanbul, Turkey, vol. XXXV, pp. 114–119 (2004)
11. Roth, G., Laganiere, R., Gilbert, S.: Robust Object Pose Estimation From Feature-Based Stereo. IEEE Transactions on Instrumentation and Measurement 55(4), 1270–1280 (2006)
12. Thies, M., Pfeifer, N., Winterhalder, D., Gorte, B.G.H.: Three-dimensional reconstruction of stems for assessment of taper, sweep and lean based on laser scanning of standing trees. Scandinavian Journal of Forest Research 19(6), 571–581 (2004)
13. Thrun, S., Burgard, W., Fox, D.: Probabilistic Robotics. The MIT press, Cambridge (2005)

Expliner – Toward a Practical Robot for Inspection of High-Voltage Lines

Paulo Debenest, Michele Guarnieri, Kenskue Takita, Edwardo F. Fukushima,
Shigeo Hirose, Kiyoshi Tamura, Akihiro Kimura, Hiroshi Kubokawa,
Narumi Iwama, Fuminori Shiga, Yukio Morimura, and Youichi Ichioka

Abstract. Preventive maintenance of high-voltage transmission power lines is a dangerous task, but the obstacles mounted on the lines have so far prevented the automation of this task. Expliner aims to overcome such obstacles by controlling actively the position of its center of mass, thus changing its configuration as needed when moving on the power lines. This work presents the design of Expliner and results of field experiments performed with very high voltages to prove the effectiveness of the proposed concept.

1 Introduction

Urban centers and industries rely heavily on electric energy provided by an electric grid. This electric energy is needed for safety, transportation, sanitation and other essential functions. If there is any problem in the electric grid linking the energy generation plants and urban or industrial centers, the supply of energy may have to be stopped, affecting in a negative way the lives of millions of people [1]. In order to avoid such problems, the preventive maintenance of electric power lines is of vital importance. However, this is a very dangerous and time-demanding job, requiring specialized people to walk on the lines, suspended several tens of meters above the ground in remote areas like mountains and deserts. In addition, usually the energy supply must be interrupted momentarily for such inspections.

Paulo Debenest, Michele Guarnieri, and Kenskue Takita
HiBot Corp.

Edwardo F. Fukushima and Shigeo Hirose
Toko Institute of Technology

Kiyoshi Tamura and Akihiro Kimura
Kansai Electric Power Corp.

Hiroshi Kubokawa, Narumi Iwama, and Fuminori Shiga
J-Power Systems Corp.

Yukio Morimura and Youichi Ichioka
Kanden Engineering

A. Howard et al. (Eds.): Field and Service Robotics 7, STAR 62, pp. 45–55.
springerlink.com   © Springer-Verlag Berlin Heidelberg 2010

There have been proposals to carry on the inspection of power lines in more efficient ways. One involves the use of helicopters to check visually the conditions of the cables [2]. However, even in the case of remotely controlled helicopters [3], it may be dangerous to fly close to power lines, and the visual inspection will provide images of only the upper side of the cables. Another approach is to have people working directly on live lines with special gear to insulate them from the very high voltages [4]. This method requires good access to the cables from the ground, which may not be possible in mountains or other remote locations. In addition, it still depends on people moving on the cables, a risky operation.

In the field of robotics, several researchers have proposed machines for remote inspection of power lines. However, the presence of obstacles on the cables (such as cable spacers and clamp suspenders connected to the towers) makes the automation of this task more difficult. Campos et al. have proposed a machine to inspect the warning spheres installed on the high-voltage lines [5], but the machine is not able to overcome any sort of obstacle. Sawada et al. have proposed a robot with several degrees of freedom in order to cross clamp suspenders [6], but this resulted in a bulky and heavy machine that is difficult to carry to the field. Tang and Zhu have proposed a different machine that moves on the ground line, above the high voltage lines and with fewer obstacles [7][8]. However, the inspection performed with this machine is similar to the one with a helicopter, since only the upper side of the high voltage cables can be inspected with a video camera. Montambault and Pouliot are developing a very practical machine for inspection of single high-voltage lines [9][10], including the defrosting of frozen wires. In Japan, where the current research is being conducted, the high-voltage lines are grouped in bundles of four cables, and this presents different challenges that have not been solved in a satisfactory way until now.

2 Proposal of Automation

Figure 1 shows Expliner, the machine proposed by the authors to perform remote inspection of high-voltage lines. Its mobility is based on pulleys, placed on the upper cables, driven by electric motors. The pulley units are connected to a single horizontal base.

The horizontal base also has a vertical element, with a 2-DOF manipulator connected to its lower end. On the tip of the manipulator there is a counter-weight, housing batteries and electronics. Therefore, by moving the manipulator, it is possible to change the position of the center of mass of the machine. Thus, it becomes possible to lift one of the two pulley units in order to overcome large obstacles, as presented for the first time in [11].

If Expliner would move only in a straight line, it would be necessary to have only 1 degree of freedom in the manipulator. However, there are times when the machine must perform rather complex motions, such as when crossing clamp suspenders, or when moving on an inclined loading pipe between the tower and the

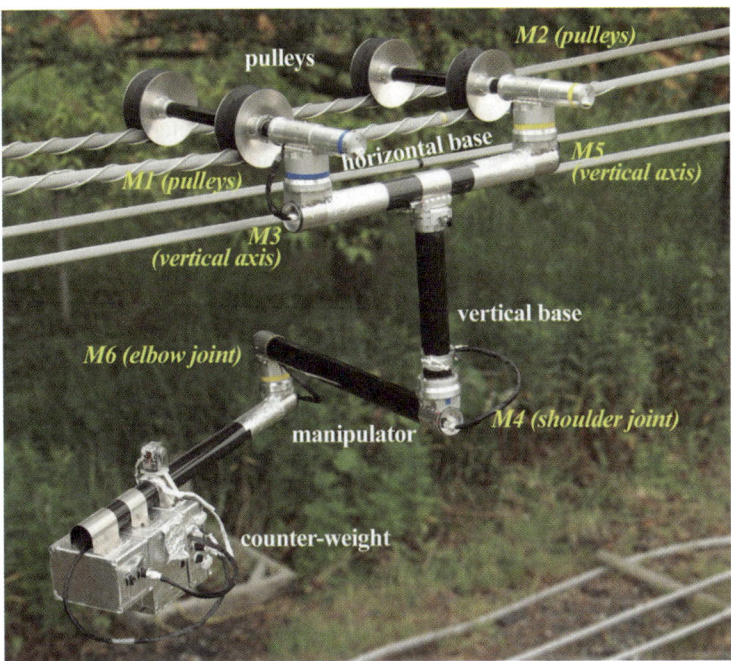

Fig. 1 Concept of Expliner with main components.

Fig. 2 Loading pipe between tower and cables (left) and clamp suspender (right).

main cables, as displayed in Figure 2. For such cases, not only the 2 DOFs of the manipulator are needed, but also the rotation around the vertical axis of each motion unit. These motions have already been described in [11].

The next sections will focus on the mechanical design of Expliner, and will also describe the main differences between the current machine and the first prototype, presented in [11].

3 Mechanical Design

Friction

In order to dimension the actuators, it was necessary to determine the friction between the rubber-coated pulleys and the power cables. In spite of the theoretical
models taking into account the deformation of the rubber around the contact area,
the lack of friction data made it necessary to determine the friction empirically, by
applying loads on the pulleys, mounted on the cables, and measuring the force required to pull them at a constant speed. The experimental setup and the data are
displayed in Figure 3, and show that the pulling force is linearly proportional to
the load applied on the pulleys, yielding a coefficient of friction that was used for
the selection of the actuators.

Actuators

All actuators were designed by the authors, so that the motors and transmissions
be assembled in the most compact and effective way. In addition, the main components of Expliner can be assembled by sliding joints, which were also incorporated in the design of the cases of the actuators.

Each motion unit is equipped with a 200W brushless motor connected to a train
of planetary gears with a total reduction ration of 60, maximum continuous torque
of 40.4Nm and speed of 60.6rpm. With the pulleys of Expliner, this represents a
linear speed of between 23m/min and 29m/min, depending on the size of the cables where the machine is moving.

The vertical axis of each motion units is powered by a 200W brushless motor
with Harmonic-Drive reduction embedded in its case, resulting in a maximum
continuous torque of 558.4Nm and 3.6rpm. The same actuator is used to drive the
first joint of the manipulator (the "shoulder"), while the second joint (the "elbow")
is driven by a 200W brushless motor with embedded Harmonic-Drive, and with a
maximum continuous torque of 349Nm and speed of 5.7rpm.

The dimensioning of the actuators of the manipulator took into account lifting
the counter-weight when the manipulator is in a vertical configuration, something
that does not happen in real applications, but which may happen during tests.

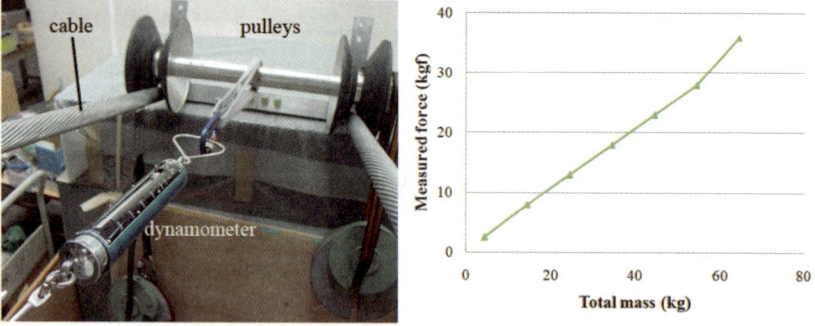

Fig. 3 Friction experimental setup (left) and results from experiments (right).

Fig. 4 Sliding joint for easy assembly.

Structure

The structure of Expliner is composed mainly of CFRP pipes, and they were designed to withstand the extreme postures required when overcoming obstacles.

In order to make Expliner easy to carry to the inspection sites, it was designed to be easily assembled with sliding joints and stopping pins, as shown in Figure 4. In places where the wiring might prevent an easy assembly, slide-in connectors were also employed.

Safety Hooks

Another improvement from the first prototype of Expliner was the introduction of the safety hooks, which keep the robot attached to the power lines even in case of sudden winds or accidents. The safety hooks must provide enough clearance to operate even with large loading pipes, but must be very close to the pulley in its engaged position to prevent the cable from sliding at the gap between hook and

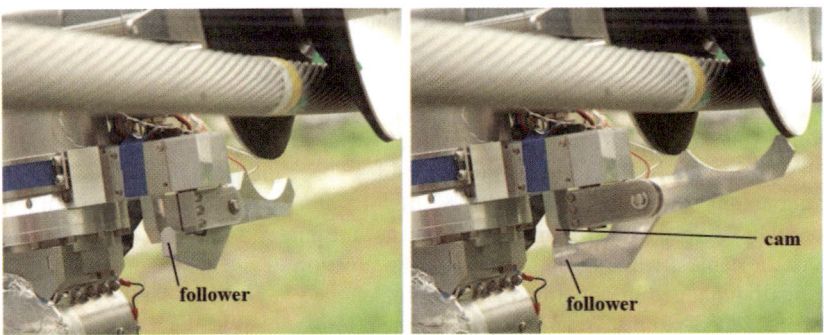

Fig. 5 Safety Hook disengaged (left) and engaged (right).

pulley. Therefore, the safety hooks were assembled with a cam mechanism that modifies the angle of the hook as it moves from the disengaged position (open) to the engaged position (closed), as shown in Figure 5. Each safety hook is driven by a worm-gear transmission, so it is self-locking.

It is necessary to disengage the safety hooks when crossing obstacles such as cable spacers or clamp suspenders, but also when moving on the loading pipe. Care is taken to insure that at all times at least one safety hook is engaged, so that Expliner would not fall from the cables in case of malfunctions or accidents.

The new version of Expliner is easier to assemble, easier to operate, and also lighter that the first prototype (60kg, against 85kg of the first version).

4   Control Architecture

The control architecture has been greatly improved since the first prototype presented in [11]. Each actuator has its motor drive (HiBot TITech Drive Version 1) positioned close to it, in a splash-proof assembly. Communication between the motor drives and the main micro-controller, positioned in the counter-weight, was implemented with CAN bus.

By installing the motor drives close to the motors, wiring in the joints of the robot was greatly reduced and simplified. In addition, the entire machine is shielded for operation at very high voltages (500kV) and with high electric currents (1400A).

The counter-weight houses not only the main micro-controller (HiBot SH2Tiny), but also the wireless communication devices, which connect to a portable control station by wireless LAN. The wireless communication diagram is presented in Figure 6.

5   Human-Machine Interface

Expliner is controlled from a portable control case shown in the lower right corner of Figure 6. From this control case it is possible to drive all joints independently, but this would result in cumbersome and time-demanding operations. Therefore, the obstacle-crossing motions were automated, but are always controlled by the operator, as described next.

When crossing a clamp suspender (Figure 7), the operator controls the position of the center of mass by moving a joystick forward or backward, while the two joints of the manipulator are driven automatically, so that the center of mass is always positioned between the two upper cables. Once the front motion unit is lifted with enough clearance from the cables, the operator starts the next step in the motion, which consists of rotating the front motion unit and bringing the pulleys out from the cables, while at the same time the position of the center of mass is automatically changed to accommodate for this change in posture. These motions are also achieved simply by moving a joystick forward or backward. Once this is completed, the operator moves the machine forward. When the front unit has passed the obstacle, it is brought back to its initial angle by a single joystick

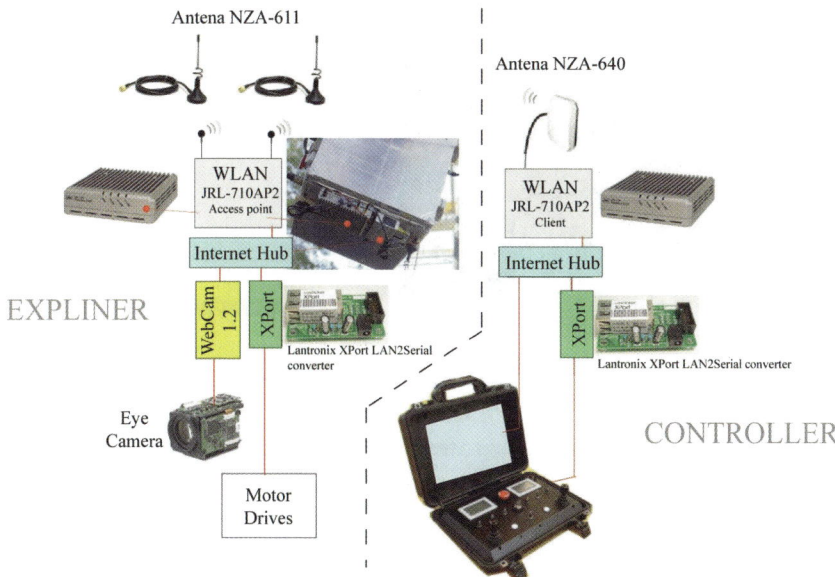

Fig. 6 Wireless communication diagram

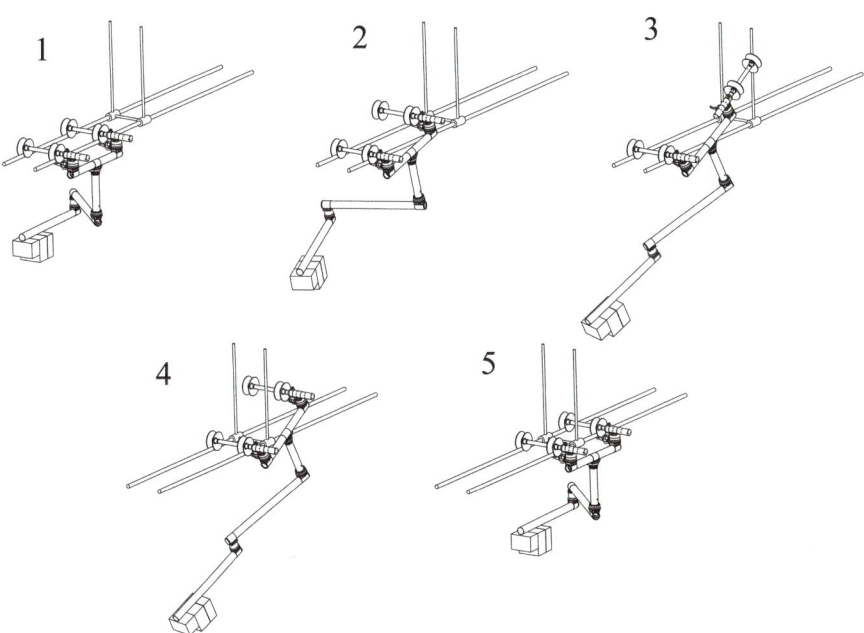

Fig. 7 Sequence of motions for overcoming clamp suspender (first half of motion)

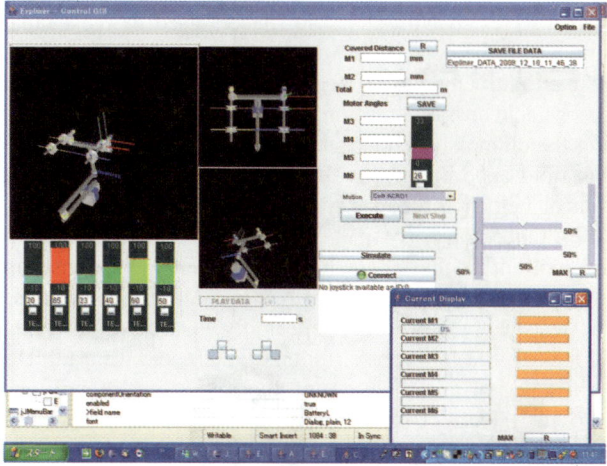

Fig. 8 Graphic Interface with data from each actuator and other sensors.

motion. Finally, when the front motion unit is aligned to the cables, the center of mass is brought back to the center position, which brings the front pulley unit back onto the cables. This process is then repeated to lift the rear motion unit.

All joints are driven in position control, with data from encoders. All motion patterns and limits are pre-defined, in order to make operation as fast and safe as possible. Therefore, the joints of Expliner will move until the defined angle, and will always go back to the initial postures with an accuracy of $0.1°$.

Similar automated patterns have been implemented for other motions, such as moving between the loading pipe and the cables, so that the center of mass is always kept in a safe position. This automated control is especially useful when Expliner is moving on the single loading pipe. However, even with this level of automation, the presence of a human operator is always required due to the complexity of the environment.

The operator is always receiving telemetry data from the robot, including the temperature of each motor, the current consumption, the voltage level, the angle of each joint, and the attitude of the machine, with its spatial configuration. The graphical interface presented in Figure 8 makes the control of the machine more intuitive and easy to understand.

6 Field Experiments

The latest prototype of Expliner has undergone intense field tests with very promising results. Speed tests performed on horizontal cables indicated an average speed of 27m/min, above the requirement of 20m/min. The required speed of 20m/min was set based on limitations in the sensors used to inspect the cables, which will be presented in the near future.

Expliner was also tested with all cable spacers used in West Japan (Figure 9), where it is supposed to be deployed. Additionally, Expliner was tested on inclined cables, with a maximum angle of 30 degrees, and even in such conditions was able to perform the obstacle crossing motion, as displayed in Figure 9.

Overcoming the clamp suspender and crossing to the other side of a tower was also verified in field experiments, as shown in Figure 10. The introduction of the automated control helped to reduce the time necessary to perform this motion, from around 12 minutes to approximately 3 minutes (with a trained operator).

Finally, the motion on the single loading pipe and the transition motion between loading pipe and cable were confirmed in real field conditions, as shown in Figure 11. All tests were performed repeatedly on test cables and also on live wires with 500kV. In the latter case, footage obtained with a ultra-violet camera revealed the existence of a corona around the robot, but no malfunction was observed, thus proving the effectiveness of the shielding.

Fig. 9 Cable spacers (left) and Expliner crossing obstacle on inclined cable (right).

Fig. 10 Expliner lifting the front motion unit (left) and rotating the motion unit after crossing the clamp suspender (right).

Fig. 11 Expliner moving on loading pipe (left) and performing an automated motion.

7 Conclusions

With the development of Expliner, the automation of inspection of high voltage power lines became one step closer to reality. The field tests presented in this paper showed that the concept of changing the position of the center of mass of the machine can be employed in dangerous and complex applications such as power lines suspended tens of meters above the ground. The developments in human-machine interface and the automation of complex motions have made the control of the machine faster and more reliable. The deployment of sensors to acquire data from the power lines is a future step, but the authors are already working on it, and plan to present concrete results in the near future.

References

1. Andersson, G., et al.: Causes of the 2003 major grid blackouts in North America and Europe, and recommended means to improve system dynamic performance. IEEE Trans. Power Systems 20(4), 1922–1928 (2005)
2. Ishino, R., Tsutsumi, F.: Detection System of Damaged Cables Using Video Obtained from an Aerial Inspection of Transmission Lines. In: IEEE Power Engineering Society General Meeting, vol. 2, pp. 1857–1862 (2004)
3. Golightly, I., Jones, D.: Visual control of an unmanned aerial vehicle for power line inspection. In: Proc. IEEE Int'l. Conference on Advanced Robotics, pp. 288–295 (2005)
4. Harano, H., Syutou, K., Ikesue, T., Kawabe, S.: The development of the Manipulator method for 20kV class overhead distribution system. In: Transmission and Distribution Conference and Exhibition, vol. 6, pp. 2112–2116 (2002)
5. Campos, M.F.M., et al.: A mobile manipulator for installation and removal of aircraft warning spheres on aerial power transmission lines. In: Proc. IEEE Int'l Conf. on Robotics & Automation, Washington D.C., USA, pp. 3559–3564 (2002)
6. Sawada, J., Kusumoto, K., Munakata, T., Maikawa, Y., Ishikawa, Y.: A Mobile Robot for Inspection of Power Transmission Lines. IEEE Trans. on Power Delivery 6(1), 309–315 (1991)

7. Tang, L., Fu, S., Fang, L., Wang, H.: Obstacle-navigation strategy of a wire-suspend robot for power transmission lines. In: Proc. IEEE Int'l Conf. on Robotics and Biomimetics, Shenyang, China, pp. 82–87 (2004)
8. Zhu, X., Wang, H., Fang, L., Zhao, M., Zhou, J.: Dual Arms Running Control Method of Inspection Robot Based on Obliquitous Sensor. In: Proc. 2006 IEEE/RSJ Int'l Conference on Intelligent Robot and Systems, Beijing, China, pp. 5273–5278 (2006)
9. Montambault, S., Pouliot, N.: The HQ LineROVer: Contributing to Innovation in Transmission Line Maintenance. In: Proc. IEEE 10th Int'l Conf. on Transmission and Distribution Construction, Operation and Live-Line Maintenance, pp. 33–40 (2003)
10. Montambault, S., Pouliot, N.: Geometric design of the LineScout, a teleoperated robot for power line inspection and maintenance. In: Proc. 2008 IEEE Intl' Conf. on Robotics & Automation, pp. 3970–3977 (2008)
11. Debenest, P., et al.: Expliner – Robot for Inspection of Transmission Lines. In: Proc. 2008 IEEE Intl' Conf. on Robotics & Automation, pp. 3978–3984 (2008)

Part II
Perception and Control

Experimental Study of an Optimal-Control-Based Framework for Trajectory Planning, Threat Assessment, and Semi-Autonomous Control of Passenger Vehicles in Hazard Avoidance Scenarios

Sterling J. Anderson, Steven C. Peters, Tom E. Pilutti, and Karl Iagnemma[*]

Abstract. This paper describes the design of an optimal-control-based active safety framework that performs trajectory planning, threat assessment, and semi-autonomous control of passenger vehicles in hazard avoidance scenarios. The vehicle navigation problem is formulated as a constrained optimal control problem with constraints bounding a navigable region of the road surface. A model predictive controller iteratively plans an optimal vehicle trajectory through the constrained corridor. Metrics from this "best-case" scenario establish the minimum threat posed to the vehicle given its current state. Based on this threat assessment, the level of controller intervention required to prevent departure from the navigable corridor is calculated and driver/controller inputs are scaled accordingly. This approach minimizes controller intervention while ensuring that the vehicle does not depart from a navigable corridor of travel. It also allows for multiple actuation modes, diverse trajectory-planning objectives, and varying levels of autonomy. Experimental results are presented here to demonstrate the framework's semi-autonomous performance in hazard avoidance scenarios.

1 Introduction

Recent traffic safety reports from the National Highway Traffic and Safety Administration show that in 2007 alone, over 41,000 people were killed and another 2.5 million injured in motor vehicle accidents in the United States [1]. The

Sterling J. Anderson, Steven C. Peters, and Karl Iagnemma
Department of Mechanical Engineering, Massachusetts Institute of Technology,
77 Massachusetts Ave. Cambridge, MA 02139, USA
e-mail: {ster,scpeters,kdi}@mit.edu

Tom E. Pilutti
Ford Research Laboratories, Dearborn, MI 48124, USA
e-mail: tpilutti@ford.com

[*] Corresponding author.

A. Howard et al. (Eds.): Field and Service Robotics 7, STAR 62, pp. 59–68.
springerlink.com   © Springer-Verlag Berlin Heidelberg 2010

longstanding presence of passive safety systems in motor vehicles, combined with the ever-increasing influence of active systems, has contributed to a decline in these numbers from previous years. Still, the need for improved collision avoidance technologies remains significant.

Recent developments in onboard sensing, lane detection, obstacle recognition, and drive-by-wire capabilities have facilitated active safety systems that share steering and/or braking control with the driver [2,3]. Among existing proposals for semi-autonomous vehicle navigation, lane-keeping systems using audible warnings [4], haptic alerts [5], steering torque overlays [6], and various combinations of these have been developed [7].

Many of the navigation systems developed in previous work address only one piece of the active safety problem. While some use planning algorithms such as rapidly-exploring random trees [3], evolutionary programming [8] or potential fields analysis [9] to plan a safe vehicle path, others simply begin with this path presumed [10]. The threat posed by a particular path is seldom assessed by the controller itself and is often only estimated by a simple threat metric such as lateral vehicle acceleration required to track the path [11]. Finally, hazard avoidance is commonly performed using one or more actuation methods without explicitly accounting for the effect of driver inputs on the vehicle trajectory. Such controllers selectively replace (rather than assist) the driver in performing the driving task. Yu addressed this problem in mobility aids for the elderly by designing an adaptive shared controller which allocates control authority between the human user and a controller in proportion to the user's current and past performance [12]. While sufficient to control low-speed mobility aids, this reactive approach to semi-autonomy is not well suited for higher-speed applications with significant inertia effects and no pre-planned trajectory.

In this paper, a framework for passenger vehicle active safety is developed that performs vehicle trajectory planning, threat assessment, and hazard avoidance in a unified manner. This framework leverages the predictive and constraint-handling capabilities of Model Predictive Control (MPC) to plan trajectories through a preselected corridor, assess the threat this path poses to the vehicle, and regulate driver and controller inputs to maintain that threat below a given threshold. The next section describes the semi-autonomous control framework and its associated trajectory prediction, control law, threat assessment, and intervention law. Experimental setup and results are then presented, and the paper closes with general conclusions and recommendations.

2 Framework Description

The navigation framework operates as follows. First, an objective function is established to capture desirable performance characteristics of a safe/"optimal" vehicle path. Boundaries tracing the edges of the drivable road surface are assumed to have been derived from forward-looking sensor data and a higher-level corridor planner. These boundaries establish constraints on the vehicle's projected position and are used together with a model of the vehicle dynamics to calculate an optimal sequence of inputs and the associated vehicle trajectory. The predicted trajectory

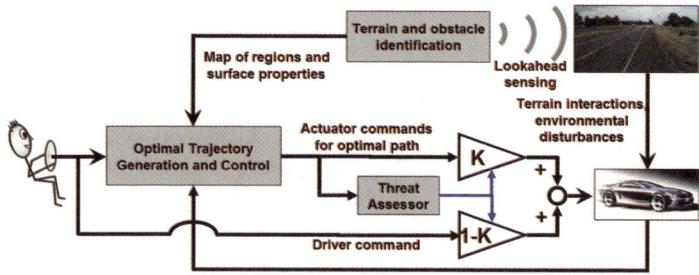

Fig. 1 Diagram of an active safety system.

is assumed to be a "best-case" scenario that poses the minimum threat to the vehicle given its current state. This threat is then used to calculate the intervention required to prevent departure from the navigable corridor and driver/controller inputs are scaled accordingly. Fig. 1 shows a block diagram of this system.

In this paper it is assumed that road lane data is available and that road hazards have been detected, located, and mapped to form the boundaries of a 2-dimensional corridor of travel. Radar, LIDAR, and vision-based lane-recognition systems [3,13], along with various sensor fusion approaches [14] have been proposed to provide the lane, position, and environmental information needed by this framework. Additionally, where multiple corridor options exist, it is assumed that a high-level path planner has selected a single corridor through which the vehicle should travel.

2.1 Vehicle Path Planning

The best-case (or baseline) path through a given region of the state space is established by a model predictive controller. Model Predictive Control is a finite-horizon optimal control scheme that iteratively minimizes a performance objective defined for a forward-simulated plant model subject to performance and input constraints. At each time step, t, the current plant state is sampled and a cost-minimizing control sequence spanning from time t to the end of a control horizon of n sampling intervals, $t+n\Delta t$, is computed subject to inequality constraints. The first element in this input sequence is implemented at the current time and the process is repeated at subsequent time steps. The basic MPC problem setup is described in [15].

The vehicle model used in this paper considers the kinematics of a 4-wheeled vehicle, along with its lateral and yaw dynamics. Vehicle states include the position of its center of gravity $[x, y]$, its yaw angle ψ, yaw rate $\dot{\psi}$, and sideslip angle β, as illustrated in Fig. 2. Table 1 defines and quantifies this model's parameters.

Tire compliance is included in the model by approximating lateral tire force (F_y) as the product of wheel cornering stiffness (C) and wheel sideslip $(\alpha$ or β for front or rear wheels respectively) as shown in (1).

$$F_y = C\alpha \tag{1}$$

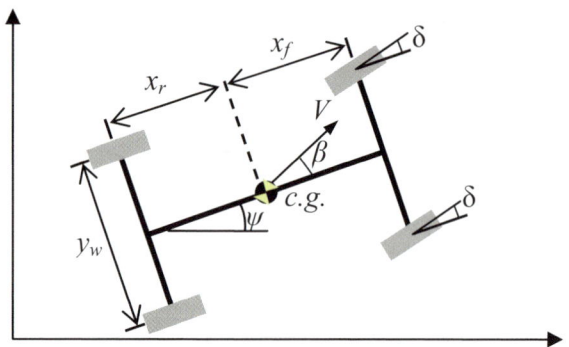

Fig. 2 Vehicle model used in MPC controller.

Table 1 Vehicle model parameters.

Symbol	Description	Value [units]
m	Total vehicle mass	2050 [kg]
I_{zz}	Yaw moment of inertia	3344 [kg·m^2]
x_f	C.g. distance to front wheels	1.43 [m]
x_r	C.g. distance to rear wheels	1.47 [m]
y_w	Track width	1.44 [m]
C_f	Front cornering stiffness	1433 [N/deg]
C_r	Rear cornering stiffness	1433 [N/deg]
μ	Surface friction coefficient	1

Linearized about a constant speed and assuming small slip angles, the equations of motion for this model are (where δ represents the steering angle input)

$$\dot{x} = V \tag{2}$$

$$\dot{y} = V(\psi + \beta) \tag{3}$$

$$\dot{\beta} = \frac{-(C_r + C_f)}{mV}\beta + \left(\frac{(C_r x_r - C_f x_f)}{mV^2} - 1\right)\dot{\psi} + \frac{C_f}{mV}\delta \tag{4}$$

$$\ddot{\psi} = \frac{(C_r x_r - C_f x_f)}{I_{zz}}\beta - \frac{(C_r x_r^2 + C_f x_f^2)}{I_{zz}V}\dot{\psi} + \frac{C_f x_f}{I_{zz}}\delta \tag{5}$$

where C_f and C_r represent the cornering stiffness of the lumped front wheels and the lumped rear wheels, and x_f and x_r are the longitudinal distances from the c.g. of the front and rear wheels, respectively.

2.1.1 Constraint Setup and Objective Function Description

As mentioned above, this framework assumes that the environment has been delineated previously. The boundaries of the navigable road surface at each timestep are then described by the constraint vectors

$$\mathbf{y}^{y}_{max}(k) = \left[y^{y}_{max}(k+1) \quad \cdots \quad y^{y}_{max}(k+p) \right]^{T}$$

$$\mathbf{y}^{y}_{min}(k) = \left[y^{y}_{min}(k+1) \quad \cdots \quad y^{y}_{min}(k+p) \right]^{T}$$

(6)

By enforcing vehicle position constraints at the boundaries of the navigable region of the road surface (i.e. the lane edges on an unobstructed road), the controller forces the MPC-generated path to remain within the constraint-bounded corridor whenever dynamically feasible. Coupling this lateral position constraint with input constraints $\mathbf{u}_{min/max}$, input rate constraints $\Delta\mathbf{u}_{min/max}$, and vehicle dynamic considerations, the navigable operating corridor delineated by $\mathbf{y}^{y}_{max}$ and $\mathbf{y}^{y}_{min}$ translates to a safe operating region within the state space.

The controller's projected path through the constraint-imposed tube is shaped by the performance objectives established in the MPC cost function. While many options exist for characterizing desirable vehicle trajectories, here, the total sideslip angle at the front wheels (α) was chosen as the trajectory characteristic to be minimized in the objective function. This choice was motivated by the strong influence front wheel sideslip has on the controllability of front-wheel-steered vehicles since cornering friction begins to decrease above critical slip angles. In [16] it is shown that limiting tire slip angle to avoid this strongly nonlinear (and possibly unstable) region of the tire force curve can significantly enhance vehicle stability and performance. Further, the linearized tire compliance model described here does not account for this decrease, motivating the suppression of front wheel slip angles to reduce controller-plant model mismatch.

The MPC objective function with weighting matrices $R_{(\cdot)}$ then takes the form

$$J_{k} = \sum_{i=k+1}^{k+p} \frac{1}{2}\alpha_{i}^{T}R_{\alpha}\alpha_{i} + \sum_{i=k}^{k+p-1} \frac{1}{2}\delta_{i}^{T}R_{\delta}\delta_{i} + \sum_{i=k}^{k+p-1} \frac{1}{2}\Delta\delta_{i}^{T}R_{\Delta\delta}\Delta\delta_{i} + \frac{1}{2}\rho_{\varepsilon}\varepsilon^{2}$$

(7)

where ε represents constraint violation and was included to soften select position constraints as $\mathbf{y}^{j}_{min} - \varepsilon\mathbf{V}^{j}_{min} \leq \mathbf{y}^{j} \leq \mathbf{y}^{j}_{max} + \varepsilon\mathbf{V}^{j}_{max}$.

2.2 Threat Assessment

The vehicle path calculated by the MPC controller is assumed to be the best-case or safest path through the environment. As such, key metrics from this prediction are used to assess instantaneous threat posed to the vehicle. By setting constraint violation weights (ρ_{ε}) significantly higher than the competing minimization weight (R_{α}) on front wheel sideslip, optimal solutions satisfy corridor constraints before minimizing front wheel sideslip. When constraints are not active, front wheel sideslip – and the corresponding controllability threat – is minimized. When the solution is constrained, predicted front wheel sideslip increases with the severity of the maneuver required to remain within the navigable corridor.

Various approaches are available to reduce the predicted front wheel sideslip vector $\boldsymbol{\alpha}$ to a scalar threat metric Φ. In this paper,

$$\Phi(k) = \max\left(\left|\alpha_{k+1} \quad \alpha_{k+2} \quad \cdots \quad \alpha_{k+p}\right|\right)^T \tag{8}$$

was chosen for its good empirical performance when used to regulate controller intervention (described in the next section).

2.3 Hazard Avoidance

Given a best-case vehicle path through the environment and a corresponding threat, desired inputs from the driver and controller are blended and applied to the vehicle. This blending is performed based on the threat assessment: a low predicted threat causes more of the driver's input and less of the controller's input to be applied to the vehicle, while high threat allows controller input to dominate that of the driver. This "scaled intervention" may thereby allow for a smooth transition in control authority from driver to controller as threat increases.

Denoting the current driver input by u_{dr} and the current controller input by u_{MPC}, the blended input seen by the vehicle, u_v, is defined as

$$u_v = K(\Phi)u_{MPC} + (1 - K(\Phi))u_{dr} \tag{9}$$

The intervention function K is used to translate predicted vehicle threat Φ (obtained from the MPC trajectory plan) into a scalar blending gain. This function is bounded by 0 and 1 and may be linear, piecewise-linear, or nonlinear. Linear and piecewise-linear forms of this function may be described by

$$K = f(\Phi) = \begin{cases} 0 & 0 \le \Phi \le \Phi_{eng} \\ \dfrac{\Phi_{aut} - \Phi}{\Phi_{aut} - \Phi_{eng}} & \Phi_{eng} \le \Phi \le \Phi_{aut} \\ 1 & \Phi \ge \Phi_{aut} \end{cases} \tag{10}$$

where Φ_{eng} and Φ_{aut} represent the threat level at which the controller engages and the level at which it is given full control authority and effectively acts as an autonomous controller.

Using predicted threat (Φ) as calculated in (8) with an appropriate cost function formulation of the form (7) ensures that 1) the threat metric regulating controller intervention is minimized in the path plan (and associated control calculation) and 2) the controller maintains full control authority when constraints are binding. Increasing Φ_{eng} widens the "low threat" band in which the driver's inputs are unaffected by the controller. Increasing the value of Φ_{aut}, on the other hand, delays complete controller intervention until more severe maneuvers are predicted. The friction-limited bounds on the linear region of the tire force curve (1) suggest a natural upper limit of $\Phi_{aut} \le 5$ degrees in order to ensure that by the time the predicted maneuver required to remain within the safe region of the state space reaches this level of severity, the controller has full control authority and can – unless unforeseen constraints dictate otherwise – guide the vehicle to safety.

3 Experimental Setup

Experimental testing was performed at 14 m/s using a test vehicle and three human drivers. Driver and actuator steering inputs were coupled via an Active Front Steer (AFS) system. An inertial and GPS navigation system was used to measure vehicle position, sideslip, yaw angle, and yaw rate while a 1 GHz dSPACE processor ran controller code and interfaced with steering actuators.

Three common scenarios were used to analyze system performance. In each scenario, obstacles, hazards, and driver targets were represented to the driver by cones and lane markings and to the controller by a constrained corridor (with onboard sensing and constraint mapping assumed to have been performed previously by "virtual sensors" and high-level planners respectively). Only results from multiple-hazard-avoidance tests are shown below. In these tests (illustrated in Fig. 3), both lanes of travel were blocked at different locations, forcing the vehicle to change lanes to avoid the first hazard, then change lanes again to avoid the second.

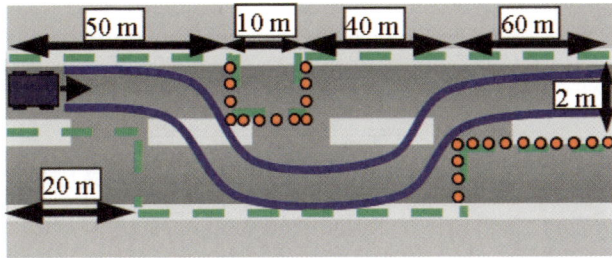

Fig. 3 Multiple hazard avoidance test setup showing hazard cone placement (circles) and lane boundaries (dashed).

Table 2 Controller parameters.

Symbol	Description	Value [units]
p	Prediction horizon	$\{35, 40\}$
n	Control horizon	$\{18, 20\}$
$R_y^{()}$	Weight on front wheel slip	0.2657
R_u	Weight on steering input	0.01
R_u	Weight on steering input rate (per t)	0.01
$u_{min/max}$	Steering input constraints	± 10 [deg]
$u_{min/max}$	steering input rate (per t) constraints	$\pm .75$ [deg] (15 deg/s)
$y^y_{min/max}$	Lateral position constraints	Scenario-dependent
	Weight on constraint violation	1×10^5
$[\;_{eng}\;\;_{aut}]$	Thresholds for controller intervention	$\{[0\ 3], [1\ 3]\}$ deg
V	Variable constraint relaxation on vehicle position	$[1.25, \cdots, 1.25, 0.01]$

Two types of human driver inputs were tested. Drowsy, inattentive, or otherwise impaired drivers were represented by a constant driver steer input of zero degrees. In these tests, the unassisted driver's path formed a straight line directly through the obstacle(s). To represent active driver steer inputs, the drivers were asked to steer either around or into obstacles.

Controller parameters are described and quantified in Table 2.

4 Experimental Results

The semi-autonomous framework proved capable of keeping the vehicle within the navigable corridor for each of the maneuvers, using various system/controller configurations, and with three different human drivers. Results from multiple hazard avoidance experiments are shown below.

Fig. 4 compares a semi-autonomous multi-hazard-avoidance maneuver to an autonomous maneuver ($K=1$).

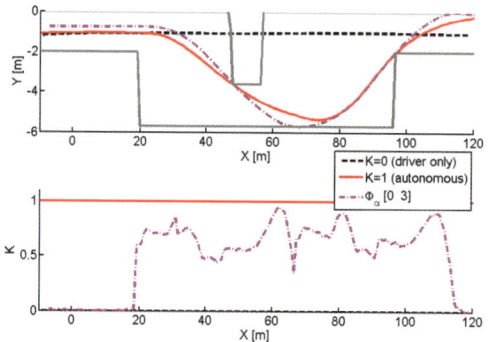

Fig. 4 Multiple hazard avoidance tests showing the similarity between semi-autonomous (dash-dot) and autonomous (solid) vehicle trajectories.

Notice that the semi-autonomous controller delayed intervention until the driver's inputs put the vehicle at risk of leaving the navigable road surface. When the framework did intervene, it allocated enough control authority to the controller to avert corridor departure or loss of control. Also notice that even with average controller intervention $K_{ave}=0.44$, the vehicle trajectory obtained using the semi-autonomous controller very closely resembles the "best case" trajectory taken by the autonomous controller. This results from the selective nature of the semi-autonomous system – it intervenes only when necessary, then relinquishes control to the driver once threat to the vehicle has been reduced.

Fig. 5 shows experiments in which the driver was instructed to swerve at the last minute to avoid hazards.

Notice that intervention by the semi-autonomous controller slightly preceded an otherwise-late driver reaction. The combined effect of both inputs was then sufficient to avoid both road hazards.

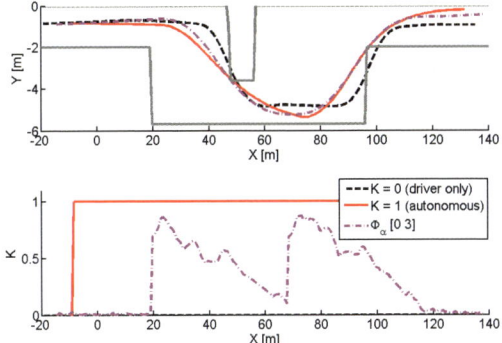

Fig. 5 Multiple hazard avoidance tests showing the vehicle trajectory with an unassisted driver input (dashed) and autonomous controller (solid), and semi-autonomous controller (dash-dot).

Finally, in each of the above experimental results, this shared-adaptive controller behaves as a stable closed-loop system. While this was also true of all of the other simulated and experimental results conducted to date, no rigorous stability proof is presented in this paper.

5 Conclusions

This paper presented an optimal-control-based framework that performs trajectory planning, threat assessment, and semi-autonomous control of passenger vehicles in hazard avoidance. This framework has been proven experimentally capable of satisfying position, input, and dynamic vehicle constraints using multiple threat metrics and intervention laws. Additionally, this framework has been shown to provide significant autonomy to a human driver, intervening only as necessary to keep the vehicle under control and within the navigable roadway corridor. Experimental results have also shown this control framework to be stable even in the presence of system-inherent time delays, though a rigorous stability proof is a topic of current investigation.

Finally, while human factors have not been studied in depth here, it is expected that with additional investigation, a best-case, or average driver-preferred intervention law may be described and intervention settings tuned accordingly. Further work is needed before this research is road-ready.

Acknowledgments. The authors would like to thank Eric Tseng, Matt Rupp, Mitch McConnell, Len Johnson, Steve Hermann, Kyle Carey, Reid Steiger, Tim Zwicky, Roger Trombley, Chris Wallis, and Jeff Rupp, all of Ford Motor Co. for their assistance in conducting the abovementioned experiments.

References

1. National Highway Traffic Safety Administration (NHTSA), 2007 Traffic Safety Annual Assessment - Highlights, NHTSA National Center for Statistics and Analysis (2008)
2. Weilkes, M., Burkle, L., Rentschler, T., Scherl, M.: Future vehicle guidance assistance - combined longitudinal and lateral control. Automatisierungstechnik 42, 4–10 (2005)
3. Leonard, J., How, J., Teller, S., et al.: A perception-driven autonomous urban vehicle. Journal of Field Robotics 25, 727–774 (2008)
4. Jansson, J.: Collision avoidance theory with application to automotive collision mitigation. Doctoral Dissertation, Linkoping University (2005)
5. Pohl, J., Birk, W., Westervall, L.: A driver-distraction-based lane-keeping assistance system. Proceedings of the Institution of Mechanical Engineers. Part I: Journal of Systems and Control Engineering 221, 541–552 (2007)
6. Mobus, R., Zomotor, Z.: Constrained optimal control for lateral vehicle guidance. In: Proceedings of the 2005 IEEE Intelligent Vehicles Symposium, June 6-8, pp. 429–434. IEEE, Piscataway (2005)
7. Netto, M., Blosseville, J., Lusetti, B., Mammar, S.: A new robust control system with optimized use of the lane detection data for vehicle full lateral control under strong curvatures. In: ITSC 2006: 2006 IEEE Intelligent Transportation Systems Conference, Piscataway, NJ, September 17-20, pp. 1382–1387. Institute of Electrical and Electronics Engineers Inc., United States (2006)
8. Vaidyanathan, R., Hocaoglu, C., Prince, T.S., Quinn, R.D.: Evolutionary path planning for autonomous air vehicles using multiresolution path representation. In: 2001 IEEE/RSJ International Conference on Intelligent Robots and Systems, October 29-November 3, pp. 69–76. Institute of Electrical and Electronics Engineers Inc. (2001)
9. Rossetter, E.J., Christian Gardes, J.: Lyapunov based performance guarantees for the potential field lane-keeping assistance system. Journal of Dynamic Systems, Measurement and Control, Transactions of the ASME 128, 510–522 (2006)
10. Falcone, P., Tufo, M., Borrelli, F., Asgari, J., Tseng, H.: A linear time varying model predictive control approach to the integrated vehicle dynamics control problem in autonomous systems. In: 46th IEEE Conference on Decision and Control 2007, CDC, Piscataway, NJ, December 12-14, 2007, pp. 2980–2985. Institute of Electrical and Electronics Engineers Inc, United States (2008)
11. Engelman, G., Ekmark, J., Tellis, L., Tarabishy, M.N., Joh, G.M., Trombley, R.A., Williams, R.E.: Threat level identification and quantifying system, U.S. Patent US 7034668 B2, April 25 (2006)
12. Yu, H., Spenko, M., Dubowsky, S.: An adaptive shared control system for an intelligent mobility aid for the elderly. Autonomous Robots 15, 53–66 (2003)
13. McBride, J.R., Ivan, J.C., Rhode, D.S., Rupp, J.D., Rupp, M.Y., Higgins, J.D., Turner, D.D., Eustice, R.M.: A perspective on emerging automotive safety applications, derived from lessons learned through participation in the DARPA Grand Challenges. Journal of Field Robotics 25, 808–840 (2008)
14. Zomotor, Z., Franke, U.: Sensor fusion for improved vision based lane recognition and object tracking with range-finders. In: Proceedings of the 1997 IEEE Conference on Intelligent Transportation Systems, ITSC, November 9-12, pp. 595–600. IEEE, Piscataway (1997)
15. Garcia, C., Prett, D., Morari, M.: Model predictive control: theory and practice-a survey. Automatica 25, 335–348 (1989)
16. Falcone, P., Borrelli, F., Asgari, J., Tseng, H.E., Hrovat, D.: Predictive active steering control for autonomous vehicle systems. IEEE Transactions on Control Systems Technology 15, 566–580 (2007)

Receding Horizon Model-Predictive Control for Mobile Robot Navigation of Intricate Paths

Thomas M. Howard, Colin J. Green, and Alonzo Kelly

Abstract. As mobile robots venture into more difficult environments, more complex state-space paths are required to move safely and efficiently. The difference between mission success and failure can be determined by a mobile robots capacity to effectively navigate such paths in the presence of disturbances. This paper describes a technique for mobile robot model predictive control that utilizes the structure of a regional motion plan to effectively search the local continuum for an improved solution. The contribution, a receding horizon model-predictive control (RHMPC) technique, specifically addresses the problem of path following and obstacle avoidance through geometric singularities and discontinuities such as cusps, turn-in-place, and multi-point turn maneuvers in environments where terrain shape and vehicle mobility effects are non-negligible. The technique is formulated as an optimal controller that utilizes a model-predictive trajectory generator to relax parameterized control inputs initialized from a regional motion planner to navigate safely through the environment. Experimental results are presented for a six-wheeled skid-steered field robot in natural terrain.

1 Introduction

Mobile robot navigation is the challenge of selecting intelligent actions from the continuum of possible actions that make progress towards achieving some goal under the constraints of limited perceptual information, computational resources, and planning time of the system. It also often viewed as the problem of balancing path following and obstacle avoidance in autonomous system architectures. Regional motion planning is the problem of planning beyond the sensor horizon.

A state-space trajectory is typically defined as a vector valued function of monotonic time (t). There are, however, circumstances where time is replaced by potentially nonmonotonic functions of distance (s) or heading (ψ) to form a path. Path

Thomas M. Howard, Colin J. Green, and Alonzo Kelly
Robotics Institute, Carnegie Mellon University, Pittsburgh, PA 15213-3980, USA
e-mail: {thoward,cjgreen,alonzo}@ri.cmu.edu

A. Howard et al. (Eds.): Field and Service Robotics 7, STAR 62, pp. 69–78.
springerlink.com © Springer-Verlag Berlin Heidelberg 2010

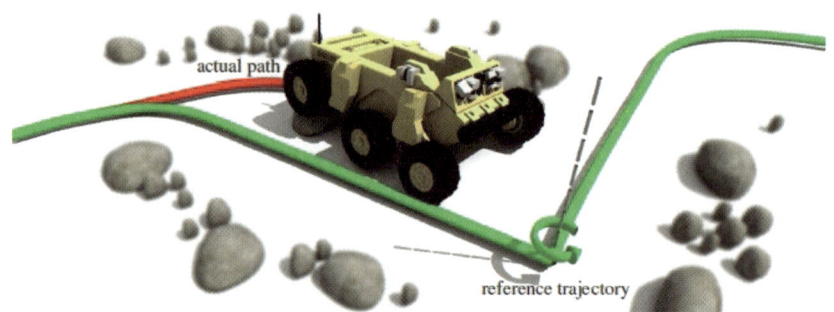

Fig. 1 An illustration of a vehicle attempting to to follow a reference trajectory with geometric singularities.

representations are used to achieve behaviors that allows velocity to remain unspecified. A cusp is a point in a trajectory where linear velocity changes sign. While cusps are discontinuous in path curvature, they are not discontinuous in state space trajectories and are perfectly feasible motions. Furthermore, the concept of forward on a path is not well-defined for cusps (and likewise for point turns) whereas forwards in time always has meaning. The capacity of a state space trajectory representation to remove discontinuities and permit a forward horizon to be defined are the basis of our preference for this representation.

A reference trajectory is the state-space trajectory $(\mathbf{x}(t)^1)$ provided by a regional motion planner (or other form of global guidance). The reference actions $(\mathbf{u}(\mathbf{x},t))$ are the inputs which cause the vehicle to follow the path perfectly in the absence of disturbances. In the presence of disturbances, the reference input signals that correspond to a disturbance free trajectory must be augmented by corrective actions to null the following error over some time horizon.

1.1 Motivation

As mobile robots navigate intricate motion plans composed of cusps, turn-in-place, and multi-turn maneuvers, the geometric singularities and discontinuities of these inflection points become problematic. Commonly applied techniques cannot generally reason about actions beyond these problematic points, which can endanger the system or impede path following performance by limiting the navigation horizon.

Consider the situation illustrated in Figure 1. In this example, the mobile robot deviates from the reference trajectory from disturbances including errors in modeling dynamics, terramechanical properties, and mobility.

The popular class of pursuit algorithms [1] will round path corners, avoid cusps, and fail for turn-in-place maneuvers where the pursuit point becomes undefined. In contexts where such intricate maneuvers were generated by a path planner in

---

[1] The state $(\mathbf{x})$ contains the vehicle position, orientation, velocity, or any other quantity of interest

order to avoid obstacles, a pursuit planner is inadequate. Sampling-based obstacle avoidance techniques [6] sometimes fail for intricate path navigation because of the computational resources required to search the entire input or state space densely enough to find an acceptable solution.

For effective intricate path navigation, a technique is needed which can exploit the reference trajectory structure to search in the local continuum for actions which minimize path deviation and avoid obstacles. This is the process of parametric relaxation, the technique of rendering a functional on a few parameters in order to permit relaxation of a trajectory (for optimization purposes) by searching a small number of degrees of freedom.

1.2   Related Work

There has been substantial research in the problem of developing effective, efficient mobile robot navigators. Early path following controllers operate on the assumption of tracking a single lookahead point and have been greatly extended in the literature [4]. In [12], effective search spaces for navigation in roads and trails were produced by generating nudges and swerves to the motion that reacquires the lane center.

An alternative approach involves sampling in the input space of the vehicle. In [6], navigation search spaces were generated by sampling in the input space of curvature. This approach also estimated the response of each action through a predictive motion model subject to the initial state constraints to more accurately predict the consequences of the actions. Egographs [8] represent a technique for generating expressive navigation search spaces offline by precomputing layered trajectories for a discrete set of initial states. Precomputed arcs and point turns comprised the control primitive sets used to autonomously guide planetary rovers for geologic exploration [2] where convolution on a cost or goodness map determined the selected trajectory. This approach was an extension of Morphin [11], an arc-planner variant where terrain shape was considered in the trajectory selection process. Another closely related algorithm is the one presented in [3], where an arc-based search space is evaluated based on considering risk and interest.

Other techniques such as rapidly-exploring random trees [7] have been effectively used to generate search spaces around the mobile robot to navigate cluttered, difficult environments and generate sophisticated maneuvers including u-turns. [9] presents a reactive path following controller for a unicycle type mobile robot built with a Deformable Virtual Zone to navigate paths without the need for global path replanning.

1.3   Discriminators

The main contribution of this work is the development of a receding horizon model-predictive controller (RHMPC) that effectively navigates intricate paths in complex environments. The algorithm leverages the capacity to generate the reference controls for a given reference trajectory. This capability exists because the sequence of reference controls can be generated by a trajectory generator that understands the association between actions and the corresponding state-space trajectory. Our particular preference is parameterized controls, but the key issue is that the controls are known, however represented, which correspond exactly to the reference trajectory. The corresponding reference trajectory inputs are already available in many regional motion planner implementations, so this simply requires that this additional information is passed to the navigator with the reference trajectory. Field experiments results demonstrate that the proposed technique can effectively navigate intricate paths composed of path singularities and discontinuities.

2 Technical Approach

This section describes the issues related to navigation of intricate paths generated by regional motion planners, the methods by which parameterized controls are generated, and the trajectory optimization techniques used to generate corrective actions. The trajectory follower is formulated as an optimal control problem:

$$
\begin{aligned}
&\text{minimize:} \quad J(\mathbf{x}, \mathbf{u}, t) \\
&\text{subject to:} \quad \dot{\mathbf{x}} = \mathbf{f_{PMM}}(\mathbf{x}, \mathbf{u}, t) \\
&\qquad\qquad\quad \mathbf{x}(t_I) = \mathbf{x_I} \\
&\qquad\qquad\quad \mathbf{u}(\mathbf{x}) \in \mathbf{U}(\mathbf{x}), \quad t \in [t_I, t_F]
\end{aligned}
\tag{1}
$$

The problem is one of determining actions from a set of functions $(\mathbf{U}(\mathbf{x}, t))$ to represent the control inputs $(\mathbf{u}(\mathbf{x}, t))$ which, when subject to the predictive motion model $(\mathbf{f_{PMM}}(\mathbf{x}, \mathbf{u}, t))$, minimize a penalty function $(J(\mathbf{x}, \mathbf{u}, t))$. An additional requirement for the trajectory follower is that the resulting control must be defined for a specific period of time or distance. This allows the optimized path to be evaluated for hazards to ensure vehicle safety.

2.1 Control Parameterization

One of the most difficult problems in motion planning involves reducing the continuum of actions to a manageable space to search. The trajectory following technique that we present uses a portion of the reference controls, which may be only piecewise continuous, as the initial guess. First, the reference trajectory is divided into the primitives used by the motion planner as shown in Figure 2(a). For each action, there exists a set of controls that, when applied to the system, produce a path segment of a certain shape. Parameterized freedom vectors $(\mathbf{p_i})$ control the shape of each set of inputs $(\mathbf{u}(\mathbf{p_i}, \mathbf{x}, t))$ that define the reference trajectory.

The initial guess for the parameterized control inputs $(\mathbf{u_{RHMPC}}(\mathbf{p_{RHMPC}}, \mathbf{x}, t))$ is defined by the sequence of trajectory segments between the nearest state and the predefined fixed control horizon (Figure 2(b)). In this example, the free

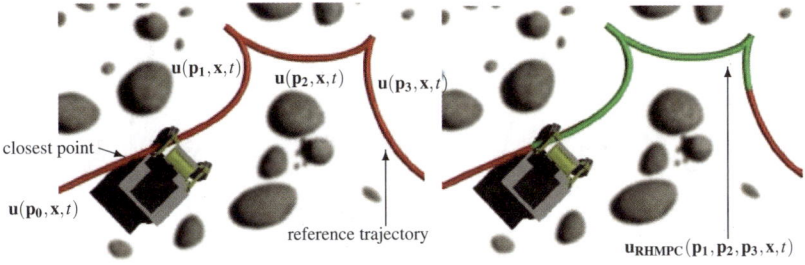

(a) Reference trajectory segmentation   (b) Horizon-limited action extraction

Fig. 2 An illustration of the parameterized action initialization in the RHMPC technique.

parameters of the receding horizon model-predictive controller (**P**RHMPC) are defined by a concatenation of free parameters in the control inputs:

$$\mathbf{P}_{\text{RHMPC}} = \begin{bmatrix} \mathbf{p}_1 & \mathbf{p}_2 & \mathbf{p}_3 \end{bmatrix}^T \qquad (2)$$

2.2 Path Deviation Optimal Control

Once the control input parameterization is determined, the next step is to modify the parameters to compensate for disturbances, approximations, and errors in the motion model. This technique seeks to minimize a cost function ($J(\mathbf{x}, \mathbf{u}, t)$) by modifying a set of control inputs:

$$J(\mathbf{x}, \mathbf{u}, t) = \Phi(\mathbf{x}(t_I), t_I, \mathbf{x}(t_F), t_F) + \int_{t_I}^{t_F} \mathcal{L}(\mathbf{x}(t), \mathbf{u}(\mathbf{p}, \mathbf{x}), t) dt \qquad (3)$$

The initial corrective action is evaluated through the predictive motion model subject to the initial state constraints to obtain a cost estimate as illustrated in Figure 3. While the gradient of the cost function with respect to the parameterized control input freedom exceeds a threshold, the algorithm adjusts the control inputs to minimize the integrated penalty function ($\mathcal{L}(\mathbf{x}, \mathbf{u}, t)$). The parameterized freedoms are modified iteratively through any standard optimization technique, such as gradient descent, as the cost function gradient is determined entirely numerically:

$$\mathbf{P}_{\text{RHMPC}_i} = \mathbf{P}_{\text{RHMPC}_{i-1}} - \alpha \nabla J(\mathbf{x}, \mathbf{u}, t), \quad i \geq 1 \qquad (4)$$

2.3 Integrating Observed Cost Information

There are several situations when a mobile robot should intentionally deviate from the reference trajectory. Navigating around recently observed static and dynamic obstacles faster than the replanning rate of the regional motion planner is important when perceptual information is frequently updated. Another reason for deviation is the suboptimality of the reference trajectory itself. One solution is to stop and

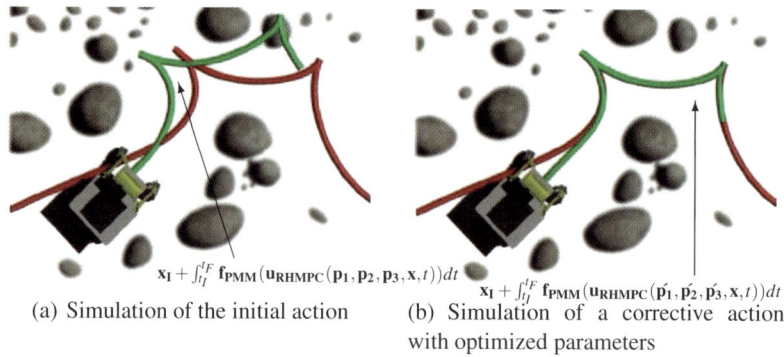

$$\mathbf{x_I} + \int_{t_I}^{t_F} \mathbf{f_{PMM}}(\mathbf{u_{RHMPC}}(\mathbf{p_1},\mathbf{p_2},\mathbf{p_3},\mathbf{x},t))dt$$

$$\mathbf{x_I} + \int_{t_I}^{t_F} \mathbf{f_{PMM}}(\mathbf{u_{RHMPC}}(\mathbf{\acute{p}_1},\mathbf{\acute{p}_2},\mathbf{\acute{p}_3},\mathbf{x},t))dt$$

(a) Simulation of the initial action   (b) Simulation of a corrective action
                                             with optimized parameters

Fig. 3 An illustration of the parameterized action correction in the RHMPC technique.

request a refined or alternative plan. A potentially better method is to include cost information into the utility functional optimized by the receding horizon model-predictive controller to determine the obstacle avoidance maneuver. The presented technique is naturally suited to deform the current action for local obstacle avoidance and path smoothing. The desired behaviors can be integrated by modifying the cost function to include a weighted penalty for obstacle cost.

3   Implementation

The regional motion planer used to generate feasible reference trajectories for these experiments runs A* on a graph composed of regularly arranged discrete nodes in a state space, similar to [10]. The connectivity between nodes in the discretized graph was provided by a motion template consisting of forward, reverse, and turn-in-place trajectories with lengths varying between $3m$ and $9m$. This particular implementation operated on a $60m$ x $60m$ vehicle-centered cost map. Reference trajectory updates were provided by the regional motion planner at a rate of $2Hz$.

The resulting reference trajectory is a series of sequential independent parameterized trajectories. Intricate paths composed of multiple cusps and/or turn-in-place actions often result from the diversity of edges in motion planning graphs and the complexity of the environment. The model-predictive trajectory generator [5] was used in both the motion template generation and the path deviation optimal control. Actions in the motion template were composed of constant linear velocities and either second-order spline curvature functions parameterized by distance or constant angular velocity functions parameterized by heading. Generic spline classes defined by sequences of individual command profiles parameters were optimized by the receding horizon model-predictive controller. The corrective actions were generated by the receding horizon model-predictive controller at a rate of $20Hz$.

4 Experiments

A set of experiments were designed as a comparison between a navigator that used the presented trajectory follower and one that directly executed the regional motion plan. Both systems used the same version of a lattice planner that searches dynamically feasible actions which was specifically designed for the test platform. Each field experiment was required to achieve a series of waypoints in an environment with updating perceptual information generated by an on-board perception system combined with limited overhead prior data.

The platform for the field experiments was Crusher (Figure 4(a)), a six-wheeled skid steered outdoor mobile robot. The multi-kilometer experiments were conducted at a test site in Pittsburgh, Pennsylvania with variable off-road terrain (Figure 4(b)).

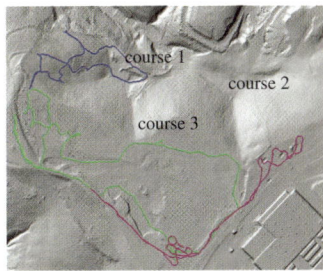

(a) Crusher   (b) The field experiment courses

Fig. 4 The mobile robot and test environment for the trajectory follower field experiments.

Integrated path cost was the main metric used to measure success for the field experiments, which is related to the risk, mobility, and traversability for the vehicles configuration in the environment. While inherently unitless and scaleless, it provides the best metric for measuring performance because both the motion planner and the trajectory follower optimize this quantity.

5 Results

This section presents the results of the three field experiments comparing the performance of the presented trajectory follower to a system directly executing the regional motion plan. Figure 5 shows several selected examples from the field experiments where the receding horizon model-predictive control was used to navigate intricate paths in varied, natural terrain. Two different situations are shown involving geometric singularities and discontinuities including cusps and turn-in-place actions where the generated RHMPC action is shown as a solid green line.

Figure 5(a) shows the receding horizon model-predictive control determined for following of a trajectory with an initial turn-in-place action with path following disturbance. The current vehicle state is off the reference trajectory and a corrective

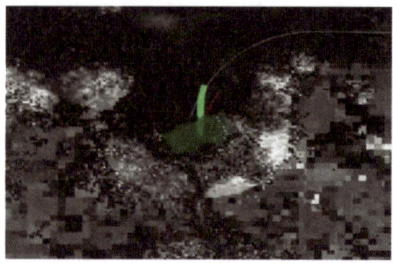
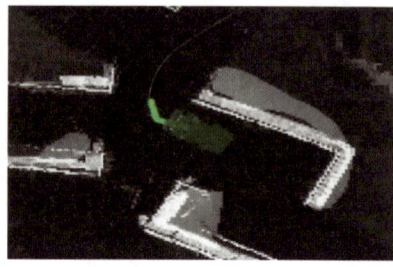

(a) Planning a control that reasons about a turn-in-place action and compensates for a path following disturbance

(b) Planning a control with a constant horizon through a future turn-in-place action

Fig. 5 Selected examples of the RHMPC navigator used in the multi-kilometer field experiments.

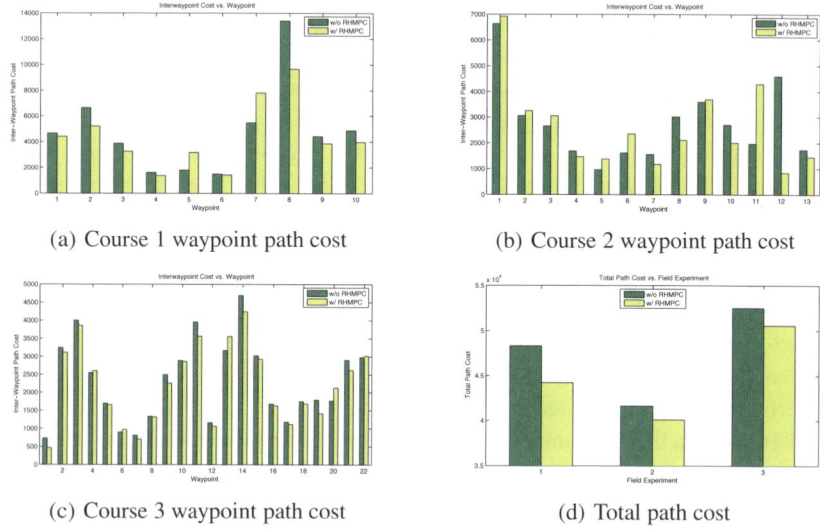

(a) Course 1 waypoint path cost

(b) Course 2 waypoint path cost

(c) Course 3 waypoint path cost

(d) Total path cost

Fig. 6 The waypoint and total path cost for a series of comparison runs on three courses

control is determined which adjusts the length of the angular velocity command and bends the straight segment to reacquire the path in a feasible manner. The last example shown in Figure 5(b) involves planning through a future turn-in-place action between two nominally straight segments. This example shows the flexibility of the technique, where the turn-in-place action does not necessarily need to start or end at a specific point in the receding horizon model-predictive control.

Figure 6 shows the integrated cost of each systems between each waypoint. It is useful to look at each waypoint-waypoint segment separately because each one can be considered to be an independent trial. On average, the system using the trajectory follower slightly outperformed or achieved a similar level of performance of the alternative system.

For portions of the course where disturbances relative to the predicted motion are uncommon or the local cost gradient was small near the disturbances very little improvement would be expected, with more improvement expected in cases where small system disturbances can quickly lead to significantly different path cost. Figure 6(d) shows the total integrated cost for each of the three field experiments. It is important to note that while the trajectory follower did not outperform the alternative system between every waypoint, it did improve the overall performance of the system by up to 7.2%. The variability in the results is expected because of the chaotic nature of outdoor mobile robots were any number of small changes can cause the robot to select a significantly different path.

6 Conclusions

The receding horizon model-predictive control algorithm enables mobile robots to navigate intricate paths by utilizing the paths by relaxing parameterized controls that correspond exactly to the path shape. This technique enables planning through inflection points and turn-in-place actions in paths to better reason about the recovery trajectory. This method makes it possible to intelligently search the local continuum for an action which minimizes path following error and/or avoids obstacles. It also enables several other important behaviors including the capacity to define a utility function in situations where pursuit planners fail and the ability to correctly follow path discontinuities like cusps which are otherwise feasible motions. Several multi-kilometer field experiments demonstrated that the inclusion of the presented trajectory follower as a mobile robot navigator improves upon the metric that the regional motion planner minimizes.

Acknowledgements. This research was conducted at the Robotics Institute of Carnegie Mellon University under contract to NASA/JPL as part of the Mars Technology Program. Field experiments on the Crusher mobile robot platform were sponsored by the Defense Advanced Research Project Agency (DARPA) under contract Unmanned Ground Combat Vehicle PerceptOR Integration (contract number MDA972-01-9-0005). The views and conclusions contained in this paper are those of the authors and should not be interpreted as representing the official policies, either expressed or implied, of the U.S. Government.

References

1. Amidi, O.: Integrated mobile robot control. Technical Report CMU-RI-TR-90-17, Carnegie Mellon, Pittsburgh (1990)
2. Besiadecki, J.J., Leger, P.C., Maimone, M.W.: Tradeoffs between directed and autonomous driving on the Mars Exploration Rovers. International Journal of Robotics Research 26(91), 91–104 (2007)

3. Bonnafous, D., Lacroix, S., Siméon, T.: Motion generation for a rover on rough terrains. In: Proceedings of the 2001 IEEE/RSJ International Conference on Intelligent Robots and Systems, October 2001, vol. 2, pp. 784–789 (2001)
4. Coulter, R.C.: Implementation of the pure pursuit path tracking algorithm. Technical Report CMU-RI-TR-92-01, Carnegie Mellon, Pittsburgh (1992)
5. Howard, T.M., Kelly, A.: Optimal rough terrain trajectory generation for wheeled mobile robots. International Journal of Robotics Research 26(2), 141–166 (2007)
6. Kelly, A., Stentz, T.: Rough terrain autonomous mobility - Part 2: An active vision and predictive control approach
7. Kuwata, Y., Fiore, A., Teo, J., Frazzoli, E., How, J.P.: Motion planning for urban driving using RRT. In: Proceedings of the IEEE/RSJ International Conference on Intelligent Robots and Systems, September 2008, pp. 1681–1686 (2008)
8. Lacze, A., Moscovitz, Y., DeClaris, N., Murphy, K.: Path planning for autonomous vehicles driving over rough terrain. In: Proceedings of the 1998 IEEE ISIC/CIRA/ISAS Joint Conference, pp. 50–55 (September 1998)
9. Lapierre, L., Zapata, R., Lepinay, P.: Combined path-following and obstacle avoidance control of a wheeled robot. International Journal of Robotics Research 26(4), 361–375 (2007)
10. Pivtoraiko, M., Knepper, R., Kelly, A.: Optimal, smooth, nonholonomic mobile robot motion planning in state lattices. Technical Report CMU-RI-TR-07-15, Carnegie Mellon, Pittsburgh (2007)
11. Simmons, R., Krotkov, E., Chrisman, L., Cozman, F., Goodwin, R., Hebert, M., Katragadda, L., Koenig, S., Krishnaswamy, G., Shinoda, Y., Whittaker, W.L., Klarer, P.: Experience with rover navigation for lunar-like terrains. In: Proceedings of the 1995 IEEE Conference on Intelligent Robots and Systems, August 1995, vol. 1, pp. 441–446 (1995)
12. Thrun, S., Montemerlo, M., Dahlkamp, H., Stavens, D., Aron, A., Diebel, J., Fong, P., Gale, J., Halpenny, M., Goffmann, G., Lau, K., Oakley, C., Palatucci, M., Pratt, V., Stang, P.: Stanley: The robot that won the DARPA Grand Challenge. Journal of Field Robotics 23(9), 661–692 (2006)

Posterior Probability Estimation Techniques Embedded in a Bayes Filter for Vibration-Based Terrain Classification

Philippe Komma and Andreas Zell

Abstract. Vibration signals acquired during robot traversal provide enough information to yield a reliable prediction of the current terrain type. In a recent approach, we combined a history of terrain class estimates into a final prediction. We therefore adopted a Bayes filter taking the posterior probability of each prediction into account. Posterior probability estimates, however, were derived from support vector machines only, disregarding the capability of other classification techniques to provide these estimates. This paper considers other classifiers to be embedded into our Bayes filter terrain prediction scheme, each featuring different characteristics. We show that the best classification results are obtained using a combined k-nearest-neighbor and support vector machine approach which has not been considered for terrain classification so far. Furthermore, we demonstrate that other classification techniques also benefit from the temporal filtering of terrain class predictions.

1 Introduction

In outdoor applications such as rescue missions or agricultural assignments the mobile robot navigates over varying ground surfaces, each possessing different characteristics. To ensure a safe traversal in outdoor environments the robot should adapt its driving style according to the presence of ground surface hazards like slippery or bumpy surfaces. These hazards are denoted as non-geometric hazards [18]. Therefore, most approaches employ a model-based prediction scheme which estimates the current terrain type from sensor readings. In a model generation phase, the model learns the correct assignment of a labeled terrain class given the respective observation. In the recall phase, that is, during terrain traversal over unknown terrain, the robot then uses this model to predict the current ground surface. For

Philippe Komma and Andreas Zell
Chair of Computer Architecture, Computer Science Department,
University of Tübingen, Sand 1, D-72076 Tübingen, Germany
e-mail: {philippe.komma,andreas.zell}@uni-tuebingen.de

A. Howard et al. (Eds.): Field and Service Robotics 7, STAR 62, pp. 79–89.
springerlink.com © Springer-Verlag Berlin Heidelberg 2010

acquiring input data, a variety of sensors such as vision [11, 1] or ladar sensors [15, 10] can be employed. Recently, several researchers considered vehicle vibrations for terrain classification as originally proposed in [8]. In this context, vibration data acquired from accelerometers have been successfully applied to planetary rovers [3], autonomous ground vehicles [7], and experimental unmanned vehicles [6]. In [16] a comparison was drawn between different base classifiers providing the model for vibration-based terrain classification. These techniques, however, estimate the terrain type using single sensor measurements only, disregarding the temporal coherence between consecutive measurements. We addressed this problem in [9]. There, we applied a Bayes filter to combine the posterior probabilities of several recent terrain class predictions into a final prediction. In our approach, posterior probability estimation was performed using a support vector machine (SVM) since this classifier was reported to yield the best classification results in a single observation-based prediction scheme [16].

To motivate our current research we first note that the performance of a classifier in the context of Bayes-filtered terrain classification does not depend on the classification quality only but also on the quality of the prediction certainty: Since the final classification is based on the posterior probability of single predictions, it benefits from a model which performs confident correct predictions and uncertain erroneous predictions. Classifiers which provide these characteristics result in a better prediction performance when embedded into our Bayes filter approach. This is because erroneous predictions obtain a lower weight in the filtering process and thus influence the final prediction less significantly. The quality of various classifiers relating to the prediction certainty is unclear and is hence investigated in this paper. Second, the SVM classifier is not an appropriate choice in all domains, especially for online learning [17] where an enduring model generation phase is not applicable. Thus, this paper focuses on the selection of an adequate classifier with regard to its limiting factors such as training and model selection time, storage requirements, and the run-time complexity of the recall phase. We further applied the SVM-KNN classifier introduced in [21] which in our terrain classification task was significantly superior to all other classifiers considered so far.

The remainder of this paper is organized as follows: Section 2 briefly describes our terrain classification model, taking both one and several recent observations into account. The posterior probability estimation techniques of the classifiers to be embedded in our temporally-filtered classification approach are introduced in Sect. 3. After summarizing our experimental setup in Sect. 4 we present and discuss experimental results in Sect. 5. Finally, a conclusion is given in the last section.

2 Terrain Classification Model

This section summarizes our terrain classification technique based on both single observations and temporal filtering of several recent terrain predictions. A detailed description is presented in [9].

The objective of our approach is to estimate the terrain type the robot is currently traversing. Predictions are model-based, assigning a certain terrain class from a set of classes to recorded observations. We represent the observations by acceleration data sampled at a frequency of 100 Hz over a period of 1.28 s. The acceleration data can be regarded as the vibration which the terrain induces to the body of the robot. For feature extraction, we applied the Fast Fourier Transform (FFT) to the raw input signal to determine its FFT amplitude spectrum in a second step. We then normalized the data by scaling each component of the preprocessed vibration signal to have a mean of 0 and a standard deviation of 1. The scaled amplitude spectrum entries constitute the inputs for the terrain classification model.

In the recall phase, the robot predicts the current terrain type, using the terrain classification model generated during training. Therefore, the same preprocessing has to be applied to the acquired vibration data. Using the posterior probability estimation techniques presented in the next section, the application of the final feature vector to the classifier does not only provide a class prediction but also an approximation of the posterior $p(x = i|u)$. This probability distribution denotes the probability that a preprocessed vibration segment u belongs to terrain class i. Next, we describe how $p(x|u)$ can be embedded into a Bayes filter framework.

Using a Bayes filter [14], the state of a dynamic system at a time t is represented by a random variable x_t. In our context, $x_t \in [1;k]$ models the uncertainty with which the robot navigates over one of the k terrain types. Given $t + 1$ preprocessed vibration segments $u_{0:t} = \{u_0, u_1, \dots, u_t\}$ recorded by accelerometer sensors, the estimated target distribution is determined by $p(x_t|u_{0:t})$. In [9] we showed that $p(x_t|u_{0:t})$ can be formally defined as:

$$p(x_i = i|u_{0:t}) = \alpha_t p(x_t = i|u_t) \sum_j p(x_t = i|x_{t-1} = j) p(x_{t-1} = j|u_{0:t-1}).$$

Here, $p(x_t|u_t)$ substitutes the measurement probability $p(u_t|x_t)$ and represents the probability that the vibration measurement u_t can be observed when navigating over a certain terrain type x_t. $p(x_t|u_t)$ is derived from the Bayes inversion $p(u_t|x_t) = p(x_t|u_t)\frac{p(u_t)}{p(x_t)}$ assuming that $p(x_t)$ is distributed uniformly. Further note that $p(u_t)$ is constant for all i and can thus be included in the normalizing constant α_t.

The transition probability $p(x_t|x_{t-1})$ denotes the probability that the robot moves from terrain type $x_{t-1} = j$ to $x_t = i$. Bayes filters model the dynamic system by a first-order Markov process assuming that the information provided by the state x_t suffices to predict future states without considering earlier observations. Our approach is based on the heuristic that the terrain class most likely does not change from one measurement to the next. Thus, we assign a relatively large value v to $p(x_t = i|x_{t-1} = i)$. $p(x_t = i|x_{t-1} = j)$, with $i \neq j$, is derived from the following two heuristics: First, the probability $p(x_t = i|x_{t-1} = j)$ should increase with the probability to confuse class i with class j. Second, a transition from state $x_{t-1} = j$ to state $x_t = i$ should be based on the probability to predict the terrain class at time t correctly. Both probabilities can directly be estimated from the confusion matrix. For further details, we refer to [9]. By dynamically changing v, the probability that the system remains in its current state, we obtain an approach being both reactive and stable enough to detect fast terrain transitions and selective misclassifications. In our implementation, v is either increased or decreased by a constant factor depending on whether the current prediction equals the system state at time $t - 1$. Upper and lower bounds for v ensure that the probability of a state transition neither becomes too large nor too small.

For the definition of the initial probability distribution $p(x_0)$, we make no assumptions that the robot is placed on a specific terrain type at time $t = 0$. Hence, $p(x_0)$ is assumed to be uniformly distributed.

3 Posterior Probability Estimation

In this section, we briefly describe all classifiers that have been embedded into our Bayes filter classification approach. Therefore, we explain how posterior probabilities $p(x = i|u)$ can be predicted for each class i under consideration. Since each classifier features different characteristics we conclude this section by indicating in which situations the choice of a certain classifier is appropriate.

k-nearest neighbor classification (KNN) [5] determines the set of k-nearest-neighbors contained in a training set to a testing instance u. Then, we calculate the frequency of occurrence of each class in the neighbor set. The class with the largest frequency becomes the predicted class for the testing instance u. The posterior probability $p(x = i|u)$ is defined as the ratio between the number of occurrences of class i in the neighbor set n_i and the number of considered neighbors k, $p(x = i|u) = \frac{n_i}{k}$.

The multilayer perceptron (MLP) [2] is an instance of an artificial neural network. It consists of artificial neurons which are interconnected in a well-defined manner. These neurons are arranged in three different layers: in an input layer, a hidden layer, and an output layer. When applying an input u to the network input, the neurons of the hidden layer perform a weighted sum of the input components: $net_l = \langle w_l, u \rangle$. Here, net_l denotes the net activation of neuron h_l and w_l is the weight vector determining the specific contribution of each input component to the final sum. We then apply an activation function f_{act}, typically chosen as $f_{act} = tanh(net_l)$, to each net activation to obtain the final output for the neurons of the hidden layer. The determination of the net activation of the output neurons is equivalent to the ones of the hidden layer except that we do not add weighted input coefficients but weighted activations of the hidden neurons. For classification problems, the activation function of the output neurons is replaced by the softmax function which takes the form $f_{act} = \exp(net_m) / \sum_{m'} \exp(net_{m'})$, where net_m is the net activation for output neuron m. Each output neuron represents a certain class to discriminate. The predicted class is the one which is represented by the neuron with the maximum activation. It can be shown [2] that the activations can directly be interpreted as posterior probabilities.

Probabilistic neural networks (PNN) [13] are another instance of artificial neural networks. In the training phase, scaled training patterns are inserted into a matrix W_c, $c \in [1;k]$, according to the class c they belong to. Each row of W_c represents a single pattern. The scaling is performed such that the L_2 norm of each training instance equals to one. In the recall phase, the same scaling is applied to the test vector u. For each class c, the inner product between each pattern w_i of the weight matrix W_c and the query u is determined yielding the net activation $net_{c,i}$. The net activations are non-linearly transformed using the activation function $f_{act}(net_{c,l}) = \exp((net_l - 1)/\sigma^2)$, where σ is a model parameter defining the size of the Gaussian window. For each class, the sum over all transformed net activations is determined, $s_c = \sum_l f_{act}(net_{c,l})$, and the predicted class becomes the one which maximizes s_c. Given that the probability of each class is distributed uniformly, posterior probabilities $p(x = i|u)$ can then be defined as $p(x = i|u) = (n_i^{-1} s_i)/(\sum_j n_j^{-1} s_j)$, where n_c is the number of training instances for class c.

Given two classes c_1 and c_2 to discriminate, a support vector machine (SVM) [4] establishes a separating hyperplane such that each instance of the first class resides in one subspace and each instance of the other class resides in the other subspace. To increase generalization we maximize the margin which is the distance from the hyperplane to the instances closest to it. In the non-separable case, that is, if no hyperplane exists which separates the two classes, instances of class c_1 are allowed to reside in the subspace representing class c_2 and vice versa. However, a penalty term is added for each non-separable training point. Problems exist, which are not linearly separable in the original space spanned by the training data but which become linearly separable when mapping the inputs u_i into a higher dimensional feature space, $z = \phi(u)$. Using the "kernel trick" the actual mapping does not have to be performed. Instead, we exploit the fact that the inner product of basis functions $\phi(x)^T \phi(y)$ is replaced by a kernel function $K(x,y)$. In our experiments, we used the radial basis function kernel defined as $K(x,y) = \exp\left(-\|x - y\|^2/\sigma^2\right)$. Multi-class classification using n classes is achieved by establishing $n(n-1)/2$ binary classifiers in a one-versus-one classification scheme. Adopting the technique of [12], a parameterized sigmoid function is applied to the decision value of each binary classification which results in posterior probabilities of both classes. Finally, we obtain the posterior for each class i, $p(x = i|u)$, using the pairwise coupling method of [19].

The SVM-KNN approach [21] combines the characteristics of both the KNN and the SVM classifiers. It does not require a training phase. Instead, predictions are performed by first pruning the training set. Therefore, the k-nearest-neighbors to a given query u are identified. Then, a multi-class SVM is trained online using the pairwise distances between all entries of the union of the query and the neighbor set. Prior to the SVM model training, these distances have to be transformed into a kernel matrix. In our approach, this is realized by applying the function $f(d_{ij}) = \exp(-d_{ij}^2/\sigma^2)$ to the pairwise distances d_{ij}. As distance function we chose the L_2 norm. The posterior probability $p(x|u)$ is then obtained by applying the query u to the trained SVM.

Classifier selection should be handled with care since each approach has different characteristics. KNNs and PNNs belong to the class of lazy learning techniques. That is, all computations are delayed until a prediction query is requested. On the one hand, this renders a time-demanding training phase unnecessary which is advantageous if the underlying phenomenon changes frequently. On the other hand, all patterns have to be available at run-time which might pose a problem if storage

Table 1 The respective model parameter(s), the number of considered candidates during model selection, model selection and training times, prediction complexity, and storage requirements of the proposed classifiers.

	KNN	MLP	PNN	SVM	SVM-KNN
model param.	k = 6	hid. = 96	$\sigma = 0.07$	$C = 9.05, \sigma = 0.02$	k = 640, C = 9.05, $\sigma = 0.02$
model sel. cand.	31	8	64	196 (14×14)	k: 30; C,σ: 196
model sel. time (h)	0:52:55	32:34:00	0:07:54	24:55:47	50:09:40
training time (h)	-	1:08:43	0:00:01	0:00:54	-
testing time (ms)	13.21	0.02	0.54	1.07	464.9
storage req. (kB)	763.5	52.5	763.5	22.5	763.5

is limited. Given that the acquired training set consists of n samples, storage requirements are $O(n \cdot d)$, where d is dimensionality of a training instance. Furthermore, if the calculating capacity is constrained in the recall phase, the desired prediction frequency might not be accomplished due to a large set of training patterns. For example, when using the KNN classifier, a naïve approach involves $O(n)$ distance calculations to determine the k-nearest-neighbors. Although accelerating data structures like M-trees [20] exist, high-dimensional nearest-neighbor search is known to be a non-trivial task suffering from the curse of dimensionality.

MLP and SVM classifiers typically provide compact models, resulting in a fast prediction performance. Model training, however, is computationally much more demanding since both methods iteratively try to minimize a given error function. The time spent on choosing a classifier with a good generalization behavior is significantly increased by the model selection process which has to consider a sufficiently large set of candidate model parameter settings.

The SVM-KNN approach is characterized by an involved model selection and testing phase. Since a class prediction also requires the determination of the k-nearest-neighbor set to a given query, the training set has to be present at run-time. We note, however, that this approach still guarantees predictions performed in real-time. Hence, we included SVM-KNN classification in our investigations.

Table 1 summarizes the key characteristics for the proposed classifiers: the respective model parameter(s) which yielded the best generalization and model selection times along with the number of tested model candidates. For the respective best classification model we further present the training time using data contained in one fold of a 5-fold cross validation scheme, the average testing time for a single query, and storage requirements (measured in kB). We performed all run-time analyses on a Pentium D 3.0 GHz desktop PC. For the storage considerations, we represented each floating point number as *double*, each 8 bytes in size.

4 Experimental Setup

In our experiments, an Xsens MTi altitude and heading reference system was mounted on an aluminum plate on top of our RWI ATRV-Jr outdoor robot to measure

Fig. 1 The employed terrain types: 1: indoor floor, 2: asphalt, 3: gravel, 4: grass, 5: boule court.

vibration signals in left-right direction at 100 Hz. During data acquisition, the robot navigated over five different terrains (Fig. 1): indoor PVC floor, asphalt, gravel, grass, and clay (the surface of a boule court). To not constrain the model to work at a certain driving speed, we varied the speed between 0.2, 0.4, and 0.6 m/s. In total, the dataset consists of 7635 patterns, corresponding to approximately 1.5 hours of robot navigation.

We performed individual terrain classifications using vibration data acquired during 1.28 s of robot travel. For two consecutive segments we permit an overlap of 28 samples to achieve a prediction frequency of 1 Hz. The combination of terrain class predictions was realized by our adaptive Bayes filter approach introduced in Sect. 2. To quantify the performance of the latter, we applied the following evaluation procedure: We assembled consecutive vibration segments representing the same terrain type to give a travel distance of constant length. Then, assembled segments of varying terrain types were grouped together yielding the final test set. In different experiments, we varied the distance covered by a robot before it reaches a new terrain class. This distance is denoted as the *travel distance d* (measured in meters) in the following. In each experiment, d was chosen from the set $d \in \{0; 4; 8; 12; 16\}$. The 0 m experiment describes the worst case scenario for approaches based on temporal filtering. Here, single segments of varying terrain classes are concatenated, each representing data acquired during 1 s of robot travel. Since the robot speed varies between 0.2, 0.4, and 0.6 m/s, this experiment includes travel distances of 0.2, 0.4, and 0.6 m. Note that according to the confusion matrix, certain terrain transitions are easier to detect than other ones. Hence, the results depend on the order in which assembled terrain segments of varying terrain type are presented. We minimized this effect by randomly permuting this order and averaging the classification results determined after 20 reruns of a particular experiment.

As quality measure we used the true positive rate (TPR). It is the ratio (measured in per cent) of the number of correct predictions for which the predicted class x_t equals the actual class $\hat{x}_t$ and the number of instances contained in the test set. We derived the prediction performance using 5-fold cross validation and averaging the true positive rate over all five folds.

5 Experimental Results

Table 2 shows the results for the proposed classifiers when using single observations (SO) and Bayes-filtered posterior probabilities of recent predictions (AB). Note that

the true positive rate for the single observation-based approach differs between varying experiments. This is due to the model evaluation procedure introduced in the
previous section which selects a varying test set for each travel distance.

Related to both the SO and the AB approach, the SVM-KNN technique yields
the best prediction performance, followed by SVM, MLP, KNN, and PNN classification (Fig. 2(a)). The differences in the true positive rates of the applied classifiers
proved to be statistically significant, using a two-tailed t-test at a significance level of
1%. The combined support vector machine and k-nearest-neighbor approach benefits from the reduced training set resulting in another configuration of the separating
hyperplane. This hyperplane results in a higher generalization as compared to the
one established by the SVM approach which uses all training patters at once. The
classification performance of each classifier is also reflected in the increase of the
true positive rate obtained when using the adaptive Bayes technique in comparison
with the single observation approach (Fig. 2(b)): the larger the true positive rate
for a given classifier, the larger the benefit of using temporally-filtered predictions.
This statement holds for all classifiers but the KNN approach: Here, the adaptive
Bayes technique results in the largest increase of the true positive rate. Investigations

Table 2 Prediction performance (in %) for varying classifiers and travel distances (dist.) using single observation-based (SO) and adaptive Bayes filter-based (AB) terrain classification.

dist. (m)	0		4		8		12		16	
approach	SO	AB	SO	AB	SO	AB	SO	AB	SO	AB
SVM-KNN	89.1	89.1	89.8	93.3	89.8	94.8	91.8	96.7	91.8	97.0
SVM	88.5	88.4	89.3	92.4	89.3	94.2	91.0	95.8	90.9	96.1
MLP	86.7	86.7	87.4	90.6	87.4	91.0	89.0	92.4	88.9	92.1
KNN	80.6	79.7	81.2	83.6	81.1	85.0	82.9	88.5	82.7	88.4
PNN	79.2	79.1	80.2	82.4	80.1	82.7	83.7	86.1	82.8	85.3

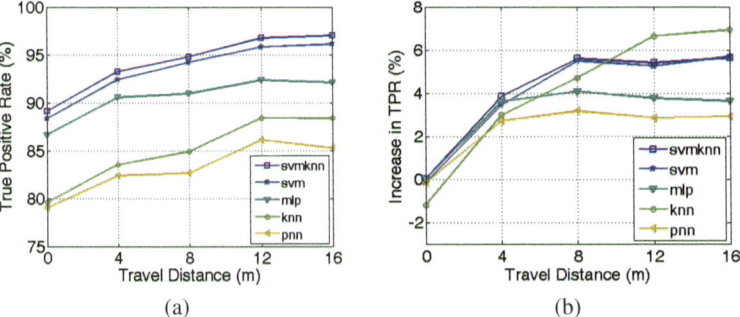

(a)   (b)

Fig. 2 (a) True positive rates for the adaptive Bayes approach and (b) the relative increase
of classification performance for the adaptive Bayes approach related to single observation-
based classification when varying the classifier and the travel distance.

revealed that in the case of a misclassification the posterior probability of the erroneously predicted class p_e was rather small on average, $p_e = 0.64 \pm 0.19$, in comparison with the ones obtained for a correctly predicted class, $p_c = 0.87 \pm 0.18$. Hence, the transition into another, that is, erroneous system state becomes less likely if the system previously resided in the correct state, no terrain transition occurred, but the classifier erroneously estimated the wrong terrain class. We obtained similar results using SVM and SVM-KNN classifiers: Here, the posterior probability of erroneous predictions was $p_e = 0.69 \pm 0.18$ and $p_e = 0.66 \pm 0.18$, respectively. PNN and MLP classifiers perform wrong predictions with higher confidences: $p_e = 0.87 \pm 0.17$ and $p_e = 0.84 \pm 0.17$, respectively. Referring to Fig. 2(b), this is another explanation for the smaller increase in TPR compared to other classifiers which provide more uncertain erroneous predictions. In addition, we observed that the posterior probabilities of correct predictions were larger than 0.87 on average for all classifiers.

Fig. 2(b) shows the true positive rates when applying the Bayes filter technique to the proposed classifiers. It reveals that all classifiers are not influenced by high-frequent terrain changes in a significant manner when embedded into our Bayes filter prediction scheme.

6 Conclusion

In this paper, we systematically investigated the applicability of several posterior probability estimation techniques in the context of terrain classification based on temporal coherence. We exploited temporal coherence using a Bayes filter approach which takes several recent terrain class predictions into account. Depending on the choice of the classifier and the distance, a robot has to navigate over a certain terrain type before a terrain transition occurs, the classification performance increased by up to 6.9%. This number denotes the increase of classification performance related to a classification approach based on individual observations only. We showed that the Bayes filtering approach was nearly always superior to the single-observation approach with the only exception of the KNN classifier at a travel distance of less than or equal to 0.6 m. The significantly best experimental results were obtained using a combined support vector machine and k-nearest neighbor approach which has not been employed in the domain of terrain classification so far. Further investigation revealed that the various classifiers did not only differ in classification performance but also in the confidence of erroneous predictions. In the context of Bayesian filtering this is an important issue since a decrease in this confidence results in a decreased influence of wrong predictions on the final classification.

As a further contribution we examined the proposed classifiers with respect to their limiting factors such as storage requirements, prediction times, model generation times, and model selection times. The results provide criteria for choosing an appropriate classifier for a variety of hardware configurations.

References

1. Angelova, A., Matthies, L., Helmick, D.M., Perona, P.: Fast terrain classification using variable-length representation for autonomous navigation. In: Proceedings of the Conference on Computer Vision and Pattern Recognition, Minneapolis, MN, USA, pp. 1–8 (2007)
2. Bishop, C.: Neural Networks for Pattern Recognition. Oxford University Press, Oxford (1995)
3. Brooks, C.A., Iagnemma, K.: Vibration-based terrain classification for planetary exploration rovers. IEEE Transactions on Robotics 21(6), 1185–1191 (2005)
4. Burges, C.J.C.: A tutorial on support vector machines for pattern recognition. Data Mining and Knowledge Discovery 2, 121–167 (1998)
5. Cunningham, P., Delany, S.: k-nearest neighbour classifiers. Technical report, UCD School of Computer Science and Informatics (2007)
6. DuPont, E.M., Moore, C., Roberts, R.G.: Terrain classification for mobile robots traveling at various speeds: An eigenspace manifold approach. In: Proceedings of the IEEE International Conference on Robotics and Automation (ICRA 2008), pp. 3284–3289 (2008)
7. DuPont, E.M., Moore, C.A., Collins Jr., E.G., Coyle, E.: Frequency response method for terrain classification in autonomous ground vehicles. Autonomous Robots 24(4), 337–347 (2008)
8. Iagnemma, K., Dubowsky, S.: Terrain estimation for high-speed rough-terrain autonomous vehicle navigation. In: Proceedings of the SPIE Conference on Unmanned Ground Vehicle Technology IV, Orlando, FL, USA (2002)
9. Komma, P., Weiss, C., Zell, A.: Adaptive bayesian filtering for vibration-based terrain classification. In: IEEE International Conference on Robotics and Automation (ICRA 2009), Kobe, Japan, May 2009, pp. 3307–3313 (2009)
10. Lalonde, J.-F., Vandapel, N., Huber, D.F., Hebert, M.: Natural terrain classification using three-dimensional ladar data for ground robot mobility. Journal of Field Robotics 23(10), 839–861 (2006)
11. Manduchi, R., Castano, A., Talukder, A., Matthies, L.: Obstacle detection and terrain classification for autonomous off-road navigation. Autonomous Robots 18, 81–102 (2005)
12. Platt, J.: Advances in Large-Margin Classifiers. chapter: Probabilities for SV Machines, pp. 61–74. MIT Press, Cambridge (2000)
13. Specht, D.: Probabilistic neural networks. Neural Networks 3(1), 109–118 (1990)
14. Thrun, S., Burgard, W., Fox, D.: Probabilistic Robotics. In: Intelligent Robotics and Autonomous Agents, MIT Press, Cambridge (2005)
15. Vandapel, N., Huber, D., Kapuria, A., Hebert, M.: Natural terrain classification using 3-d ladar data. In: Proceedings of the IEEE International Conference on Robotics and Automation (ICRA 2004), New Orleans, LA, April 2004, pp. 5117–5122 (2004)
16. Weiss, C., Fechner, N., Stark, M., Zell, A.: Comparison of different approaches to vibration-based terrain classification. In: Proceedings of the 3rd European Conference on Mobile Robots (ECMR 2007), Freiburg, Germany, pp. 7–12 (2007)
17. Weiss, C., Zell, A.: Novelty detection and online learning for vibration-based terrain classification. In: Proceedings of the 10th International Conference on Intelligent Autonomous Systems (IAS 2008), Baden-Baden, Germany, pp. 16–25 (2008)

18. Wilcox, B.H.: Non-geometric hazard detection for a Mars microrover. In: Proceedings of the AIAA/NASA Conference on Intelligent Robots in Field, Factory, Service and Space, Houston, TX, USA, vol. 2, pp. 675–684 (1994)
19. Wu, T.-F., Lin, C.-J., Weng, R.C.: Probability estimates for multi-class classification by pairwise coupling. Journal of Machine Learning Research 5, 975–1005 (2004)
20. Yianilos, P.N.: Data structures and algorithms for nearest neighbor search in general metric spaces. In: Proceedings of the fourth annual ACM-SIAM Symposium on Discrete algorithms (SODA 1993), Philadelphia, PA, USA, pp. 311–321 (1993)
21. Zhang, H., Berg, A., Maire, M., Malik, J.: SVM-KNN: Discriminative nearest neighbor classification for visual category recognition. In: Proceedings of the 2006 IEEE Computer Society Conference on Computer Vision and Pattern Recognition (CVPR 2006), Washington, DC, USA, pp. 2126–2136 (2006)

Towards Visual Arctic Terrain Assessment

Stephen Williams and Ayanna M. Howard

Abstract. Many important scientific studies, particularly those involving climate change, require weather measurements from the ice sheets in Greenland and Antarctica. Due to the harsh and dangerous conditions of such environments, it would be advantageous to deploy a group of autonomous, mobile weather sensors, rather than accepting the expense and risk of human presence. For such a sensor network to be viable, a method of navigating, and thus a method of terrain assessment, must be developed that is tailored for arctic hazards. An extension to a previous arctic terrain assessment method is presented, which is able to produce dense terrain slope estimates from a single camera. To validate this methodology, a set of prototype arctic rovers have been designed, constructed, and fielded on a glacier in Alaska.

1 Introduction

An important aspect of autonomous field robotic navigation is terrain assessment. When an autonomous agent is deployed in unstructured, natural environments, the exact condition of the environment cannot be known ahead of time. Instead, the agent must assess the terrain condition locally, then revise its navigation plan as necessary. Much of the literature in the area of terrain assessment focuses on desert environments, arising from the needs of NASA's Mars rovers and the first two DARPA Grand Challenge events [2, 4, 6].

In contrast, little work has focused on navigating in arctic environments, despite the scientific importance of such areas. Though many scientists believe the condition of the giant ice sheets in Greenland and Antarctica are a key to understanding global climate change, there is still insufficient data to accurately predict the future

Stephen Williams and Ayanna M. Howard
School of Electrical and Computer Engineering
Georgia Institute of Technology
Atlanta, GA 30332
e-mail: swilliams8@gatech.edu, ayanna.howard@ece.gatech.edu

A. Howard et al. (Eds.): Field and Service Robotics 7, STAR 62, pp. 91–100.

behavior of those ice sheets. While satellites are able to map the ice sheet elevations with increasing accuracy, data about general weather conditions (i.e. wind speed, barometric pressure, etc.) must be measured at the surface.

In order to obtain measurements, human expeditions must be sent to these remote and dangerous areas. Alternatively, a group of autonomous robotic rovers could be deployed to these same locations, mitigating the cost, effort, and danger of human presence. For this to be a viable solution, a method for navigating in the arctic, and thus of assessing arctic terrain, must be developed. This paper extends the work presented in [10], creating dense slope estimates of the terrain from a single camera. Sect. 2 briefly describes the types of terrain likely to be encountered in the arctic regions of Greenland or Antarctica. Sect. 3 details the slope assessment algorithm. A set of prototype arctic rovers have been designed and constructed. A description of the units and the field tests conducted is presented in Sect. 4. The slope estimate results from the field tests are shown in Sect. 5. Finally, conclusions and future work are discussed in Sect. 6.

2   Environment

Despite being covered by snow, arctic regions present a large assortment of terrain challenges, a small sample of these are shown in Fig. 1. Large quantities of fresh surface snow can be present during certain times of the year. This fresh snow is soft, creating a potential sinking hazard for wheeled vehicles. The soft snow is also more readily melted, causing a dimpling of the surface, referred to as "sun cups," which can span 0.5 meters or more. Over time the winds harden the snow surface making it more amenable to locomotion. However, these same winds also sculpt the snow into dune-like structures that can be as large as one meter, again impeding motion. The underlying ice sheet is also responsible for several types of terrain hazards. As the ice sheet flows, forces build due to differential velocities of different ice sections. These forces can cause nearly vertical fractures in the ice known as crevasses. Crevasses can be as deep as 30 meters and are often covered with snow, making their detection all the more difficult. A narrow crevasse is shown in Fig. 1(c), which becomes obscured by snow toward the top of the image. In the thinner regions of the ice sheet, the surface is affected by the underlying mountains, causing significant local-scale elevation changes. Even on seemingly flat terrain, the actual snow depth can change drastically, with the ice sheet exposed in some locations, and covered by several meters of snow in others.

3   Slope Assessment

The slope estimation technique presented in [10] divided the image into large blocks in which the surface texture was analyzed. A single slope estimate was produced which was aligned with the predominate surface texture direction. The resulting estimates, although noisy, were shown to be representative of the actual slope within a simulated environment, and sufficient input for a slope-avoidance control scheme.

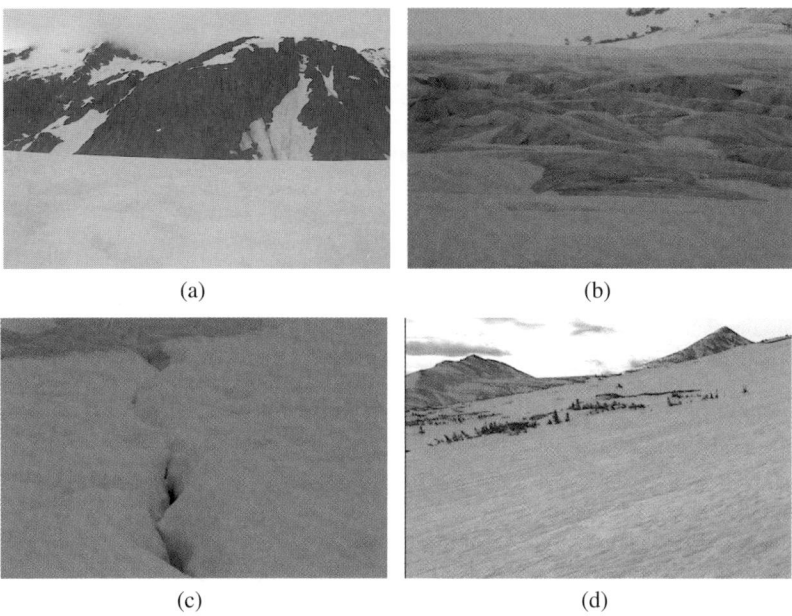

(a)   (b)

(c)   (d)

Fig. 1 Images from the top of Mendenhall Glacier, Alaska showing (a) visible sun cups, (b) a large section of exposed ice sheet, and (c) a small crevasse visible through the snow. (d) An image from an analogous site on the Arikaree Glacier, Colorado showing the potential steep slopes in glacial terrain.

Presented below is an improvement upon this algorithm which results in a set of dense slope estimates for the scene.

In images of arctic terrain, the surface texture has very low contrast. In order to analyze this texture, the foreground contrast must first be boosted. An adaptive, nonlinear preprocessing stage has been introduced, originally formulated to enhance x-ray images and CT scans [9]. Contrast limited adaptive histogram equalization (CLAHE) separates the image into different contextual regions. Within each region, a histogram equalization procedure is calculated. To prevent over-enhancement of local areas, a contrast limit is imposed. In effect, this applies an upper bound to the slope of the gradient at a specific location, resulting in smoothly varying contrast.

However, the presence of image distractors, such as background mountains, have an adverse effect on both the contrast enhancement and the subsequent slope estimates. A method of histogram thresholding, presented in [10] has been applied here. It is assumed that the majority of the image is filled with the snowy region. Consequently, in the histogram of the image, the largest peak should be associated with the grayscale values of the snow. An adaptive threshold based on the boundaries of this peak produces an image mask which can effectively separate the region of interest from unwanted objects and areas. Fig. 2 shows the results of the mask and contrast enhancement on a single exemplar glacial image. For the first time, the

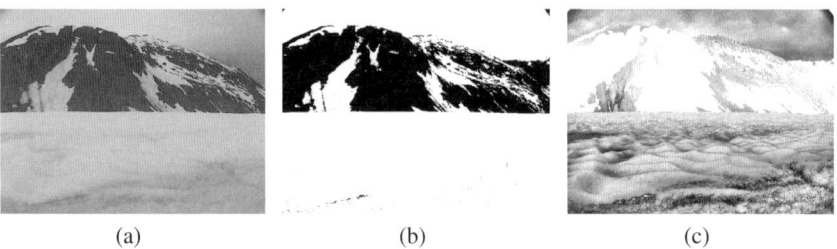

(a)   (b)   (c)

Fig. 2 (a) An image from Mendenhall Glacier, Alaska, (b) the mask produced by the histogram threshold procedure which separates the region of interest from the background objects, and (c) the results of the CLAHE contrast enhancement of the masked image. For the first time, the underlying structure of the scene is clearly visible.

underlying scene structure is clearly visible. Although, the image noise has clearly been amplified as well.

The enhanced snow texture exhibits small-scale directional details, which are visually similar to those of fingerprints. In the area of fingerprint enhancement, where it is desired to find and follow the small ridge details of a print, it is common to create an orientation image to aid in the processing [3, 5]. A least square estimate procedure for calculating this orientation is presented in [3]. In a similar fashion, the final slope estimate is produced by finding the least square estimate of the dominant Fourier spectrum direction in a neighborhood around each pixel.

To calculate the orientation of a given pixel, (i, j), the image gradient within a neighborhood of that pixel is first calculated. Then the two component vectors, v_x and v_y, are generated, as described in Equations 1 and 2. The orientation, θ, is then defined as the least squares solution to Equation 3. The entire slope calculation process can be processed in real-time.

$$v_x(i,j) = \sum_{neighborhood} 2\partial_x(u,v)\partial_y(u,v) \tag{1}$$

$$v_y(i,j) = \sum_{neighborhood} \partial_x^2(u,v) - \partial_y^2(u,v) \tag{2}$$

$$\theta(i,j) = \frac{1}{2}\tan^{-1}\left(\frac{v_y(i,j)}{v_x(i,j)}\right) \tag{3}$$

4   Field Tests

To validate the slope assessment algorithm, three prototype mobile weather sensor nodes were constructed. The rovers, referred to as "Sno-motes", were subsequently fielded on a frozen lake near Wapakoneta, Ohio and on Mendenhall Glacier in Juneau, Alaska.

4.1 Sno-mote Mk1

A 1/10 scale snowmobile chassis was selected for the prototype platform, endowing the rover with an inherent all-terrain drive system. The platform was modified to include an ARM-based processor running a specialized version of Linux. The motherboard offered several serial standards for communication, in addition to wifi and bluetooth. A daughterboard provided an ADC unit and PWM outputs for controlling servos. The drive system was modified to accept PWM motor speed commands, and the steering control was replaced with a high-torque servo. For ground truth position logging, a GPS unit connects to the processor via the bluetooth interface, while robot state and camera images are sent directly to an external control computer via the wifi link. To simulate the science objectives of the mobile sensor, a weather-oriented sensor suite was added to the rover. The deployed instrument suite includes sensors to measure temperature, barometric pressure, and relative humidity.

4.2 Alaska Test Site

The three "Sno-mote" platforms and related equipment were shipped to Juneau, Alaska for field testing. Two potential test sites were selected based on the relevance of weather data, the similarity of the terrain to arctic conditions, and logistics. Site 1, Lemon Creek Glacier, has been the subject of annual mass balance measurements since 1953 as part of the Juneau Icefield Research Program (JIRP) [8], making weather measurements in this area particularly relevant. The second site, Mendenhall Glacier, is one of Alaska's most popular tourist attractions [7]. The current public interest of this particular site makes additional information valueable. Both sites are only accessible via helicopter.

Helicopter travel to glacial areas is heavily dependent on the weather conditions, particularly low cloud deck heights. This presents a dangerous "white out" situation for the helicopter pilot in which the snow-covered peaks, the ground of the landing site, and the sky are all indistinguishably white. During the course of our flight to Lemon Creek such a situation was determined to exist, forcing the group to abandon the site for the day. A few images were acquired from this site before departure, a sample of which is shown in Fig. 3(a).

The weather conditions preventing travel to either of the test site finally lifted on Day 4, allowing travel to Mendenhall. The site surface is visually flat and covered with snow, though there are sections of the terrain where the underlying ice sheet is exposed. Despite the flat appearance, the snow varied in depth from a few centimeters to over a meter. This snow was deposited recently and was quite soft. Upon arrival at the site, a test area was explored with ice-axes to ensure it was safe. Cracks in the underlying ice, called crevasses, are often completely concealed by surface snow.

The rovers were driven manually to assess the mobility performance in the different snow conditions present. During these traverses it was discovered that the platform suffered from stability issues. Due to the narrow track footprint in the rear, the chassis would often roll sideways when attempting to navigate perturbations in the snow surface. Additionally, the snowmobile would sink in the fresh snow, causing

(a)   (b)

Fig. 3 (a) A sample of images acquired from Lemon Creek Glacier, Alaska before weather conditions forced the site to be abandoned for the day. Some of the underlying mountain range is visible though the glacier surface, but the terrain is predominately white, with the slope characteristics almost invisible. (b) A Sno-motes deployed at Mendenhall Glacier in Juneau, Alaska.

the DC drive motor to stall from excess torque. Due to the chassis limitations, a set of short traverses were performed in selected locations. During these traverses, the local temperature, barometric pressure, relative humidity, GPS location, and camera images were all logged at 2 Hz and timestamped to ensure proper off-line reconstruction and analysis. Fig. 1(a) - 1(c) show some sample images acquired at the Mendenhall test site.

4.3   Sno-mote Mk2

The main reason tracked vehicles are used for snow traversal is the large area of the track distributes the vehicle weight, allowing it to "float" on the surface. Possibly the most capable snow vehicle is the "Alpina Sherpa" [1], which was designed with two tracks to further reduce the applied pressure. Due to the discovered mobility issues with the original platform, a set of chassis modifications were designed and implemented, with inspiration taken from the "Sherpa". The original front suspension mechanism was replaced by a passive double-wishbone system, increasing the ski-base over 30%. The rear track system was replaced with a custom, dual-track design, which both widened the rear footprint and effectively doubled the snow contact surface area. A 500 W brushless motor and high-current speed controller drive the new track system. The overall increase in the platform width drastically improved the platform's stability and role characteristics. Fig. 4 shows the modified chassis, as deployed on a frozen lake near Wapakoneta, Ohio.

4.4   Ohio Test Site

A test site near Wapakoneta, Ohio was selected to verify the performance of the new chassis. The site was blanketed with 8-12 inches of fresh snow next to a frozen lake.

Fig. 4 The modified chassis of the second generation "Sno-mote" deployed near Wapakoneta, Ohio. Clearly visible is the new dual-track drive system.

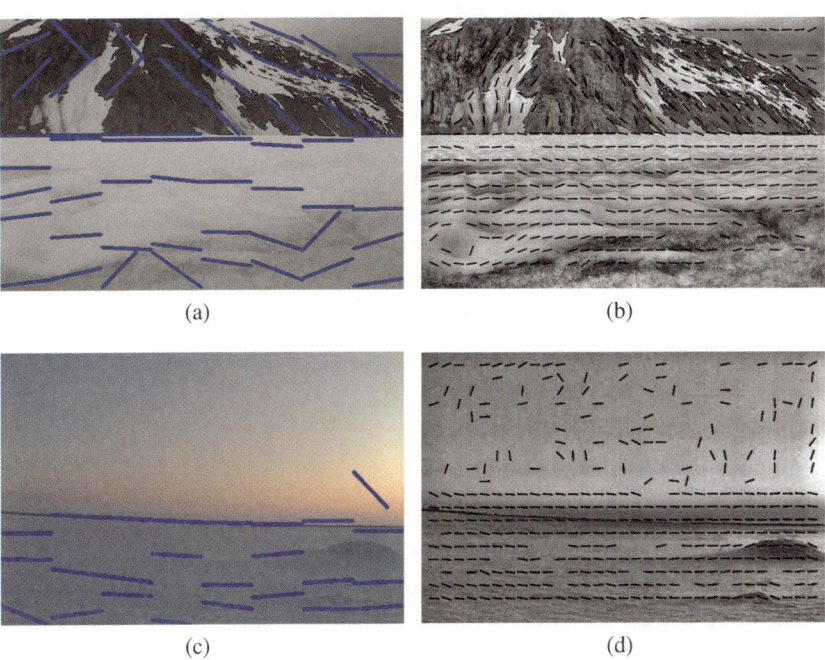

(a)   (b)

(c)   (d)

Fig. 5 (a, b) An image taken during the field tests at Mendenhall Glacier, Alaska has been processed by the original slope estimation procedure and the proposed method. Similarly, (c, d) an image from the Ohio field tests has been processed by both methods. The proposed method produces much denser estimates that are better able to capture smaller scale surface details.

Several long traverses were conducted, which transitioned from land to lake several times. During these traverses, the GPS location and camera image were logged at 15 Hz and timestamped. The lake bank consisted of irregularly spaced large rocks, between which large amounts of snow had collected, forming a drivable incline between $10°$ and $30°$.

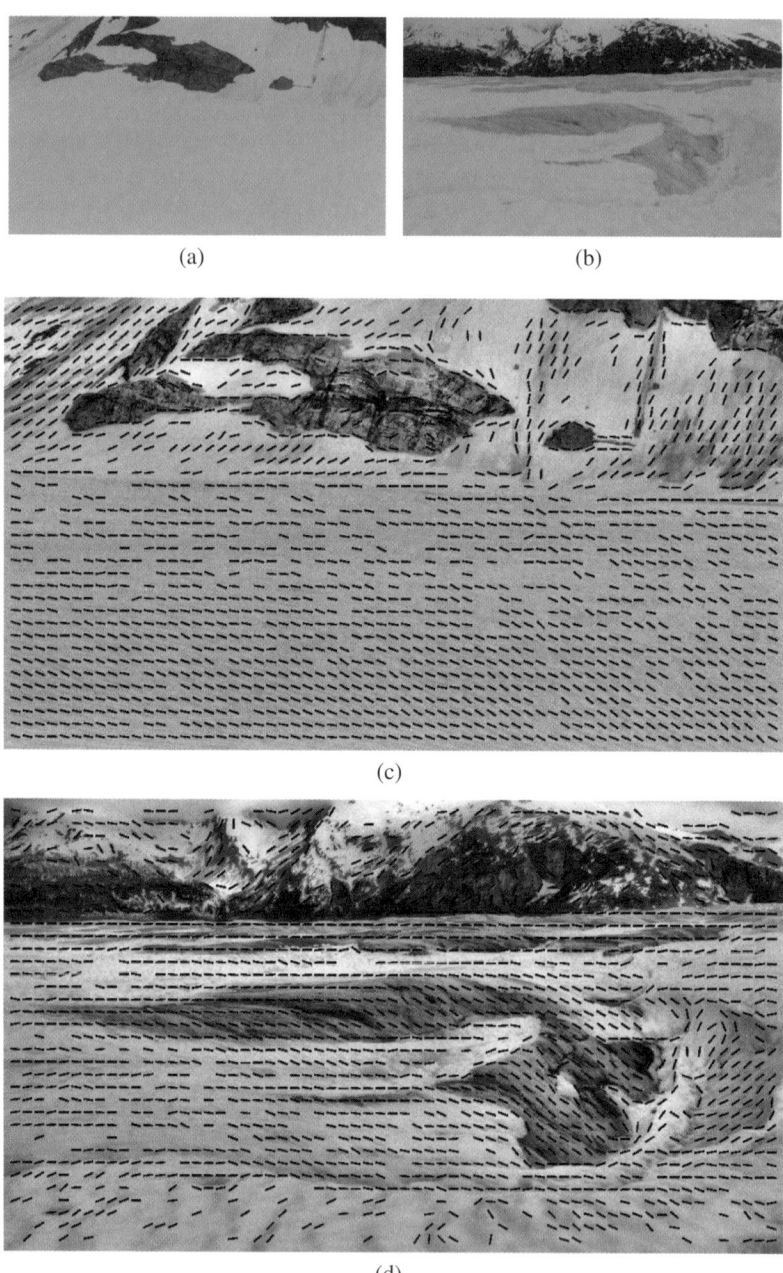

(a)   (b)

(c)

(d)

Fig. 6 Images (a) and (c) show an original image from Lemon Creek glacier, and the enhanced and processed version, respectively. Despite the surface texture being nearly invisible in the original image, the slope estimation process is able to produce dense, visually consistent slope measurements. Images (b) and (d) show an image of a large crevasse at Mendenhall glacier and the resulting slope estimates. The slope profiles in (d) clearly show the elevation changes at the edge of the crevasse.

The improved chassis performed well during the tests, never rolling, even when negotiating a path between rocks up a $20°$ slope. While it was still possible for the chassis to loose traction, especially in very soft snow or up steep inclines, the new drive motor was never forced into a stall condition. However, one unexpected observation from these test was that the control computer, which consisted of a consumer-grade laptop, ceased to operate when its temperature dropped below $20°F$. The "Sno-mote" control computer and hardware were unaffected by the cold.

5   Results

The slope estimation algorithm, presented in Sect. 3, has been applied to the images acquired during the field tests. For comparison, the original slope estimate technique, presented in [10], has also been calculated. The results of both techniques are shown in Fig. 5. While both methods show the general regions in front of the camera to be flat, the denser information of the proposed method is better able to capture the smaller scale surface trends. This is particularly evident in the lower left of the images in Fig. 5(a) and 5(b). The proposed method accurately indicates the slopes around a depression in the snow, whereas the original method provides only a single, slightly upward slope indication. Also, the new method is able to handle the ice as well as the snow image textures. The original method provides spurious measurements in the ice regions that do not reflect the true terrain grade. In the Ohio images, Fig. 5(c) and 5(d), the original method completely ignores the small dune structures, whereas the new method does indicate the sloping regions on either side of both structures.

Examples of processed arctic terrain are provided in Fig. 6. In the first image set from Lemon Creek Glacier, the terrain grade in the original image is virtually invisible. Yet, the estimate process is able to provide dense estimates, even in the areas that originally seemed uniformly white. The second pair of images illustrates a large crevasse on Mendenhall Glacier. The slope estimation process is able to handle both the snow and exposed ice textures without modification. The estimates provided clearly show the snow and ice sloping into the mouth of the crevasse, while a relatively safe area exists in the far left.

6   Conclusions

When navigating in arctic terrain, the local terrain slope is an important factor when determining traversabilty. Vehicle limitations may impose terrain grade limits, or local areas of steep decent may imply hazards. A purely visual slope estimation technique has been presented which creates dense slope estimates from a single image, even in the inherently low contrast environment of the arctic.

A set of prototype rovers have been constructed, based upon a snowmobile design, and fielded on a frozen lake in Ohio, as well as on Mendenhall Glacier in Juneau, Alaska. A sample of the slope estimation results from these field tests have been included. Qualitatively, the results appear consistent with human perceived slope determinations, and are an improvement over a previously presented method, both in terms of estimate density and slope misclassification.

In the future, these slope estimates will be developed into a full traversabilty assessment, were drivable terrain is classified as safe and terrain hazards are labeled. This would, in turn, be used by the navigation and path planning system to plot safe trajectories.

Acknowledgements. This work was supported by the National Aeronautics and Space Administration under the Earth Science and Technology Office, Applied Information Systems Technology Program. The authors would also like to express their gratitude to our collaborators Dr. Derrick Lampkin, Pennsylvania State University, for providing the scientific motivation for this research, Dr. Magnus Egerstedt, Georgia Institute of Technology, for providing his experience in multi-agent formations, and Dr. Matt Heavner, Associate Professor of Physics, University of Alaska Southeast, for providing his expertise in glacial field work.

References

1. Alpina snowmobiles (2009), http://www.alpina-snowmobiles.com/
2. Dang, T., Kammel, S., Duchow, C., Hummel, B., Stiller, C.: Path planning for autonomous driving based on stereoscopic and monoscopic vision cues. In: 2006 IEEE International Conference on Multisensor Fusion and Integration for Intelligent Systems, pp. 191–196 (2006)
3. Hong, L., Wan, Y., Jain, A.: Fingerprint image enhancement: Algorithm and performance evaluation. IEEE Transactions on Pattern Analysis and Machine Intelligence 20(8), 777–789 (1998)
4. Howard, A., Seraji, H.: Vision-based terrain characterization and traversability assessment. Journal of Robotic Systems 18, 577–587 (2001)
5. Kawagoe, M., Tojo, A.: Fingerprint pattern classification. Pattern Recogn 17(3), 295–303 (1984)
6. Lee, J., Crane, C.: Road following in an unstructured desert environment based on the em(expectation-maximization) algorithm. In: International Joint Conference SICE-ICASE, pp. 2969–2974 (2006)
7. Mendenhall glacier visitor center, juneau (2008), http://www.alaskageographic.org/static/847/ mendenhall-glacier-visitor-center-juneau
8. Miller, M., Pelto, M.: Mass balance measurements on the lemon creek glacier, juneau icefield, alaska 1953-1998. Geografiska Annaler: Series A, Physical Geography 81(11), 671–681 (1999)
9. Reza, A.M.: Realization of the contrast limited adaptive histogram equalization (CLAHE) for Real-Time image enhancement. The Journal of VLSI Signal Processing 38(1), 35–44 (2004)
10. Williams, S., Howard, A.M.: A single camera terrain slope estimation technique for natural arctic environments. In: ICRA, pp. 2729–2734. IEEE, Los Alamitos (2008)

Part III
Tracking and Servoing

Pedestrian Detection and Tracking Using Three-Dimensional LADAR Data

Luis E. Navarro-Serment, Christoph Mertz, and Martial Hebert

Abstract. The approach investigated in this work employs three-dimensional LADAR measurements to detect and track pedestrians over time. The sensor is employed on a moving vehicle. The algorithm quickly detects the objects which have the potential of being humans using a subset of these points, and then classifies each object using statistical pattern recognition techniques. The algorithm uses geometric and motion features to recognize human signatures. The perceptual capabilities described form the basis for safe and robust navigation in autonomous vehicles, necessary to safeguard pedestrians operating in the vicinity of a moving robotic vehicle.

1 Introduction

The ability to avoid colliding with other objects is essential in autonomous vehicles, especially in cases where they operate in close proximity to people. The timely detection of a pedestrian makes the vehicle aware of a potential danger in its vicinity, and allows it to modify its course accordingly. There is a large body of work done using laser line scanners as the primary sensor for pedestrian detection and tracking. In our group, we have developed detection and tracking systems using SICK^{TM} laser line scanners; these implementations work well in situations where the ground is relatively flat [5]. However, a 3D LADAR (i.e. one who produces a set of 3D points, or point cloud) captures a more complete representation of the environment and the objects within it. In [6], we presented an algorithm that detects pedestrians from 3D data. Its main improvement over the version with 2D data was that it constructs a ground elevation map, and uses it to eliminate ground returns. This allows pedestrian detection even when the surrounding ground is uneven. To

Luis E. Navarro-Serment, Christoph Mertz, and Martial Hebert
The Robotics Institute, Carnegie Mellon University, Pittsburgh, PA 15213
e-mail: lenscmu@ri.cmu.edu, cmertz@andrew.cmu.edu, hebert@ri.cmu.edu

A. Howard et al. (Eds.): Field and Service Robotics 7, STAR 62, pp. 103–112.

classify the humans the algorithm uses motion, size, and noise features. Persons are classified well as long as they are moving. However, there are still too many false positives when classifying stationary humans.

In this paper, we describe a strategy to detect and classify humans using the full 3D point cloud of the object. This will improve the classification of both moving and static pedestrians. However, the improvement will be most significant for static humans. The algorithm quickly detects the objects that have the potential of being humans using a subset of the point cloud, and then classifies each object using statistical pattern recognition techniques. We present experimental results of detection performance using 3D LADAR, which were obtained from field tests performed on a Demo III XUV [7].

2   Related Work

Some researchers have applied classification techniques to the detection and tracking problem. The approach reported in [1] applies AdaBoost to train a strong classifier from simple features of groups of neighboring points. This work focuses on 2D range measurements. Examples using three-dimensional data include [4], where 3D scans are automatically clustered into objects and modeled using a surface density function. A Bhattacharya similarity measure is optimized to register subsequent views of each object enabling good discrimination and tracking, and hence detection of moving objects. In [3], the authors describe a pedestrian detection system which uses stereo vision to produce a 3D point cloud, and then classifies the cloud according to the point shape distribution considering the first two central moments of the 2D projections using a naive Bayes classifier. Motion is also used as a cue for human detection.

In [8] the authors report an algorithm capable of detecting both stationary and moving humans. Their approach uses multi-sensor modalities including 3D LADAR and long wave infrared video (LWIR). Similarly, in [9] the same research group presents a technique for detecting humans that combines the use of 3D LADAR and visible spectrum imagery. In both efforts the authors employ a 2D template to extract features from the shape of an object. Among other differences, as opposed to our work, they extract a shape template from the projection in only one plane, and compute a measure of how uniformly distributed the returns are across the template.

3   Algorithm Description

In this section, the algorithm for pedestrian detection and classification is described. In our approach, since operation in real time is a chief concern, we do object detection and tracking in a 2D data subset first, and then use the object's position and size information to partition the set of 3D measurements into smaller groups, for further analysis. We describe these steps in the following sections.

3.1 Projection into 2D Plane

To reduce the computational cost of processing the entire point cloud, we initially isolate a 2D virtual slice, which contains only points located at a certain height above ground. As shown in Fig. 1, a 3D scanner produces a point cloud, from which a "slice" is projected onto the 2D plane, resulting in a virtual scan line. This scan line is a vector of range measurements coming from consecutive bearings, which resembles the kind of data reported directly by a line scanner such as the SICK^{TM} laser scanner. This is done by collapsing into the plane all the points residing within the slice, which is defined by its height above the ground.

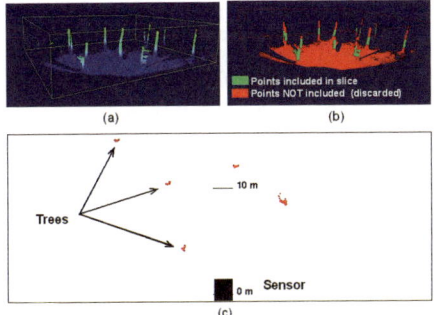

(a)   (b)

Points included in slice
Points NOT included (discarded)

Trees

10 m

0 m   Sensor

(c)

Fig. 1 Projection of virtual scan line. (a) A point cloud is collected from the environment shown. (b) The points located within a certain height above ground are projected into a 2D plane, and processed as if it were a single scan line. The resulting projection is shown in (c), top view

The ground elevation is stored in a scrolling grid that contains accumulated LADAR points and is centered at the vehicle's current position. The points are weighted by age, more recent points have a heigher weight. The mean and standard deviation of the heights of all scan points that are inside each cell are computed, and the elevation is then calculated by subtracting one standard deviation from the average height of all the points in the cell. The key properties of this simple algorithm are that mean and standard deviations can be calculated recursively, and that the elevation is never below the lowest point while still having about 80% of the points above ground.

The system adapts to different environments by varying the shape of the sensing plane i.e., by adjusting the height of the slice from which points are projected onto a two-dimensional plane. Spurious measurements produced by ground returns are avoided by searching for measurements at a constant height above the ground. Since our research was done in an open outdoor environment, we did not encounter overhanging structures like overpaths or ceilings. These might be topics of future research.

3.2 Motion Features

After detecting and tracking objects using the virtual scan line we can compute a *Motion Score* (MS). The MS is a measure of how confident the algorithm is that the detected object is a human, based on four motion-related
variables: the object's size, the distance it has traveled, and the variations in
the object's size and velocity. The size test discriminates against large objects
like cars and walls. The distance traveled test discriminates against stationary objects like barrels and posts. The variation tests discriminate against
vegetation, since their appearance changes a lot due to their porous and flexible nature. The individual results of these tests are scored, and then used
to calculate the MS. A detailed description of each test and all parameters
involved is presented in [6].

3.3 Geometric Features

To discriminate against static structures, we compute a group of distinguishing geometric features from the set of points belonging to each object being
tracked in 2D, and then feed these features to a classifier, which determines
whether the object is a human or not. This concept is depicted in Fig. 2.

As shown in Fig. 2(a), the process starts when a point cloud is read from
the sensor. We define $Z_j = \{\mathbf{x}_1, \mathbf{x}_2, \ldots, \mathbf{x}_N\}$ as the set of N points contained
in a frame collected at time t_j, whose elements are represented by Cartesian
coordinates $\mathbf{x} = (x, y, z)$. The points corresponding to one frame are shown,
and are colored according to their height above ground. To avoid the computational cost of processing the entire point cloud, we extract a 2D virtual
slice, as described in Section 3.1 (Fig. 2(b)). For each object being tracked,

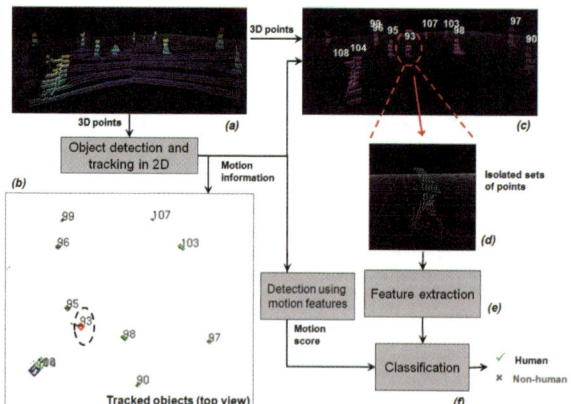

Fig. 2 Improved pedestrian detection. Geometric features present in subsets of the
point cloud are used by a classifier to distinguish pedestrians from static objects.

its position, velocity, and size are estimated using the algorithm described in [6]. These values are used to compute the MS. The object's position and size information are used to isolate, from the original point cloud, only those points corresponding to potential humans, as shown in Fig. 2(c). In this way, the three-dimensional information corresponding to each object is recovered in the form of smaller sets of points. At this point, we have a collection of M sets $\{S_1, S_2, \ldots, S_M\}$, where $S_{i \in \{1,2,\ldots,M\}} \subset Z_j$. A feature vector is computed from each of these sets (Figs. 2(d) - (e)), and then fed to a classifier that determines for each object whether it is a human or not, Fig. 2(f). This decision is made for each object, and is based on the most recent set of points collected from the sensor. The classifier also takes into account the information used to calculate the MS; this is described in a subsequent section.

A set of features is computed with the purpose of extracting the most informative signatures of a human in an upright posture from the 3D data. The legs are particularly distinctive of the human figure, so the algorithm computes statistical descriptions from points located around the legs. Similar descriptions are computed from the trunk area, representing the upper body. Additionally, the moment of inertia tensor is used to capture the overall distribution of all points. Finally, to include the general shape of the human figure, we compute the normalized 2D histograms on two planes aligned with the gravity vector.

3.3.1 Feature Extraction

Let $S_k = \{\mathbf{x}_1, \mathbf{x}_2, \ldots, \mathbf{x}_n\}$ be the set of points belonging to the object k, whose elements are represented by Cartesian coordinates $\mathbf{x} = (x, y, z)$. A set of suitable features is computed from S_k, as depicted in Fig. 2(d), which constitutes a profile of the object.

We begin by performing Principal Component Analysis (PCA) using all the elements of S_k, to identify the statistical patterns in the three-dimensional data (see Fig. 3). This involves the subtraction of the mean $\mathbf{m}$ from each of the three data dimensions. From this new data set with zero mean, we calculate the covariance matrix $\mathbf{\Sigma} \in \Re^{3 \times 3}$, and the normalized moment of inertia tensor $\mathbf{M} \in \Re^{3 \times 3}$, treating all points as unit point masses:

$$\mathbf{\Sigma} = \frac{1}{n-1} \sum_{k=1}^{n} (\mathbf{x}_k - \mathbf{m})(\mathbf{x}_k - \mathbf{m})^T$$

$$\mathbf{M} = \begin{bmatrix} \sum_{k=1}^{n} (y_k^2 + z_k^2) & -\sum_{k=1}^{n} x_k y_k & -\sum_{k=1}^{n} x_k z_k \\ -\sum_{k=1}^{n} x_k y_k & \sum_{k=1}^{n} (x_k^2 + z_k^2) & -\sum_{k=1}^{n} y_k z_k \\ -\sum_{k=1}^{n} x_k z_k & -\sum_{k=1}^{n} y_k z_k & \sum_{k=1}^{n} (x_k^2 + y_k^2) \end{bmatrix}$$

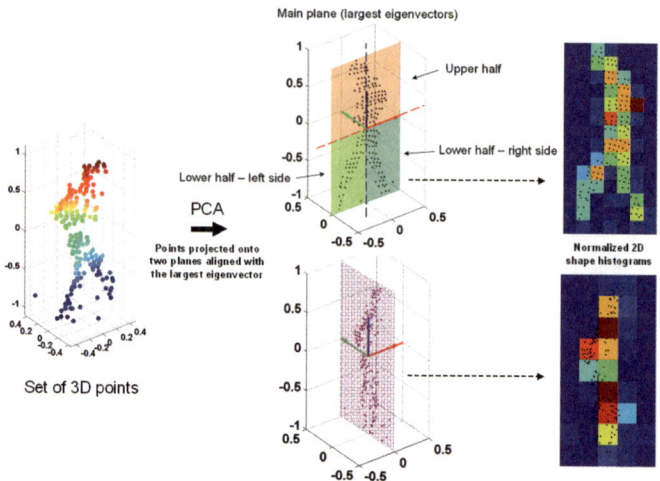

Fig. 3 A set of geometric features is computed from a the set of points belonging to an object.

Since both Σ and M are symmetric, we only use 6 elements from each as features.

Resulting from the PCA are three pairs of eigenvectors and eigenvalues, sorted according to decreasing eigenvalue. Call these eigenvectors e_1, e_2, e_3, with their corresponding eigenvalues $\lambda_1 > \lambda_2 > \lambda_3$. We assume that a pedestrian is in an upright position, so the principal component e_1 is expected to be vertically aligned with the person's body[1]. Together with the second largest component e_2, it forms the main plane (Fig.3, center top), and also forms the secondary plane with the smallest component, e_3 (Fig.3, center bottom). We then transform the original data into two representations using each pair of components e_1, e_2 and e_1, e_3, from which we proceed to compute additional features (the third possible representation, i.e. using the two smallest components e_2, e_3, is not used).

We focus on the points included in the main plane, to analyze the patterns that would correspond to the legs and trunk of a pedestrian, as shown in Fig 3, center top. These zones are the upper half, and the left and right lower halves. After separating the points into these zones, we calculate the covariance matrix (in 2D) over the transformed points laying inside each zone. This results in 9 additional features (3 unique values from each zone).

Finally, we compute the normalized 2D histograms for each of the two principal planes (Fig. 3, right), to capture the shape of the object. We use 14×7 bins for the main plane, and 9×5 for the secondary plane. Each bin is used as a feature, so there are 143 features representing the shape. A total of 164 geometric features are determined for each object.

---

[1] Dealing with the violation of this assumption is the focus of current research.

3.4   Human Detection

A classifier (Fig. 2(f)), composed of two independent Support Vector Machines (SVM) [2], determines for each object whether it is human or not. The first classifier is a SVM that receives the vector of 164 geometric features computed directly from S_k, and scores how closely the set matches a human shape. We call this the *Geometric Score* (GS). The GS is particularly effective for detecting static pedestrians. Similarly, the features used to determine the MS (i.e. object's size, the distance it has traveled, and the corresponding size and motion noises) contain valuable information about the motion of the target. Together with the GS, these features are fed to a second SVM, whose output represents the distance to the decision surface of the SVM. The *Strength of Detection* (SOD), the total measure of how strongly the algorithm rates the object as being a human, is calculated as the logistic function of the distance to the decision surface. This number is reported for each object. If the GS cannot be computed (e.g. insufficient data from a distant target, or violation of the upright position assumption), then the MS is reported as the SOD for that object.

3.4.1   Training

 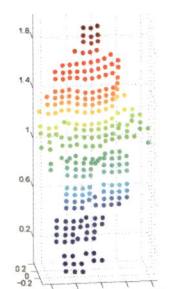

We trained the GS classifier using a combination of simulated and real examples. Because it is impossible to collect enough real data to evaluate perception algorithms in all possible situations, we have created a simulator capable of producing synthetic examples of sensor data. The simulator uses a ray tracing engine to generate a set of ray intersections between sensor and the objects in the scene to simulate. This information is then used to

Fig. 4 Simulated target (left), and its corresponding point cloud (right).

produce synthetic LADAR measurements according to a set of parameters for a particular sensor, as shown in Fig. 4. We trained the GS classifier using over 3500 examples (27.4% humans, 72.6% non-humans). The human set included 62% of simulated examples. The second classifier was trained using only real examples, since the motion and size noises used to determine the MS are of a dynamic nature and consequently harder to simulate efficiently (over 46000 examples: 6% humans, 94% non-humans).

We trained both SVMs using a five-fold cross validation procedure. We found that both radial basis function (RBF) and polynomial kernels resulted in similar levels of classification performance. After multiple tests, we determined that a RBF kernel was the best for the calculation of the GS, while a polynomial kernel was preferred for the second classifier.

4   Experimental Results

This section presents the results of several experimental runs. These results were obtained from field tests performed on a Demo III XUV [7]. The data comes from 14 different runs, where the variations include static and moving vehicles, pavement and off-road driving, and pedestrians standing, walking, or jogging. The data was taken at 17 Hz, and the average duration of each run was about 1 minute. There were altogether 48 humans and 1075 non-human objects, where those who came in and out of the field-of-view were counted twice. The ground truth was produced by labelling the data by hand.

In the upper part of Fig. 5 the ROC curve and the precision-recall curves are shown. Each human in one cycle is a positive example and each non-human object in one cycle is a negative example. There are about 6300 positive and 60000 negative examples. These plots illustrate the current performance of our system. The blue traces indicate the MS score, which is our previous detection algorithm. The red traces indicate the geometric score, i.e. the classification using the geometric features computed directly from the object's set of 3D points, but without any motion clues. As seen in the plots, neither algorithm by itself outperforms the other throughout the entire operational range. For low false positive rates the GS is better and at high false positive rates MS is better. As we mentioned earlier, the MS only works for static humans at high false positive rates. The synergistic combination of both results has significantly better performance, as indicated by the black traces.

An alternative representation of ROC and precision-recall is shown in the lower part of Fig. 5, where each object is counted per run. The score of an object is the mean of the score over all cycles, with a minimum of 10 cycles. As mentioned above, there are 48 humans (= positive examples) and 1075 other objects (negative examples). A noteworthy operating point is where there are basically no false positives (rate is 10^{-3}) and still the true positive rate is 0.75.

We identified several cases where performance is decreased. Typically, partial views of a human (e.g. only the upper body and part of the legs are seen by the sensor) result in false negative detections. Also, false detections occur when a target is only partially inside the sensor's field of view. Similarly, pedestrians in non-upright positions usually result in false negative detections. This is expected, because of the particular attention paid to legs and torso when extracting geometric features. An exception to this is the case of kneeling humans, which we have been able to detect consistently, though they are usually borderline classified as humans. Solving these problems is the focus of current research.

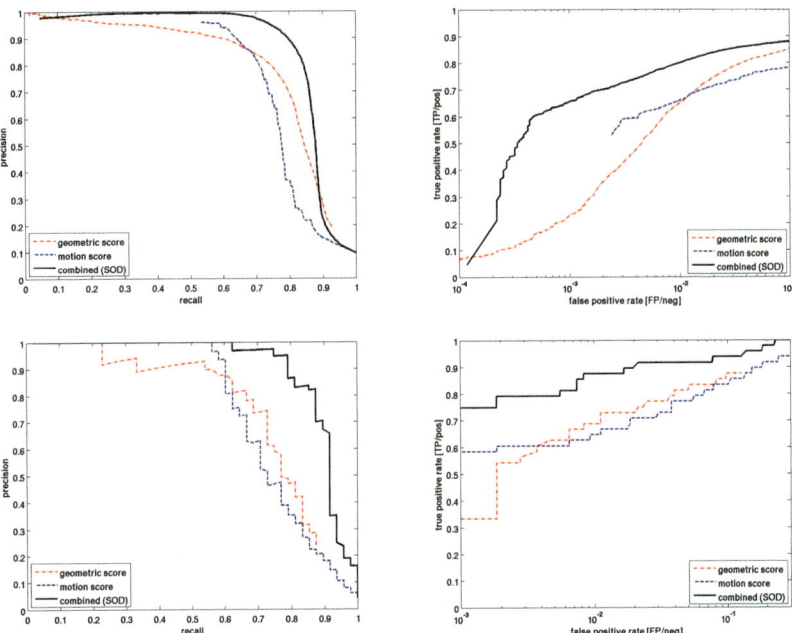

Fig. 5 The plots on the left show the precision-recall and the ones on the right the ROC curves. Shown are the curves produced by using the geometric score (red dashed), the motion score (blue dot-dash) and the combined (black solid). For the upper curves each object in each cycle and in the lower curves each object in the full run is counted as one example.

5 Conclusion

We described a pedestrian detection and tracking system using only three-dimensional data. The approach uses geometric and motion features to recognize human signatures, and clearly improves the detection performance achieved in our previous work. The set of features used to determine the human and motion scores was designed to detect humans in upright positions. To increase the robustness of detection of humans in other postures, in future research we will investigate ways of extracting signatures from the point cloud that are highly invariant to deformations of the human body.

Acknowledgements. We thank General Dynamics Robotic Systems for their support. This work was conducted through collaborative participation in the Robotics Consortium sponsored by the U. S. Army Research Laboratory under the Collaborative Technology Alliance Program, Coop. Agreement DAAD19-01-209912.

References

1. Arras, K.O., Mozos, O.M., Burgard, W.: Using Boosted Features for the Detection of People in 2D Range Data. In: Proc. of the 2007 IEEE Int. Conf. on Robotics and Automation, Roma, Italy, April 10-14, pp. 3402–3407 (2007)
2. Burges, C.J.C.: A Tutorial on Support Vector Machines for Pattern Recognition. In: Data Mining and Knowledge Discovery, vol. 2, pp. 121–167. Kluwer Academic Pub., Boston (1998)
3. Howard, A., Matthies, L.H., Huertas, A., Bajracharya, M., Rankin, A.: Detecting Pedestrians with Stereo Vision: Safe Operation of Autonomous Ground Vehicles in Dynamic Environments. In: Proc. of the 13th. International Symposium of Robotics Research, November 26-29 (2007)
4. Morris, D., Colonna, B., Haley, P.: Ladar-based Mover Detection from Moving Vehicles. In: Proc. of the 25th Army Science Conference (November 2006)
5. Navarro-Serment, L.E., Mertz, C., Hebert, M.: Predictive Mover Detection and Tracking in Cluttered Environments. In: Proc. of the 25th. Army Science Conference, November 27-30 (2006)
6. Navarro-Serment, L.E., Mertz, C., Vandapel, N., Hebert, M.: LADAR-based Pedestrian Detection and Tracking. In: IEEE Workshop on Human Detection from Mobile Platforms, Pasadena, California, May 20 (2008)
7. Shoemaker, C.M., Bornstein, J.A.: The Demo III UGV Program: a Testbed for Autonomous Navigation Research. In: Proc. of the IEEE Int. Symposium on Intelligent Control, Gaithersburg, MD, September 1998, pp. 644–651 (1998)
8. Thornton, S., Hoffelder, M., Morris, D.: Multi-sensor Detection and Tracking of Humans for Safe Operations with Unmanned Ground Vehicles. In: 1st. IEEE Workshop on Human Detection from Mobile Platforms, Pasadena, California, May 20 (2008)
9. Thornton, S., Patil, R.: Robust Detection of Humans Using Multi-sensor Features. In: Proc. of the 26th. Army Science Conference, December 1-4 (2008)

Passive, Long-Range Detection of Aircraft: Towards a Field Deployable Sense and Avoid System[*]

Debadeepta Dey, Christopher Geyer, Sanjiv Singh, and Matt Digioia

1 Introduction

Unmanned Aerial Vehicles (UAVs) typically fly blind with operators in distant locations. Most UAVs are too small to carry a traffic collision avoidance system (TCAS) payload or transponder. Collision avoidance is currently done by flight planning, use of ground or air based human observers and segregated air spaces. US lawmakers propose commercial unmanned aerial systems access to national airspace (NAS) by 30th September 2013. UAVs must not degrade the existing safety of the NAS, but the metrics that determine this have to be fully determined yet. It is still possible to state functional requirements and determine some performance minimums. For both manned and unmanned aircraft to fly safely in the same airspace UAVs will need to detect other aircraft and follow the same rules as human pilots.

Debadeepta Dey
Carnegie Mellon University, 5000 Forbes Avenue, Pittsburgh, USA
e-mail: debadeep@cs.cmu.edu

Christopher Geyer
Carnegie Mellon University, 5000 Forbes Avenue, Pittsburgh, USA
e-mail: cgeyer@cs.cmu.edu

Sanjiv Singh
Carnegie Mellon University, 5000 Forbes Avenue
e-mail: ssingh@ri.cmu.edu

Matthew Digioia
The Penn State Electro-Optics Center, 222 Northpointe Blvd Freeport, Freeport, USA
e-mail: mdigioia@eoc.psu.edu

[*] This material is based upon work supported by the Unique Missions Division, Department of the Army, United States of America under Contract No. W91CRB-04-C-0046. Any opinions, findings and conclusions, or recommendations expressed in this material are those of the author(s) and do not necessarily reflect the views of the Department of Army or Penn State EOC.

A. Howard et al. (Eds.): Field and Service Robotics 7, STAR 62, pp. 113–123.
springerlink.com　　　　　　　　　© Springer-Verlag Berlin Heidelberg 2010

Fig. 1 Selection of 11×11 subwindows showing the image of the Piper Archer II which was used as the intruder aircraft for collecting imagery, at a range of 1.5 miles. The camera and lens used had 0.41 milliradian/pixel resolution and a field of view of $30°$(H) $\times$ $21°$(V).

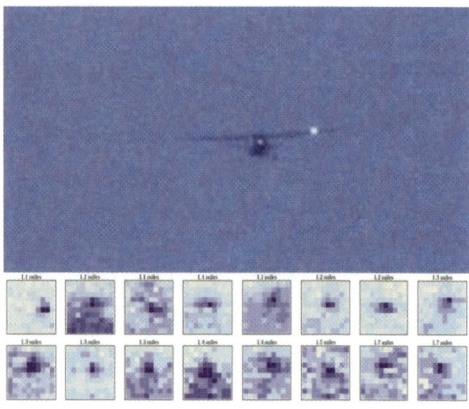

Key specifications of the international committee F38 on UAS systems standard F2411-04 (1) proposed requirements which include a field of regard of 220° (horizontal) $\times$ 30°(vertical), minimum detection range of 3 statute miles under visual flight rules and a required miss distance of 500 ft. Without this capability, widespread utilization of UAVs will not be possible.

This paper focuses on the sensing of aircraft with passive vision. Small size, low weight and power requirement make cameras attractive for this application. Multiple cameras can be used to cover the wide field-of-regard. A typical image of an aircraft at a range of 3 miles is a few pixels in diameter. Fig.1 shows a $11x11$ window around the image of the intruder aircraft at various ranges. Part of the challenge in detecting aircraft in such a wide field of regard reliably is the low signal to background ratio. Active sensors like radar are not feasible because of their prohibitive power and size requirements (2) for UAVs. Passive vision provides a low cost, low power solution albeit at the cost of a relatively high false positive rate. Using an approach based on morphological filters augmented with a trained classifier we have been able to obtain 98% detection rate out to 5 statute miles and a false positive rate of 1 every 50 frames.

In section 2 the related work in sense and avoid systems is outlined. In section 3 we discuss the details of the vision based aircraft detection algorithm. In section 4 we outline our efforts to collect imagery of flying aircraft and details about the result of our algorithm on the corpus of real ground truthed imagery of aircraft. Finally in section 5 we discuss the path forward towards a field deployable sense and avoid system.

2   Related Work

Utt et al. (20) describe a fielded vision-based sensory and perception system. McCandless (14) proposes an optical flow method for detecting moving aircraft. The use of morphological filtering is popular in computer vision based sense and avoid systems (11; 4). This approach generates a significant number of false positives.

Fig. 2 The minimum detection distance required to guarantee collision avoidance for varying intruder speeds and ownship speeds of 40, 50 and 60 knots. For the best case scenario the minimum detection distance is 700 meters and for the worst case scenario the minimum detection distance is 2100 m

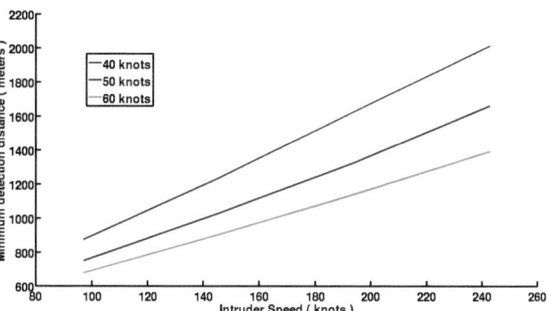

Petridis et al. use AdaBoost (17) to detect aircraft in low resolution imagery. Track-Before-Detect (TBD) (9) is an approach used mainly on infrared imagery. Defence Research Associates have implemented a vision based sense and avoid system on a Predator UAV system (13).

Efforts to directly model the range of atmospheric conditions under VFR remain untouched. A field deployable sense and avoid system must be able to operate in a variety of atmospheric conditions including fog, haze and directly against the glare of the sun. The operation of the system must not degrade beyond an acceptable level under all these conditions. We have developed an image formation model which accounts for the various atmospheric conditions. We used this model to predict the signal to background ratio of the image of the aircraft. The image formation model is described in detail in (12). The model also allows the determination of the suitability of any sensor combination before using the sensor and also to determine the minimum derived resolution for achieving a specified performance. The performance of the image formation model has been validated by the vast corpus of real imagery of flying aircraft that we collected during the course of this project.

2.1 Requirements

The range requirements of an aircraft detection system are influenced by two main factors: regulations developed by the FAA and the maneuvering capabilities of the UAV.

Duke et al.(6) and Schaeffer et al.(18) outline core competencies needed by a human equivalent system. The human equivalence mandated by OSD (15) and ACC (7) require vehicles to avoid non-cooperative vehicles without such systems. Shakernia et al. (19) leverage work of Utt et al. (20) on using maneuvers to reduce the intrinsic uncertainty about range when using an image based detector.

In order to decide the range requirements of the system we opted for a collision avoidance system by Frazzoli et al. (10) that enables an UAV to aggressively maneuver without breaching its envelope. Further details are in (12).

3   Approach

We experimented with a number of different approaches to detecting small targets with low signal to background ratios with an emphasis on methods that have both high detection rates and low computational complexity.

We have developed a multi-stage method that starts with a large number of candidates and winnows these down. We start with a morphological filter that looks for high contrast regions in the image that are most likely to be aircraft. Next we use a classifier that has been trained on positive and negative examples and finally we track the candidates over time to remove false positives. Below we discuss each "stage" of detection in detail.

3.1   Stage 1: Morphological Filtering

In the first stage, we apply a morphological filter that detects deviations from the background intensity. We use two types, one favors dark targets against lighter backgrounds (positive), and the other favors light targets against darker backgrounds (negative). The positive morphological filter takes the form:

$$\mathcal{M}^+(x,y) = \mathcal{I}(x,y) - \max\{\max_{|i|\leq w}\min_{|j|\leq w}\rangle(x+i+j,y),$$
$$\max_{|i|\leq w}\min_{|j|\leq w}\mathcal{I}(x,y+i+j)\} \tag{1}$$

As long as no $2w+1$ sub-window (we used $w=2$) contains all target pixels (higher intensity) and no background pixels (lower intensity), then all sub-windows will contain at least one (darker) background pixel. Since the background could be

Fig. 3 The image on the left shows part of the image of the Piper Archer II at a range of 2.87 miles. The image on the right shows the result of the morphological operation of the left image in Stage 1 of the processing pipeline. The dark aircraft image shows up as bright white spot.

noisy, the max's have the effect of finding a conservative greatest lower-bound for the background intensity. The difference, then, yields an estimate of the difference between signal and background for the pixel. The negative morphological filter, $\mathscr{M}^-$, swaps min's for max's, and negates the expression. From $\mathscr{M}^+$ we choose the top n_+ pixels above a threshold T_+, while suppressing local non-maxima, and construct a list of detections. We do the same for $\mathscr{M}^-$. Fig.3 shows an example aircraft image and the result of the morphological filtering on the example image.

3.2   Stage 2: Construction of a Shape Descriptor

For each detection we fit a Gaussian function to its $(2r+1) \times (2r+1)$ sub-window (we settled on $r=7$) and construct a shape descriptor for the detection. Through trial and error we found a descriptor that was a good discriminator in test sequences. The descriptor encodes the parameters of the fitted Gaussian, as well as statistics computed from the residual image. We use an axis-aligned Gaussian, parameterized as follows:

$$\mathscr{G}(x,y;\sigma_x,\sigma_y,b,s) = b + \frac{s}{2\pi\sigma_x\sigma_y} e^{-\frac{x^2}{2\sigma_x^2} - \frac{y^2}{2\sigma_y^2}} \tag{2}$$

We center the Gaussian at the the pixel with the largest absolute deviation from the window's mean intensity. We use gradient descent to minimize the sum of square errors between the input sub-window and $\mathscr{G}(\cdot;\xi)$, minimizing over $\xi = (\sigma_x, \sigma_y, b, s)$. To do this efficiently, we avoid repeated calls to the exponential function by pre-computing both a set of template $\mathscr{G}$'s over a range of σ_x and σ_y pairs, with $(b,s) = (0,1)$, and a set of finite difference approximations to the partial derivatives of $\mathscr{G}$ with respect to σ_x and σ_y. Fig.4 shows an example aircraft image and the fitted two dimensional Gaussian window centered on the image of the aircraft.

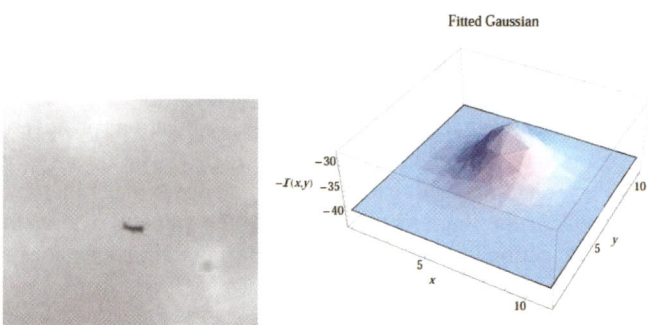

Fitted Gaussian

Fig. 4 The image on the left shows an image of the aircraft at 2.87 miles. The figure on the right shows the two dimensional Gaussian function fitted on the 15×15 subwindow centered on the image of the aircraft in the right image in Stage 2 of the processing pipeline.

Using the best fitting Gaussian $\mathscr{G}^*$, we compute a shape descriptor from the residual difference between the input image and $\mathscr{G}^*$ in upper-left (UL), upper-right (UR), lower-left (LL), lower-right (LL), and center (C) regions. We construct both positive and negative half-sign sums. For example:

$$S_{\mathrm{UL}}^+ = \sum_{\substack{1 \le x \le w \\ 1 \le y \le w}} \max\left[0, \mathscr{G}^*(x,y) - \mathscr{I}(x,y)\right]$$

$$\dots$$

$$S_{\mathrm{C}}^- = -\sum_{\substack{w/2 < x < 3w/2 \\ w/2 < x < 3w/2}} \min\left[0, \mathscr{G}^*(x,y) - \mathscr{I}(x,y)\right]$$

Then, we construct min's and max's of positive and negative half-sign sums, e.g., $S_{\max}^+ = \max\left(S_{\mathrm{UL}}^+, \dots, S_{\mathrm{C}}^+\right)$, and for each statistic we take its log normalized by the background intensity b, e.g., $s_{\max}^+ = \log\left(S_{\max}^+/b\right)$. We also compute the estimated signal to background ratio:

$$\mathrm{SBR} = \frac{|b| + |s|/2\pi\sigma_x\sigma_y}{|b|} \tag{3}$$

Finally, the shape descriptor we use is:

$$\mathbf{d} = \big(b, s, \sigma_x, \sigma_y, \mathrm{SBR}, s_{\min}^+, s_{\max}^+, s_{\min}^-, s_{\max}^-, \\ s_{\mathrm{UL}}^+, s_{\mathrm{UR}}^+, s_{\mathrm{LL}}^+, s_{\mathrm{LR}}^+, s_{\mathrm{C}}^+, s_{\mathrm{UL}}^-, s_{\mathrm{UR}}^-, s_{\mathrm{LL}}^-, s_{\mathrm{LR}}^-, s_{\mathrm{C}}^-\big)$$

We associate this 19-dimensional vector with each detection.

3.3 Stage 3: SVM-Based Classification of Potential Targets

The next stage of the algorithm is to pass the computed descriptor $\mathbf{f}$ for each detection through a *support vector machine* (SVM) (5). We trained the SVM using shape descriptors computed for positive and negative examples taken from a sequence of hand-labeled images. For negative examples we used the false negatives produced by the morphological filter. We used OpenCV's (3) implementation of an SVM (3), and chose to use radial basis functions for the classifier.

OpenCV's SVM implementation returns a hard classification: positive or negative depending on the sign of a summation, e.g. $y = \mathrm{sign}\, x$, where $x = \sum_i f_i(\mathbf{d})$. We want to associate a probability with each detection, so as to make them comparable, and so we modified the implementation to use the value of x to estimate the probability that the detection is from a true target. During training we construct empirical densities of x for positive (p_x^+) and negative (p_x^-) classes using a mixture of Gaussians, and store a log-likelihood ratio function $\ell(x) = \log p_x^+/p_x^-$ in a look-up table keyed on x. We choose the kernel bandwidth just large enough to make the odds monotonic in x. We keep only those detections whose odds exceed a minimum value of $p_{\min}$.

Fig. 5 A manned aircraft equipped with a GPS was flown in a series of flights such that it was in the field of view of the ground based cameras. The circles show the distance to the cameras in miles.

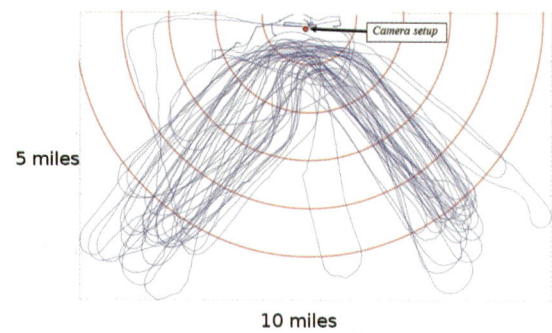

3.4   Stage 4: Tracking

The purpose of this stage is to track detections over time, associating detections to a list of tracked targets. Since many of the false positives are intermittent, we also use tracking to reduce the false positive rate. We arrived at a simple procedure for target tracking that provides a full screen tracking system for high definition imagery.

First, we always maintain a list of targets, and in steady state, it is the job of the tracker to associate to every existing target a detection. With any remaining detections, it also decides whether to create new targets.

For each existing target we consider a set of candidate detections, which are chosen from a wide search area around the predicted position of the target. For each potential matching detection, we evaluate the likelihood that the target and detection are associated given their respective descriptors.

Then, given a list of the likelihoods for the possible pairings we construct a graph with a node for each target and each detection, and edges between possible pairings, where the weights are the log likelihoods of pairings.

We construct a cost matrix, whose rows correspond to targets, columns to detections, and entries are the log likelihoods of the potential pairing, with $-\infty$ given to non-candidate pairings. The goal is to choose entries from the matrix, no more than one from every row and no more than one from every column such that the sum of probabilities is a maximum. We use the Hungarian algorithm (16) to find this matching. For the number of targets we typically have, usually less than 200, this computation can be computed in less than 8 milliseconds.

4   Data Collection and Results

Collaborating with the Penn State Electro-Optics Center (EOC), we collected ground to air imagery of aircraft with ten different camera/lens combinations. Four different infra-red cameras were also used to acquire imagery. The EOCs Payload Development Center, a comprehensive aerial systems integration lab (located on the field at the Jimmy Stewart Airport, Indiana, PA), was utilized to provide hardware, testing and technical support, and flight operations.

Table 1 Shows the number of false positives per frame for Stage 1 and Stage 3 as a function of the true positive percentage. Stage 3 reduces the false positive rate by a factor between 6 and 17 times as Stage 1

TP %	Stage 1 FP/frame	Stage 3 FP/frame	FP Reduction Factor
95%	120	20	5.9×
90%	66	3.9	17×
80%	14	1.0	14×
70%	8	0.66	12×
60%	6.2	0.56	11×

We tracked a Piper Archer II flying in the air from the ground with cameras mounted on pan tilt unit and synchronised with the geolocation of the aircraft so that it always remained in field of view of the cameras.

In Fig.5 we show the pattern the intruder aircraft flew as we gathered imagery. Till date we have collected 2.5 terabytes of imagery of which in 2 terabytes the position of the aircraft has been picked out manually for ground truth purposes. This corpus of real imagery has been used to analyze the performance of our algorithm.

We evaluated the performance of each stage of the algorithm using receiver operator characteristic curves (ROC) curves, which measure specificity (ability to reject outliers) and sensitivity (ability to detect true target) of a detector on about 2 terabytes of imagery of above the horizon flying aircraft.

Stage 3 improves the false positive rate by a factor between 6 and 17 depending on the detection rate over Stage 1. Refer Table.1 . We get a vast improvement with tracking in Stage 4. In the case of both Stage 1 and Stage 3 the variable affecting rates is a threshold. For Stage 1, the threshold is the value returned by the morphological filter at the detection. For Stage 3, the threshold is the probability according to the SVM classifier, that the detection is a target.

Fig.6 on the left shows the ROC curve for Stage 1, Stage 3 and Stage 4 of the algorithm. Whereas before the value affecting performance was a threshold on the output of a filter or classifier, in this case the threshold is the number of frames for which a target has been tracked. It is to be noted that the best overall detection rate of Stage 4 is higher than the best overall detection rate of Stage 3, even though it is based on the output of Stage 3. We believe that this is a temporal effect, in that detections that are intermittently below threshold, are picked up by the tracker. The detection rate decreases slightly at closer ranges. This is due to the fact that the algorithms were not optimized for close ranges.

Fig.6 on the right shows the effects of the variance of the minimum number of frames that a potential target has to be tracked for before being declared as a target. The points on the curve are the number of frames that a target has to have been tracked for, for it to be declared a possible target. In our experiments we let this threshold go up to 30 frames, at which point the false positive rate was 0.014

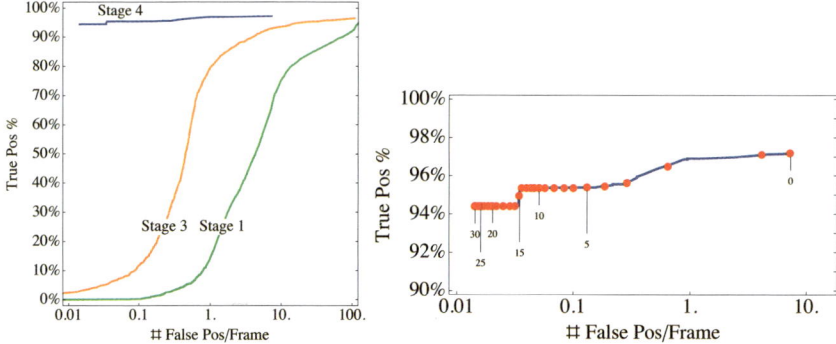

Fig. 6 Figure on the left shows the ROC curve for true positive and false positive for the 3 main stages of the algorithm on 4 mp imagery. The curve for Stage 4 shows almost perfect detection rate with a false positive rate of as low as 0.02 per frame. Figure on the right shows the effect of varying the minimum number of frames that a potential target has to be tracked for before being declared as a target in Stage 4. This curve is very flat as most true positives have long tracks and false positives have short tracks.

FP/frame and detection rate was 92%. It is to be noted that this curve is very flat. Most of the true positives have long tracks and almost all outliers have short tracks.

Overall there is a significant decrease in the number of false positives per frame. We add to Table.1 the results of Stage 4 and present them in Table.2. The entries for detection rates below 95% are not filled in because we chose not to evaluate the threshold frames beyond 30. If we had, the detection rate would have eventually fallen. We find that tracking in Stage 4 improves the false positive rate by a factor of over 500 times over Stage 3.

We found a reasonable compromise in false positive and true positive rate when we insisted that targets be tracked for at least 10 frames. Then the overall detection rate was 95%, the false positive rate was 0.05 false positives per frame. The detection rate is nearly 100% between 2.5 and 3.75 miles.

Table 2 Performance of various stages of the algorithm. Stage 4 achieves a false positive reduction rate of 571 times over Stage 1

TP %	Stage 1 FP/frame	Stage 3 FP/frame	Stage 4 FP/frame	FP Reduction Factor
97%	-	-	7.3	-
95%	120	20	0.035	571×
90%	66	3.9	-	-
80%	14	1.0	-	-
70%	8	0.66	-	-
60%	6.2	0.56	-	-

About 80% of the false positives that made it through the tracking of at least 10 frames were items that are of interest to collision avoidance. Most of the false positives were birds or landmarks on the ground that were not segmented out by the horizon detector (e.g. an antenna in the distance). These targets are of interest and could be considered useful.

We have developed and demonstrated a vision based algorithm that achieves a reasonable true positive rate of approximately 98% out to 5 statute miles and a false positive rate of 1 in every 50 frames which exceeds the FAA (8) regulatory requirement of reliable detection out to 3 statute miles.

5   Future Work

Currently, our system detects bearing to targets that must be avoided. An important extension will be to estimate range to the target so that precise maneuvers can be planned. We are currently investigating active ranging systems that can be pointed at potential targets, to estimate range and further reduce false positives.

Fusing infrared imagery with visible spectrum imagery and collecting below horizon imagery are areas we need to address.

Although the current algorithm takes about 0.8 seconds per 4 mp frame on an AMD Athlon X2 3800+ processor, most of the computation is image processing and hence amenable to parallelization. Specialized hardware like Digital Signal Processors are promising.

All of the above issues affect how a collision detection and warning system should be designed so as to cover the desired field-of-regard.

References

1. Standard specification for design and performance of an airborne sense-and-avoid system, f2411-04. ASTM International, West Conshohocken, PA (2004)
2. Bernier, R., Bissonnette, M., Poitevin, P.: Dsa radar - development report. In: UAVSI 2005 (2005)
3. Bradski, G.: The OpenCV Library. Doctor Dobbs Journal 25(11), 120–126 (2000)
4. Carnie, R., Walker, R., Corke, P.: Image processing algorithms for UAV "sense and avoid". In: Proceedings 2006 IEEE International Conference on Robotics and Automation, ICRA 2006., pp. 2848–2853 (2006)
5. Cristianini, N., Shawe-Taylor, J.: An Introduction to Support Vector Machines and other kernel-based learning methods. Cambridge University Press, Cambridge (2000)
6. Duke, E., Vanderpool, C., Duke, W.: Turning pinoccio into a real boy: A turing test for UAV operations. In: AIAA Infotech at Aerospace 2007 Conference and Exhibit (2007)
7. Ebdon, M.D., Regan, M.J.: Sense-and-avoid requirement for remotely operated aircraft, roa (2004)
8. Federal Aviation Administration: Order 7610.4 (2006)
9. Fernandez, M.F., Aridgides, T., Bray, D.: Detecting and tracking low-observable targets using IR. In: Proceedings of SPIE, vol. 1305, p. 193 (1990)

10. Frazzoli, E., Dahleh, M.A., Feron, E.: Real-time motion planning for agile autonomous vehicles. Journal of Guidance Control and Dynamics 25(1), 116–129 (2002)

11. Gandhi, T., Yang, M.-T., Kasturi, R., Camps, O.I., Coraor, L.D., McCandless, J.: Detection of obstacles in the flight path of an aircraft. In: CVPR, pp. 2304–2311. IEEE Computer Society, Los Alamitos (2000)

12. Geyer, C., Dey, D., Singh, S.: Prototype Sense-and-Avoid System for UAVs. Technical Report CMU-RI-TR-09-09, Robotics Institute, Carnegie Mellon University, Pittsburg (2009)

13. McCalmont, J., Utt, J., Deschenes, M.: Detect and Avoid Technology Demonstration. In: Proceedings of the American Institute of Aeronautics and Astronautics Infotech (2002)

14. McCandless, J.: Detection of aircraft in video sequences using a predictive optical flow algorithm. Optical Engineering 38, 523 (1999)

15. Office of the Secretary of Defense: Airspace integration plan for unmanned aviation (2004)

16. Papadimitriou, C., Steiglitz, K.: Combinatorial Optimization: Algorithms and Complexity. Courier Dover Publications (1998)

17. Petridis, S., Geyer, C., Singh, S.: Learning to detect aircraft at low resolutions. In: Gasteratos, A., Vincze, M., Tsotsos, J.K. (eds.) ICVS 2008. LNCS, vol. 5008, pp. 474–483. Springer, Heidelberg (2008)

18. Schaeffer, R.: A standards-based approach to sense-and-avoid technology. In: AIAA 3rd Unmanned Unlimited Technical Conference, Workshop and Exhibit (2004)

19. Shakernia, O., Chen, W., Raska, V.: Passive ranging for UAV sense and avoid applications. In: AIAA's Infotech at Aerospace (2005)

20. Utt, J., McCalmont, J., Deschenes, M.: Development of a sense and avoid system. In: AIAA Infotech at Aerospace (2005)

Multiclass Multimodal Detection and Tracking in Urban Environments*

Luciano Spinello, Rudolph Triebel, and Roland Siegwart

Abstract. This paper presents a novel approach to detect and track pedestrians and cars based on the combined information retrieved from a camera and a laser range scanner. Laser data points are classified using boosted Conditional Random Fields (CRF), while the image based detector uses an extension of the Implicit Shape Model (ISM), which learns a codebook of local descriptors from a set of hand-labeled images and uses them to vote for centers of detected objects. Our extensions to ISM include the learning of object sub-parts and template masks to obtain more distinctive votes for the particular object classes. The detections from both sensors are then fused and the objects are tracked using an Extended Kalman Filter with multiple motion models. Experiments conducted in real-world urban scenarios demonstrate the usefulness of our approach.

1 Introduction

One research area that has turned more and more into the focus of interest during the last years is the development of driver assistant systems and (semi-)autonomous cars. In particular, such systems are designed for operation in highly unstructured and dynamic environments. Especially in city centers, where many different kinds of transportation systems are encountered (walking, cycling, driving, etc.), the requirements for an autonomous system are very high. One key prerequisite for such systems is a reliable detection and distinction of dynamic objects, as well as an accurate estimation of their motion direction and speed. In this paper, we address this problem focusing on the detection and tracking of pedestrians and cars. Our system is a robotic car equipped with cameras and a 2D laser range scanner. As we will show, the use of different sensor modalities helps to improve the detection results.

Luciano Spinello, Rudolph Triebel, and Roland Siegwart
Autonomous Systems Lab, ETH Zurich, Switzerland
e-mail: {luciano.spinello,rudolph.triebel}@mavt.ethz.ch,
rsiegwart@ethz.ch

* This work was funded within the EU Projects BACS-FP6-IST-027140 and EUROPA-FP7-231888.

A. Howard et al. (Eds.): Field and Service Robotics 7, STAR 62, pp. 125–135.

The system we present here employs a variety of different methods from machine learning and computer vision, which have been shown to provide good detection rates. We extend these methods obtaining substantial improvements and combine them into a complete system of detection, sensor fusion and object tracking. We use supervised-learning techniques for both kinds of sensor modalities, which extract relevant information from large hand-labeled training data sets. In particular, the major contributions of this work are:

- Several extensions to the vision based object detector by Leibe *et al.* [13] using a feature based voting scheme denoted as Implicit Shape Models (ISM). Our major improvements to ISM are the subdivision of objects into sub-parts to obtain a more differentiated voting, the use of *template masks* to discard unlikely votes, and the definition of *superfeatures* that exhibit a higher evidence of an object's occurrence and are more likely to be found.
- The application and combination of boosted Conditional Random Fields (CRF) for classifying laser scans with the ISM based detector using vision. We use an Extended Kalman Filter (EKF) with multiple motion models to fuse the sensor information and to track the objects in the scene.

This paper is organized as follows. The next section describes work that is related to ours. Sec. 3 gives a brief overview of our overall object detection and tracking system. In Sec. 4, we introduce the implicit shape model (ISM) and present our extensions. Sec. 5 describes our classification method of 2D laser range scans based on boosted Conditional Random Fields. Then, in Sec. 6 we explain our EKF-based object tracker. Finally, we present experiments in Sec. 7 and conclude the paper.

2   Related Work

Several approaches can be found in the literature to identify a person in 2D laser data including analysis of local minima [19, 23], geometric rules [24], using maximum-likelihood estimation to detect dynamic objects [10], using AdaBoost on a set of geometrical features extracted from segments [1], or from Delaunay neighborhoods [20]. Most similar to our work is that of Douillard *et al.* [5] who use Conditional Random Fields to classify objects from a collection of laser scans. In the area of vision-based people detection, there mainly exist two kinds of approaches (see [9] for a survey). One uses the analysis of a *detection window* or *templates* [8, 4], the other performs a *parts-based* detection [6, 11]. Leibe *et al.* [13] present a people detector using *Implicit Shape Models* (ISM) with excellent detection results in crowded scenes. In earlier works, we showed already extensions of this method with a better feature selection and an improved nearest neighbor search [21, 22].

Existing people detection methods based on camera *and* laser data either use hard constrained approaches or hand tuned thresholding. Zivkovic and Kröse [25] use a learned leg detector and boosted Haar features from the camera images and employ a parts-based method. However, both their approach to cluster the laser data using Canny edge detection and the use of Haar features to detect body parts is hardly suited for outdoor scenarios due to the highly cluttered data and the larger variation of illumination. Schulz [18] uses probabilistic exemplar models learned from training data of both sensors and applies a Rao-Blackwellized particle filter (RBPF) to track a person's appearance in the data. However, in outdoor scenarios illumination changes often and occlusions are very likely, which is why contour matching is not appropriate. Also, the RBPF is computationally demanding, especially in crowded environments. Douillard *et al.* [5] also use image features to enhance the object detection but they do not consider occlusions and multiple image detection hypotheses.

3 Overview of Our Method

Our system consists of three main components: an appearance based detector that uses the information from camera images, a 2D-laser based detector providing structural information, and a tracking module that uses the combined information from both sensor modalities and provides an estimate of the motion vector for each tracked object. The laser based detection applies a Conditional Random Field (CRF) on a boosted set of geometrical and statistical features of 2D scan points. The image based detector extends the multiclass version of the Implicit Shape Model (ISM)[13]. It only operates on a region of interest obtained from projecting the laser detection into the image to constrain the position and scale of the detected objects. Then, the tracking module applies an Extended Kalman Filter (EKF) with two different motion models, fusing the information from camera and laser. In the following, we describe the particular components in detail.

4 Appearance Based Detection

Our vision-based people detector is mostly inspired by the work of Leibe *et al.* [13] on scale-invariant Implicit Shape Models (ISM). In summary, an ISM consists in a set of local region descriptors, called the *codebook*, and a set of displacements and scale factors, usually named *votes*, for each descriptor. The idea is that each descriptor can be found at different positions inside an object and at different scales. Thus, a vote points from the position of the descriptor to the center of the object as it was found in the training data. To obtain an ISM from labeled training data, all descriptors are clustered, usually using agglomerative clustering, and the votes are computed by adding the scale and the displacement of the objects' center to the descriptors in the codebook. For the detection, new descriptors are computed on a test image and matched against the descriptors in the codebook. The votes that are cast by each matched descriptor are collected in a 3D *voting space*, and a maximum density estimator is used to find the most likely position and scale of an object.

In the past, we presented already several improvements of the standard ISM approach (see [21, 22]). Here, we show some more extensions of ISM to further improve the classification results. These extensions concern both the learning and the detection phase and are described in the following.

4.1   ISM Extensions in the Learning Phase

Sub-Parts: The aim of this procedure is to enrich the information from the voters by distinguishing between different object subparts from which the vote was cast. We achieve this by learning a circular histogram of interest points from the training data set for each object class. The number of bins of this histogram is determined automatically by using k-means clustering. The final number of clusters, here denoted as q, is obtained using the Bayesian Information Criterion (BIC). Note that this subpart extraction does not guarantee a semantical subdivision of the object (i.e.: legs, arms, etc. for pedestrians) but it is interesting to see that it nevertheless resembles this automatically without manual interaction by the user (see Fig. 1, left and center).

Template Masks: In the training data, labeled objects are represented using a binary image named *segmentation mask*. This mask has the size of the object's bounding box and is 1 inside the shape of the object and 0 elsewhere. By overlaying all these masks for a given object class so that their centers coincide and then averaging over them, we obtain a *template mask* of each object class (see Fig. 1, left and center). This method is more robust against noise than, e.g., Chamfer matching [3], and does not depend on an accurate detection of the object contours. We use the template mask later to discard outlier votes cast from unlikely areas.

Superfeatures: The original ISM maintains all features from the training data in the codebook as potential voters and does not distinguish between stronger and weaker votes. This has the disadvantage that often too many votes are cast, even if an occurance of the object is not likely given the training data, and leads to many false positive detections. To overcome this, we propose to extract *superfeatures* from the training data, i.e. descriptor vectors that cast a stronger vote than standard features. We keep these superfeatures in a separate codebook to avoid clutter in the implementation. A superfeature is defined by a local density maximum in descriptor space, where only feature vectors are considered that correspond to interest points from a dense area in the image space (in x, y, and scale). This definition ensures that for superfeatures a high evidence of the occurrence of the object is combined with a high probability to encounter an interest point. We compute superfeatures by first employing mean shift estimation on all interest points found in the training data set for each class, and then clustering the feature vectors in descriptor space that correspond to the interest points from the found areas of high density. This clustering is done agglomeratively. In the end, we select the 50% of the cluster centers that correspond to the biggest clusters. The right part of Fig. 1 shows an example. Note that the superfeatures inherently reflect the skeleton of the object.

4.2   ISM Extensions in the Inference Phase

Sub-Parts and Template Masks: After collecting all the votes for a given set of extracted input features from a test image, we first discard the ones that are implausible by placing the template mask at the potential object centers and removing the votes

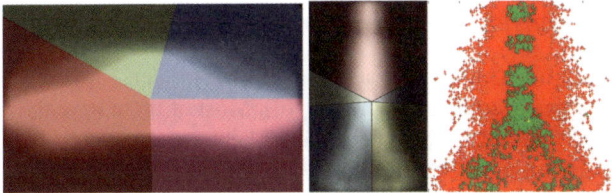

Fig. 1 Left and Center: Sub-parts, depicted in colored slices, and template masks, in white. They are computed from the training set. Note that even though the subparts are computed unsupervised, they exhibit some semantic interpretation. **Right**: Superfeatures are stable features in image and descriptor space. This figure shows Shape Context descriptors at Hessian interest points (in red) for the class 'pedestrian'. The position of the superfeatures are depicted in green.

that are cast from outside the mask. For the remaining ones we find the maximum density point **m** using mean shift and insert all votes for **m** into a circular histogram with q bins: one per sub-part of the object. We denote each such histogram as a *hypothesis* $\mathbf{h} = (h_1, \dots, h_q)$ of an object's position. The *strength* σ of a hypothesis is defined as the sum of all bins, i.e. the number of all voters for the object center. To find the best hypothesis we define a partial order $\prec$ based on a function Δ_h:

$$\mathbf{h}_i \prec \mathbf{h}_j \Leftrightarrow \Delta_h(\mathbf{h}_i, \mathbf{h}_j) < 0 \quad \text{where} \quad \Delta_h(\mathbf{h}_i, \mathbf{h}_j) := \sum_{k=1}^{q} \text{sign}(h_k^i - h_k^j). \quad (1)$$

Using this, we select the hypothesis with the highest order (in case of ambiguity we use the one with the highest strength) for each class. Then, we find the best hypothesis *across* all classes as described below, remove all its voters and recompute the ordering. This is done until a minimum hypothesis strength σ_{min} is reached. Thus, the parameter σ_{min} influences the number of false positive detections.

Superfeatures: Superfeatures and standard features vote for object centers in the same voting space, but the votes from superfeatures are weighted higher (in our case by a factor of 2). Thus, the score of a hypothesis is higher if the fraction of superfeatures voting for it is higher. In some cases where an object's shape visibility is low only superfeatures might be used to obtain a very fast detection.

Best Inter-Class Hypothesis: As mentioned above, we need to rate the best object hypotheses from all classes. To be independent on an over- or under-representation of a class in the codebooks, we do this by comparing the relative areas covered by the voters from all class hypotheses. More precisely, we define a square area γ around each voter that depends on the relative scale of the descriptor, i.e. the ratio of the test descriptor's scale and that of the found descriptor in the codebook. The fraction of the area covered by all voters of a hypothesis and the total area of the object (computed from the template mask) is then used to quantify the hypothesis. Care has to be taken in the case of overlapping class hypotheses. Here, we compute the set intersection of the interest points in the overlapping area and assign their corresponding γ values alternately to one and the other hypothesis.

5   Structure Based Detection

For the detection of objects in 2D laser range scans, several approaches have been presented in the past (see for example [1, 16]). Most of them have the disadvantage that they disregard the conditional dependence between data points in a close neighborhood. In particular, they can not model the fact that the label l_i of a given scan point $\mathbf{z}_i$ is more likely to be l_j if we know that l_j is the label of $\mathbf{z}_j$ and $\mathbf{z}_j$ and $\mathbf{z}_i$ are neighbors. One way to model this conditional independence is to use Conditional Random Fields (CRFs) [12], as shown by Douillard *et al.* [5]. CRFs represent the conditional probability $p(\mathbf{y} \mid \mathbf{z})$ using an undirected cyclic graph, in which each node is associated with a hidden random variable l_i and an observation $\mathbf{z}_i$. In our case, the l_i is a discrete label that ranges over 3 different classes (pedestrian, car and background) and the observations $\mathbf{z}_i$ are 2D points in the laser scan. At this point we omit the mathematical details about CRFs and refer to the literature (e.g. [5, 17]). We only note that for training the CRF we use the L-BFGS gradient descent method [14] and for the inference we use max-product loopy belief propagation.

We use a set of statistical and geometrical features $\mathbf{f}_n$ for the nodes of the CRF, e.g. height, width, circularity, standard deviation, kurtosis, etc. (for a full list see [20]). We compute these features in a local neighborhood around each point, which we determine by jump distance clustering. However, we don't use this features directly in the CRF, because, as stated in [17] and also from our own observation, the CRF is not able to handle non-linear relations between the observations and the labels. Instead, we apply AdaBoost [7] to the node features and use the outcome as features for the CRF. For our particular classification problem with multiple classes, we train one binary AdaBoost classifier for each class against the others. As a result, we obtain for each class k a set of M weak classifiers u_i (decision stumps) and corresponding weight coefficients α_i so that the sum

$$g_k(\mathbf{z}) := \sum_{i=1}^{M} \alpha_i u_i(\mathbf{f}(\mathbf{z})) \tag{2}$$

is positive for observations assigned with the class label k and negative otherwise. We apply the inverse logit function $a(x) = (1 + e^{-x})^{-1}$ to g_k to obtain a classification likelihood. Thus, the node features for a scan point $\mathbf{z}_i$ and a label l_i are computed as $\mathbf{f}_n(\mathbf{z}_i, l_i) = a(g_{l_i}(\mathbf{z}_i))$. For the edge features $\mathbf{f}_e$ we compute two values, namely the Euclidean distance d between the points $\mathbf{z}_i$ and $\mathbf{z}_j$ and a value g_{ij} defined as

$$g_{ij}(\mathbf{z}_i, \mathbf{z}_j) = \text{sign}(g_i(\mathbf{z}_i) g_j(\mathbf{z}_j))(|g_i(\mathbf{z}_i)| + |g_j(\mathbf{z}_j)|). \tag{3}$$

This feature has a high value if both $\mathbf{z}_i$ and $\mathbf{z}_j$ are equally classified (its sign is positive) and low otherwise. Its absolute value is the sum of distances from the decision boundary of AdaBoost where $g(\mathbf{z}) = 0$. Thus, we define the edge features as

$$\mathbf{f}_e(\mathbf{z}_i, \mathbf{z}_j, l_i, l_j) = \begin{cases} \left(a(d(\mathbf{z}_i, \mathbf{z}_j)) \quad a(g_{i,j}(\mathbf{z}_i, \mathbf{z}_j)) \right)^T & \text{if } l_i = l_j \\ (0 \qquad\qquad 0)^T & \text{otherwise.} \end{cases} \tag{4}$$

The intuition behind Eq. (4) is that edges that connect points with equal labels have a non-zero feature value and thus yield a higher potential.

6 Object Tracking and Sensor Fusion

To fuse the information from camera and laser and for object tracking we use an Extended Kalman Filter (EKF) as presented in [21]. In our implementation, we use two different motion models – Brownian motion and linear velocity – in order to cope with pedestrian and car movements. The data association is performed in the camera frame: we project the detected objects from the laser scan into the camera image. Assuming a fixed minimal object height, we obtain a rectangular search region, in which we consider all hypotheses from the vision based detector for the particular object class. Using a previously calibrated distance r_0 of an object at scale 1.0 (using the normalized training height), we can estimate the distance r_{est} of a detected object in the camera image by multiplying r_0 with the scale of the object. Then, r_{est} is compared to the measured distance r_{meas} from the laser and both detections are assigned to each other if $|r_{meas} - r_{est}|$ is smaller than a threshold τ_d (in our case 2m).

We track cluster centers of gravity in the 2D laser frame using two system states:

$$\mathbf{x}_{m1} = \langle (x^{cog}, y^{cog}), (v_x^{cog}, v_y^{cog}), (c_1, \dots, c_n) \rangle \text{ and } \mathbf{x}_{m2} = \langle (x^{cog}, y^{cog}), (c_1, \dots, c_n) \rangle,$$

one for each motion model. Here, (v_x^{cog}, v_y^{cog}) is the velocity of the cluster centroid (x_x^{cog}, y_y^{cog}) and $c_1, \dots, c_n$ are the probabilities of all n classes. We use a static state model where the observation vector $\mathbf{w}$ consists of the position of the cluster and the class probabilities for each sensor modality:

$$\mathbf{w} = \langle \hat{x}^{cog}, \hat{y}^{cog}, (c_1, \dots, c_n)^1, \dots, (c_1, \dots, c_n)^s \rangle. \tag{5}$$

Here, $(\hat{x}^{cog}, \hat{y}^{cog})$ is a new observation of a cluster center and s denotes the number of sensors. The matrix H models the mapping from states to the predicted observation and is defined as $H = (P^T S_1^T \dots S_s^T)^T$, where P maps to pose observations and the S_i map to class probabilities per sensor. For example, for one laser, one camera and constant velocity we have

$$P = \begin{pmatrix} 1 & 0 & 0 & 0 & 0 & 0 \\ 0 & 1 & 0 & 0 & 0 & 0 \end{pmatrix} \qquad S_1 = S_2 = \begin{pmatrix} 0 & 0 & 0 & 0 & 1 & 0 \\ 0 & 0 & 0 & 0 & 0 & 1 \end{pmatrix}. \tag{6}$$

7 Experimental Results

To acquire the data, we used a car equipped with two CCD cameras and a 2D laser range finder mounted in front (see Fig. 2, right). The 3D transform between the laser and the camera coordinate frame was calibrated beforehand. We acquired training data sets for both sensor modalities. For the camera, we collected images of pedestrians and cars that we labeled by hand. The pedestrian data set consists of 400 images of persons with a height of 200 pixels in different poses and with different

Fig. 2 Left: For car classification, we use codebooks from 7 different views. For training, mirrored images are included for each view to obtain a wider coverage. **Center:** For pedestrians we use 2 codebooks of side views with mirroring. Lateral views have sufficient information to generalize frontal/back views. **Right:** Setup used for the city data set. Only a small overlap of the cameras' field of view is used to cover a larger part of the laser scans. No stereo vision is used in this work.

clothing and accessories such as backpacks and hand bags in a typical urban environment. The class 'car' was learned from 7 different viewpoints as in [13] (see also Fig. 2, left). Each car data set consists of 100 pictures from urban scenes with occlusions. Car codebooks are learned using Shape Context (SC) descriptors [2] at Hessian-Laplace interest points [15]. The pedestrian codebook uses lateral views and SC descriptors at Hessian-Laplace and Harris-Laplace interest points for more robustness. Experience shows [13] that lateral views of pedestrians also generalize well to front/back views. Our laser training data consists of 800 annotated scans with pedestrians, cars and background. There is no distinction of car views in the

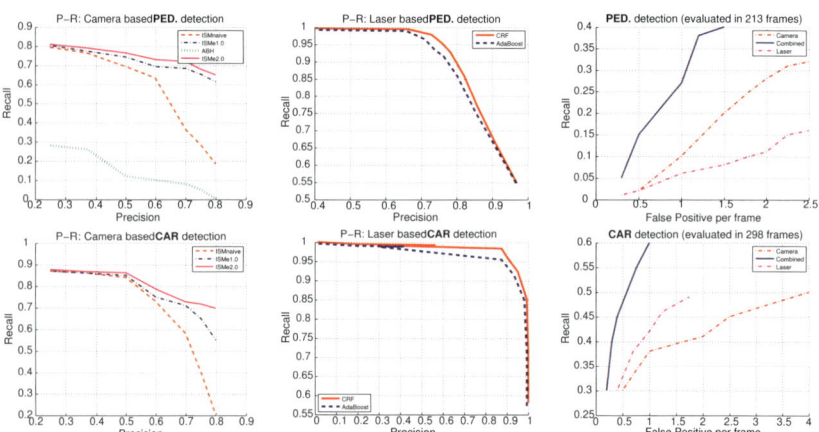

Fig. 3 Quantitative evaluation. **Upper row:** pedestrian detection, **Lower row:** car detection. From left to right we show the results only using camera, only using laser, and both. As we can see, our approach outperforms the other methods for both sensor modalities. The image based detection is compared with standard ISM, our first extension of ISM (ISMe1.0) and AdaBoost with Haar features. Our CRF-based laser detector is compared with AdaBoost. We can also see that the combination of both sensors improves the detection result of both single sensors.

Fig. 4 Cars and pedestrian detected and tracked under occlusion, clutter and partial views. In the camera images, upper row, blue boxes indicate car detections, orange boxes pedestrian detections. The colored circle on the upper left corner of each box is the track identifier. Tracks are shown in color in the second row and plotted with respect to the robot reference frame.

laser data as the variation in shape is low. The range data consists in 4 layers where each has an angular resolution of $0.25°$ and a maximum range of $15m$.

To quantify the performance of our detector we acquired two datasets containing cars and pedestrians. The results of our detection algorithm are shown in Fig. 3. Our vision based detecion named ISMe2.0 is compared to the standard ISM, our previous extension ISMe1.0, and for the pedestrian class, with AdaBoost trained on Haar features (ABH). For the class 'car', we averaged the results over all different views. We can see that our method yields the best results with an Equal Error Rate (EER) of 72.3% for pedestrians and 74% for cars. The improvements are mainly due to a decreased rate of false positive detections. The results of our laser based detection are shown in the middle column of Fig. 3. We can see that our approach using boosted CRFs performs better than standard AdaBoost. The right column of Fig. 3 depicts the results for the combined detection using laser and vision. These graphs clearly show that using both sensors the number of false positive detections decreases and the hit rate increases. Some qualitative results are shown in Fig. 4 where a passing car and a crossing pedestrian are correctly detected and tracked.

In addition, we evaluated our algorithm on a third, more challenging dataset acquired in the city of Zurich. It consists of 4000 images and laser scans. The equal error rates of this experiment resulted in 64.1% (laser-only), 64.1% (vision-only) and 68% (combined) for pedestrians, and in $(72.2\%, 73.5\%, 75.7\%)$ for cars. As a comparison, we evaluated the state-of-the-art pedestrian detector based on Histogram of Oriented Gradients [4] and ABH obtained an EER of 36.4 and 8.9.

8 Conclusions

We presented a method to reliably detect and track multiple object classes in outdoor scenarios using vision and 2D laser range data. We showed that the overall performance of the system is improved using a multiple-sensor system. We presented several extensions to the ISM based image detection to cope with multiple classes. We showed that laser detection based on CRFs performs better than a simpler AdaBoost classifier and presented tracking results on combined data. Finally, we showed the usefulness of our approach through experimental results on real-world data.

References

1. Arras, K.O., Mozos, Ó.M., Burgard, W.: Using boosted features for the detection of people in 2d range data. In: IEEE Int. Conf. on Rob. and Autom, ICRA (2007)
2. Belongie, S., Malik, J., Puzicha, J.: Shape matching and object recognition using shape contexts. IEEE Trans. on Pattern Analysis & Machine Intelligence 24(4), 509–522 (2002)
3. Borgefors, G.: Hierarchical chamfer matching: A parametric edge matching algorithm. IEEE Trans. on Pattern Analysis & Machine Intelligence 10(6), 849–865 (1988)
4. Dalal, N., Triggs, B.: Histograms of oriented gradients for human detection. In: IEEE Conf. on Comp. Vis. and Pat. Recog., CVPR (2005)
5. Douillard, B., Fox, D., Ramos, F.: Laser and vision based outdoor object mapping. In: Robotics: Science and Systems (RSS), Zurich, Switzerland (June 2008)
6. Felzenszwalb, P., Huttenlocher, D.: Efficient matching of pictorial structures. In: IEEE Conf. on Comp. Vis. and Pat. Recog (CVPR), pp. 66–73 (2000)
7. Freund, Y., Schapire, R.E.: A decision-theoretic generalization of on-line learning and an application to boosting. Journal of Computer and System Sciences 55(1), 119–139 (1997)
8. Gavrila, D., Philomin, V.: Real-time object detection for "smart" vehicles. In: IEEE Int. Conf. on Computer Vision, ICCV (1999)
9. Gavrila, D.M.: The visual analysis of human movement: A survey. Comp. Vis. and Image Und (CVIU) 73(1), 82–98 (1999)
10. Hähnel, D., Triebel, R., Burgard, W., Thrun, S.: Map building with mobile robots in dynamic environments. In: IEEE Int. Conf. on Rob. and Autom, ICRA (2003)
11. Ioffe, S., Forsyth, D.A.: Probabilistic methods for finding people. Int. Journ. of Comp. Vis. 43(1), 45–68 (2001)
12. Lafferty, J., McCallum, A., Pereira, F.: Conditional random fields: Probabilistic models for segmentation and labeling sequence data. In: Int. Conf. on Machine Learning, ICML (2001)
13. Leibe, B., Cornelis, N., Cornelis, K., Gool, L.V.: Dynamic 3d scene analysis from a moving vehicle. In: IEEE Conf. on Comp. Vis. and Pat. Recog, CVPR (2007)
14. Liu, D., Nocedal, J.: On the limited memory bfgs method for large scale optimization. Math. Programming 45(3, (Ser. B)) (1989)
15. Mikolajczyk, K., Schmid, C.: A performance evaluation of local descriptors. IEEE Trans. on Pattern Analysis & Machine Intelligence 27(10), 1615–1630 (2005)
16. Premebida, C., Monteiro, G., Nunes, U., Peixoto, P.: A lidar and vision-based approach for pedestrian and vehicle detection and tracking. In: ITSC (2007)
17. Ramos, F., Fox, D., Durrant-Whyte, H.: CRF-matching: Conditional random fields for feature-based scan matching. In: Robotics: Science and Systems, RSS (2007)
18. Schulz, D.: A probabilistic exemplar approach to combine laser and vision for person tracking. In: Robotics: Science and Systems, RSS (2006)

19. Schulz, D., Burgard, W., Fox, D., Cremers, A.: People tracking with mobile robots using sample-based joint probabilistic data ass. filters. Int. Journ. of Rob. Res (IJRR) 22(2) (2003)
20. Spinello, L., Siegwart, R.: Human detection using multimodal and multidimensional features. In: IEEE Int. Conf. on Rob. and Autom, ICRA (2008)
21. Spinello, L., Triebel, R., Siegwart, R.: Multimodal detection and tracking of pedestrians in urban environments with explicit ground plane extraction. In: IEEE Int. Conf. on Intell. Rob. and Systems, IROS (2008)
22. Spinello, L., Triebel, R., Siegwart, R.: Multimodal people detection and tracking in crowded scenes. In: Proc. of the AAAI Conf. on Artificial Intelligence (July 2008)
23. Topp, E.A., Christensen, H.I.: Tracking for following and passing persons. In: IEEE Int. Conf. on Intell. Rob. and Systems, IROS (2005)
24. Xavier, J., Pacheco, M., Castro, D., Ruano, A., Nunes, U.: Fast line, arc/circle and leg detection from laser scan data in a player driver. In: IEEE Int. Conf. on Rob. and Autom, ICRA (2005)
25. Zivkovic, Z., Kröse, B.: Part based people detection using 2d range data and images. In: IEEE Int. Conf. on Intell. Rob. and Systems (IROS), San Diego, USA (November 2007)

Vision-Based Vehicle Trajectory Following with Constant Time Delay

Hien K. Goi, Timothy D. Barfoot, Bruce A. Francis, and Jared L. Giesbrecht

Abstract. A convoy problem is formulated and solved for two four-wheeled ve-hicles. The task is for the second vehicle to follow the leader's trajectory with a constant time delay. This delayed trajectory can be viewed as the trajectory of a *delayed leader*. This novel constant-time-delay concept allows for the estimation of the delayed leader's speed and heading using the vehicle kinematics. Decoupled longitudinal and lateral controllers are developed based on the constant-time-delay approach. The lateral controller includes a look-ahead feature to compensate for steering delays. Successful field trials were conducted with full-sized military vehi-cles on a 1.3-kilometre test track. The tracking errors from differential global posi-tioning system (DGPS) ground truth covering 13 kilometres are presented.

1 Introduction

Motivating our research is a military scenario in which a vehicle convoy traverses hostile territory to deliver supplies. Naturally, equipping every vehicle in the con-voy with armour that will protect its occupants is expensive. To reduce the cost, autonomous unarmoured supply vehicles may be used, whereby each autonomous

Hien K. Goi and Bruce A. Francis
Department of Electrical and Computer Engineering, University of Toronto
Toronto, Ontario, Canada
e-mail: hien.goi@utoronto.ca, bruce.francis@utoronto.ca

Timothy D. Barfoot
University of Toronto Institute for Aerospace Studies
Toronto, Ontario, Canada
e-mail: tim.barfoot@utoronto.ca

Jared L. Giesbrecht
Defence Research and Development Canada-Suffield
Medicine Hat, Alberta, Canada
e-mail: jared.giesbrecht@drdc-rddc.gc.ca

A. Howard et al. (Eds.): Field and Service Robotics 7, STAR 62, pp. 137–147.
springerlink.com   © Springer-Verlag Berlin Heidelberg 2010

vehicle would follow the trajectory of the vehicle ahead of it. To follow the vehicle ahead, an autonomous vehicle can sometimes take advantage of a global positioning system (GPS), inter-vehicle communications, and/or lane markers/magnets. However, since the vehicle convoy is in hostile territory, GPS signals may be jammed, inter-vehicle communications may be intercepted, the roads may be unstructured.

Based on this motivating example, our project goal is to design and test a control system to allow a convoy of full-sized autonomous vehicles with large inter-vehicle spacing to follow the lead vehicle's trajectory without cutting corners on turns. The control system should use only on-board sensors, avoiding the use of GPS, inter-vehicle communications, and lane markers/magnets. This paper reports our preliminary experimental results performed on two full-sized vehicles where the leader vehicle is manually driven and the follower vehicle is autonomously controlled. In our field trials to date, we do not use lane marker/magnets or radio communications between vehicles. However, we do use GPS to measure the follower's position due to a poor odometry system.

To our knowledge, there have been only a few prior experimental works relevant to our project goal. The most relevant is Gehrig and Stein [4], who tested a path-following strategy that, with the addition of autonomous speed control, could potentially follow the leader's trajectory at large distances without cutting corners. Their 'Control Using Trajectory' (CUT) algorithm stores the time history of the leader's path and steers towards the leader's position that is a constant distance ahead of the follower's current position. Although Gehrig and Stein's experiments showed improvements in tracking the leader's path over a system without CUT, the experimental data was limited to less than 15 seconds.

Other experimental works that use only on-board sensors include Benhimane et al. [1], Franke et al. [3], and Kehtarnavaz et al. [6]. Benhimane et al. developed a vehicle-following system with the objective of tracking a virtual leader a constant distance behind the leader. However, since the trajectory of the leader is not stored, the follower may cut corners on turns. Furthermore, their experimental data was limited to 2 minutes, and the maximum follower speed was 1 m/s. Both Franke et al. and Kehtarnavaz et al. had vehicle-following systems that were based on the follower's range and bearing to the leader. As such, both implemented steering controls that simply steered toward the leader. Such steering controls are known to deviate from the leader's path [4].

Daviet and Parent [2] also performed vehicle-following experiments without using inter-vehicle communications. Their vehicles travelled up to 10 m/s, but the corresponding distance separation was only 4.5 metres, and the experimental data shown was for only 30 seconds. They also used the strategy of simply steering towards the leader, but they do suggest storing the leader's trajectory in a future implementation. Although Schneiderman et al. [8] used radio communications, the radio link was used to interact with the follower's computer system. They demonstrated a path-following system with a 33-kilometre traverse at speeds of 13.9 to 20.8 m/s and at following distances of 5 to 15 metres. However, only the steering was autonomous, and the follower did not store the leader's path.

1.1 Problem Formulation

To meet our project goal, we take a novel approach to tracking the leader's trajectory. Our objective is for the follower to track the planar trajectory of the leader delayed by a constant time, τ. Specifically, if $(x(t), y(t))$ is the position of the follower with respect to an inertial frame and $(x_0(t), y_0(t))$ is the position of the leader with respect to the same frame, then we want $(x(t), y(t))$ to track $(x_0(t - \tau), y_0(t - \tau))$. For brevity, we define the delayed leader position, $(x_0(t - \tau), y_0(t - \tau))$, as $(x_d(t), y_d(t))$. The leader, delayed leader, and follower are shown in Fig. 1a. It is important to note that our definition is different from the constant time headway [9] definition. The tracking error in our definition is with respect to the leader's delayed position, while the tracking error in constant time headway is with respect to the leader's current position.

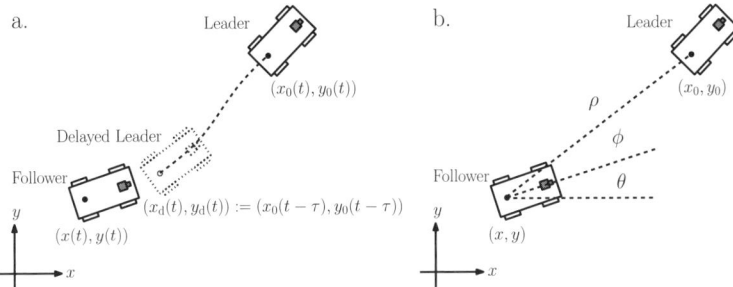

Fig. 1 (**a**) Leader, delayed leader, and follower in an inertial frame. (**b**) The leader's and follower's positions are related by the follower's heading, θ, and the range, ρ, and bearing, ϕ, to the leader.

There are two main advantages to our approach: 1) tracking the delayed leader provides us with 'future' delayed-leader positions since we have measurements up to the leader's current position; and 2) the following distance varies based on the leader's speed. The first advantage allows us to track the leader's trajectory without having to measure the leader's speed or heading. Instead, the delayed leader's speed and heading are estimated using the delayed leader's future positions. Having future position measurements also allows our system to use interpolation to handle the occasional data dropout, which is to be expected with a vehicle-following system on bumpy roads. Due to space limitations, the details of our interpolation technique are not discussed in this paper. The second advantage naturally causes the following distance to be smaller when the leader slows down on difficult portions of the road, e.g., turns and rough terrain. The smaller following distance allows for more accurate measurements of the leader's position, which will help the tracking during those difficult portions.

2 System Architecture and Design

Given the problem formulation, the follower requires a means to localize its position, (x, y), and heading, θ, relative to an inertial frame. This localization can be done using GPS or wheel odometry. Since the convoy is to operate in hostile territory, our preference is to use wheel odometry. To measure the leader's relative position, we use a pan-tilt-zoom monocular camera system with a colour tracker to servo around a coloured target on the back of the leader [5]. Knowing the offsets between the camera and the follower's rear axle and between the target and the leader's rear axle, we can obtain the range, ρ, and bearing, ϕ, to the leader from the camera's output, as shown in Fig. 1b.

A top level diagram of our vehicle control system is shown in Fig. 2. The camera system outputs the range and bearing to the leader. The odometry measures the follower's speed and steering, (v, γ), and provides estimates of the follower's position and heading, $(\hat{x}, \hat{y}, \hat{\theta})$. The range, bearing, and odometric estimates are fed into a nonlinear observer to produce estimates of the delayed leader's position, heading, and speed, $(\hat{x}_d, \hat{y}_d, \hat{\theta}_d, \hat{v}_d)$, along with estimates of a look-ahead point's position and heading, $(\hat{x}_l, \hat{y}_l, \hat{\theta}_l)$. These estimates are used by the control laws to produce the commanded speed and steering, (v_c, γ_c), which are the inputs to the follower. The details of the vehicle model for the follower, the control laws, and the nonlinear observer are provided in the following subsections.

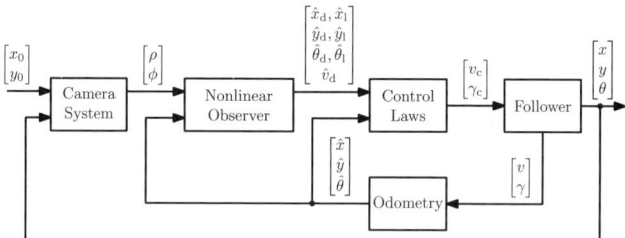

Fig. 2 Top level diagram of vehicle control system.

2.1 Follower Vehicle Model

For the vehicle kinematics, we chose the bicycle model, which is given by

$$\dot{x} = v\cos\theta \ , \quad \dot{y} = v\sin\theta \ , \quad \dot{\theta} = \frac{v}{d}\tan\gamma \ ,$$

where (x, y) is the position of the rear axle, θ is the vehicle's heading, d is the distance between the front and rear axles, v is the vehicle's speed, and γ is the vehicle's steering angle. We derived a local linear model by examining the longitudinal and lateral tracking errors, (e_1, e_2), in the follower's frame. The tracking errors are defined to be

$$\begin{bmatrix} e_1 \\ e_2 \end{bmatrix} := \begin{bmatrix} \cos\theta & \sin\theta \\ -\sin\theta & \cos\theta \end{bmatrix} \begin{bmatrix} x_d - x \\ y_d - y \end{bmatrix} .$$

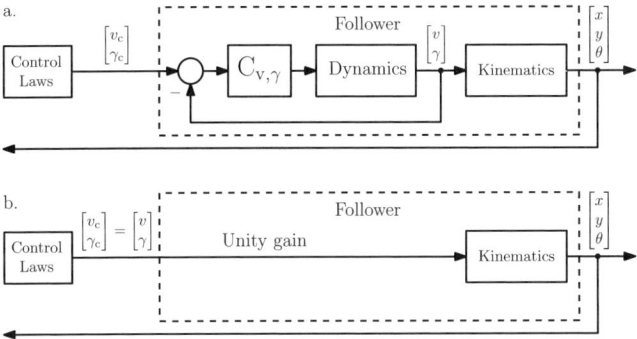

Fig. 3 (**a**) Inner/outer loop architecture for the follower. (**b**) The outer-loop controller is designed by assuming the inner loop is a unity gain.

Linearizing the tracking errors and the bicycle model along a constant-velocity trajectory, we obtain a local kinematic model for the follower given by

$$\dot{e}_1 = v_d - v, \ \dot{e}_2 = v_d e_3, \ \dot{e}_3 = -\frac{v_d}{d}\gamma,$$

where v_d is the speed of the delayed leader and $e_3 := \theta_d - \theta$ is the heading error.

For the vehicle dynamics, we assume the follower has an inner-loop controller, $C_{v,\gamma}$, that stabilizes its throttle and steering dynamics[1]. This assumption creates an inner/outer loop architecture where $C_{v,\gamma}$ stabilizes the vehicle dynamics in the inner loop and our controller controls the vehicle kinematics in the outer loop, as shown in Fig. 3.

A common practice with the above architecture is to design the outer-loop controller by treating the inner loop as a unity gain [7]. This assumption works well if the gains of the outer-loop controller are kept low enough such that the bandwidth of the outer loop is approximately 5 to 10 times smaller than the bandwidth of the inner loop. As a result, the kinematic model of the follower is

$$\dot{e}_1 = v_d - v_c, \ \dot{e}_2 = v_d e_3, \ \dot{e}_3 = -\frac{v_d}{d}\gamma_c. \tag{1}$$

In our implementation, we validated the bandwidth separation between our inner and outer loops through simulation and through the actual tuning of the gains in experimental trials.

2.2 Control Laws

Since the longitudinal and lateral directions of (1) are decoupled, it can be shown that choosing

---

[1] This was the case for the vehicles that we employed.

$$v_c = v_d + k_{p,1}e_1 , \quad k_{p,1} > 0$$
$$\gamma_c = k_{p,2}e_2 + k_{p,3}e_3 , \quad k_{p,2} , \; k_{p,3} > 0$$

will regulate the tracking errors to zero for the linearized model.

After some initial field trials, we discovered that the follower was turning late, which caused a large lateral error. We hypothesized that the late turning was caused by the low gains in our controller and the delays in the vehicle's steering dynamics. To compensate, we added a look-ahead feature for the lateral controller. We define the look-ahead point as

$$(x_l(t), y_l(t)) := (x_0(t - \tau + l), y_0(t - \tau + l)) , \quad 0 \leq l \leq \tau ,$$

where l is a constant look-ahead time. With a look-ahead time defined, the lateral and heading errors are computed by

$$e_2 = -(x_l - x)\sin\theta + (y_l - y)\cos\theta , \; e_3 = \theta_l - \theta ,$$

where θ_l is the heading of the look-ahead point.

2.3 Nonlinear Observer

From the control laws, it is obvious that we need estimates of the tracking and heading errors, (e_1, e_2, e_3), and the delayed leader's speed, v_d. The tracking and heading errors are calculated from the state of the follower, (x, y, θ), the state of the delayed leader, (x_d, y_d, θ_d), and the state of the look-ahead point, (x_l, y_l, θ_l). The delayed leader's speed can be calculated from the delayed leader's instantaneous change in position, $(\dot{x}_d, \dot{y}_d)$.

The follower's odometry provides an estimate of its state. The delayed leader's position is simply the leader's current position delayed by τ. From Fig. 1b, the leader's position can be computed by

$$x_0 = x + \rho\cos(\phi + \theta) , \; y_0 = y + \rho\sin(\phi + \theta) .$$

We use a data buffer to emulate the constant time delay in our implementation. The position of the look-ahead point is calculated in the same manner.

From the bicycle kinematics, the delayed leader's speed and heading can by calculated by

$$v_d = \sqrt{\dot{x}_d^2 + \dot{y}_d^2} , \; \theta_d = \mathrm{atan2}(\dot{y}_d, \dot{x}_d) .$$

We obtain an estimate of $\dot{x}_d$ by fitting a line to an n-second window of x_0 measurements centred around $t - \tau$, where n is configurable and a multiple of the data rate for x_0. A depiction is shown in Fig. 4, where the line is fitted using least squares. The slope of the line is then used as the estimate of $\dot{x}_d$. The estimate of $\dot{y}_d$ is obtained in the same manner, thus allowing us to calculate $\hat{\theta}_d$. A similar approach is used to estimate the look-ahead point's heading. Plots of the estimated and actual

Fig. 4 Estimating $\dot{x}_d$ using a line fitting window centred around $t - \tau$. The window size is n seconds, where n is configurable and a multiple of the data rate for x_0. The line is fitted using least squares, and the slope of the line is used as the estimate of $\dot{x}_d$.

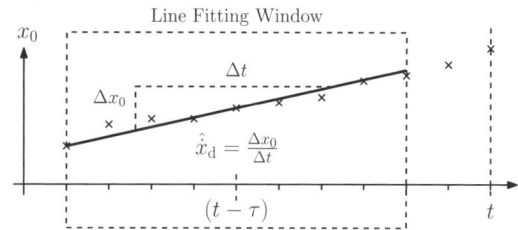

delayed leader's speeds and headings during a field trial are given in the next section to validate this windowing technique.

3   Field Trials

Field trials were conducted at Defence Research and Development Canada (DRDC)-Suffield, Alberta, Canada, in November, 2008, with two MultiAgent Tactical Sentry (MATS) vehicles. A picture of the MATS leader vehicle is shown in Fig. 5. The coloured target is used by the follower's camera system to measure the range and bearing to the leader. Each MATS vehicle is equipped with an on-board computer, a pan-tilt-zoom monocular camera, a GPS antenna, and a data link to a ground station to receive DGPS corrections. The DGPS data serves to provide ground truth for the trials. Each MATS is also equipped with odometric sensors that provide the vehicle's current speed and steering angle.

The test track is a 1.3-kilometre loop shown in Fig. 6. The track is a gravel road and is approximately 7 metres wide. The most difficult portion of the track is the 60-degree hairpin turn located at the north-west corner of the track.

3.1   Odometry Localization

Tracking the delayed leader using odometry localization proved to be difficult with the follower's current odometric sensors. The problem stemmed from an encoder

Fig. 5 The MATS leader vehicle. The coloured target is used by the follower's camera system to measure the range and bearing to the leader. Each MATS vehicle is equipped with an on-board computer, a pan-tilt-zoom monocular camera, a GPS antenna, and a data link to a ground station to receive DGPS corrections.

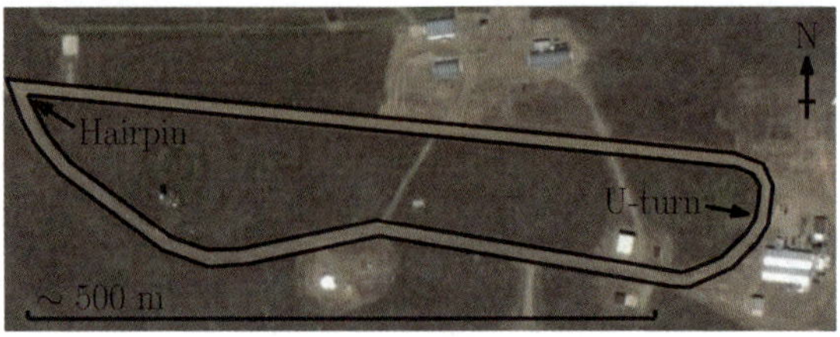

Fig. 6 The 1.3-kilometre test track used for field trials. The track is a gravel road and is approximately 7 metres wide. The most difficult portions of the track are the U-turn and the hairpin turn.

Fig. 7 The follower's actual path in comparison with its path estimated from odometric sensors. A crowned road caused the odometry to produce a circular path estimate when the follower was actually traveling straight.

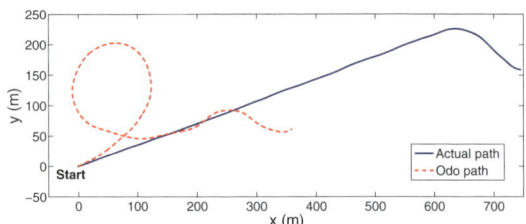

located on the steering column used to measure the steering angle. Since there was significant 'play' between the steering wheel and the front wheels, the steering measurement did not accurately represent the angle of the front wheels and was highly sensitive to road slope. As a result, the follower's heading estimate was very inaccurate, resulting in poor path estimates. An example is shown in Fig. 7, when tracking with odometry localization was performed off the test track. In this case, a crowned road caused the odometry to produce a circular path estimate when the follower was actually traveling straight.

3.2 DGPS Localization

Using DGPS for localization of its position, the follower was able to successfully track the delayed leader for 10 laps of the 1.3-kilometre track. A summary of the test results is shown in Table 1. The constant time delay was set to 8 seconds, and the look-ahead time was set to 3 seconds. The mean follower speed for the entire traverse was 2.2 m/s (7.9 km/h), and the mean following distance was 19 metres.

Table 1 Summary of results from 10 laps of 1.3-kilometre track. The constant time delay was set to 8 seconds, and the look-ahead time was set to 3 seconds. The controller gains $(k_{p,1}, k_{p,2}, k_{p,3}) = (0.08 \ s^{-1}, 0.04, 0.04)$. The lateral error, ε_2, is calculated in the delayed leader's frame, and t_f is the finishing time for the entire 13-kilometre traverse.

Description	Symbol	Value
Mean Follower Speed	v	2.2 m/s (7.9 km/h)
Mean Following Distance		19 m
Mean Lateral Error $\pm$Standard Deviation	$\frac{1}{t_f} \int_0^{t_f} \varepsilon_2(q)dq$	0.07 m $\pm$0.46 m
Maximum Absolute Lateral Error	$\max_t |\varepsilon_2(t)|$	2.73 m

Fig. 8 The longitudinal and lateral errors, $(\varepsilon_1, \varepsilon_2)$, are in the delayed leader's frame, and the longitudinal and lateral errors, (e_1, e_2), are in the follower's frame.

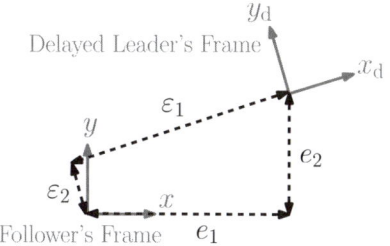

The mean lateral error was 0.07 metres with a standard deviation of 0.46 metres. The maximum absolute lateral error was 2.73 metres, which occurred during one of the turns at the hairpin. Since it is more natural to calculate an error with respect to the reference, the lateral errors here are calculated in the delayed leader's frame. The difference between tracking errors in the delayed leader's frame and tracking errors in the follower's frame is shown in Fig. 8.

Plots of DGPS ground truth data for a typical lap around the track are shown in Fig. 9. Figure 9a shows the leader's and follower's paths, while a close-up of the hairpin turn is shown in Fig. 9b. The longitudinal and lateral errors in the delayed leader's frame are shown in Fig. 9c, along with the delayed leader's and follower's speeds. The simultaneous large error increases around the 50-second and 400-second marks correspond to the U-turn and the hairpin turn, respectively. It should be noted that the longitudinal error did not get to zero during the 13-kilometre traverse. We have fixed this issue by adding an integral gain on the longitudinal error in the control law for the commanded speed. Unfortunately, due to time and weather

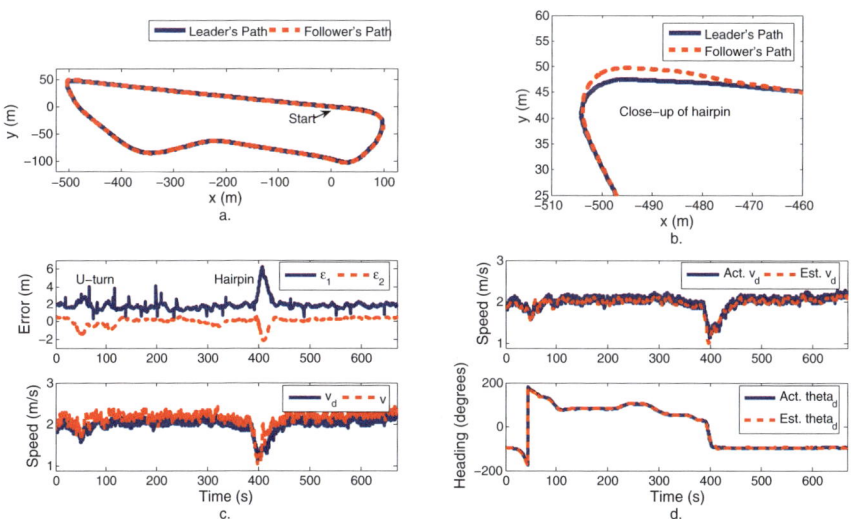

Fig. 9 (**a**) The leader's and follower's paths. (**b**) A close-up of the paths during the hairpin (**c**) The longitudinal and lateral tracking errors, and the delayed leader's and follower's speeds. (**d**) The delayed leader's speed and heading compared with their estimates from windowing.

constraints, we have not been able to properly tune and test the follower with the improved controller. Since the longitudinal error was always positive and our control law is $v_c = v_d + k_{p,1}e_1$, the follower's speed was always slightly larger than the delayed leader's speed. However, because the follower deviated from the leader's path, it was never able to catch up to the delayed leader, resulting in its inability to reduce the longitudinal error to zero. In Fig. 9d, the delayed leader's actual speed and heading are compared with their estimates. The similarities of the plots suggest that windowing around $t - \tau$ yields accurate speed and heading estimates.

4 Summary and Future Work

We have introduced the novel concept of tracking the trajectory of a vehicle ahead delayed by a constant time. This constant time delay forms the basis for our controller design and allows us to use 'future' position measurements to estimate the delayed leader's speed and heading. Successful field trials were conducted with two full-sized vehicles over a 13-kilometre traverse in which the follower vehicle achieved a mean lateral error of 0.07 metres with a standard deviation of 0.46 metres.

For future work, we would like to perform vehicle-following experiments with odometry localization. We are confident that odometry localization will work with our approach as long as the odometric estimates are reasonably accurate over τ (the constant time delay) seconds. To fix our poor heading estimate, we plan to implement a heading gyro on the follower. We would also like to conduct tests with multiple followers and at higher speeds. Testing multiple followers will provide us with important data on how tracking errors propagate in our system. To test at higher speeds, we plan to implement gain scheduling for our lateral controller since our lateral closed-loop system is dependent on the delayed leader's speed. These tests will further validate the feasibility of our approach and will bring us closer to an operational autonomous convoy.

References

1. Benhimane, S., Malis, E., Rives, P., Azinheira, J.R.: Vision-based control for car platooning using homography decomposition. In: Proceedings of the IEEE International Conference on Robotics and Automation, pp. 2161–2166 (2005)
2. Daviet, P., Parent, M.: Longitudinal and lateral servoing of vehicles in a platoon. In: Proceedings of the IEEE Intelligent Vehicles Symposium, pp. 41–46 (1996)
3. Franke, U., Bottiger, F., Zomotor, Z., Seeberger, D.: Truck platooning in mixed traffic. In: Proceedings of the Intelligent Vehicles Symposium, pp. 1–6 (1995)
4. Gehrig, S.K., Stein, F.J.: A trajectory-based approach for the lateral control of car following systems. In: IEEE International Conference on Systems, Man, and Cybernetics, vol. 4, pp. 3596–3601 (1998)
5. Giesbrecht, J.L., Goi, H.K., Barfoot, T.D., Francis, B.A.: A vision-based robotic follower vehicle. In: Proceedings of the SPIE Defence, Security, and Sensing, vol. 7332, pp. 14–17 (2009) (to appear)

6. Kehtarnavaz, N., Griswold, N.C., Lee, J.S.: Visual control of an autonomous vehicle (BART)—the vehicle-following problem. IEEE Trans. on Veh. Tech. 40(3), 654–662 (1991)
7. Marshall, J., Barfoot, T., Larsson, J.: Autonomous underground tramming for center-articulated vehicles. Journal of Field Robotics 25(6-7), 400–421 (2008)
8. Schneiderman, H., Nashman, M., Wavering, A.J., Lumia, R.: Vision-based robotic convoy driving. Mach. Vis. Appl. 8, 359–364 (1995)
9. Swaroop, D., Rajagopal, K.R.: A review of constant time headway policy for automatic vehicle following. In: Proceedings of the IEEE Intelligent Transportation Systems, pp. 65–69 (2001)

Part IV
Localization

Radar Scan Matching SLAM Using the Fourier-Mellin Transform

Paul Checchin, Franck Gérossier, Christophe Blanc,
Roland Chapuis, and Laurent Trassoudaine

Abstract. This paper is concerned with the Simultaneous Localization And Mapping (SLAM) problem using data obtained from a microwave radar sensor. The radar scanner is based on Frequency Modulated Continuous Wave (FMCW) technology. In order to meet the needs of radar image analysis complexity, a trajectory-oriented EKF-SLAM technique using data from a 360° field of view radar sensor has been developed. This process makes no landmark assumptions and avoids the data association problem. The method of egomotion estimation makes use of the Fourier-Mellin Transform for registering radar images in a sequence, from which the rotation and translation of the sensor motion can be estimated. In the context of the scan-matching SLAM, the use of the Fourier-Mellin Transform is original and provides an accurate and efficient way of computing the rigid transformation between consecutive scans. Experimental results on real-world data are presented.

1 Introduction

Environment mapping models have been studied intensively over the past two decades. In the literature, this problem is often referred to as simultaneous localization and mapping (SLAM). For a broad and quick review of the different approaches developed to address this problem, one can consult [2], [8], [9] and [25]. Localization and mapping in large outdoor environments are applications related to the availability of efficient and robust perception sensors, particularly with regard to the problem of maximum range and the resistance to the environmental conditions. Most approaches to map learning generate 2D models from range sensor data. Even though lasers and cameras are well suited sensors for indoor environments, their

Paul Checchin, Franck Gérossier, Christophe Blanc, Roland Chapuis,
and Laurent Trassoudaine
Clermont Université, Université Blaise Pascal, LASMEA, BP 10448,
F-63000 CLERMONT-FERRAND

CNRS, UMR 6602, LASMEA, F-63177 AUBIERE.
e-mail: firstname.name@univ-bpclermont.fr

A. Howard et al. (Eds.): Field and Service Robotics 7, STAR 62, pp. 151–161.
springerlink.com   © Springer-Verlag Berlin Heidelberg 2010

strong sensitivity to atmospheric conditions has created an interest for doing SLAM with radars and sonars [21]. Microwave radar provides an alternative solution for environmental imaging and overcomes the shortcomings of laser, video and sonar sensors. In this paper, a trajectory-oriented SLAM technique is presented using data from a 360° field of view radar sensor. This radar is based on Frequency Modulated Continuous Wave (FMCW) technology [16].

In Section 2, a review of articles related to our research interests is carried out in order to position our work in relation to existing methods. Section 3 presents the microwave radar scanner developed by a Cemagref research team (in the field of agricultural and environmental engineering research) [22]. The way to obtain a radar image (i.e. the power spectra with polar coordinates) is briefly presented. Section 4 gives the SLAM formulation used in this paper. There, the Fourier-Mellin Transform is applied to register images in a sequence and to estimate the rotation and translation of the radar system (see Section 5). This process makes no landmark assumptions, and avoids the data association problem by storing a detailed map instead of sparse landmarks. Finally Section 6 shows experimental results of this work, which were implemented (and tested on recorded real data) in Matlab and C/C++. Section 7 concludes and introduces future work.

2   Related Work

2.1   In the Field of Radar Mapping

In order to perform outdoor SLAM, laser sensors have been widely used [19] [11] [4]. A recent application with Velodyne HDL-64 3D LIDAR is presented in [13]. To provide localization and map building, the input range data is processed using geometric feature extraction and scan correlation techniques. Less research exists using sensors such as underwater sonar [21] and Frequency Modulated Continuous Wave (FMCW) radar. Interestingly, this last kind of sensor was already used by Clark in [6] at the end of the last century. In an environment containing a small number of well separated, highly reflective beacons, experiments were led with this sensor to provide a solution to the SLAM problem [8] using an extended Kalman filter framework and a landmark based approach. Finally, in [17], methods were presented for building a map with sensors that return both range and received signal power information. An outdoor occupancy grid map related to a 30 m vehicle's trajectory is analyzed. So far, there seems to have been no trajectory-oriented SLAM work based on radar information over important distances. However, vision-based, large-area SLAM has already been carried out successfully for underwater missions, using information filters over long distances [10] [15].

2.2   In the Field of Scan Matching SLAM

Since Lu and Milios presented their article [14] in search of a globally consistent solution to the 2D-SLAM problem with three degrees of freedom poses, many techniques have been proposed in the literature concerning robotics as well as computer vision. A common method of pose estimation for mobile robots is scan matching. By solving the rigid transformation between consecutive scans from a range sensor, the robot's motion in the time period between the scans can be inferred. The sensor used is most often a scanning laser range finder. One of the most popular approaches for scan matching is the Iterative Closest Point (ICP) algorithm [3]. In ICP, the transformation between scans is found iteratively by assuming that every point in the first scan corresponds to its closest point in the second scan, and by calculating a closed form solution using these correspondences. However, sparse and noisy data, such as that from an imaging radar, can cause an ICP failure. A single noisy reading can significantly affect the computed transformation, causing the estimated robot pose to drift over time. Other recent trends in SLAM research are to apply probabilistic methods to 3D mapping. Cole et al. [7] use an extended Kalman filter on the mapping problem. Olson et al. [18] have presented a novel approach to solve the graph-based SLAM problem by applying stochastic gradient descent to minimize the error introduced by constraints.

In its current version, our algorithm is close to the method suggested by Cole et al. [7]. However, the Fourier-Mellin Transform for registering images in a sequence is used to estimate the rotation and translation of the radar sensor motion (see Section 5). In the context of scan-matching SLAM, the use of the Fourier-Mellin Transform is original and provides an accurate and efficient way of computing the rigid transformation between consecutive scans. It is a global method that takes into account the contributions of both range and power information of the radar image.

3   A Microwave Radar Scanner

The exploited radar uses the frequency modulation continuous wave (FMCW) technique which has been known for several decades [23][16]. Frequency modulation presents two advantages for mobile robotics application, where distances are hundreds of meters [22]. First, it permits a low transmission power, which is safer for the user (the mean power determines the range). Second, a transposition of temporal variables into the frequency domain allows to obtain the measure more easily (a very short delay time Δt is switched to a broad variation of frequency Δf).

The FMCW radar is called K2Pi (2π for panoramic - in K band). A general view of the radar is presented in Figure 1 and its main characteristics are listed in Table 1. The radar is equipped with a rotating antenna in order to achieve a complete 360° per second monitoring around the vehicle, with an angular resolution of 3°, in the 3-100 m range. The image construction is based on the classical Plan Position Indicator (PPI) representation, i.e. the power spectra with polar coordinates. An example of radar images is presented in Figure 1. Variations of shading indicate variations of amplitude in the power spectra. These images are "radar referenced": the heading indications are related to the internal encoder of the radar and not to the earth's magnetic field.

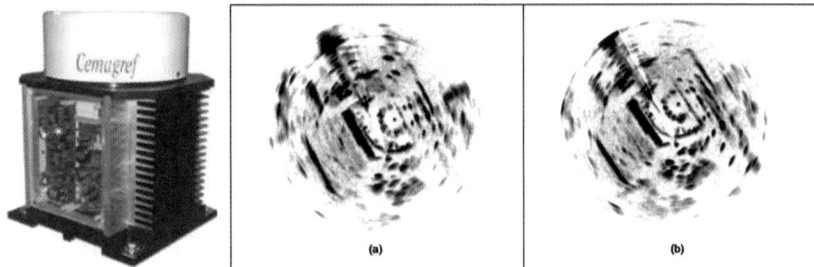

Fig. 1 Left side. The K2Pi FMCW radar. All the radar components are implemented in the same housing: microwave components, electronic devices for emission and reception, and the data acquisition and signal processing unit. The radar is mono-static: a single antenna, protected by a radome, is used for both transmitting and receiving. **Right side.** Two consecutive radar images ((a) & (b)) that are fairly similar.

Table 1 Characteristics of the K2Pi FMCW radar.

Carrier frequency $F0$	24 GHz
Transmitter power Pt	20 dBm
Antenna gain G	20 dB
Bandwidth	250 MHz
Angular resolution	3°
Angular precision	0.1°
Range Min/Max	3 m/100 m
Distance resolution	0.6 m
Distance precision (canonical target at 100 m)	0.05 m
Size (length-width-height)	27-24-30 cm
Weight	10 kg

4 Problem Formulation

4.1 SLAM Process

The used formulation of the SLAM problem is to estimate the vehicle trajectory defined by the estimated state $\mathbf{x}_k = \left[\mathbf{x}_{v_k}^T, \mathbf{x}_{v_{k-1}}^T, \ldots, \mathbf{x}_{v_1}^T \right]^T$. $\mathbf{x}_{v_i} = [x_i, y_i, \phi_i]^T$ is the state vector describing the location and orientation of the vehicle at time i. There is no explicit map; rather each pose estimate has an associated scan of raw sensed data that can be next aligned to form a global map.

4.2 Radar Scan Matching SLAM

The developed approach for a SLAM process is based on the following observation: two consecutive radar images are very similar to the "eye" point of view. For that reason a matching approach based on cross correlation function was selected [1].

Scan matching is the process of translating and rotating a radar scan such that a maximal overlap with another scan emerges. Assuming this alignment is approximately Gaussian, a new vehicle pose is added to the SLAM map by only adding the pose to the SLAM state vector. So, as described previously, observations are associated to each pose. They are compared and registered to offer potential constraints on the global map of vehicle poses. This is not only useful for odometry based state augmentation, but it is also an essential point for loop closing.

The estimator used here is the EKF, but it is not a limitation: algorithms like those presented in Section 2 could be tested too. Given a noisy control input $\mathbf{u}(k+1)$ at time $k+1$, upon calculation of the new vehicle pose, $\mathbf{x}_{v_{n+1}}(k+1|k)$, and a corresponding covariance matrix, $\mathbf{P}_{v_{n+1}}(k+1|k)$, the global state vector, $\mathbf{x}$, and corresponding covariance matrix, $\mathbf{P}$, can be augmented as follows:

$$\mathbf{x}(k+1|k) = \begin{bmatrix} \mathbf{x}(k|k) \\ \mathbf{x}_{v_n} \oplus \mathbf{u}(k+1) \end{bmatrix} \tag{1}$$

$$\mathbf{P}(k+1|k) = \begin{bmatrix} \mathbf{P}(k|k) & \mathbf{P}(k|k)\frac{\partial(\mathbf{x}_{v_n} \oplus \mathbf{u}(k+1))^T}{\partial \mathbf{x}_{v_n}} \\ \frac{\partial(\mathbf{x}_{v_n} \oplus \mathbf{u}(k+1))}{\partial \mathbf{x}_{v_n}}\mathbf{P}(k|k) & \mathbf{P}_{v_{n+1}}(k+1|k) \end{bmatrix}. \tag{2}$$

The operator $\oplus$ is the well-known displacement composition operator. $\mathbf{P}_{v_{n+1}}(k+1|k)$ is the covariance of the newly added vehicle state. Let us assume that two scans, $\mathbf{S}_i$, $\mathbf{S}_j$, have been registered. So, an observation $\mathbf{T}_{i,j}$ of the rigid transformation between poses in the state vector exists. Therefore a predicted transformation between the two poses can be found from the observation model as follows:

$$\mathbf{T}_{i,j}(k+1|k) = \mathbf{h}(\mathbf{x}(k+1|k)) = \ominus(\ominus\mathbf{x}_{vj}(k+1|k) \oplus \mathbf{x}_{vi}(k+1|k)) \tag{3}$$

where the operator $\ominus$ is the inverse transformation operator. This is then used as the initial estimate for our registration algorithm as follows:

$$\mathbf{T}_{i,j}(k+1) = \mathbf{\Psi}(\mathbf{T}_{i,j}(k+1|k), \mathbf{S}_i, \mathbf{S}_j) \tag{4}$$

where $\mathbf{\Psi}$ represents a registration algorithm. The state update equations are then the classical EKF update equations. The search for a transformation $\mathbf{T}_{i,j}$ is achieved by maximizing a cross correlation function [1].

5 Fourier-Mellin Transform for Automatic Image Registration

5.1 Principle

The problem of registering two scans in order to determine the relative positions from which the scans were obtained, has to be solved. The choice of an algorithm is strongly influenced by the need for real-time operation. A FFT-based algorithm was chosen to perform scan matching.

Algorithm 1. Steps of the Fourier-Mellin Transform algorithm applied to FMCW radar images

---

1. Get radar images I_k and I_{k-1}.
2. Apply thresholding filter to eliminate the speckle noise in both images.
3. Apply FFT to images $I_k \rightarrow \hat{I}_k$ and $I_{k-1} \rightarrow \hat{I}_{k-1}$.
4. Compute the magnitudes $M_k = |\hat{I}_k|$, $M_{k-1} = |\hat{I}_{k-1}|$
5. Transform the resulting values from rectangular to polar coordinates. $M() \rightarrow MP()$.
6. Apply the FFT to polar images, a bilinear interpolation is used. $MP() \rightarrow \widehat{MP}()$.
7. Compute $\widehat{Corr}(w_\rho, w_\theta)$ between $\widehat{MP}_k(w_\rho, w_\theta)$ and $\widehat{MP}_{k-1}(w_\rho, w_\theta)$ using Eq. 6.
8. Compute the inverse FFT $Corr(\rho, \theta)$ of $\widehat{Corr}(w_\rho, w_\theta)$.
9. Find the location of the maximum of $Corr()$ and obtain the rotation value.
10. Construct a new image Ir by applying reverse rotation to I_{k-1}.
11. Apply FFT to image Ir_{k-1}.
12. Compute the correlation $\widehat{Corr}(w_x, w_y)$ using Eq. 6.
13. Take inverse FFT $Corr(x,y)$ of $\widehat{Corr}(w_x, w_y)$.
14. Obtain the values $(\Delta x, \Delta y)$ of the shift.

Fourier-based schemes are able to estimate large rotations, scalings, and translations. Let us note that the scale factor is irrelevant in our case. Most of the DFT-based approaches use the shift property [20] [12] [24] of the Fourier transform. To match two scans which are translated and rotated with respect to each other, the phase correlation method is used, stating that a shift in the coordinate frames of two functions is transformed in the Fourier domain as a linear phase difference. To deal with the rotation as a translational displacement, the images are previously transformed into an uniform polar Fourier representation.

It is known that if two images I_1 and I_2 differ only by a shift, $(\Delta x, \Delta y)$, (i.e., $I_2(x,y) = I_1(x - \Delta x, y - \Delta y)$), then their Fourier transforms are related by:

$$\hat{I}_1(w_x, w_y).e^{-i(w_x \Delta x + w_y \Delta y)} = \hat{I}_2(w_x, w_y). \tag{5}$$

Hence the normalized cross power spectrum is given by

$$\widehat{Corr}(w_x, w_y) = \frac{\hat{I}_2(w_x, w_y)}{\hat{I}_1(w_x, w_y)} = \frac{\hat{I}_2(w_x, w_y)\hat{I}_1(w_x, w_y)^*}{|\hat{I}_1(w_x, w_y)\hat{I}_1(w_x, w_y)^*|} = e^{-i(w_x \Delta x + w_y \Delta y)} \tag{6}$$

where $*$ indicates the complex conjugate. Taking the inverse Fourier transform $Corr(x,y) = F^{-1}(\widehat{Corr}(w_x, w_y)) = \delta(x - \Delta x, y - \Delta y)$, which means that $Corr(x,y)$ is nonzero only at $(\Delta x, \Delta y) = \arg\max_{(x,y)}\{Corr(x,y)\}$. If the two images differ by rotational movement (θ_0) with translation $(\Delta x, \Delta y)$, then

$$I_2(x,y) = I_1(x\cos\theta_0 + y\sin\theta_0 - \Delta x, -x\sin\theta_0 + y\cos\theta_0 - \Delta y). \tag{7}$$

Converting from rectangular coordinates to polar coordinates makes it possible to represent rotation as shift: The Fourier Transform in polar coordinates is $\hat{I}_2(\rho, \theta) = e^{-i(w_x \Delta x + w_y \Delta y)}\hat{I}_1(\rho, \theta - \theta_0)$. Let M_1 and M_2 denote the magnitudes of $\hat{I}_1$ and $\hat{I}_2$

$(M_1 = |\hat{I}_1|, M_2 = |\hat{I}_2|)$. So, M_1 and M_2 are related by $M_1(\rho, \theta) = M_2(\rho, \theta - \theta_0)$. The shift between the two images can now be resolved using Eq. 6.

5.2   Scan Registration

In order to perform a scan registration algorithm, the Fourier-Mellin Transform (FMT) has been chosen [5] [20]. The FMT is a global method that takes the contributions from all points in the images into account in order to provide a way to recover all rigid transformation parameters, i.e. rotation, translation. It is an efficient and accurate method to process a couple of images that are fairly similar (see Fig. 1). The steps of the scan registration algorithm are described in Alg. 1.

6   Experimental Results

This section provides experimental results of the Scan SLAM application using the radar sensor previously described. The radar and the proprioceptive sensors were mounted on a utility car moving at a speed ranging from 0 to 25 km/h. Here, two experimental runs are presented. They were performed in an outdoor field, Blaise Pascal University campus, with a complex environment (buildings, cars, trees, roads, road signs, etc.). The radar was on top of the vehicle, 3 meters above the ground. The estimated trajectories obtained with the Scan SLAM process are presented in Figures 2 and 5. The successive positions of the radar are separated by an interval of one second. The photograph (see Fig. 2) is an aerial image of the experimental zone. The trajectory of the vehicle simultaneously measured with a centimetrically-precise GPS is overlayed. For these experiments, all data acquisitions have been realized in real time but SLAM processing has been realized off-line. One step of the process (scan registration, prediction and update) is achieved in less than one

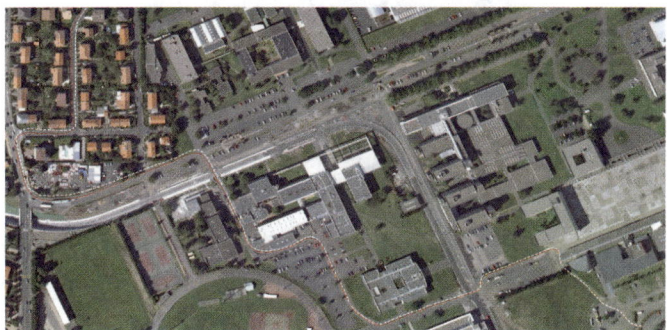

Fig. 2 Overlay of the estimated trajectory and the aerial image of Blaise Pascal University campus. The total traveled distance is around 1,135 m. The thin red line shows the trajectory of the vehicle measured with a centimetrically-precise GPS. The vehicle estimates are in thick white dashes.

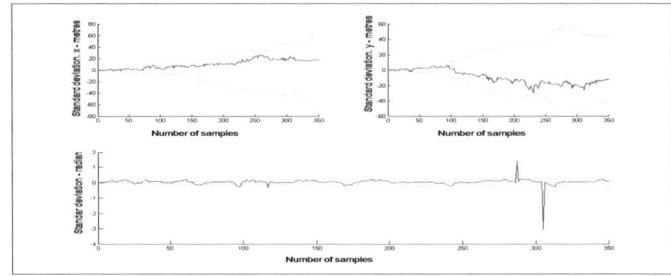

Fig. 3 Error and standard deviation (lower and upper bounds) related to the trajectory depicted in Fig. 2. Near sample 300, there is a GPS loss.

Fig. 4 Global map related to the trajectory depicted in Fig. 2.

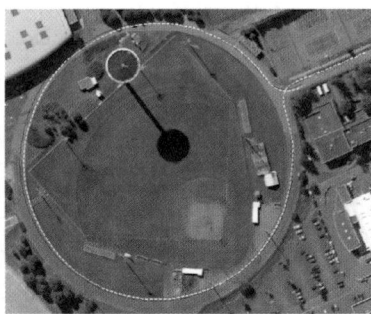

Fig. 5 The total traveled distance is around 700 m. The thin red line shows the trajectory of the vehicle measured with a centimetrically-precise GPS. The vehicle estimates after the loop closing are in thick white dashes.

second with Matlab on a dual-core 2 GHz laptop. A quantitative evaluation of the localization performances of the implemented process has been achieved. The position errors are calculated using the estimates and GPS data, assumed to be the ground truth (see Fig. 3 and Fig. 6). The first experiment was made on a distance

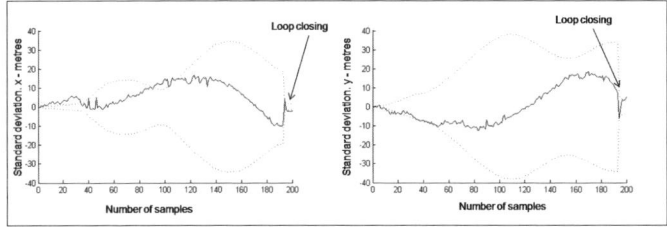

Fig. 6 Standard deviation along the north (x) and west (y) axes between pose estimation and GPS and influence of the loop closing.

of 1,135 m without loop closing. Figure 2 shows the trajectory of the vehicle. In Figure 3, error and standard deviation (lower and upper bounds) are presented. The global map that is obtained is shown in Figure 4. The second experiment was made on a distance of 700 m with loop closing (a circular trajectory around the campus sports-ground). In Figure 5, the corrected trajectory after loop closing is presented. In Figure 6, error and standard deviation (lower and upper bound) are presented.

7 Conclusion and Future Work

This paper presented results of SLAM using a microwave radar sensor. Due to the complexity of radar target detection, identification, tracking and association, a trajectory-oriented SLAM process based on the Fourier-Mellin Transform was developed; in this way, target assumptions about their position and nature were avoided.

Currently, this work considers only a static environment, assuming that there are no mobile elements around the radar. However, in order to develop a perception solution for high velocity robotics applications, future work will be devoted to the enhancement of the global map using methods such as the one described in [18]. Once the sensor delivers the measurement of Doppler frequency to take the relative velocity of mobile targets into account, integration of SLAM with Mobile Object Tracking (SLAMMOT) will be considered [25].

Acknowledgements. This work is supported by the Agence Nationale de la Recherche (ANR - the French national research agency) (ANR Impala PsiRob 2006 – ANR-06-ROBO-0012). The authors would like to thank M-O. Monod, R. Rouveure, P. Faure, J. Morillon and all other members of the Cemagref and THALES OPTRONIQUE SA who contributed to this project.

References

1. Aschwanden, P., Guggenbuhl, W.: Experimental results from a comparative study on correlation-type registration algorithms. In: Forstner, W., St. Ruwiedel (eds.) Robust computer vision: quality of vision algorithms, pp. 268–289. Wichmann (1992)

2. Bailey, T., Durrant-Whyte, H.F.: Simultaneous Localization and Mapping (SLAM): Part II - State of the Art. Robotics and Automation Magazine, 10 (2006)
3. Besl, P.J., McKay, N.D.: A method for registration of 3-d shapes. IEEE Trans. Pattern Anal. Mach. Intell. 14(2), 239–256 (1992)
4. Bosse, M., Zlot, R.: Map Matching and Data Association for Large-Scale Two-dimensional Laser Scan-based SLAM. Int. J. Robotics Research 27(6), 667–691 (2008), doi:10.1177/0278364908091366
5. Chen, Q., Defrise, M., Deconinck, F.: Symmetric Phase-Only Matched Filtering of Fourier-Mellin Transforms for Image Registration and Recognition. IEEE Trans. Pattern Anal. Mach. Intell. 16(12), 1156–1168 (1994)
6. Clark, S., Dissanayake, G.: Simultaneous localization and map building using millimeter wave radar to extract natural features. In: IEEE Int. Conf. on Robotics and Automation, Detroit, Michigan, May 1999, pp. 1316–1321 (1999)
7. Cole, D., Newman, P.: Using Laser Range Data for 3D SLAM in Outdoor Environments. In: IEEE Int. Conf. on Robotics and Automation, Florida (2006)
8. Dissanayake, G., Newman, P., Durrant-Whyte, H.F., Clark, S., Csobra, M.: A solution to the simultaneous localization and map building (SLAM) problem. IEEE Trans. Robotics and Automation 17(3), 229–241 (2001)
9. Durrant-Whyte, H.F., Bailey, T.: Simultaneous Localization and Mapping (SLAM): Part I - The Essential Algorithms. Robotics and Automation Magazine, 9 (2006)
10. Eustice, R., Singh, H., Leonard, J., Walter, M., Ballard, R.: Visually Navigating the RMS Titanic with SLAM Information Filters. In: Proceedings of Robotics: Science and Systems (RSS), Cambridge, MA, USA (June 2005)
11. Howard, A., Wolf, D.F., Sukhatme, G.S.: Towards 3D Mapping in Large Urban Environments. In: Proc. IEEE/RSJ Int. Conf. on Intelligent Robots and Systems, pp. 419–424 (2004)
12. Kuglin, C.D., Hines, D.C.: The Phase Correlation Image Alignment Method. In: Proc. IEEE Conf. Cybernetics and Soc., September 1975, pp. 163–165 (1975)
13. Leonard, J., et al.: A Perception Driven Autonomous Urban Vehicle. Journal of Field Robotics, Special Issue on the 2007 DARPA Urban Challenge, Part III 25(10), 727–774 (2008)
14. Lu, F., Milios, E.: Robot pose estimation in unknown environments by matching 2d range scans. Journal of Intelligent and Robotics Systems 18, 249–275 (1997)
15. Mahon, I., Williams, S.B., Pizarro, O., Johnson-Roberson, M.: Efficient View-Based SLAM Using Visual Loop Closures. IEEE Trans. Robotics 24(5), 1002–1014 (2008)
16. Monod, M.O.: Frequency modulated radar: a new sensor for natural environment and mobile robotics. Ph.D. Thesis, Paris VI University, France (1995)
17. Mullane, J., Jose, E., Adams, M.D., Wijesoma, W.S.: Including Probabilistic Target Detection Attributes Into Map Representations. Int. Journal of Robotics and Autonomous Systems 55(1), 72–85 (2007)
18. Olson, E., Leonard, J., Teller, S.: Fast iterative alignment of pose graphs with poor initial estimates. In: Proc. of the IEEE Int. Conf. on Robotics and Automation (2006)
19. Pfaff, P., Triebel, R., Stachniss, C., Lamon, P., Burgard, W., Siegwart, R.: Towards Mapping of Cities. In: IEEE Int. Conf. on Robotics and Automation, Rome, Italy (2007)
20. Reddy, B.S., Chatterji, B.N.: An FFT-based Technique for Translation, Rotation, and Scale-Invariant Image Registration. IEEE Trans. Image Processing 3(8), 1266–1270 (1996)
21. Ribas, D., Ridao, P., Tardós, J.D., Neira, J.: Underwater SLAM in a marina environment. In: IEEE/RSJ Int. Conf. on Intelligent Robots and Systems, San Diego, USA (October 2007)

22. Rouveure, R., Faure, P., Checchin, P., Monod, M.O.: Mobile Robot Localization and Mapping in Extensive Outdoor Environment based on Radar Sensor - First Results. In: PSIP 2007 - Physics in Signal and Image Processing, 31 January-2 February (2007)
23. Skolnik, M.I.: Introduction to radar systems. In: Electrical Engineering Series. Ed. McGraw-Hill International Editions, New York (1980)
24. Stone, H., Orchard, M., Chang, E.-C., Martucci, S.: A Fast Direct Fourier-Based Algorithm for Subpixel Registration of Images. IEEE Trans. Geoscience and Remote Sensing 39(10), 2235–2243 (2001)
25. Wang, C.-C.: Simultaneous Localization, Mapping and Moving Object Tracking. Doctoral dissertation, Tech. report CMU-RI-TR-04-23, Robotics Institute, Carnegie Mellon Univ. (2004)

An Automated Asset Locating System (AALS) with Applications to Inventory Management

Thomas H. Miller, David A. Stolfo, and John R. Spletzer

Abstract. In this work, we present a proof-of-concept Automated Asset Locating System (AALS) for enhancing inventory management. AALS integrates LIDAR and RFID sensor measurements into a Rao-Blackwellized particle filter for simultaneously localizing its pose with the positions of assets in the environment. We present significant experimental results where the proof-of-concept system successfully traveled a total distance of 1.4 km autonomously, while detecting and mapping all 143 available assets in real-time, and with a mean position error of < 80 cm.

1 Introduction

Radio Frequency Identification (RFID) systems use radio frequency to identify, locate and track features of interest. The technology sees widespread use in commercial applications to include baggage handling, passport readers, and toll collection to name but a few [1]. There are several RFID variants: passive, semi-passive, and active. In this work, we limit our discussion to the former. A passive RFID system is composed of three primary components: a reader (RF transmitter/receiver), a passive tag, and a host computer. The tag is composed of an antenna coil and an integrated circuit that contains both modulation circuitry and non-volatile memory. The tag is energized by the RF carrier signal transmitted by the reader. Using this scavenged energy, the information stored on the tag – to include a unique identifier for that tag instance – can be transmitted back to the reader [2]. The strength of RFID is that it explicitly solves the data association problem. As each tag is associated with a unique identifier, false correspondences across tag detections are eliminated.

In this work, we investigate the potential for applying RFID and robotics technologies to inventory management tasks. Manual intervention in material tracking systems is labor intensive, costly, and error-prone [3]. Furthermore, low-frequency "scheduled scanning" approaches cannot ensure that inventory remains up-to-date.

Thomas H. Miller, David A. Stolfo, and John R. Spletzer
Lehigh University, Bethlehem, PA, USA
e-mail: {thm204,das611,josa}@lehigh.edu

A. Howard et al. (Eds.): Field and Service Robotics 7, STAR 62, pp. 163–172.
springerlink.com   © Springer-Verlag Berlin Heidelberg 2010

The ability to automate the material tracking task can dramatically enhance asset visibility. To this end, we demonstrate an Automated Asset Locating System (AALS) that integrates LIDAR and RFID sensing on a mobile robot base for enhanced inventory management. The RFID system's role is dual purpose. First, the tags serve to identify assets to be tracked. Second, they are integrated into the environment as correspondence-free landmarks. In this role, they effectively introduce dramatic, artificial asymmetries into the environment. This enables reliable robot localization indoors even in largely symmetric environments, and where the scale of the environment was large compared to the range of the robot's sensors – conditions which could be problematic for traditional SLAM and localization approaches. The RFID tag's extremely compact size ($\approx$10-30 cm^2 stickers) and low cost (\$0.1-1.0) allows them to be discretely integrated into the environment. The net result is an automated system capable of reliably locating assets in the environment.

2   Related Work

Several researchers have investigated the convergence of robotics and RFID technologies. Most related to our work is that of Hähnel et al [4], where a Pioneer 2 Robot equipped with a Sick LMS200 and an RFID reader was manually steered through the environment. Using a map generated a priori, the authors employed Monte-Carlo localization (MCL) to estimate the position of RFID tags in the environment. Formal results on tag localization accuracy were not provided. However, they demonstrated that using these same tags as landmarks, robot localization could be achieved using only RFID measurements (although not to the same level of accuracy as when the LIDAR system was used). Schneegans et al [5] built on this to demonstrate a system for robot localization using a more sophisticated sensor model, and whereby an RFID snapshot was associated with a database of learned features. They compared their approach with those from [4], and found comparable accuracy in the end position estimate of the robot, but a significantly faster filter convergence rate. This work was also done off-line.

There is also significant work that has emphasized using RFID to assist in localization tasks. Kulyukin et al incorporated RFID into a robotic assistant for the visually impaired [6]. Tsukiyama demonstrated a limited implementation where RFID tags served as topological landmarks enabling the robot to correctly follow a path [7]. Mapping the position of assets was not considered. Chae and Han used a topological approach with RFID and a vision sensor [8]. Experimental results were again off-line. Miah and Gueaieb examined using tag received power (TRP) to estimate the distance from the robot to the tag [9]. However, their implementation was limited to simulations. Milella et al developed an RFID-assisted mobile robot system for mapping and surveillance using fuzzy inference methods [10]. In terms of asset tracking task, Ehrenberg et al investigated the use of a LibBot to locate books in a library environment [11]. They localized densely packed, short range tags by again employing a probabilistic RFID antenna model. The actual implementation was rather limited however, with experiments only over a single library shelf.

Our work differs from these efforts in several ways. First, we employ a Rao-Blackwellized particle filter for the simultaneous localization of the robot pose and mapping of asset positions in the environment. Second and more significantly, unlike these efforts we provide significant experimental results with AALS operating on-line. In our experiments, AALS is completely responsible for its own navigation as it self-localizes and maps the positions of assets in the environment. These results show that AALS is capable of reliably detecting and mapping the position of assets in the environment in real-time.

3   The Development Platform

The AALS proof-of-concept system was built upon an iRobot Create robotics development platform. The Create is an excellent low-cost research platform, combining a robust mobile chassis with a higher level motor control interface through RS-232 communication, odometry feedback, limited sensing, and 5V DC power output. The other primary components of AALS are:

Computing. With the exception of motor control which ran on the Create's embedded computer, all computing was done on a Lenovo X200 laptop with a 2.4 GHz Core 2 Duo processor and 2 GB memory.

LIDAR. The primary exteroceptive sensor for AALS was a Hokuyo URG-04LX LIDAR. The URG-04LX provides a 240° field of view with an angular resolution of 0.36° . It offers an advertised range of up to 5.6 meters, although in this application we found a more accurate estimate to be $\leq$4.5 meters.

RFID. The RFID transceiver used in this work was a Skyetek M9 operating at 862-955 MHz. We deliberately chose an UHF module to maximize range. The reader was multiplexed to a pair of antennae oriented to maximize detection coverage to the front and sides of the robot. To date, all development has been done using the Alien Technology ALN-9534 Gen 2 tag. In an evaluation of available Gen 2 tags, this model provided acceptable detection ranges (up to 4.0 meters) while exhibiting fairly good omnidirectional performance in a compact footprint. Images of AALS,

 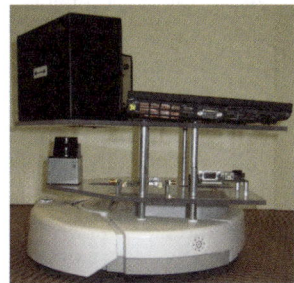

Fig. 1 Top and side profiles of AALS showing the integration of Hokuyo URG-04LX LIDAR, RFID reader, and on-board computing.

showing the integration of on-board computing, the URG-04LX, the M9, multiplexer, and antennae are at Figure 1.

4   Robot Localization and Asset Tracking

For robot localization and asset tracking, we employed a Rao-Blackwellized Particle Filter (RBPF). Such approaches were first introduced to the robotics community by Doucet *et al* [12], who observed that the simultaneous localization and mapping (SLAM) problem could be factored into two sub-problems

$$p(x_{1..t}, l_{1..m} | z_{1..t}, u_{1..t-1}) = p(x_{1..t} | z_{1..t}, u_{1..t-1}) \prod_{i=1}^{m} p(l_i | x_{1..t}, z_{1..t}) \qquad (1)$$

where $x_{1..t}$ denotes the robot pose over time, $l_{1..m}$ the m landmark positions, z the sensor measurements, and u the control inputs. The left term on the right side of (1) corresponds to the robot localization problem, and the right term to estimating the position of m conditionally independent landmarks in the map. This partitioning enabled the robot localization problem to be solved using a traditional particle filtering approach, while allowing the mapping problem to be estimated through analytical methods. The significance of this factorization was that it mitigated the otherwise exponential increase of particle samples with increases in state space dimension (*i.e.*, the number of landmarks). This result was leveraged by Montemerlo *et al* in developing FastSLAM [13], where mapping was accomplished by associating m Extended Kalman Filters (EKFs) with each particle to independently track the m landmarks $l_{1..m}$. We employed a similar approach, using Monte-Carlo Localization (MCL) to estimate the robot pose, and Kalman Filters for asset tracking.

4.1   Sensor Model Development

The LIDAR Sensor Model. AALS relies heavily upon the Hokuyo URG-04LX for localization. The URG-04LX is extremely compact and lightweight compared to the ubiquitous Sick LMS2xx series LIDARs, which made it well suited for our proof-of-concept. However, they are also myopic, demonstrating an effective range of ≤ 4.5 meters in our experiments. Such limited range can be a challenge for MCL approaches, which solves a data association problem relating the robot pose $[x(t), y(t), \theta(t)]^T$ vs. time through asymmetries in the environment. Our development site consisted of two building wings connected by a corridor ≈ 40 meters in length with little asymmetry. To mitigate the potential for the filter converging too quickly (and likely incorrectly), the conditional density functions which model the uncertainty in LIDAR measurements (and are used to weight the individual particles) were dramatically smoothed. As a result, even a relatively improbable measurement was unlikely to penalize a particle dramatically. We found that such a PDF would ensure that the robot's pose would eventually converge to the correct position/orientation even without input from the RFID sensors and regardless of the initial robot pose.

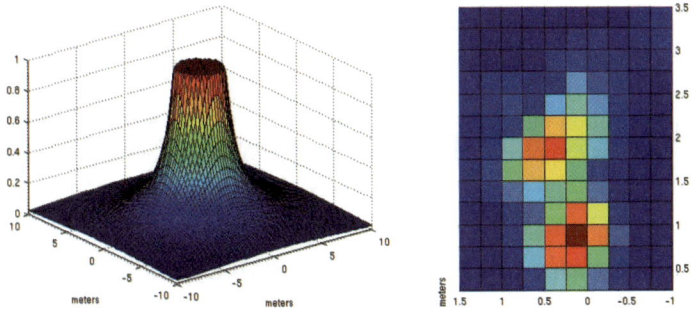

Fig. 2 (Left) Weighting function for landmark tags used in the robot localization process. (Right) PDF for asset tag detection generated empirically.

The RFID Sensor Model. Two different models were used for the RFID sensor depending upon its given role. The primary purpose of landmark tags (with positions known *a priori*) was to provide a low-cost mechanism for enhancing localization robustness, as there was no potential for data association errors. We considered their ability to assist in pose estimates and improve filter convergence as demonstrated in [4] of secondary importance. Therefore, we assumed no relative orientation information was available and a symmetric scaling function S was used to reflect the likelihood of landmark detection by the robot. To model this, we defined a critical radius r^* around each landmark where detection was expected based upon empirical results. With r^* so defined, the weight function used was

$$S(i) = \left[(d(i) < r^*) + (d(i) > r^*)\frac{r^*}{d(i)^2} \right] \tag{2}$$

where $d(i) = ||(x,y)^T - (x_i,y_i)^T||$ was the Euclidean distance from the robot to the ith landmark. When used in conjunction with the MCL process, particles within the critical radius of a detected landmark are unaffected, while the weights of those outside are scaled inversely proportional to the squared distance to the landmark. This is illustrated at Figure 2 (left). The motivation for the quadratic model is the Friis Transmission Equation, which shows that the power ratio between receiving and transmitting antennae are inversely proportional to their distance squared [14]. The placement of only several landmark tags in the environment dramatically accelerated particle filter convergence during our experiments.

For asset detection, we assumed that the estimated robot pose was approximately correct. As such, the sensor model was directional to reflect the relative robot/asset tag orientation. We initially generated a discrete PDF model empirically by collecting detection data as a function of tag position, orientation, and height as in [5]. The resulting two-dimensional PDF estimate in the antenna frame is shown at Figure 2 (right). The PDF is highly non-Gaussian, and does not lend itself to a Kalman filter implementation. However, in reality this model – as well as those typically used in related work – is ad-hoc. Antenna performance is strongly environment specific.

Signal is strongly tied to reflections from the floor, walls, ceiling, obstacles, signal absorption, the amount of metal in the environment, tag line-of-sight, the object to which the tag is affixed, *etc.*. In fact, in preliminary testing we compared a voting approach based upon our discrete PDF model with a pure Kalman filter using an overly conservative approximation of this PDF. The latter demonstrated equal or better performance, and as such we ultimately employed such an approach.

4.2   Robot Localization and Asset Position Estimation

For the most part, robot localization was accomplished using a traditional MCL approach [15]. The time update phase corresponded to the transformation of the particles' poses using a unicycle model for robot motion. Measurement updates using the LIDAR were also straightforward. However, an additional measurement update stage was integrated for whenever a landmark tag was detected. In this event, samples were re-weighted based upon $w_{k+1}(i) = S(j)w_k(i)$ where $w_k(i)$ denotes the current weight of the ith particle at time-step k, and $S(j)$ the scaling function defined by (2). After re-weighting, the particle set was re-sampled. The net effect was that particles far away from landmark j were quickly killed off.

With the ability to reliably localize the robot, we turn to the case of mapping assets. To this end, each particle p_i, $i = 1 \ldots n$, in our RBPF maintains a Kalman filter that propagates an estimate for the position and positional covariance $\{\mathbf{x}(i,j), \Sigma(i,j)\}$, $j = 1 \ldots m$, for each of the m assets detected. Note that RFID asset detections are *not* used to refine the robot pose estimate, so the asset position estimates remain uncorrelated. As a result, only n of the mn total Kalman filters need be updated for a given asset detection.

We model each RFID asset detection as a direct estimate of the asset's position, *i.e.*, $z = {}^W T_A \mathbf{x}_A$ where $\mathbf{x}_A$ is the tag position estimate in the antenna frame, and ${}^W T_A$ maps points from the antenna frame to world frame. The associated measurement covariance is then $\Sigma_R = R(\theta_R + \theta_A)\Sigma_A R(\theta_R + \theta_A)^T$ where Σ_A denotes the estimated uncertainty in the antenna frame, and R is a 2-D rotation matrix associated with the robot θ_R and antenna θ_A orientations in the world and robot frames, respectively. The measurement update is then textbook Kalman Filter, and since the asset position is assumed static there is no process update.

5   Experimental Results

5.1   Component Level Testing

As part of the proof-of-concept, we performed component level testing to determine the robustness of tag detection as a function of tag density. Of concern was the potential for message collisions if multiple irradiated tags attempted to transmit at the same time. To this end, we examined both linear arrays of 5-15 tags, and grid arrays of 12 tags (3×4) with inter-tag spacings ranging from 0-45 cm. This also included different heights above the ground plane. A representative linear array

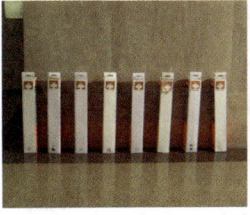 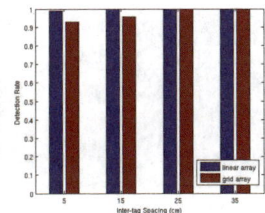 

Fig. 3 (Left-Center) Sample RFID linear array used during component level testing. All configurations demonstrated at least a 93% success rate. (Right) Sample asset configuration during system level testing.

configuration with 10 cm spacing is shown at Figure 3 (left). For each test geometry, AALS was driven multiple times past the tag array at standoff distances consistent with an expected detection based upon the sensor model derived in Section 4.1. A tag was considered detected if it was successfully identified at least one time while AALS traversed the array. Summary statistics are shown at Figure 3 (center).

There were 908 true positives, 17 false negatives, and 0 false positives. Sixteen of the 17 false negatives were with grid arrays with inter-tag spacing of 5 cm (14) and 15 cm (2). These corresponded to tag densities of 100 and 30 tags/m^2, and detection rates were 93% and 96%, respectively. These results indicate that the anti-collision protocols employed by the system worked very well for the range of geometries tested even under very high tag densities.

5.2 System Level Testing

To demonstrate the system level proof-of-concept, we conducted a series of experiments using the fourth floor of Packard Laboratory at Lehigh University as the development site. This constituted a region $\approx 48 \times 14$ meters. Our map M representation was an occupancy grid with a cell resolution of 10 cm, and was provided to AALS *a priori*. The map was constructed from digital blue prints. While nominally correct, there were significant inconsistencies between this map and the actual floorplan. Only the most serious of these were corrected. One final alteration to M included the introduction of 4 landmark tags with positions also known *a priori* by the robot. These were spaced approximately every 15 meters in our corridor set. Finally, 10-15 assets (*i.e.*, card board boxes and plastic bins with tags affixed) were placed in random locations throughout the environment. A representative configuration is at Figure 3 (right).

For global path planning, AALS was provided a route network graph $G(V,E)$ that delineated in continuous space the intended paths for navigation. Waypoints in the route network corresponded to vertices $v_i \in V$ of G, and the edge set $E \subseteq G$ corresponded to path segments where each $e_{ij} \in E$ connected a pair of waypoints (v_i, v_j). The desired path for a given mission was then specified via a waypoint sequence $(v_i, v_j, \ldots, v_n)$. For motion planning, AALS relied upon 2 modes: obstacle avoidance, and path following. Prior to particle filter convergence or in the event that the

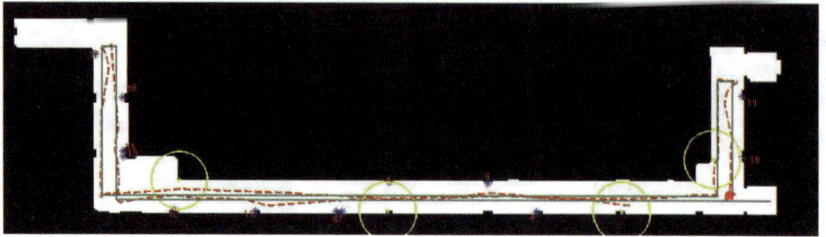

Fig. 4 Mission results showing the actual (blue "*") and estimated (red "numbers") asset locations. The mean position error in this trial was 54 cm.

specified route segment was blocked, AALS would operate in obstacle avoidance mode. For path-following, a PD controller was used where the normal distance to the current route segment was employed as an error metric. The typical mission for AALS entailed a complete circuit of the test area. This corresponded to a mission length of $\approx$ 125 meters.

After preliminary testing to characterize the system, a total of 12 missions were conducted. During these trials, the starting point was varied, as were the position and orientation of assets. This ensured that asset detection and mapping was possible with tag orientations parallel and orthogonal to the robot path. The geometry changes were also done to ensure that the sensor model for the Kalman filter was not deliberately biased. For each mission, AALS drove at a nominal linear velocity of 0.3 m/s. At the initiation of each trial, 10,000 particles were used to instantiate the prior for the robot pose. This number was reduced dynamically to as few as several hundred particles using the second-order statistics to infer convergence of the particle set. To further support real-time computation, LIDAR range measurements were sub-sampled to an angular resolution of $1.08°$. The target update rate for AALS was 2 Hz. At the conclusion of a given mission, the estimate for the position of assets was determined from

$$\begin{bmatrix} x \\ y \end{bmatrix}_i = \sum_{j=1}^{n} w(j) \begin{bmatrix} x \\ y \end{bmatrix}_{ij}, \ i = 1 \ldots m \tag{3}$$

where $[x,y]_i^T$ denotes the position of the ith landmark, $[x,y]_{ij}^T$ the ith landmark position as estimated by the Kalman filter of the jth particle, and w_j is the corresponding sample weight at mission completion. Results from a representative mission are at Figure 4. This shows the route network (green lines), the path as estimated by the robot (red dashed lines), the position of landmark tags (yellow circles), and the actual (blue "*") and estimated (red "numbers") positions of assets.

Of the 12 missions, 11 were completed successfully. The one failure occurred when an asset was deliberately placed across the path. The motion planner incorrectly determined the path was not traversable, and aborted the mission. The motion planner was subsequently modified, and this same configuration was successfully re-tested. The 11 completed missions constitute a total distance traveled of 1.4 km. During this time, all 143 assets that were placed in the environment were detected.

Table 1 Mean Position Error (MPE) for detected assets as a function of geometry.

Asset Configuration	Number Samples	Number Detected	MPE (cm)	σ (cm)
All	143	143	79.2	49.5
Border	130	130	79.4	49.8
Interior	13	13	77.1	48.3
Parallel to path	91	91	86.8	55.5
Normal to path	52	52	65.8	33.3

The estimated asset positions were then compared with hand-measured ground-truth values. Statistics for the different configurations are shown at Table 1. Border and interior configurations discriminate as to whether the asset was located on the map border or in the interior. Parallel/normal to path refers to the antenna orientation with respect to the robot's primary direction of travel.

From these, we see that the average position error was <80 cm. There was little difference between assets that were located within the interior or along the border of the map (we should note that no optimizations were done to asset location estimates that were outside the boundary of the map, which would have improved results). We do note a fairly significant difference between tag orientations that were normal vs. parallel to the robot's direction of travel. This appears to be attributed to the normal antennae being detected at longer standoff distances, and the associated Kalman filters seeing a larger number of measurement updates. However, further analysis is needed to support this hypothesis. We should note that in a warehouse or similar environment where such a system would be used, tag orientation would typically be parallel to the direction of travel and as such these errors are more representative.

For portions of three trials, we also estimated robot position using a Sick LMS291-S14 to track a retro-reflector affixed to the robot. Using this as ground truth, the mean absolute position error of the robot localization system was 53.3 cm (σ_x = 49.3 cm, σ_y=19.1 cm). The bias was not surprising due to the strong symmetry and limited configuration space in the x and y directions, respectively. Taking these findings into consideration, a more accurate estimate of tag localization performance would be a MAE of ≈60 cm.

6 Discussion

In this work, we demonstrated a proof-of-concept Automated Asset Locating System (AALS) that integrates LIDAR and RFID sensing on a mobile robot base. The RFID system's role was dual purpose in this application – identifying both asset and landmarks tags in close proximity to the robot platform. These measurements enabled the position of asset tags in the environment to be estimated with a mean error of <80 cm. Furthermore, they were able to augment the limited range of the Hokuyo URG-04LX by not only accelerating the filter's convergence rate, but also ensuring against divergence in areas of low feature asymmetry. A natural question regarding this approach is the use of MCL vs. SLAM. This decision was made so that landmark tags with known "absolute" positions in the map could readily be integrated to protect against localization failures (*e.g.*, incorrect loop closures). We are currently investigating a hybrid approach which integrates both aspects, and working with members of the NSF Center for Engineering Logistics and Distribution to evaluate AALS in a larger scale, representative environment.

Acknowledgments. Special thanks to Mr. Sean Kelly (Lehigh University) for his efforts. This work was funded in part as a National Science Foundation Center for Engineering Logistics and Distribution (CELDi) center designated project. Any opinions, findings, and conclusions or recommendations expressed in this material are those of the author(s) and do not necessarily reflect the views of NSF.

References

1. Federal Trade Commission, RFID: Applications and Implications for Consumers. Workshop Report (March 2005)
2. Sorrells, P.: Passive rfid basics. Tech. Rep. AN680, Microchip Technology, Inc. (2002)
3. RFID Solutions & Applications (2009),
   http://www.gaorfidassettracking.com
4. Hähnel, D., Burgard, W., Fox, D., Fishkin, K., Philipose, M.: Mapping and localization with RFID technology. In: ICRA (2004)
5. Schneegans, S., Vorst, P., Zell, A.: Using RFID snapshots for mobile robot self-localization. In: Proceedings of the 3rd European Conference on Mobile Robots (ECMR 2007), Freiburg, Germany, September 19-21, pp. 241–246 (2007)
6. Kulyukin, V., Gharpure, C., Nicholson, J., Pavithran, S.: RFID in Robot-Assisted Indoor Navigation for the Visually Impaired. In: IROS (2004)
7. Chae, H., Han, K.: World map based on rfid tags for indoor mobile robots. In: Proceeding of SPIE - The International Society for Optical Engineering, vol. 6006 (2005)
8. Chae, H., Han, K.: Combination of rfid and vision for mobile robot localization. In: Proceeding of the 2005 International Conference on Intelligent Sensors, Sensor Networks and Information Processing Conference, pp. 75–80 (2005)
9. Miah, M. S., Gueaieb, W.: Mobile robot navigation using custom-made RFID tag system. In: Proceedings of the 5th Scientific Research Outlook, Fes, Morocco, October 26–30 (2008)
10. Milella, A., Cicirelli, G., Distante, A.: Rfid-assisted mobile robot system for mapping and surveillance of indoor environments. Industrial Robot: An International Journal 35(2), 143–152 (2008)
11. Floerkemeier, C., Ehrenberg, I., Sarm, S.: Inventory management with an rfid-equipped mobile robot. In: Proceedings of the 3rd Annual IEEE Conference on Automation Science and Engineering, Scottsdale, AZ, USA, September 2007, pp. 1020–1026 (2007)
12. Doucet, A., de Freitas, N., Murphy, K., Russell, S.: Rao-blackwellised particle filtering for dynamic bayesian networks. In: Proceedings of the Sixteenth Conference on Uncertainty in Artificial Intelligence, pp. 176–183 (2000)
13. Montemerlo, M., Thrun, S., Koller, D., Wegbreit, B.: Fastslam: A factored solution to the simultaneous localization and mapping problem. In: Proceedings of the AAAI National Conference on Artificial Intelligence, pp. 593–598. AAAI, Menlo Park (2002)
14. Balanis, C.: Antenna Theory: Analysis and Design. Wiley, Chichester (2005)
15. Thrun, S., Fox, D., Burgard, W., Dellaert, F.: Robust monte carlo localization for mobile robots. Artificial Intelligence 128(1-2), 99–141 (2000)

Active SLAM and Loop Prediction with the Segmented Map Using Simplified Models

Nathaniel Fairfield and David Wettergreen

Abstract. We previously introduced the SegSLAM algorithm, an approach to the simultaneous localization and mapping (SLAM) problem that divides the environment up into segments, or submaps, using heuristic methods. We investigate a real-time method for Active SLAM with SegSLAM, in which actions are selected in order to reduce uncertainty in both the local metric submap and the global topological map. Recent work in the area of Active SLAM has been built on the theoretical basis of information entropy. Due to the complexity of the SegSLAM belief state, as encoded in the SegMap representation, it is not feasible to estimate the expected entropy of the full belief state. Instead, we use a simplified model to heuristically select entropy-reducing actions without explicitly evaluating the full belief state. We discuss the relation of this heuristic method to the full entropy estimation method, and present results from applying our planning method in real-time onboard a mobile robot.

1 Introduction

The tasks of mapping, localization, and planning lie at the core of mobile robotics, and to a large degree have been solved for small, two dimensional, structured environments. To make robots useful in the broader world, they need to move beyond such simple environments into large, 3D, unstructured environments.

In response, there has been recent work in the SLAM field in two areas: submap SLAM and active SLAM. Most submap SLAM methods use a combination of metric and topological maps, in which the relationships between submaps are

Nathaniel Fairfield
Robotics Institute, Carnegie Mellon University, Pittsburgh PA
e-mail: than@timbrel.org

David Wettergreen
Robotics Institute, Carnegie Mellon University, Pittsburgh PA
e-mail: dsw@cmu.edu

A. Howard et al. (Eds.): Field and Service Robotics 7, STAR 62, pp. 173–182.
springerlink.com   © Springer-Verlag Berlin Heidelberg 2010

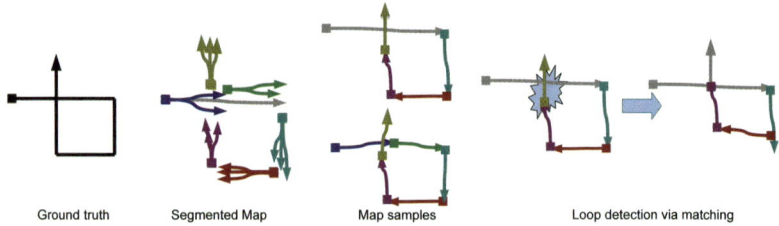

Ground truth   Segmented Map   Map samples   Loop detection via matching

Fig. 1 This figure illustrates the process of segmentation, map sample generation, and matching. The segmented map stores the particle submaps (shown as different arrows) for each color-coded segment. The segmented map also stores the relationships between segments, loosely illustrated here by the segment placements relative to each other. Note that after matching, the breadth-first map sampling algorithm does not enforce global consistency between the red and turquoise segments.

represented by the edges of a graph, and the nodes of the graph represent the submaps. The submap segmentation is usually designed such that their scale is well within the capabilities of a SLAM approach. Thus the scaling problem is addressed, but the submap algorithm must manage the graph of submaps, deciding when to create a new submap, when to re-enter an old submap, and how to represent different hypotheses about the topological relationships between submaps. In prior work, we have presented a robust, real-time, submap-based approach called SegSLAM [6]. In our SegSLAM algorithm, individual submaps are accurate 3D metric evidence grid-based maps. Using an extension of the Rao-Blackwellized Particle Filter (RBPF) formulation [5], SegSLAM maintains a stochastic graph of the transformations between submaps, called the segmented map or SegMap (Figure 1). The particles of a regular RBPF are a discrete approximation to the distribution over poses and the metric maps; the particles of SegSLAM are a discrete approximation to the distribution over poses and submaps, where the poses are in the local coordinate frame of the submaps. Since each SegSLAM particle has its own copy of each submap, we use the noun *segment* to refer to the collection of particle submaps that are temporally compatible: unlike RBPF particles, SegSLAM particles do not encode a complete trajectory hypothesis, instead the trajectory must be reconstructed by stitching together compatible segments.

Active SLAM determines the robot's actions by planning based on the SLAM uncertainty. Uncertainty, which can be quantified as *information entropy*, depends on the route that the robot uses to explore the environment. A well-known approach for balancing the need to explore new regions against the need to localize is to estimate the expected information gain in the SLAM state distribution resulting from different actions [8, 2, 3]. However, it is rarely possible to perform the estimates in closed form, so computationally expensive Monte-Carlo simulations are used.Rather than attempting to estimate the expected information gain of the full SLAM state, we show how a simplified sensor model can be used to probe the SLAM belief state, and how the results of these probes can be then used to select plans that reduce the SLAM uncertainty.

A goal of Active SLAM is to detect and close loops: when the robot returns to a place that it has previously mapped and recognizes that it is back in that place, it can correct for all of the error that has accumulated since it left. For example, [14] perform active loop closure by searching for discrepancies between the metric distance (which is short near a loop closure) and topological distance (which is long just before a loop closure). However, active loop closure doesn't necessarily predict loops, it just takes advantage of them once the robot has already observed a possible loop. Our approach extends to true loop prediction, attempting to exploit the local structure of the environment by using the submaps of SegSLAM's segmented map (SegMap) as a predictive model. This is related to the work of [11], who use "hallucination" to fill in gaps in a sparse map to assist in estimating path costs for navigation over large-scale terrains. Also related is the work of [4], which relies heavily on repetitive structure in the environment to match snippets from the current map to partially observed areas, with the idea of not exploring regions that can be predicted or of using the prediction for localization, a questionable strategy. Although not used for prediction, [9] have the most robust approach to building a model of the world, learning a Dirichlet prior over structural models from a library of previously explored environments, and using samples from this distribution for estimating the probability that the robot is inside or outside the known map.

We begin with a description of the information entropy formulation for Active SLAM, as well as some of the difficulties of explicitly reasoning about information gain. We then describe our method for using simplified models to probe the SegMap to find actions than reduce entropy, including the use of the segmented map as a predictive model for generating routes that extend into unknown regions. Finally, we demonstrate Active SLAM with predictive loop closure in a real world experiment.

2 Information Gain-Based Active SLAM

Information entropy methods for Active SLAM involves several steps. To select an action, an information gain planner evaluates different possible actions based on their expected information gain. For SLAM, this information gain is over the SLAM belief distribution, $p(x, \theta)$, which includes both the pose x and map θ. The RBPF SLAM belief distribution is represented by a set of particles, each particle representing both a map and a position within that map. As a result, the RBPF SLAM entropy can be factorized as

$$H[p(X, \Theta)] = H(p(X)) + E_X[H(p(\Theta|X))]$$

To estimate $E_X[H[p(\Theta|X)]]$, we take the average entropy of all the particle maps. We use an efficient 3D evidence grid data-structure called the Deferred Reference Count Octree [7], and under the independence assumptions of evidence grids, the information entropy H of of a map θ is the sum of the entropies of each voxel $\theta[x, y, z]$,

$$H(\theta) = \sum_{\forall x,y,z} -\rho \log(\rho) - (1 - \rho)\log(1 - \rho),$$

where $\rho = p(\theta[x,y,z])$. Since the octrees are sparsely populated, ignoring unknown regions changes from a requirement to a computational advantage. A further speed improvement arises because the octree can represent homogeneous regions at a variety of leaf node scales.

Computing the entropy of the particle poses is approximated with heuristics. The method described by [13] approximates the cloud with a multivariate normal distribution, and then computes the differential entropy of the distribution. This approximation is only accurate in cases in which the pose cloud is nearly unimodal, although it will usefully report high entropy for multi-modal distributions. [12] and [13] refine this heuristic by computing its average value over multiple points in time, though [1] describes how even this improved metric can become undefined after closing a loop. This is also more complicated for SegSLAM, because particles may be distributed over different segments, each of which has its own coordinate frame. A more mundane problem is that the entropy of the maps dominates the entropy of the poses, meaning that a small amount of exploration will result in a greater change in entropy than a convergence in the particle poses.

A reasonable simplification is to restrict the problem to finding the best of a set of candidate actions rather than finding the best of all possible actions. The set of candidate actions can be generated to have the actions differ significantly and be distributed evenly over the space of all possible actions. Unfortunately, even when only considering a few candidate actions, this process is computationally intractable for a RBPF with hundreds of particles and complex map structures. For example, the expected entropy estimation process is further complicated for SegSLAM, since the average is computed over map samples, and there are many more possible map samples than particles.

Further, a fundamental limitation of this process is that measurements can't be predicted for unmapped areas. It is also questionable whether updating the (cloned) particle maps with the simulated measurements yields a better estimate of the SLAM state. As a result, the accuracy of the entropy estimates will degrade over time: in particular, they will degrade completely for actions that enter unmapped areas.

In the next two sections we address the two limitations of information gain-based Active SLAM: computational intractability and limited horizon. First, we describe our simplified method, and then we describe how to use prediction to usefully extend the planning horizon.

3   Simplified Entropy Heuristic

In order to be computationally tractable, our method eliminates the filter simulation and entropy evaluation and instead, directly examines the variance of the simulated measurements. Under certain constraints, this variance is a proxy for the full entropy and can be used for Active SLAM.

As before, the simulated measurements $\hat{z}_{ap}$ are generated by simulating the motion of particle p while executing action a, and casting rays in the map. The simulated measurements are determined by the current SLAM belief state, and are correlated with the particle weights at least until the particle filter would resample (if we were simulating the full particle filter).

For example, assuming a 1D state and that the particle pose distribution can be approximated by a Gaussian, we can estimate the entropy of the particle poses for action a as the differential entropy of the Gaussian

$$H_a = \log(\sigma_a \sqrt{2\pi e}),$$

where σ_a is the weighted variance of the particle positions

$$\sigma_a = \sqrt{\left(\sum_p^{\#_p} w_{ap}(x_p - \mu)^2\right)}.$$

Note that the $(x_p - \mu)^2$ are constant for all actions, because all particles move in lock-step under the simplified model. Using a Gaussian model for a single range sensor, the particle weights are given as

$$w_{ap} = \frac{1}{\sqrt{2\pi\sigma_z^2}} e^{\frac{-(\hat{z}_{ap} - \bar{z})^2}{2\sigma_z^2}}.$$

Plugging in the expression for w_{ap} and discarding constants, we have

$$H_a \propto \log\left(\sum_p e^{-(\hat{z}_{ap} - \bar{z})^2}\right)$$

and since both log and exp are monotonic functions, if we can minimize H_a by choosing the action that maximizes $(\hat{z}_{ap} - \bar{z})^2$, or equivalently to choosing the action that maximizes the variance of the simulated measurements: the planning algorithm only needs to estimate the relative variances to select between different actions. Note that variance in the simulated measurements arises from the uncertainty in the SLAM belief state and is indicative of the SLAM entropy, rather than the likelihood of any particular set of measurements under the sensor model.

Within the particle filter's resampling horizon, choosing the action that maximizes the sum-squared measurement difference is equivalent to choosing the action that minimizes the entropy. Even beyond the resampling horizon, when some of the assumptions above break down, the measurement variance criteria leads to good action recommendations. Because we do not need to precisely simulate the full state of the SLAM algorithm, we can use simplified sensor models and larger motion step sizes, yielding a smoothly degrading, computationally fast, proxy for the full SLAM entropy estimate.

4 Useful Map Prediction

The overall goal of Active SLAM is to find actions that reduce the uncertainty in both mapping and localization. Closing loops is crucial for SLAM, because when the algorithm detects that it has returned to a previously mapped area, the accumulated error can be canceled. It would be very beneficial if Active SLAM could select actions that were expected to close loops: the difficulty is that making any non-trivial loop-closure predictions involves simulating measurements in unmapped regions. Rather than assuming all unknown space is empty, or occupied, or using some other prior model, our method directly predicts the structure of the environment in the unmapped area by using nearby previously constructed map segment.

In the SegMap there are multiple particle maps for each segment, or submap (Figure 2). Different particle maps can be stitched together to yield metric map samples. The SegMap represents the SLAM belief distribution, in the sense that for regions where SegSLAM has low uncertainty different map samples will be very similar, but for regions where SegSLAM is uncertain, variation in the metric submaps and the transformations between them will yield significant variation in the map samples.

To simulate measurements we generate metric map samples that include each particle's current segment, simulate the motion of the vehicle within the map sample, and cast rays. We can extend map sample generation to predict unmapped regions by making the assumption that nearby segments are good predictors of unobserved areas (Figure 2). We use the same process as for generating map samples, with the addition of a step that randomly selects a nearby segment, and grafts its start

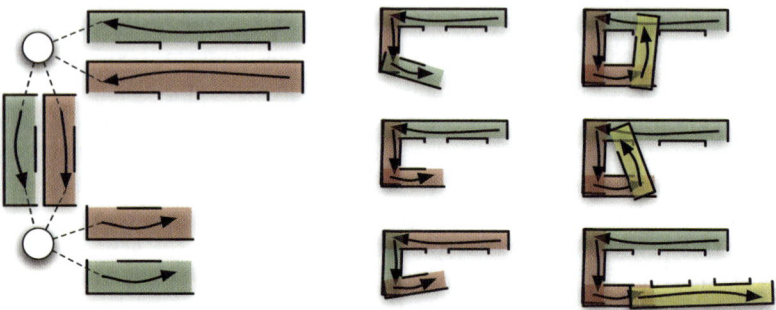

Fig. 2 Left, a diagram of a SegMap with two particles and three segments: the transforms between segments are represented by circles. Center, three different map samples generated from the SegMap. Right, a map sample is extended by three guesses (in yellow), which are transformed segments attached at the current vehicle position with a small random perturbation.

position onto the current position with some random perturbation or refinement from map matching. We will call this grafted segment a "guess," because it is a weaker hypotheses about the map structure.

The question is then how to make use of the "guess" yielded by the grafted segment. If the guess conflicts with the "known" portions of the map sample due to the regular segments, it is useless. If the guess extends into unmapped regions, it can be used to generate informed plans (for example allowing plans to extend down a hallway). But the real value of a guess is when it connects known regions while bridging across unknown regions: this allows Active SLAM to close *predicted* loops.

To accelerate the process of finding plans, our planner also uses the following simplifications. It first generates a local map sample and fuses the submaps into a single evidence grid map. It then generates a guess by selecting a random segment to append to the vehicle's current position, as well as a small random perturbation (Figure 2). This guess segment includes the vehicle's original trajectory through the segment, so to quickly check the plausibility of the guess, our planner queries the fused map along the (transformed) trajectory to verify that the guessed trajectory starts in known empty space, goes into unknown space, and returns to known empty space. If the guess passes this quick check, it is grafted into the fused map, and the planner uses the vehicle model to see if the vehicle can pass through the resulting map. If the vehicle can go from known empty space to grafted empty space and back to known empty space without collision, the plan is considered a success for the map sample.

In the next section, we present from using this approach to Active SLAM with loop prediction in a real-world experiment.

5 Active SLAM Experiment

Beneath the Field Robotics Center highbay is a network of tunnels known as the catacombs. The navigable area of the catacombs forms a square figure eight shape (Figure 3). We used Cave Crawler [10] to explore the tunnels, which are are just wide enough: in some cases the wheels rub on both sides, so we manually drove Cave Crawler in a close approximation of the planned trajectory. We limited the laser ranges to a maximum of 4 m so that returns would not reach the end of the tunnel segments. As the simplified sensor model for this Active SLAM experiment, we used a simple binary collision model that indicated traversability.

Cave Crawler started at one end of the figure eight, and used an RRT planner to find exploration actions with a bias toward moving in straight lines. After driving down one side, past the crossing tunnel and around the bottom of the loop, Cave Crawler encountered the other end of the crossing tunnel (Figure 4). At this point, it had three or four segments (varying between different runs) with which to make guesses, and 40 particle submaps for each segment. In this experiment, we evaluated guesses as described above, and checked the collision sensor model in 10 map samples, considering the maximum variance criteria to be satisfied if the binary

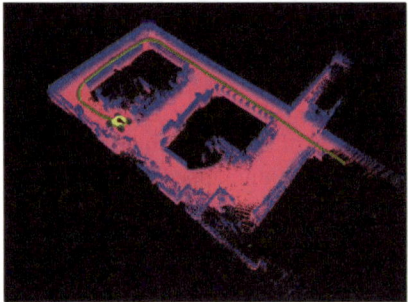

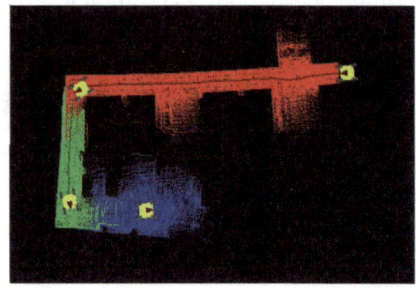

Fig. 3 Overview of the catacombs environment – the long axis of the figure eight is about 20 m, and the short axis is about 10 m. In several places Cave Crawler can barely fit through the narrow tunnels.

Fig. 4 In both experiments, Cave Crawler started at one end of the figure eight (upper right) and then proceeded along one side and around the bottom. The colored pointclouds correspond to the three segments.

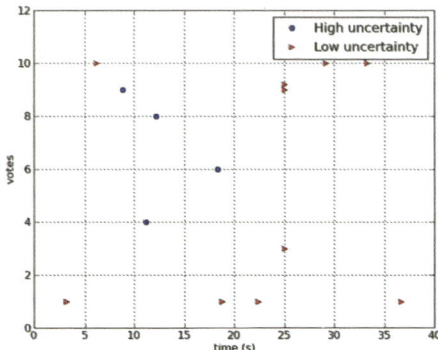

Fig. 5 Action votes over the period of time when Cave Crawler approached the crossing tunnel, for both the high and low uncertainty runs.

outcome of the collision model indicated traversability in 4-7 samples (Figure 5). Most guesses result in collisions and rarely have success rates above 2/10. But some guesses successfully bridged the middle segment of the figure eight, and the success rate of actions across these bridges depended on the SegSLAM uncertainty.

To control the SLAM uncertainty, we artificially increased the motion model noise. When the SegSLAM uncertainty was high the bridging actions had success rates of 4/10 or 6/10, and following the max variance criteria (relative to simply continuing straight) Cave Crawler took the action and closed the loop (Figure 6:top). In the runs with low SegSLAM uncertainty, the bridging actions had either high or low success rates, and Cave Crawler did not cross the center tunnel, continuing to follow the usual exploration actions up and around the figure eight to close the loop (Figure 6:bottom).

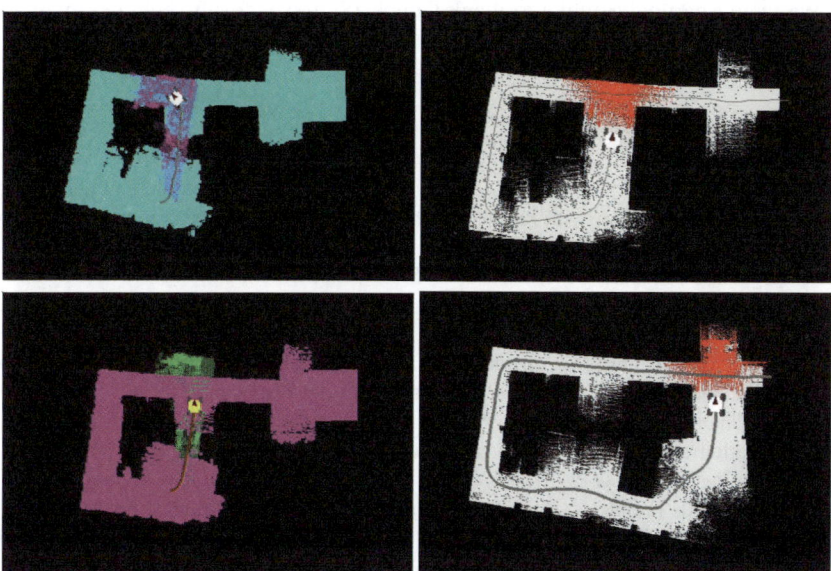

Fig. 6 Top: High uncertainty run: the guessed path through the center tunnel (left) had a success rate of 6/10, which satisfied the maximum variance criteria. After taking this action, SegSLAM closed the inner loop (right). Bottom: Low uncertainty run: the guessed path through the center tunnel (left) had a low variance success rate of 9/10. The vehicle continued to follow the outside of the figure eight, and SegSLAM closed the outer loop (right).

Using our method, Cave Crawler was able to actively predict and close a loop in real-time at full safe vehicle velocity, about 0.2 m per second in this constrained environment.

6   Conclusions

Our approach to Active SLAM can be summarized as using simplified models to heuristically estimate the entropy of the SegMap. In particular, the method probes the SegMap by analyzing the measurements generated by a simple sensor model applied to multiple map samples. We have shown that this approach is equivalent to information-gain estimation when the planning horizon is within the resampling period of the particle filter, and we have experimentally found that produces reasonable plans even beyond this horizon. We also showed how to use the SegMap as a predictive model to hypothesize about unseen regions by using nearby segments as priors. Integrating all these ideas, we demonstrated active SLAM in real-time onboard Cave Crawler.

In future work, we would like to compare our method with other information-gain heuristics over a range of structured and unstructured 3D environments.

We would particularly like to thank Chuck Whittaker for his help getting Cave Crawler into the catacombs.

References

1. Blanco, J.-L., Fernández-Madrigal, J.-A., González, J.: A novel measure of uncertainty for mobile robot slam with rao-blackwellized particle filters. International Journal of Robotics Research 27(1), 73–81 (2008)
2. Bourgault, F., Makarenko, A., Williams, S., Grocholsky, B., Durrant-Whyte, H.: Information based adaptive robotic exploration. In: IEEE/RSJ Intl. Conf. on Intelligent Robots and Systems (2002)
3. Burgard, W., Stachniss, C., Grisetti, G.: Information gain-based exploration using rao-blackwellized particle filters. In: Proc. of the Learning Workshop (2005)
4. Jacky Chang, H., George Lee, C.S., Lu, Y.-H., Charlie Hu, Y.: P-slam: Simultaneous localization and mapping with environmental-structure prediction. IEEE Transactions on Robotics 23(2), 281–293 (2007)
5. Doucet, A., de Freitas, N., Murphy, K., Russell, S.: Rao-blackwellised particle filtering for dynamic bayesian networks. In: Proc. of the Sixteenth Conf. on Uncertainty in AI, pp. 176–183 (2000)
6. Fairfield, N.: Localization, Mapping, and Planning in 3D Environments. PhD thesis, Carnegie Mellon University, Pittsburgh, PA, USA (2009)
7. Fairfield, N., Kantor, G., Wettergreen, D.: Real-time slam with octree evidence grids for exploration in underwater tunnels. Journal of Field Robotics (2007)
8. Feder, H., Leonard, J., Smith, C.: Adaptive mobile robot navigation and mapping. Int. J. Rob. Res. 18(7), 650–668 (1999)
9. Fox, D., Ko, J., Konolige, K., Stewart, B.: A hierarchical bayesian approach to the revisiting problem in mobile robot map building. In: Intl. Symp. of Robotic Research (2003)
10. Morris, A., Ferguson, D., Omohundro, Z., Bradley, D., Silver, D., Baker, C., Thayer, S., Whittaker, W., Whittaker, W.: Recent developments in subterranean robotics. Journal of Field Robotics 23(1), 35–57 (2006)
11. Nabbe, B., Kumar, S., Hebert, M.: Path planning with hallucinated worlds. In: IEEE/RSJ Intl. Conf. on Intelligent Robots and Systems, vol. 4, pp. 3123–3130 (2004)
12. Roy, N., Thrun, S.: Coastal navigation with mobile robots. In: Advances in Neural Processing Systems, vol. 12, pp. 1043–1049 (1999)
13. Stachniss, C.: Exploration and Mapping with Mobile Robots. PhD thesis, University of Freiburg (2006)
14. Stachniss, C., Haehnel, D., Burgard, W.: Exploration with active loop-closing for Fast-SLAM. In: Proc. of the IEEE/RSJ Int. Conf. on Intelligent Robots and Systems, IROS (2004)

Outdoor Downward-Facing Optical Flow Odometry with Commodity Sensors

Michael Dille, Ben Grocholsky, and Sanjiv Singh

Abstract. Positioning is a key task in most field robotics applications but can be very challenging in GPS-denied or high-slip environments. A common tactic in such cases is to position visually, and we present a visual odometry implementation with the unusual reliance on optical mouse sensors to report vehicle velocity. Using multiple kilometers of data from a lunar rover prototype, we demonstrate that, in conjunction with a moderate-grade inertial measurement unit, such a sensor can provide an integrated pose stream that is at times more accurate than that achievable by wheel odometry and visibly more desirable for perception purposes than that provided by a high-end GPS-INS system. A discussion of the sensor's limitations and several drift mitigating strategies attempted are presented.

1 Introduction

Accurate knowledge of position is critical to successful completion of field robotics tasks. In known or highly structured environments, localization relative to a pre-determined or progressively-refined map is typically performed using sensors appropriate for registering map features to observations. In general outdoor scenarios, absolute positioning using Global Positioning and Inertial Navigation systems (GPS-INS) is frequently performed, often in conjunction with input from odometry integration, and may augmented further by continuous registration of local terrain or obstacle maps. Indeed, recent GPS-INS devices advertise errors as low as a few centimeters in position and hundredths of a degree in attitude after alignment [13].

Many applications present significant challenges for these positioning strategies. GPS may be unavailable or ineffective if too few satellites are visible during urban or subterranean operations. Wheel-based odometry depends on an accurate kinematic model and can degrade greatly in the presence of wheel slip typical of low-friction

Michael Dille, Ben Grocholsky, and Sanjiv Singh
The Robotics Institute, Carnegie Mellon University,
5000 Forbes Avenue, Pittsburgh, PA 15213, USA
e-mail: {mdille3,grouch,ssingh}@andrew.cmu.edu

A. Howard et al. (Eds.): Field and Service Robotics 7, STAR 62, pp. 183–193.
springerlink.com   © Springer-Verlag Berlin Heidelberg 2010

surfaces and skid-steered vehicles. In poorly lit or very low-texture environments, systems based on vision will see reduced performance.

Planetary rover missions pose particularly challenging cases. GPS is unavailable and alternatives such as star tracking do not offer comparable accuracy. Loose surface dust easily impedes wheel odometry. Further, visual odometry and terrain mapping methods are frustrated by poor lighting, few visual surface features, and tight computational constraints limiting their implementation complexity.

The specific need we seek to fulfill is that of a pose estimation system for Scarab, the lunar rover prototype developed at Carnegie Mellon University pictured in Figure 1 [4] [20]. This 280kg skid-steered vehicle is designed to explore permanently shadowed lunar polar craters, which combines the greatest weaknesses of most positioning methods: the unavailability of GPS, constant near-total darkness, occlusion of most stars by the high crater walls, soft lunar regolith, and strictly limited computing and power facilities. Exclusive reliance on wheel odometry was first planned but then abandoned in recognition of the errors that would be induced by surface slip, the eventual implementation of softer wheels that would thereby be of varying radius, and the passive differenced rocker suspension, which while excellent for maintaining stability on bumpy terrain, makes estimating heading changes from odometry inadequate. Certainly strategies can be applied for detecting slip some proportion of the time—and [16] provides several methods for doing just this on planetary rovers—but the design of Scarab's suspension frustrates this approach and arguably more complicated physical modeling might best be replaced by an alternative sensing modality.

Conventional forward-facing visual odometry as used on the Mars Exploration Rovers (MERs) [11] was considered but deemed impractical because lighting the surrounding area would require more power than the rover can provide and induce complex shadows. Instead, we decided upon using downward-facing visual odometry. This provides tractable lighting requirements, however these forward-facing techniques cannot be directly applied because the situation is ill-posed to compute the 3-D incremental pose differences and the terrain beneath the rover is likely to be too homogeneous for point-based feature tracking to work effectively. Rather, we chose to rely on optical flow to provide an estimate of vehicle speed, which could then be integrated to estimate incremental distance-traveled as part of a broader odometry framework. Various methods, such as the Horn & Schunck algorithm [7], have long existed for computing so-called dense optical flow over regions between images. Such algorithms have been used, for instance, for autonomous heading control for obstacle avoidance for fixed-wing aircraft [21], estimating distance to the ground and canyon walls from unmanned aerial vehicles [6], and autonomous landing for helicopters [17]. In the odometry realm, downward-facing cameras have been successfully used for positioning in pre-explored environments by correlating the visible area against an existing database [8], however such methods are inapplicable to planetary rovers observing most patches for the first time.

Although an optical flow implementation using a typical camera would have been straightforward, an intriguing alternative presented itself in the form of a custom-built optical flow sensor from AirRobot GmbH [1] used for stabilization on its

Fig. 1 The Scarab lunar rover. The glow beneath is from LED lighting for the optical flow sensor.

Fig. 2 The optical sensor used on Scarab, in a custom ruggedized enclosure.

quad-rotor helicopters. This device, shown in Figure 2, contains four commodity optical mouse sensors each attached to a lens of a different focal length and reports 2-D velocity (in m/s) along the planar surface at which it is pointed and an estimate of distance to this plane. The precise details of its operations are proprietary to the manufacturer, but given that a scalar "tracking quality" value is known to be returned by mouse sensors, we conjecture that height is derived by interpolating across the quality values returned by each of the four sensors, with speed similarly interpolated across the four focal-distance-compensated speed values. The sensor nominally reports velocity with a resolution of 0.3mm/s within a range of ± 10m/s and height up to 2.5m with a resolution of 1cm.

The use of optical mouse sensors for measuring ground velocity has a number of benefits leveraging over a decade of commercial refinement. Key among these are remarkably robust operation on a wide variety of surfaces, significant lighting insensitivity, and extremely low cost (several US dollars) due to the volume in which they are produced. These typically contain a 15 to 100 pixels-square camera sensor and are believed to implement a fast hardware version of Horn & Schunck [14] reporting sub-pixel flow rates at on the order of 1kHz. Designs based on laser interferometry are becoming prevalent, however they are not as adaptable as they do not use simple lenses. Due to their low cost and simplicity, there exists wide interest in the hobbyist robotics community in these sensors. Several examples of their use in limited indoor scenarios exist in the robotics literature (eg [15], [18], and [19]), but they have yet to see use in a field robot application. A method similar to that described here using a webcam and image correlation matching was presented in [12].

In this paper, we describe our odometry method using this sensor, present results from extensive field testing, and draw conclusions on the effectiveness of commodity optical flow sensors for pose estimation.

2 Odometry Method

We first assume that the vehicle can only instantaneously move along its current heading vector, or equivalently assume the non-holonomic constraint of no wheel side-slip. We believe this to be valid because except for rare cases such as sliding laterally down a slope, wheel slip should occur only when skidding in place during turns or failing to grip the ground during forward travel. Position is integrated based on the current linear speed and heading vector at all times. Linear speed is estimated from optical sensor readings, and the heading vector comes from the attitude determined by integrating IMU angular rate values. Uncertainty in the integrated pose is propagated by modeling uncertainty for the optical sensor and the IMU. The next section formalizes this procedure followed by methods for improved accuracy.

2.1 Basic Odometry Model

We begin by defining the vehicle position $\mathbf{x}$ and orientation θ in a chosen world frame:

$$\mathbf{x}_t = [x, y, z]^T, \quad \theta_t = [\theta_x, \theta_y, \theta_z]^T \equiv \mathbf{q}_t = [q_s, q_x, q_y, q_z]^T \tag{1}$$

where the latter equivalence denotes that we may refer to the orientation as the Euler angles θ_t or a unit quaternion $\mathbf{q}_t$.

At each time-step, the IMU provides angular rates and linear accelerations in the body frame:

$$\omega_t = [\omega_x, \omega_y, \omega_z], \quad \mathbf{a}_t = [a_x, a_y, a_z]. \tag{2}$$

For a given state $\mathbf{x}_t$ we may then write the vehicle unit heading vector $\hat{\mathbf{u}}_t$ in the world frame. The non-holonomic velocity constraint can then be embedded in the definition of the vehicle's linear velocity as

$$\mathbf{v}_t = v_{body}\hat{\mathbf{u}}_t \tag{3}$$

where v_{body} is the scalar linear velocity, which is assumed to be directly provided by the optical sensor. In actuality, it reports velocity in two dimensions, but to avoid having to very precisely calibrate for any rotational offset, we simply take the norm of the velocity vector it reports to be the speed and determine the sign from that of the velocity reported along the axis most nearly aligned with the body.

At time $t + 1$, given $\mathbf{v}_{t+1}$ and the previous values $\mathbf{x}_t$ and $\mathbf{v_t}$, the new position may be estimated as

$$\mathbf{x}_{t+1} = \mathbf{x}_t + \frac{1}{2}\Delta t(\mathbf{v}_{t+1} + \mathbf{v}_t), \tag{4}$$

where Δt is the time elapsed between t and $t + 1$, computed via trapezoidal integration of the velocity vector. A similar procedure for orientation is used. Namely,

$$\dot{\mathbf{q}}_t = \frac{1}{2}q_t * [0, \frac{1}{2}(\omega_t + \omega_{t+1})]^T$$
$$\mathbf{q}_{t+1} = \mathbf{q}_t + \Delta t \dot{\mathbf{q}}_t \tag{5}$$

estimates the new orientation, where * denotes quaternion multiplication [10].

Dead-reckoning error uncertainty is propagated by the linearized covariance prediction equation[1]

$$\mathbf{P}_{k+1} = \mathbf{F}_k\mathbf{P}_k\mathbf{F}_k^T + \mathbf{G}_k\mathbf{Q}_k\mathbf{G}_k^T, \tag{6}$$

where

$$\mathbf{F} = \begin{bmatrix} 1 & 0 & -V\sin\psi\Delta t \\ 0 & 1 & V\cos\psi\Delta t \\ 0 & 0 & 1 \end{bmatrix} \text{ and } \mathbf{GQG}^T = \begin{bmatrix} \cos^2\psi\sigma_V^2\Delta t + \frac{1}{3}V^2\sin^2\psi\sigma_\omega\Delta t^3 & \cos\psi\sin\psi(\sigma_V^2\Delta t - \frac{1}{3}V^2\sigma_\omega^2\Delta t^3) & -\frac{1}{2}V\sin\psi\sigma_\omega^2\Delta t^2 \\ \cos\psi\sin\psi(\sigma_V^2\Delta t - \frac{1}{3}V^2\sigma_\omega^2\Delta t^3) & \sin^2\psi\sigma_V^2\Delta t + V^2\cos^2\psi\sigma_\omega^2\Delta t^3 & \frac{1}{2}V\cos\psi\sigma_\omega^2\Delta t^2 \\ -\frac{1}{2}V\sin\psi\sigma_\omega^2\Delta t^2 & \frac{1}{2}V\cos\psi\sigma_\omega^2\Delta t^2 & \sigma_\omega^2\Delta t \end{bmatrix}.$$

Approximate trends in state uncertainty growth over time can be observed by reducing Equation 6 to straight line constant speed motion with inital covariance $\mathbf{P}_k = diag(\mathbf{P}_x \ \mathbf{P}_y \ \mathbf{P}_\psi)$. For x-axis-aligned motion, Equation 7 indicates along-track and heading error increases proportional to the square-root of elapsed time. Cross-track error exhibits faster error growth, linear in time due to initial heading uncertainty plus growth at three-halves power of time due to heading rate noise. This highlights the importance of low heading-rate uncertainty in achieving accurate dead-reckoning.

$$\mathbf{P}_{k+\Delta t} = \begin{bmatrix} \mathbf{P}_{x,k} & 0 & 0 \\ 0 & \mathbf{P}_{y,k}+V^2\Delta t^2\mathbf{P}_{\psi,k} & V\Delta t\mathbf{P}_{\psi,k} \\ 0 & V\Delta t\mathbf{P}_{\psi,k} & \mathbf{P}_{\psi,k} \end{bmatrix} + \begin{bmatrix} \Delta t\sigma_V^2 & 0 & 0 \\ 0 & \frac{1}{3}V^2\Delta t^3\sigma_\omega^2 & \frac{1}{2}V\Delta t\sigma_\omega^2 \\ 0 & \frac{1}{2}V\Delta t\sigma_\omega^2 & \Delta t\sigma_\omega^2 x \end{bmatrix} \tag{7}$$

Distance and heading error drift rates for the system have been determined through ground-truthed experimental trials. Figure 3 shows the difference in changes observed between the dead-reckoning solution and ground-truth for varying length-time ensembles. Twenty thousand samples were compared over a two hour period at thirty second ensemble increments. This trial indicates drift rates of $1.2m/\sqrt{hour}$ in distance traveled and $2.5^o/\sqrt{hour}$ in heading.

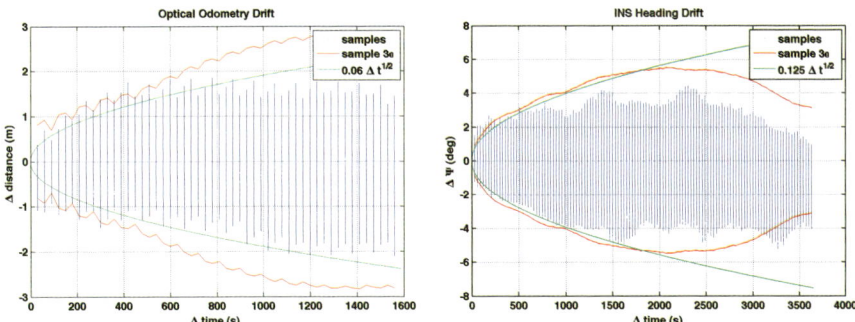

Fig. 3 Estimated drift rate for distance traveled *(left)* and heading *(right)* determined by comparing changes over increasing time ensembles between the dead-reckoning solution and ground-truth.

---

[1] For brevity, we here present uncertainty propagation for planar location and heading. The equations for the full 6-D model as we have implemented are a straightforward extension.

2.2   Improved Attitude Estimation

While the error in the pose estimate provided by the above integration procedure will necessarily grow without bound, several simple measures can be taken to greatly reduce attitude error. Many of these are intuitive and even commonplace in commercial navigation systems, however they receive rare mention in the robotics literature and are too often ignored in implementation.

A first strategy critical for long-term operation is to remove the approximately 15 degrees per hour on Earth (or 0.56 on the moon [2]) angular velocity the gyroscope will inherently pick up due to the rotation of the earth, which is done by rotating the known angular velocity of the planet into the local coordinate system and subtracting. When stationary, this measured angular velocity may even be used to perform gyrocompassing. The gyroscope bias may be computed by averaging the measured angular velocity vector while stationary (after subtracting the planetary rotation rate). With some IMUs this may needed as often as every ten minutes for even short-term performance [5]. This step was not required in our case since this value typically averaged to be negligible.

Another common technique is to use the gravity vector (weaker but still useful at about $0.16g$ on the moon [3]) as measured by the IMU's accelerometers when stationary to estimate roll and pitch by taking arctangents of the accelerations along the axes. We made use of this to reset drift in roll and pitch during stationary periods. Given now two complementary sources of roll and pitch–this direct computation most accurate when the attitude is slowly changing and integration of angular rates just the opposite–we implemented a matched complementary Butterworth filter pair to continuously merge these two streams, with a cutoff frequency of 0.05Hz found to be most appropriate given the very slow motion of the rover. This provides excellent results with errors in roll and pitch of typically much less than a degree at all times.

Finally, anticipating missions consisting of long stationary periods (primarily intended for battery recharging) followed by short periods of motion, we additionally clamped the measured angular velocity vector to zero when the vehicle is known to be stationary to avoid blatantly unnecessary noise integration.

2.3   Velocity Scale Calibration

Ideally, the optical flow sensor returns a correctly scaled lateral speed regardless of the distance between the sensor and the ground. We model the actual corrupted scaled velocity returned as

$$v_{measured} = s(h) \cdot v_{actual} + b + v, \tag{8}$$

where $s(h)$ is a scaling factor that varies with the height h, and v is noise. At most a trivial bias term b was observed, and this was effectively dealt with by clamping the measured velocity to zero when the vehicle is known to be stationary.

The height of the sensor is not fixed as Scarab is equipped with an active suspension used to lower the body so that a core-drilling apparatus may operate. Calibration trials at varying body heights were performed. Figure 9 shows a number of such runs over distances varying between 10 and 30 meters. As it indicates, rather different scaling values were found for daytime and nighttime operation, but within each class, a simple quadratic fit provides reasonable compensation.

The sensor also exhibited a scale dependence on the surface type with values tightly clustered for a given surface. An online auto-calibration scheme proved quite effective by relying on the observation that during consistent straight-line motion, wheel odometry can be very accurate. Every three seconds a battery of heuristics tests whether wheel odometry is trustworthy, including whether each wheel's velocity agrees closely with the average, all agree on direction, the reported velocity is above a noise floor, and that the IMU is reporting minimal yaw rates. Primarily, this eliminates periods including heavy slip or turning (implying likely wheel odometry error). Each period is added as a learned data point over which a variation of a recursive locally-weighted linear least squares algorithm is run to compute the scaling factor as a function of body height. Each time a new velocity value from the optical sensor is received, the best estimate of the scaling factor for that height is used. If insufficient calibration data points are available for the region surrounding that height the scaling factor determined by the original manual calibration trials is used as a fall-back.

An example run using the online scale calibration procedure is shown in Figures 10 and 11, which indicates that this process performs well, though unsurprisingly not as well as a post-hoc batch method computing the scaling function from all the data points at once. However, it performed well enough and eliminated most difficulties associated changing surfaces. All data presented uses it unless otherwise indicated.

3   Experimental Results

During the course of system development and verification, we collected tens of hours of data from field testing over several kilometers of traversal, including extended simulated lunar terrain at Moses Lake, Washington and Mauna Kea, Hawaii. Field tests were conducted both during daytime and in total darkness, for which high-intensity LED bars were mounted to the vehicle's underbelly to illuminate the area seen by the optical sensor.

Anecdotally, the optical sensor worked remarkably well across a wide variety of surfaces, including dirt, mud, grass, asphalt, and sand. Tracking was poor on poured concrete surfaces and nearly useless on painted concrete. Most of the extended field testing took place over sand to best emulate lunar terrain, and it was noted that typical auto-calibrated scaling factors were somewhat higher than for other surfaces, likely explainable by somewhat poorer tracking on this relatively featureless terrain.

As the Honeywell HG1700 IMU used for odometric integration is part of a NovaAtel SPAN GPS-INS system, a convenient source of approximate ground truth

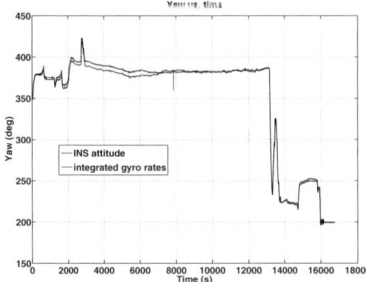

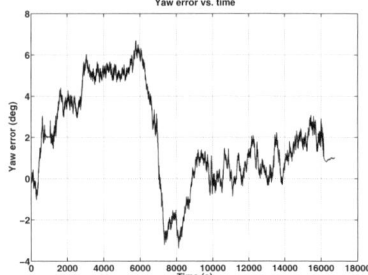

Fig. 4 Comparison of INS-reported and integrated yaw angle after application of described attitude integration improvements.

Fig. 5 Error in integrated yaw angle relative to INS-reported value after application of describe attitude integration improvements.

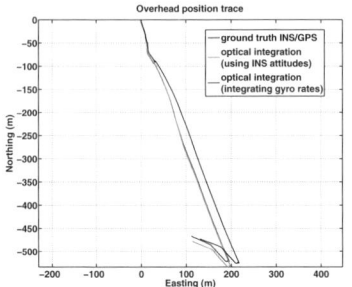

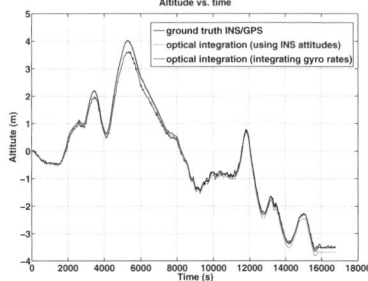

Fig. 6 Overhead comparison of odometry results against INS-reported position using both integrated angular rates and INS-reported attitude.

Fig. 7 Altitude comparison of odometry results against INS-reported position using both integrated angular rates and INS-reported attitude.

was also available. At the slow speeds the rover moves, INS position errors on the order of tens of centimeters are large relative to short-term distances traveled, however over long distances, these relatively static uncertainties are small compared to accumulated odometry error.

In the plots below, full odometry results are provided as well as the result of using INS-provided attitude (the output of the INS Kalman filter that uses GPS data) in place of integrated attitude. The purpose of this is to independently demonstrate the effectiveness of the optical sensor (accumulating just distance-traveled error in the odometry) and because further methods of improving heading accuracy would be a priority in any future implementation.

An early observation was that even when the odometry-derived pose drifted significantly, as shown in Figure 8 having atypically high gyro drift, the smoothness of the integrated solution was much more useful to the on-board perception system than that from the INS. Laser scanners are used to build local terrain and obstacle maps. Pose jumps present false obstacles without resorting to complicated registration algorithms precluded by limited on-board computing. While a map built

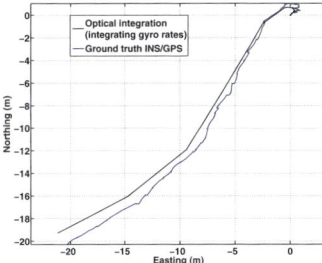

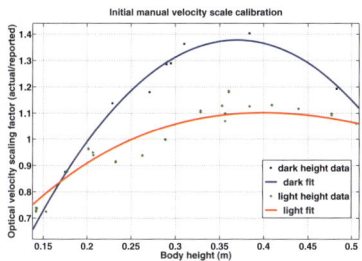

Fig. 8 Example run in which even the significantly drifting integrated estimate is preferable to a very jumpy GPS signal for perception purposes.

Fig. 9 Optical velocity scale calibration data over a number of short straight-line runs.

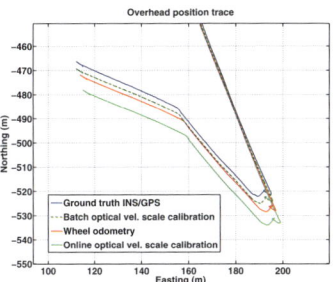

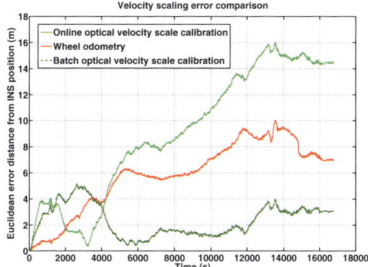

Fig. 10 Overhead comparison of optical velocity scaling methods with wheel odometry and INS position, zoomed in to the end of the run.

Fig. 11 Comparison of optical velocity scaling methods with wheel odometry and INS position.

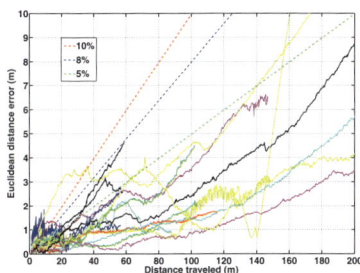

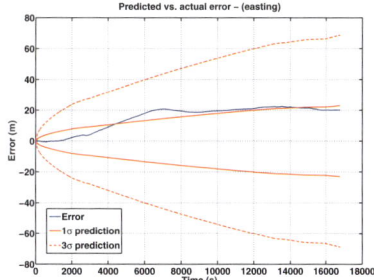

Fig. 12 Euclidean error growth across a variety of test runs.

Fig. 13 Easting position error and prediction.

from the integrated solution will not be globally correct, the resulting higher-fidelity short-term maps are more valuable for local motion planning.

As odometry error is inherently path dependent, error accumulated over different trials varied, however the results shown in Figure 6 are representative. Data from this

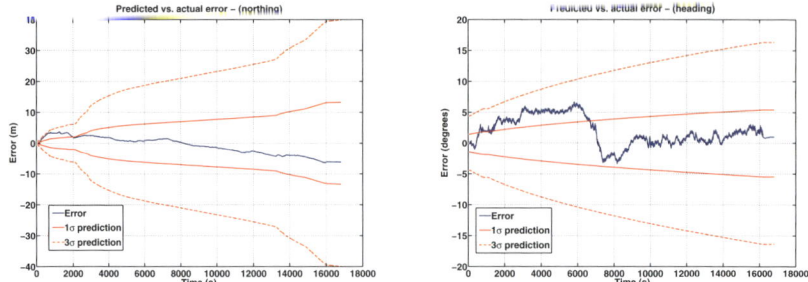

Fig. 14 Northing position error and predic- **Fig. 15** Heading error and prediction.
tion.

run, a nearly 700m nighttime traverse at Moses Lake, is also used as the examples for
the described attitude improvement and scale calibration methods. Attitude comparisons between the INS-reported values and integrated results are given in Figure 5.
Traces of position and heading uncertainty propagated using the model and parameters described are provided in Figures 13-15. Though just one example, the errors
lie within the uncertainties, lending credence to our propagation model and process
covariances. Error growth across many runs is given in Figure 12, suggesting this
method typically provides an error bounded by 5-8% of distance traveled.

4   Conclusions and Future Work

In this paper, we presented an implementation of vehicle odometry for a lunar rover
prototype using an optical mouse sensor to provide vehicle velocity. Results show
that with a moderate-grade IMU, errors can be small over long distances. Clearly the
weakest point in such a design is the accumulation of heading error, the reduction
of which would be a key focus of future implementation. The lunar application
may present some such reduction opportunities, perhaps via the stop-start nature of
motion during missions or observation from an orbiter. Other avenues of exploration
include using multiple sensors to track heading or relax the kinematic assumptions
[9] and designing a new sensor with focal lengths tuned for the mounting height.

Acknowledgements. This research was supported in part by NASA under grants,
NNX07AE30G, John Caruso, Project Manager, and NNX08AJ99G, Robert Ambrose, Program Scientist. The authors wish to thank the Scarab project team for expressing interest in
this idea and tolerating related experimentation.

References

1. AirRobot GmbH: AirRobot website (2009), http://www.airrobot.de
2. Allen, C.W., Cox, A.N.: Allen's Astrophysical Quantities. Springer, Heidelberg (2000)
3. Baldwin, R.: A Fundamental Survey of the Moon. McGraw-Hill, New York (1965)

4. Bartlett, P., Wettergreen, D., Whittaker, W.: Design of the scarab rover for mobility and drilling in lunar cold traps. In: International Symposium on Artificial Intelligence, Robotics and Automation in Space (iSAIRAS), Los Angeles, CA (2008)

5. Chung, H., Ojeda, L., Borenstein, J.: Accurate mobile robot dead-reckoning with a precision-calibrated fiber optic gyroscope. Trans. on Robotics and Automation 17(1), 80–84 (2001)

6. Griffiths, S., et al.: Maximizing miniature aerial vehicles. Robotics & Automation Mag. (2006)

7. Horn, B.K., Schunck, B.G.: Determining optical flow. Artificial Intelligence 17, 185–203 (1982)

8. Kelly, A.: Mobile robot localization from large-scale appearance mosaics. International Journal of Robotics Research 19(11), 1104–1125 (2000)

9. Kim, S., Lee, S.: Robust mobile robot velocity estimation using redundant number of optical mice, pp. 107–112 (2008), doi:10.1109/ICINFA.2008.4607977

10. Kuipers, J.B.: Quaternions and Rotations Sequences. Princeton University Press, Princeton (1999)

11. Maimone, M., Cheng, Y., Matthies, L.: Two years of visual odometry on the Mars Exploration Rovers: Field reports. J. Field Robot. 24(3), 169–186 (2007)

12. Nourani-Vatani, N., Roberts, J., Srinivasan, M.V.: IMU aided 3D visual odometry for car-like vehicles, Canberra, Australia (2008)

13. NovAtel Inc.: HG 1700 SPAN-58 Specifications (2008),
http://www.novatel.com/Documents/Papers/HG1700_SPAN58.pdf

14. Owens, R.L.: Optical mouse technology (2006),
http://web.archive.org/web/20080112045313/,
http://www.mstarmetro.net/~rlowens/OpticalMouse/

15. Palacin, J., Valgaon, I., Pernia, R.: The optical mouse for indoor mobile robot odometry measurement. Sensors and Actuators A: Physical 126(1), 141–147 (2006)

16. Reina, G., Ojeda, L., Milella, A., Borenstein, J.: Wheel slippage and sinkage detection for planetary rovers. IEEE/ASME Transactions on Mechatronics 11(2), 185–195 (2006)

17. Saripalli, S., Sukhatme, G.S.: Landing on a moving target using an autonomous helicopter. Field and Service Robotics, 277–286 (2003)

18. Sekimori, D., Miyazaki, F.: Precise dead-reckoning for mobile robots using multiple optical mouse sensors. Informatics in Control, Automation and Robotics II, 145–151 (2007)

19. Sorensen, D.: Online optical flow feedback for mobile robot localization/navigation. Master's thesis, Texas A&M University (2003)

20. Wettergreen, D., et al.: Design and experimentation of a rover concept for lunar crater resource survey. In: AIAA Aerospace Sciences, Orlando, FL (2009)

21. Zufferey, J.C., Floreano, D.: Fly-inspired Visual Steering of an Ultralight Indoor Aircraft. IEEE Transactions on Robotics 22(1), 137–146 (2006), doi:10.1109/TRO.2005.858857

Place Recognition Using Regional Point Descriptors for 3D Mapping

Michael Bosse and Robert Zlot

Abstract. In order to operate in unstructured outdoor environments, globally consistent 3D maps are often required. In the absence of a absolute position sensor such as GPS or modifications to the environment, the ability to recognize previously observed locations is necessary to identify loop closures. Regional point or keypoint descriptors are a way to encode the structure within a small local region as a fixed-sized vector, though individually do not include enough context to fully identify a previously seen place. Multiple queries to a database of descriptor vectors can quickly identify similar features, and places can be recognized from a consistent set of descriptor matches. We investigate the problem of designing informative keypoint descriptors for 3D laser maps. Several models are considered and evaluated, with a particular focus on the optimal descriptor scale and keypoint sampling density. The approach is evaluated on 3D laser point cloud data collected from a vehicle driving in unstructured off-road environments. Consistent 3D maps constructed from this data without assistance from any other sensor (such as wheel encoders, GPS, or IMU) demonstrate the effectiveness of our approach.

1 Introduction

Building globally consistent 3D maps is an essential requirement for many autonomous systems operating in unstructured, off-road environments, including earth-moving, mining, construction, and agriculture. In previous work [3], we introduce an approach for incrementally estimating the 6 DoF trajectory of a vehicle using 3D measurements taken from a continuously spinning 2D laser mounted on a vehicle (Figure 1). The resulting trajectory can be used to construct 3D maps of the environment; however these maps are only locally accurate. In order to correct global errors, constraints can be introduced into the map structure if place recognition events can be reliably detected. In this paper, we investigate the place recognition problem for

Michael Bosse and Robert Zlot
Autonomous Systems Laboratory, CSIRO ICT Centre, Brisbane, Australia
e-mail: Mike.Bosse@csiro.au, Robert.Zlot@csiro.au

A. Howard et al. (Eds.): Field and Service Robotics 7, STAR 62, pp. 195–204.

(a) Bobcat S185   **(b)** Toyota Prado 4WD

Fig. 1 The spinning laser is mounted on a variety of ground vehicles for data collection.

producing globally consistent 3D laser point cloud maps given locally consistent 3D maps as input.

For our purposes, the place recognition problem can be stated as follows: Given a 3D point cloud map of a small local region or *place*, determine whether or not this place has been previously observed, and if so, find the relative coordinate transformation aligning the matching places. These relative alignments can be used as constraints for building a global map. Place recognition arises in several applications including loop closure detection in SLAM, global localization in a known map, and fusing several maps collected over time or simultaneously on multiple vehicles.

Although point cloud maps are metric by nature, the place recognition process enforces topological constraints which correct the accumulation of registration errors over time. Typically, point cloud maps are constructed using incremental methods for aligning consecutive observations (*e.g.*, Iterated Closest Point (ICP)), which require a reasonable prior on the alignment transformation. Therefore, the ability of a place recognition algorithm to not only detect a match, but to estimate a rough alignment, is also critical.

Several general approaches have been explored for 3D place recognition, mainly in the context of loop detection for SLAM. One type of solution performs pairwise comparisons between the current place and all places within a region of uncertainty around the estimated robot pose [2, 5, 9]. However, as noted by Newman *et al.* [8], this approach can fail in some situations where the uncertainty region does not contain the matching place. In addition, at large scales, uncertainty can become so large that the exhaustive pairwise search cannot be performed in a reasonable time [12]. Silver *et al.* [11] describe a system for topologically mapping underground mines, where the tunnel structure allows the search for loop closures to be limited to intersections, which can be detected using 2D laser scans. Ryde and Hu [10] use a coarse-to-fine occupancy grid matching algorithm which initially samples the environment uniformly for match candidates, then further considers only the top matches when refining at a higher resolution. Our approach uses a global database of regional descriptors or *keypoints* extracted from all previously observed places. Place recognition is accomplished by finding those places which have a significant number of keypoints in common with the query place. As the database can be searched in sublinear time with respect to the number of keypoints, this approach can be implemented efficiently and independent of current pose uncertainty or loop scale.

The use of regional shape descriptors is also a common approach for object recognition (*e.g.*, [6, 7]). However, for the object recognition problem, descriptors from a scene are typically matched against a set of candidate objects in a predefined library to detect the existence of a particular type and variant of an object. In addition, when attempting to recognize places, it is not always obvious how to partition the environment, so boundary effects can occur if the segments are not in precisely in the same location. While in object recognition, occlusion is an important consideration, in place recognition changes in the environment—such as parked cars changing location, changing vegetation, movable furniture, doors, people, *etc.*—are also important factors.

This paper contributes an evaluative study of regional shape descriptor models for outdoor place recognition. In addition to a comparison of several models, we also investigate the optimal descriptor scale and keypoint sampling density, which are often arbitrarily or unsystematically chosen. We extend a descriptor model first described for use in two dimensional maps [1] for use in three dimensional maps. This extension is not trivial, as we must be able to consistently identify three degrees of rotational freedom in the descriptor coordinate frame, as opposed to one rotational DoF in 2D maps. Existing approaches tend to fix two of the three dimensions, and make the descriptors rotationally invariant to the third, at a performance cost [6, 7]. The regional keypoint descriptors are used in a nearest neighbor voting scheme to identify and recognize previously observed places. The ability of our approach to autonomously generate registered outdoor maps is demonstrated in challenging unstructured environments.

The remainder of the paper is organized as follows. In Section 2 we describe our framework and compare several keypoint selection strategies and descriptor models. We present mapping results generated continuously from a spinning laser range sensor in both industrial and off-road environments in Section 3. Finally, in Section 4, we present our conclusions and future work.

2   Approach

In this section, we compare several keypoint selection heuristics and descriptor models in terms of their suitability and effectiveness in our place recognition framework. We partition our maps by considering the data from each five-second segment of vehicle trajectory to be a "place". For each place, we extract a set of keypoints, compute their descriptors, and then query a global keypoint database for each new keypoint's ten nearest neighbors[1]. Each returned neighbor votes for the local map that generated it in a positional voting scheme. The local maps with the highest vote scores, if over a threshold, are considered as candidate place matches. The

---

[1] We use the ANN *kd*-tree implementation to find approximate ($\varepsilon = 1$) nearest neighbors: http://www.cs.umd.edu/~mount/ANN

matches can be subsequently verified using a geometric consistency check among the matched keypoints, followed by a global graph optimization step.

Raw descriptor vectors are initially processed in order to embed them in a Euclidean space and reduce the number of dimensions. The embedding occurs in two steps, and requires training data containing pairs of matched keypoints from different passes through the environment. First, each individual dimension is transformed by applying the empirical cumulative distribution function to the values, resulting in a 1D uniform distribution for that dimension. Second, the resulting descriptor is linearly transformed to maximize the separability between matching and non-matching pairs. Dimension reduction is then used to prevent overfitting to noise in addition to improving the computation efficiency of nearest neighbor searches. A detailed description of this descriptor processing technique is available in a previous publication [1].

The first step in representing a place by a set of keypoints is to select the locations about which to build descriptors. To ensure a high place recognition rate, keypoints should be selected such that the likelihood of selecting keypoints in a similar locations on a subsequent pass is high. The simplest selectors either sample randomly from the point cloud or distribute the keypoints in space. With an intelligent selection scheme there is the potential to use much fewer points as compared to more simplistic sampling approaches, while still ensuring they are stable and salient. Additionally, using too many points will unnecessarily increase the computation and memory requirements of the system. However, we found little advantage in any of the intelligent selection heuristics we experimented with, and intend to further study this problem in future work.

Our selection heuristic therefore simply subsamples the original point cloud such that no two points are within a minimum distance, effectively limiting the maximum density to a predetermined value. For computing the orientation at a keypoint, we first determine the first- and second-order moments of the point cloud within a fixed distance of the keypoint. By taking the eigenvectors of the covariance matrix, we are able to determine local coordinate axes, up to a sign ambiguity. To resolve this ambiguity, we ensure that the z-axis has a positive component in the local "up" direction, and the x-axis has a positive dot product with the direction to the center of mass.

2.1 Keypoint Description

Typically, 3D keypoint descriptors divide the space around the keypoint into a set of bins, then compute statistics on the points that fall into each bin to produce a descriptor vector. We consider several descriptor types, some of which are taken from the existing literature while others are of novel design. We consider three key properties as design parameters for any descriptor:

Shape.   The main defining characteristic of a keypoint descriptor is the arrangement and number of the bins around the keypoint. There is a trade-off between the number of bins and the potential descriptiveness of the feature. When there are too many bins, many of the bins will be undersampled, thereby introducing non-Gaussian quantization errors to the descriptors, which are difficult to model.

Scale. There is a trade-off in determining the size of the local support region around a keypoint from which to build the descriptor. Larger scales have more context from which to make a salient descriptor, while smaller scales are not as sensitive to boundary effects, missing data, and transient objects. However, computing larger-scale keypoints require significantly more computation.

Statistics. The statistics gathered for each bin are also important in defining the information contained in the descriptor vector. Many existing descriptor models [6, 7] include only a count of the number of points in each bin; however, higher-order moments [12] or even complete orientation histograms [4] are also possible. Care must be taken in the design of the number and shape of the bins to ensure that enough points fall into each bin in order for the higher order statistics to be reliable.

For the experiments presented in this paper, we evaluate the effect of varying the keypoint scale for three main descriptor shape types: spin images, shape contexts, and moment grids. As the descriptors have multiple parameters related to their shape (*e.g.*, number and size of bins), determining the optimal shapes is a difficult high-dimensional optimization problem. The parameters used for each descriptor are based upon a non-exhaustive empirical study in which a sampling of reasonable configurations was explored.

2.1.1 Base Dimensions

To each descriptor model, we include seven rotationally invariant shape-related statistics derived from the eigenvalues of the covariance matrix of the local point cloud. These include the three eigenvalues (where $\lambda_1 > \lambda_2 > \lambda_3$), the planarity of the region $p = 2(\lambda_2 - \lambda_3)/s$, the cylindrical-ness of the region $c = (\lambda_1 - \lambda_2)/s$, as well as $p + c$, and $p - c$, where $s = \lambda_1 + \lambda_2 + \lambda_3$. When used on its own as a descriptor, the seven base dimensions are reduced to three dimensions by the preprocessing algorithm.

2.1.2 Spin and Shell Images

Spin images [7] are cylindrical descriptors divided into bins both radially and along the cylinder axis. The resulting support region therefore consists of a set of 3D annular bins. The descriptor vector consists of a count of the number of points in each of these bins. For the implementation used in this paper, we choose 10 radial and 10 height divisions yielding (with the seven base dimensions) a 107-dimensional descriptor vectors that typically (depending on the scale) reduce down to 9-dimensions after preprocessing.

Shell images are a generalization of spin images that are rotationally invariant as the bins are defined as spherical shells about the keypoint. We choose 20 quadratically spaced radial divisions which after preprocessing typically results in a 5-dimensional feature vector.

2.1.3   3D Shape Contexts

The support region of 3D shape contexts [6] are spheres, with the bins arranged by dividing space quadratically along the radius, and linearly in the azimuth and elevation directions. As with spin images, the descriptor vector simply maintains a count of the number of points in each bin. In contrast to the original authors' approach, we do not normalize either the shape context or spin image by volume or density, as these effects are accounted for in our descriptor normalization procedure. Since the average point density is much lower in our data set as compared to the original authors', we choose 5 radius, 8 azimuth, and 4 elevation divisions. The resulting 167-dimensional descriptors is typically reduced to 18 dimensions after preprocessing. We also do not need to repeat the descriptor for every rotation of the azimuth bins since we can reliably compute all 3 rotational degrees of freedom of the keypoint frame.

2.1.4   Moment Grids

3D moment grids use a rectilinear voxelization of the space around the keypoint. They are extended from a 2D descriptor previously used for two-dimensional place recognition [12]. For each cube-shaped bin, we compute moments up to second-order, for a total of ten per voxel to include in the descriptor vector. We use $2 \times 2 \times 2$ and $3 \times 3 \times 3$ grids, as well as a $2 \times 2 \times 2$ cylindrical version (radius, azimuth, height). Including the base dimensions, the sizes of these descriptors are 87, 277, and 87 dimensions respectively, but are typically reduced by the processing algorithm to 16, 20, and 12 dimensions.

3   Results

In order to assess ideal keypoint scales and densities, a registered map is required for training and testing purposes. An earlier untuned version of this approach was used with a separate training dataset to create a globally consistent point cloud and trajectory from which matching keypoint pairs could be extracted. The training data was collected in both off-road and industrial environments. To generate the match set, the dataset is evenly sampled at 0.5 m resolution and matching keypoint pairs are identified as those that are within 0.5 m and 15 degrees, and observed more than 30 seconds apart. A set of unmatched pairs is generated by selecting keypoints at random. The retrieval rates are measured using half of the matched and unmatched pairs as training, with the remainder as a validation set in a repeated random sub-sampling cross-validation scheme.

3.1   Keypoint Scale

Our first aim is to determine an appropriate scale for the region encoded by each keypoint descriptor. For each descriptor type, we generate keypoints at a variety of

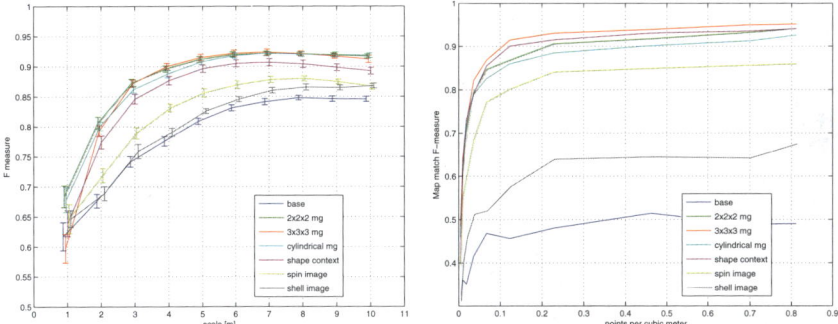

(a) Keypoint scale vs keypoint match accu- (b) Keypoint density vs map match accuracy
racy

Fig. 2 Results from experiments examining the F-measure at various keypoint scales and densities. Each datapoint in (a) represents an average over 30 trials, with 3σ error bars. Note that (a) measures accuracy relative to keypoint retrieval, whereas (b) is with respect to overall map match detections.

scales from 1 m to 10 m and measure the retrieval accuracy over the known matches. Note that the keypoint orientations are dependent on the scale, so must be computed separately for each experiment. To quantify the accuracy at each descriptor scale, we measure the precision and recall rates to compute the F-measure, which is defined as the harmonic mean of precision and recall.

Figure 2a plots the F-measure versus the scale for each descriptor type considered. As expected, we observe an improvement in F-measure as the scale increases; however, after about 7 m the improvement tapers off for all descriptor types. We interpret this result as demonstrating a general improvement as more points are contained within the keypoint region; eventually, however, sensitivity to boundary effects and misregistrations decrease the utility of the descriptors. We also note that the three moment grid descriptors perform best, though the number of bins and their arrangement seem to be less important. This suggests that the inclusion of higher-order moments adds meaningful information, as long as the bins contain a sufficient number of support points for meaningful statistics to be computed. The 3D shape context descriptor performs nearly as well as the moment grid descriptors.

3.2 Keypoint Density

Given an appropriate descriptor scale, we now evaluate the overall map matching performance resulting from the nearest neighbor voting procedure described in Sections 1 and 2. For the purposes of the voting, we aggregate the votes over five-second intervals, with the associated place represented by the trajectory point at the center of that interval. The choice of five seconds is made to provide sufficient time to observe enough context around the vehicle without being overly influenced by drift accrued in the local scan-matching. For every query place, we retain the twenty

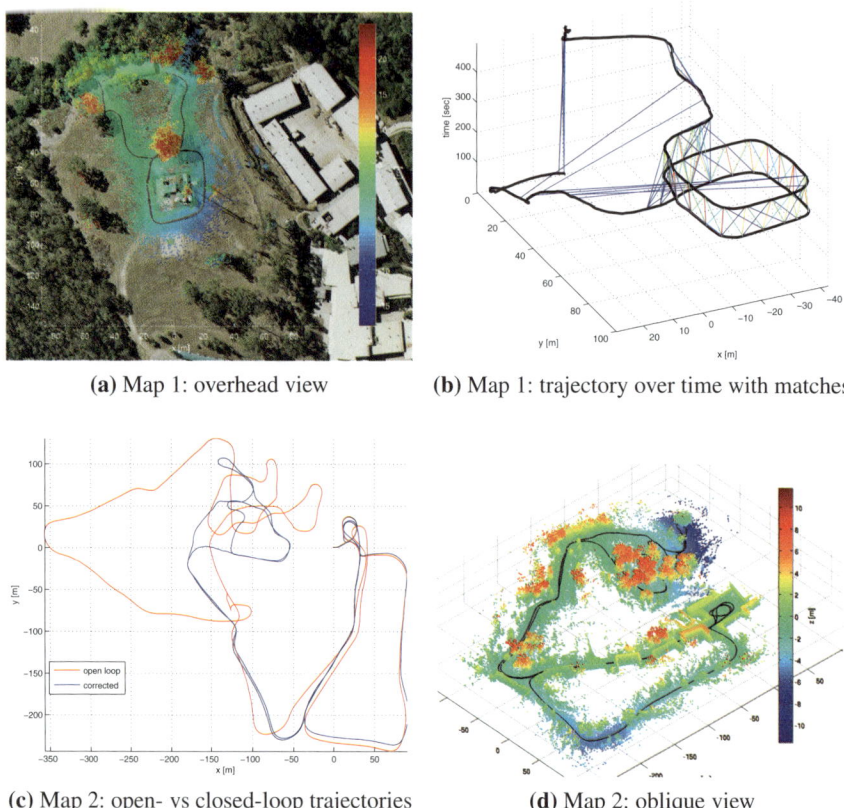

(a) Map 1: overhead view   **(b)** Map 1: trajectory over time with matches

(c) Map 2: open- vs closed-loop trajectories   **(d)** Map 2: oblique view

Fig. 3 Two maps constructing using the keypoint voting scheme. (a) Map of a lightly wooded off-road environment overlaid on a Google Maps image. (b) The trajectory of map 1 with time indicated on the vertical axis. The links between places correspond to matches that pass the vote score threshold and geometric consistency checks. Color indicates the strength of the score after the consistency check (blue low, red high). The loop optimization step re-weights the edges and is robust to the incorrect matches. (c) The open-loop and closed-loop (after loop closures) trajectories from a second map. (d) Oblique view of the second map with color indicating terrain height.

places with the highest vote scores. For each potential map match pair, we also perform a rigidity test on the keypoint matches and reduce the vote score by removing any pairs not geometrically consistent with the largest cluster of keypoint transformations (RANSAC could also have been used for this step). Any scores higher than a threshold are then taken as a map match. Note that additional consistency checks (*e.g.*, ICP) could be made to further invalidate false positives.

In this set of experiments, the map matching performance is evaluated for different keypoint densities for all the descriptor types and a fixed scale of 7 m. The map matching F-measure is defined similarly to the keypoint match F-measure using the precision and recall rates resulting from thresholding the vote scores. Note that when determining ground truth for map match pairs it is often difficult to decide whether there is enough overlap to warrant a match; therefore we evaluate precision and recall based on the examples where this decision is clear. The true positive map matches are defined as places that are temporally distinct (visited more than 30 seconds apart), spatially near (the trajectories are within 10 m), and which observe points overlapping with (within 0.5 m of) another place in the globally registered point cloud. True negatives are defined as temporally distinct, spatially distant (trajectories are more than 20 m apart), and non-overlapping.

Figure 2b illustrates the map match F-measures for various keypoint densities. The results demonstrate that fairly low keypoint densities (0.1 to 0.3 points per cubic meter) are sufficient to achieve F-measures close to the maximum, as the F-measure curve essentially flattens after this density. Again, the moment grid and 3D shape context descriptors are observed to perform significantly better than the other alternatives tested.

3.3   Mapping Results

Two maps constructed using our approach are illustrated in Figure 3. In both maps, loop closures are detected using the described place recognition framework with 3×3 moment grid descriptors at a scale of 7 m and density of 0.3 per cubic meter. A globally consistent trajectory is computed by a robust minimization of the loop closure constraints (details of which are beyond the scope of this paper), and the 3D point cloud is updated from the corrected trajectory. While we do not have ground truth for these maps, the accuracy can be verified qualitatively by comparison with satellite imagery (as in Figure 3a).

4   Conclusions

A systematic investigation into the design of regional point descriptors has led to a model and associated parameters suitable for 3D place recognition. Though several of the models considered are sufficiently capable of finding loop closures in our data, moment grid descriptors perform best in terms of accuracy and efficiency at this task. While the calculation of higher-order moments for these descriptors requires about 13% more time, this increase in computation cost is not highly significant as keypoint generation can easily be done in real-time. A limitation of our approach is that the descriptors must be calibrated using a registered training set and it can be challenging to obtain a sufficiently large number of true positive examples. However, since the training step can be performed on a separate dataset, the proposed place recognition algorithm can be run online. Future work will focus on developing intelligent keypoint selection heuristics which will reduce computation load while still maintaining a high probability of keypoint detection.

Acknowledgements. The authors thank Lennon Cork, Paul Flick, Fabien Molliner, Julian Ryde, John Whitham and the rest of the CSIRO Autonomous Systems Lab team for their assistance.

References

1. Bosse, M., Zlot, R.: Keypoint design and evaluation for place recognition in 2D lidar maps. In: Robotics: Science and Systems Conference, "Inside Data Association" Workshop (2008)
2. Bosse, M., Zlot, R.: Map matching and data association for large-scale 2D laser scan-based SLAM. International Journal of Robotics Research 27(6), 667–692 (2008)
3. Bosse, M., Zlot, R.: Continuous 3D scan-matching with a spinning 2D laser. In: Proceedings of the IEEE International Conference on Robotics and Automation (2009)
4. Cole, D.M., Harrison, A.R., Newman, P.M.: Using naturally salient regions for SLAM with 3D laser data. In: IEEE International Conference on Robotics and Automation SLAM Workshop (2005)
5. Cole, D.M., Newman, P.M.: Using laser range data for 3D SLAM in outdoor environments. In: Proceedings of the IEEE International Conference on Robotics and Automation (2006)
6. Frome, A., Huber, D., Kolluri, R., Bülow, T., Malik, J.: Recognizing objects in range data using regional point descriptors. In: Pajdla, T., Matas, J(G.) (eds.) ECCV 2004. LNCS, vol. 3023, pp. 224–237. Springer, Heidelberg (2004)
7. Johnson, A., Hebert, M.: Using spin images for efficient object recognition in cluttered 3D scenes. IEEE Transactions on Pattern Analysis and Machine Intelligence 21, 433–449 (1999)
8. Newman, P., Cole, D., Ho, K.: Outdoor SLAM using Visual Appearance and Laser Ranging. In: Proceedings of the IEEE International Conference on Robotics and Automation, Florida (2006)
9. Nüchter, A., Lingemann, K., Hertzberg, J., Surmann, H.: 6D SLAM—3D mapping outdoor environments. Journal of Field Robotics 24(8/9), 699–722 (2007)
10. Ryde, J., Hu, H.: Mobile robot 3D perception and mapping with multi-resolution occupancy lists. In: Proceedings of the IEEE International Conference on Mechatronics and Automation (2007)
11. Silver, D., Ferguson, D., Morris, A., Thayer, S.: Topological exploration of subterranean environments. J. of Field Robotics Special Issue on Field and Service Robotics 23(6/7) (June/July 2006)
12. Zlot, R., Bosse, M.: Place recognition using keypoint similarities in 2D lidar maps. In: International Symposium on Experimental Robotics (2008)

Part V
Mapping

Scan-Point Planning and 3-D Map Building for a 3-D Laser Range Scanner in an Outdoor Environment

Keiji Nagatani, Takayuki Matsuzawa, and Kazuya Yoshida

Abstract. During search missions in disaster environments, an important task for mobile robots is map building. An advantage of three-dimensional (3-D) mapping is that it can provide depictions of disaster environments that will support robotic teleoperations used in locating victims and aid rescue crews in strategizing. However, the 3-D scanning of an environment is time-consuming because a 3-D scanning procedure itself takes a time and scan data must be matched at several locations. Therefore, in this paper, we propose a scan-point planning algorithm to obtain a large scale 3-D map, and we apply a scan-matching method to improve the accuracy of the map. We discuss the use of scan-point planning to maintain the resolution of sensor data and to minimize occlusion areas. The scan-matching method is based on a combination of the Iterative Closest Point (ICP) algorithm and the Normal Distribution Transform (NDT) algorithm. We performed several experiments to verify the validity of our approach.

Keywords: Search and Rescue, Scan points planning.

1 Introduction

Recently, requests for the development of robotic systems for search-and-rescue operations have been increasing rapidly. After the Hanshin-Awaji earthquake in 1995 (Japan) and the World Trade Center attack in 2001 (U.S.A.), large research projects for search and rescue were kicked off in Japan.

One of the important tasks for mobile robots in search-and-rescue missions is map building. Three-dimensional (3-D) mapping is used to provide representations of disaster environments that will support robotic teleoperations used in locating victims and aid rescue crews in strategizing. To realize 3-D mapping in disaster environments, we constructed the 3-D scanner device shown in Figure 1-(a). It consists of a laser range finder (SICK), rotation

Keiji Nagatani, Takayuki Matsuzawa, and Kazuya Yoshida
Tohoku University

A. Howard et al. (Eds.): Field and Service Robotics 7, STAR 62, pp. 207–217.
springerlink.com

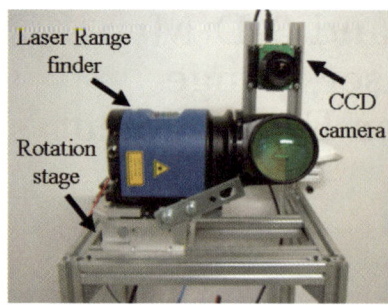 

(a) Developed sensor device   (b) Acquisition example

Fig. 1 Development of a sensor system and acquisition example of an environment

stage, and CCD camera to obtain information in color. It takes one minute to scan an environment at one place, and the device generates 3-D information to create a remote display of the target environment (e.g., Figure 1-(b)), which helps the rescuers get a complete picture.

Search-and-rescue missions require rapid, accurate mapping. However, to map a large-scale environment, rescue robots must obtain 3-D information in different locations, and the scan data obtained at different locations should be merged to represent a large-scale environment. In such cases, reasonable scan-point planning is very important; however, measurements in different locations generate gaps of objects in the map caused by positioning errors at each scan point. Therefore, we propose a scan-point planning algorithm to obtain a large scale 3-D map, and we have applied a scan-matching method to improve map accuracy. In the scan-point planning algorithm, the resolution of sensor data and occlusion area allow as much of a target area as possible to be covered. The scan-matching method is based on the Iterative Closest Point (ICP) algorithm [1] and the Normal Distribution Transform (NDT) algorithm [2]. In this paper, we discuss the construction of 3-D maps in outdoor environments, and we report the results of the environment mapping of our campus buildings.

2 Related Works

Two major approaches are used to obtain a 3-D map. One, stereo matching, involves the use of two or more cameras (e.g., [3]), while the other involves a 3-D laser range scanner (e.g., [4]). The merits of the former approach are fast measurement and simultaneous acquisition of texture information. However, there are some disadvantages: (1) stereo matching requires brightness and feature information, (2) its measurement area is narrow, and (3) distance accuracy decreases as the distance from the objects increases. Therefore, in this study, we use a 3-D laser range scanner with a CCD camera to produce the 3-D color map shown in Figure 1.

In regard to scan-point planning, a classic problem is the Art Gallery Problem [5], which involves the number of guards required to monitor the target floor completely. Basically, the problem assumes that the shape of the floor is known and the target environment is 2-dimensional (2-D). Recently, some novel researches for view point planning were performed in 3-D. Sequeira et al. proposed a view planning method for automatic acquisition of environment information with consideration of sampling density [6]. In our scan-point planning, we propose a scan-point planning method that combines frontier-based navigation [7] and the Art Gallery Problem.

Recently, scan matching has been used for the adjustment of the relative position of the scan data and particularly for simultaneous localization and mapping (SLAM) in mobile robot navigation[8]. In this research, we apply the conventional Iterative Closest Points (ICP) algorithm [1] and the Normal Distribution Transform (NDT) algorithm [2] to our scan data. Because both algorithm have advantages and disadvantages, we have combined them to obtain a robust and accurate 3-D map.

3   Scan-Point Planning

To obtain a large scale environment by repeated 3-D scanning, we have established the following procedures:

1. Conducting a 3-D scan
2. Representing the scan information on a Multi-Level Surface map (MLS-map)
3. Determining the movable region
4. Planning the next scan point in the region
5. Moving the 3-D scanner to the designated point
6. Repeating steps 1 through 5 until the target region is completely covered

3.1   *Representation of Scan Information*

To perform scan-point planning in large-scale environments, point cloud representation is unsuitable because of the massive volume of data and the non-constant density. Therefore, at the beginning of this study, we applied a Digital Elevation Map (DEM) to represent target environments [9] for scan-point planning. In this method, each scan point is registered into one cell of the lattice domain on a 2 dimensional (2-D) x-y plane as height information from a base level. The DEM advantageously represents a non-flat environment. However, the method can not be used to represent spaces under objects because only the highest scan point is effective. An example of point cloud representation is shown in Figure 2-(a). It is a side view of a four-story building (left) and a tree (right). A DEM representation of Figure 2-(a) is shown in Figure 2-(b) from the same viewpoint. A space under the tree is not visible

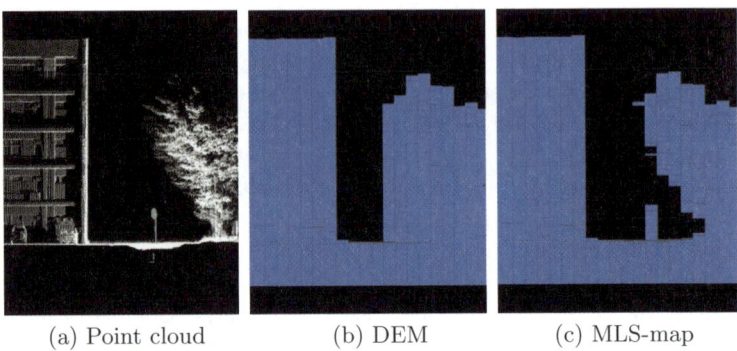

(a) Point cloud   (b) DEM   (c) MLS-map

Fig. 2 Comparison of DEM and MLS-map

in the DEM representation, which is undesirable for the determination of the movable region of mobile robots.

A Multi-Level Surface map (MLS-map) [10] is one solution to the representation of such an environment. In this method, some edge-point positions of objects are stored in each cell of the lattice domain on a 2-D x-y plane. Figure 3 shows a concept of the MLS-map. Information about two objects is stored in one cell, which represents a space between the objects. Based on the above method, the point cloud representation in Figure 2-(a) is represented by the MLS-map in Figure 2-(c). The space under the tree is now visible.

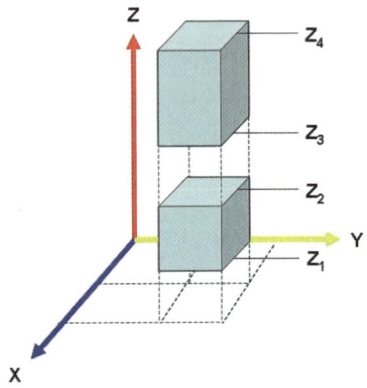

Fig. 3 Introduction to MLS map

3.2 Determination of Movable Region

For mobile robots, the path between the current scan point and the next scan point must be connected. Therefore, we define a movable region as an area (1) which is connected to a current scan point and (2) whose differential height between adjacent cells in the MLS-map is smaller than the threshold. Of course, the movable region depends on the mobility of the target mobile robot. In our implementation, the threshold was set at 0.15 [m].

3.3 Region Segmentation

To maintain the resolution of scan data, we divide the target region into three regions, (1) scan-completed region, (2) low-resolution region, and

(3) unscanned region. In reality, map resolution depends on not only the scan range
but also the orientation; however, to simplify the planning of the next scan points,
we use the following definitions: (1) The
scan-completed region (C) is a set of cells
in the target region which has scan data
and whose distance from the closest scan
point is less than a fixed value. In our
implementation, the distance value is set
at 20 [m]. (2) The low-resolution region
(M) is a set of cells in the target region
which has scan data but does not belong
to the scan-completed region. (3) The un-

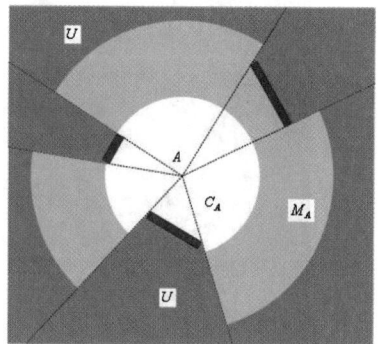

Fig. 4 Region segmentation

scanned region (U) is a region which does not belong to (1) and (2) because
a cell in the region is too far from the scan points or is occluded by objects.

Figure 4 shows an example of the region segmentation of an initial scan
from scan point A. The white region is the scan-completed region(C_A) from
point A, the gray region is the low-resolution region (M_A), and the blue
rectangles are obstacles. The dark-gray region is the unscanned region (U),
where the blue rectangles occlude or are far from point A.

3.4 Evaluation Function for Next Scan Point

To minimize the scan procedure while maintaining the resolution of the scan
data, the unscanned region or the low-resolution region should be changed,
as much as possible, into a scan-completed region by the next scan. Two
different indices are included, therefore, we defined an evaluation function as
follows:

$$F_X = area((U \cap C_B) \cup (U \cap M_B)) \cdot \alpha +$$
$$area(M_A \cap C_B) \cdot (1 - \alpha) \qquad (1)$$

where $area(\cdot)$ is an area value in the bracket, $C_(B)$ is a prospective area
which becomes a scan-completed region when the next scan is conducted at
point B, $M_(B)$ is a prospective area which becomes a low-resolution region
when the next scan is conducted at point B, and α is the weight value. If
exploration in the unscanned region is not important, the α should be very
small or zero.

The calculation of Equation (1) at one cell in the movable region requires
a ray-tracing scan in 3-D virtual space, which is time-consuming. Therefore,
we applied a hill-climbing search from several randomized initial locations.

Figure 5 shows a top view of an example of a planned result in an MLS-
map in the case of an initial scan. In this figure, the blue circle depicts an
initial scan point, the red circle is the planned scan point in the case of

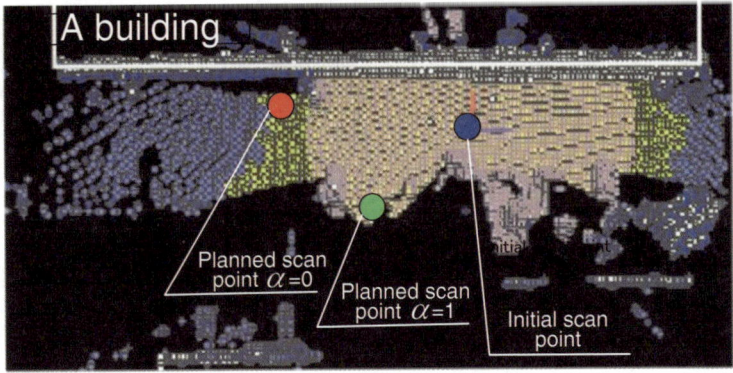

Fig. 5 Top view of an example of target region

(a) A view from the initial   (b) A view from the next   (c) A view from the next
    scan point   scan point ($\alpha = 0.0$)   scan point ($\alpha = 1.0$)

Fig. 6 An example of a scan point

$\alpha = 1.0$, the green circle is the planned scan point in the case of $alpha = 0.0$, the flesh-colored region is the scan-completed area, the yellow region is a low-resolution movable area, and the blue region is a low-resolution but not movable area. Figure 6 shows virtual views, (a) is a view from the initial scan point corresponding to the blue circle in Figure 5, (b) is a view from the planned scan point in the case of $\alpha = 0.0$ corresponding to the red circle in Figure 5, and (c) is a view from the planned scan point in the case of $\alpha = 1.0$ corresponding to the green circle in Figure 5. In the virtual views, scanning in unknown area has priority in the case of $\alpha = 1.0$.

3.5 Examples of Planning Results in Outdoor Environments

We tried the above scan-point planning method in several different outdoor environments. The motion of the 3-D scan unit to the next scan position was performed by a human operator, not by a mobile robot platform. Termination judgement was made by the operator, who checked the coverage region. Due to space limitations, only one experimental result will be discussed here.

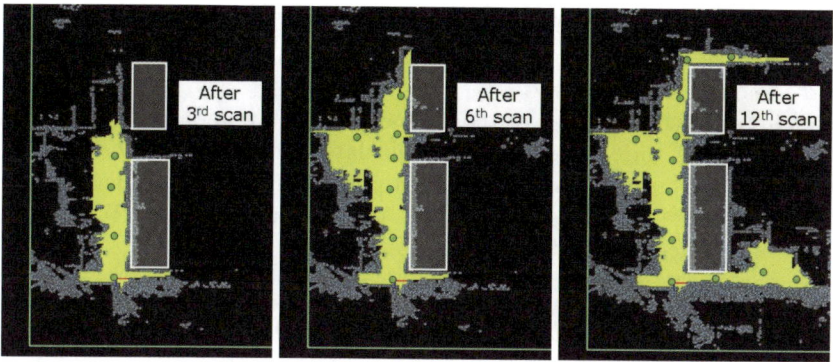

Fig. 7 A result of scan-point planning in an outdoor environment (our campus)

The target environment is our campus (Mechanical Department, Aoba-Yama, Tohoku University, Japan), which includes buildings, trees, and parking slots. The weighting factor of α is set at 0.5, and the cell size of the MLS-map is set at 1.0 [m] square. Figure 7 shows a transition of the scan position and scan area in the first environment. In the 13 scan and movement motions, the detection area (the map itself) was expanded step by step. The total detection area is about 16,500 grids (that is equal to square meters). The result shows that our proposed algorithm worked well for the large-scale outdoor environment.

4 Scan Matching for Improved Map Accuracy

In outdoor environments, global positioning systems (GPS), electromagnetic sensors, and inclination sensors can be used to obtain a scan point. However, because of sensing errors or noise, each scan-point location may include a positioning error. To compensate for such errors, scan matching is used to adjust the location and to construct an accurate map, particularly in simultaneous localization and mapping (SLAM). For scan matching, we apply the Iterative Closest Point (ICP) algorithm [1] and the Normal Distribution Transform (NDT) algorithm [2].

4.1 Summary of the ICP Algorithm

In the ICP algorithm, two given point sets are registered in Cartesian coordinates. In each iteration step, the algorithm selects the closest points as correspondences and calculates the rotation matrix $\boldsymbol{R}$ and the translation matrix $\boldsymbol{t}$ to minimize the following equation:

$$E(\boldsymbol{R}, \boldsymbol{t}) = \sum_{i=1}^{N_m} \sum_{j=1}^{N_d} \omega_{i,j} \left\| \boldsymbol{m}_i - (\boldsymbol{R}\boldsymbol{d}_j + \boldsymbol{t}) \right\|^2 \tag{2}$$

where N_m and N_d are the number of points in the reference data set M and the matching data set D respectively. $\omega_{i,j} = 1$ when m_i is the closest point to d_j, and $\omega_{i,j} = 0$ otherwize.

To improve the accuracy of the ICP algorithm, we applied the ICP algorithm of points and segments, which calculates, not a distance between the closest points, but a distance between a point in the reference data set and a segment between two closest points in the matching data set. We call this algorithm a line-segment ICP algorithm.

4.2 Summary of the NDT Algorithm

In the NDT algorithm, a target space is divided into grids. Then the distribution of scan points in the reference data set in one grid is represented by a normal distribution. An average in the grid i is represented by q_i, and a covariance matrix in the grid is represented by Σ_i. Based on the above data, an evaluation function is defined as the sum of the matching level between a point x_i' in the matching data set and the normal distribution i, which corresponds to point x_i', as follows:

$$E(p) = \sum_{i}^{N-1} exp\frac{-(x_i' - q_i)^t \Sigma_i^{-1}(x_i' - q_i)}{2} \qquad (3)$$

Detailed explanations and equations are given in [2].

The accuracy of the result depends greatly on the size of each grid. To improve the robustness of the NDT algorithm, we applied an algorithm in which the size of each grid is dynamically changed. At first, grid size is large for global matching. After that, the matching sequence is repeated with progressively smaller grid sizes. In our implementation, the initial size of the grid is 20 [m], the second size is 15 [m], and the final size is 10 [m]. We call this algorithm the Narrower NDT algorithm.

4.3 Combination of the ICP and NDT Algorithms

To construct an accurate 3-D map of an outdoor environment, we applied the above scan-matching algorithms to our experimental results, as shown in section 3.5. We mounted an inclination sensor on the 3-D scanner, so the adjustment parameters in this scan matching are the position (x, y, z) and orientation θ of an obtained environment.

Through the above application experience, we identified the following comparative qualitative features: (1) The ICP algorithm is more accurate than the NDT algorithm when the matching is successful. (2) The ICP algorithm becomes stuck in the local minima much more easily than the NDT algorithm does. Based on the above features, to pursue both accuracy and robustness for scan matching, we propose a combination of the two algorithms, or the

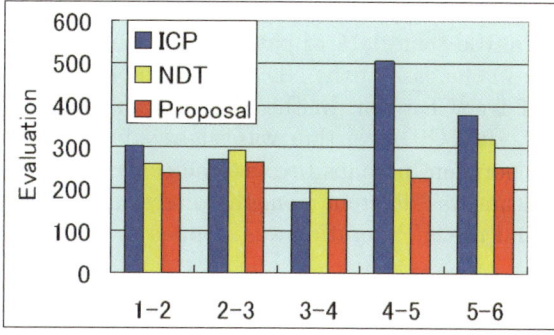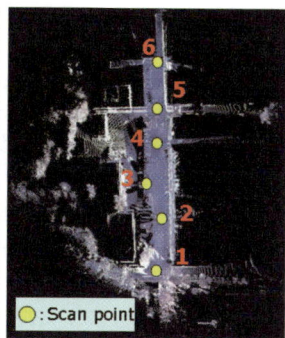

Fig. 8 A comparison of matching algorithms (our campus)

NDT-ICP Combination. First, the Narrower NDT algorithm is applied, and then the line-segment ICP algorithm is applied.

Figure 8 shows a graph comparing the above three matching algorithms (left) and scan locations (right). Table 1 shows the numerical evaluation values of the same result. A "number-number" represents the scan location indices for matching; for example, "2-3" represents a match between scan point 2 and scan point 3. Each evaluation value is the ICP score calculated by Equation 2. The smaller of the evaluation values denotes better matching.

Based on the comparison results, the NDT-ICP Combination algorithm works well. In the case of scan matching between scan points 4 and 5, the solo ICP algorithm failed to

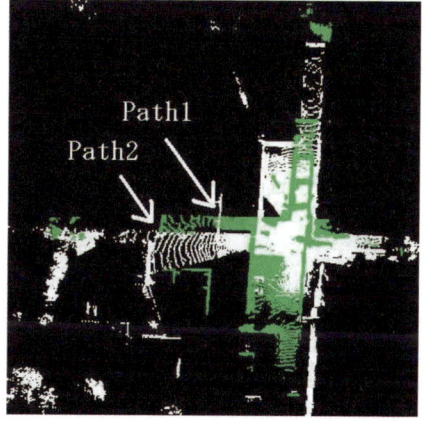

Fig. 9 Matching result "4-5" by solo ICP algorithm

match. The failed example is shown in detail in Figure 9. White dots are scan data obtained at scan point 4, and green dots are those obtained at scan point 5, where the scan points of path 1 appeared. The scan data of

Table 1 Scan Matching Comparison : Proposal

	1-2	2-3	3-4	4-5	5-6
Line-segment ICP	301.0	269.2	169.2	503.9	378.4
Narrower NDT	258.7	292.1	201.9	245.6	317.7
Proposal (NDT-ICP Comb.)	236.1	262.7	173.9	226.5	254.3

path 2, obtained at scan point 4, disappeared. However, the scan data of path 1 seemed to be pulled by the scan data of path 2. Finally, matching failed, as shown in the figure. In the case of the NDT-ICP Combination algorithm, such mismatching did not happen. In the case of scan matching between scan points 3 and 4, the ICP algorithm was slightly better than the NDT-ICP Combination algorithm, perhaps because the convergence direction of the solo ICP algorithm was different from the direction after the NDT algorithm was applied and the ICP iteration was stopped in the different situations.

5 Conclusions and Future Work

In this paper, we have proposed a scan-point planning algorithm to obtain a large scale 3-D map efficiently and a combination of scan-matching algorithms (NDT and ICP) to improve mapping accuracy. Finally, we have offered an example of mapping in an outdoor environment to confirm the validity of the above approach. We have also applied the approach to two different environments, a small natural field at Mt. Aosasa in Sendai City and a park on our campus (without large buildings). In both environments, the proposed approach worked well. In the former case, the target environment included small trees, and the solo ICP algorithm became stuck in the local minima for scan matching. These results are not included here due to space limitations.

In our current implementation of scan-point planning, we have not considered the moving cost of mobile robots. Therefore, the next scan point may be far from the current scan point, as happened in the planning procedure shown in this paper. In the future works, it is required to discuss optimality of the viewpoint selection deeply. Furthremore, although we have defined scan resolution simply as the distance to an object, it should be considered in the orientation of the targets. By solving the above problems, we aim to obtain a feasible and accurate 3-D map in an outdoor environment.

References

1. Besl, P.J., McKay, N.D.: A method for registration of 3-d shapes. IEEE Tran. on Pattern Analysis and Machine Intelligence 14(2), 239–256 (1992)
2. Biber, P., Straber, W.: The normal distributions transform: A new approach to laser scan matching. In: IEEE/RSJ International Conference on Intelligent Robots and Systems, pp. 2743–2748 (2003)
3. Ayache, N.: Artificial Vision for Mobile Robots. MIT Press, Cambridge (1991)
4. Ikeuchi, K., Sato, Y.: Modeling from Reality. Kluwer Academic Publishers, Dordrecht (2001)
5. Rourke, J.O.: Art Gallery Theorems and Algorithms. Oxford University Press, Oxford (1987)

6. Klein, K., Sequeira, V.: View planning for the 3d modelling of real world scenes. In: Proc. of IEEE/RSJ International Conference on Intelligent Robots and Systems (2000)
7. Yamauchi, B.: A frontier-based approach for autonomous exploration. In: IEEE International Symposium on Computational Intelligence in Robotics and Automation, pp. 146–151 (1997)
8. Thrun, S., Burgard, W., Fox, D.: Probabilistic Robotics. MIT Press, Cambridge (2005)
9. Fong, E.H.L., Adams, W., Crabbe, F.L., Schultz, A.C.: Representing a 3-d environment with a 2 1/2-d map structure. In: IEEE/RSJ Conference on Intelligent Robots and Systems, pp. 2986–2991 (2003)
10. Triebel, R., Pfaff, P., Burgard, W.: Multi-level surface maps for outdoor terrain mapping and loop closing. In: IEEE/RSJ International Conference on Intelligent Robots and Systems, pp. 2276–2282 (2006)

Image and Sparse Laser Fusion for Dense Scene Reconstruction

Alastair Harrison and Paul Newman

Abstract. This paper is concerned with reconstructing the metric geometry of a scene imaged with a single camera and a scanning laser. Our aim is to assign each image pixel with a range value using both image appearance and sparse laser data. We pose the problem as an optimization of a cost function encapsulating a spatially varying smoothness cost and measurement compatibility. In particular we introduce a second order smoothness term. We derive cues for discontinuities in range from changes in image appearance and reflect this in the objective function. We show that our formulation distills down to solving a large linear system which can be solved swiftly using direct methods. Results are presented and analyzed using synthetic cases to demonstrate salient behaviours and on real data to highlight real-world applicability.

1 Introduction and Motivation

This paper is about dense mapping of workspaces using common place cameras and scanning lasers. Cameras provide near instantaneous capture of the workspace's appearance (texture and colour) but, from a single view, little geometrical information. On the other hand, scanning lasers produce comparatively slow, sparse metric sampling and beyond reflectance, capture little of the scene's appearance. This motivates us to consider how we might fuse sparse laser data and images to infer a range for every pixel in the image, allowing us to reconstruct a 3D scene with all the texture, colour and appearance information captured in the original image. The heart of the problem is how to sensibly infer ranges for pixels which are not near any laser measurements without introducing intolerable distortions. Our method is general in that

Alastair Harrison
University of Oxford, Oxford, OX1 3PJ
e-mail: arh@robots.ox.ac.uk

Paul Newman
University of Oxford, Oxford, OX1 3PJ
e-mail: pnewman@robots.ox.ac.uk

A. Howard et al. (Eds.): Field and Service Robotics 7, STAR 62, pp. 219–228.
springerlink.com © Springer-Verlag Berlin Heidelberg 2010

it is not tied to any particular 3D laser scanner mechanism or geometry. Note also that we aim to recover the dense geometry of a scene over scales which prohibit the use of other direct methods such as stereo unless a truly large baseline is used.

2 Related Work

The problem of inferring 3D surface models of a scene using laser or camera sensors has been studied extensively over many years (see, for example [1, 2, 3, 4]). However, limitations in hardware and a requirement for speedy data gathering in mobile robotics typically results either in high resolution optical images only allowing inference of very basic 3D geometry, or, alternatively, low resolution range images which often sample the scene too sparsely to allow for faithful reconstruction. Multiple view reconstruction provides an attractive alternative due to a near instantaneous gathering of dense 3D data leading to dense scene reconstructions from image data alone [5, 6]. Unfortunately, stereo reconstruction fidelity is limited in range by the baseline and the image resolution. This seriously impedes accurate reconstruction beyond a few meters from the camera. Another alternative can be found in the exploitation of the complementary nature of vision and range sensing. While optical images and range images represent different quantities, they share "similar second order statistics and scaling properties" [7].

Only a relatively small body of work exists on the inference of surfaces by fusing laser data and camera images. Usually, these techniques exploit the fact that edges in the optical image often correspond to discontinuities in depth, and that smooth surfaces tend to correspond to areas of similar colour and texture. In [8], depth values for pixels in an image are inferred using belief propagation in a Markov Random Field (MRF) framework. The technique requires that the supplied range measurements contain some high density areas from which to seed the solution, and is unable to assign depth values outside of those already in the measurements. The techniques described in [9], [10] and [7] are able to fuse the information from both sources to significantly improve the resolution of low quality range images. The method of [9] is particularly relevant to this work. It employs an MRF formulation with a first-order smoothness prior. The technique favours fronto-parallel surfaces, but does not suffer too greatly from this because the range measurements are sufficiently regular and dense, coming from a special range camera sensor. This 'pins' the estimates to lie near the true surface.

In contrast to [9] the method presented here is targeted at any combination of commonly available monocular camera and scanning laser. In particular, this requires inference of range measurements based on sparse, inhomogeneous range data. In such cases, the fronto-parallel tendency of inferred surfaces induced by only considering a first-order smoothness prior leads to increasingly inaccurate reconstructions. We address that issue by introducing a second-order smoothness prior while still framing the problem as a well-understood optimization of a linear system of equations.

3 Problem Formulation

In this section we shall show how a general description of the problem can be formulated in such a way that in the end, only the solution of a single linear system is required. We begin by introducing our notation.

We are given a u by v pixel image $\mathscr{I}$ and a $3D$ point cloud of k laser measurements $\mathscr{L} = \{l_1 \cdots l_k\}$. We shall use the notation $\mathbf{I}_i$ to represent the ith pixel in a vectorized image (all pixels stacked in a single vector of length $N = u \times v$). For each $\mathbf{I}_i$ we associate a range x_i. Our task is to use both $\mathscr{I}$ and $\mathscr{L}$ to find a vector $\mathbf{x} = [x_1, x_2 \cdots x_N]^T$ - a range for every pixel in the image. We shall also refer to x_i as a "range node". Each point in $\mathscr{L}$ can be projected into $\mathscr{I}$ under a distortion correcting camera model and associated to the nearest pixel. Each laser point then yields a range measurement z_i tied to pixel $\mathbf{I}_i$. Note the laser measurements are sparse so not every pixel will have a range measurement — in fact very few will. We use the notation $i \in \mathscr{L}$ to imply the index variable i ranges over all pixels which have an associated range measurement.

We shall pose the problem as one of finding the optimal range vector $\mathbf{x}^*$ such that

$$\mathbf{x}^* = \underset{\mathbf{x}}{\operatorname{argmin}}\{\lambda_1\lambda_2\Theta_s(\mathbf{x}, \mathbf{I}) + \lambda_1(1 - \lambda_2)\Theta_c(\mathbf{x}, \mathbf{I}) + (1 - \lambda_1)\Theta_d(\mathbf{x}, \mathbf{z})\} \qquad (1)$$

where $\Theta_s(\mathbf{x}, \mathbf{I})$ is a first order cost penalizing depth discontinuities, $\Theta_c(\mathbf{x}, \mathbf{I})$ is a second order cost penalizing curvature and $\Theta_d(\mathbf{x}, \mathbf{z})$ is a data cost penalizing errors between inferred ranges and observed range measurements. The scalars $\lambda_1, \lambda_2 \in [0, 1]$ are weightings between the three terms. We shall now consider these terms in more detail.

3.0.1 Data Cost

The data cost is defined as a squared error between assigned range, x_i and measured range, z_i

$$\Theta_d(\mathbf{x}, \mathbf{z}) = \sum_{i \in \mathscr{L}} \sigma_i(x_i - z_i)^2 \qquad (2)$$

$$= ||\mathbf{W}(\mathbf{x} - \mathbf{z})||^2 \qquad (3)$$

where $\mathbf{W}$ is a diagonal matrix with entries

$$\mathbf{W}_{i,i} = \begin{cases} \sigma_i & \text{if } i \in \mathscr{L} \\ 0 & \text{otherwise} \end{cases} \qquad (4)$$

and σ_i is a measure of our confidence in measurement z_i.

3.0.2 Discontinuity Cost

As in [9], we use a depth smoothness or *first-order* prior of the form

$$\Theta_s(\mathbf{x}, \mathbf{I}) = \sum_i \sum_{j \in \mathcal{N}(i)} e_{i,j}(x_i - x_j)^2 \tag{5}$$

where $\mathcal{N}(i)$ are the horizontal and vertical neighbours of i. As edge strength between nodes we use an exponentiated L_2 norm of the difference in pixel appearance

$$e_{i,j} = \exp -\frac{||\mathbf{c}_i - \mathbf{c}_j||^2}{\sigma_d^2} \tag{6}$$

where $\mathbf{c}_i$ is the RGB colour vector of pixel i and σ_d is a tuning parameter (small σ_d increases sensitivity to changes in the image). Equation 5 may be written in matrix form as

$$\Theta_s(\mathbf{x}, \mathbf{I}) = ||\mathbf{Sx}||^2 \tag{7}$$

where each row of $\mathbf{S}$ represents a weighted average of a pair of adjacent range nodes.

3.0.3 Smoothness/Curvature Cost

In contrast to [9] we make the further assumption that in the absence of cues to the contrary, such as discontinuities in appearance, the gradient of surfaces varies smoothly. Under this *second order* smoothness assumption, given a neighbourhood $\mathcal{N}(i)$ of node x_i we may make a range prediction $\hat{x}_i$ as a linear combination of neighbouring ranges x_j for $j \in \mathcal{N}(i)$. This allows us to write simply

$$\hat{\mathbf{x}} = \mathbf{Px} \tag{8}$$

where $\mathbf{P}$ is a suitably formed prediction matrix. We define curvature cost $\Theta_c(\mathbf{x}, \mathbf{I})$ in the form

$$\Theta_c(\mathbf{x}, \mathbf{I}) = ||\hat{\mathbf{x}} - \mathbf{x}||^2 \tag{9}$$

$$= ||(\mathbf{P} - \mathbf{1})\mathbf{x}||^2 \tag{10}$$

Here, $\mathbf{1}$ is the identity matrix. While details of how $\mathbf{P}$ is created will be postponed until Section 4 we may proceed by understanding this cost as the L_2 norm of the deviation of $\mathbf{x}$ from the prediction based on modeling surfaces as locally continuous and smooth.

3.1 *Reduction to* $Ax = b$

We may further expand Equation 3 to the form

$$\Theta_d(\mathbf{x}, \mathbf{z}) = \mathbf{x}^T \mathbf{W}^T \mathbf{Wx} - 2\mathbf{z}^T \mathbf{W}^T \mathbf{Wx} + \mathbf{z}^T \mathbf{W}^T \mathbf{Wz} \tag{11}$$

and Equations 10 and 5 to

$$\Theta_c(\mathbf{x}, \mathbf{I}) = \mathbf{x}^T \mathbf{R}^T \mathbf{R} \mathbf{x}, \quad \Theta_s(\mathbf{x}, \mathbf{I}) = \mathbf{x}^T \mathbf{S}^T \mathbf{S} \mathbf{x} \tag{12}$$

where $\mathbf{R} = \mathbf{P} - \mathbf{1}$.

Substituting Equations 11, 12 into 1 and solving for $\mathbf{x}$ reduces the problem to

$$\mathbf{A}\mathbf{x} = \mathbf{b} \tag{13}$$

with

$$\mathbf{b} = \mathbf{W}^T \mathbf{W} \mathbf{z} \tag{14}$$

$$\mathbf{A} = \frac{\lambda_1 \lambda_2 \mathbf{R}^T \mathbf{R} + \lambda_1 (1 - \lambda_2) \mathbf{S}^T \mathbf{S} + (1 - \lambda_1) \mathbf{W}^T \mathbf{W}}{1 - \lambda_1} \tag{15}$$

Equations 13 to 15 imply that all we need to do to perform the optimization is to solve a large sparse linear system.

4 Constructing the Prediction Matrix

In this section we detail how the prediction matrix $\mathbf{P}$ is created. For simplicity we show only 1D cases but it should be noted that $\mathbf{P}$ contains elements to penalize curvature in *both horizontal and vertical* directions.

We decompose $\mathbf{P}$ into a weighted sum of three prediction operators - extrapolation from left and right, and interpolation.

$$\mathbf{P} = \mathbf{W}_L \mathbf{P}_L + \mathbf{W}_M \mathbf{P}_M + \mathbf{W}_R \mathbf{P}_R \tag{16}$$

where subscripts *L,M,R* imply left-extrapolation, mean (interpolation) and right-extrapolation respectively. The $\mathbf{W}$'s are suitably constructed weighting matrices derived from image appearance which we shall expand upon shortly in Section 4.1. The use of extrapolation and interpolation can be understood graphically with reference to Fig. 1 which shows a simplified 1D case.

4.1 Anticipating Depth Discontinuities from Image Cues

The image $\mathscr{I}$ can be used to provide cues about the behaviour of the surface we hope to reconstruct. Our basic assumption is one that has been used before [9] — sharp changes in range tend to appear as changes in appearance (edges) in an image. We have a range node for each pixel (see Equation 16) and its value can be predicted by a weighted sum of extrapolation and interpolation from its neighbours. We describe only the horizontal case for simplicity, but our method is applied in the vertical case too. For each node x_i the weighting is determined by the properties of pixel i and its neighbourhood. Broadly speaking, if a pixel is identical to its left and right neighbours then pure interpolation will occur. If however there is a discontinuity in pixel appearance then interpolation will be down weighted and either left or right extrapolation emphasized.

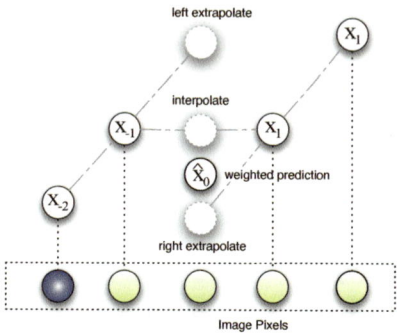

Fig. 1 Depth prediction via weighted interpolation and extrapolation in 1D. The predictions of the range x_0 by left and right extrapolation and interpolation are shown in faded grey. The discontinuity in the image shown at the bottom of the figure (each range node has a single pixel attached to it) causes the left extrapolation to be down-weighted — the image edge is a cue for a possible discontinuity in range between node x_{-1} and x_0. The final prediction, $\hat{x}_0$ is shown in the center.

To explain how the weighting matrices $\mathbf{W}_{L,M,R}$ are created we shall consider the simple 1D case shown in Fig. 2. Interpolation is preferable to extrapolation. With this preference in mind and considering node x_0 in Fig. 2, we can write the importance weights of left / right extrapolation and interpolation as $w_{l,m,r}$

$$w_m = e_{(-1,0)}e_{(0,1)} \tag{17}$$

$$w_r = e_{(-2,-1)}e_{(-1,0)}(1 - w_m) \tag{18}$$

$$w_l = e_{(2,1)}e_{(1,0)}(1 - w_m) \tag{19}$$

with $e_{i,j}$ as defined in Equation 6. The above relationships can be understood by noting that if the pixel attached to range node x_0 is identical to its neighbours ($e_{(-1,0)}$ and $e_{(0,1)}$ are unity) then $w_m = 1$ and $w_r = w_l = 0$ - interpolation has 100% of the weighting. As the pixels $\mathbf{I}_{-1}$ and $\mathbf{I}_1$ become increasingly different, the left and right extrapolations receive more weight. In the limit, if two pixels are entirely different, the edge weight between them tends to zero and the attached range nodes will have no direct link between them. It does not make the two nodes independent - there may be other dependencies via long circuitous routes through other nodes. It does

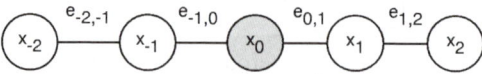

Fig. 2 A 1D chain of range nodes (a section of $\mathbf{x}$) and the edges between neighbours. Considering x_0, right extrapolation uses only nodes to the right and left extrapolation uses the two left hand nodes. Interpolation uses nodes x_{-1} and x_1. The edges between nodes are a function of the difference in pixel appearance between adjacent range nodes (each range node is associated with a single pixel in the image).

however mean that range discontinuities across this boundary are not penalized because the range prediction made by multiplication by **P** is based on an extrapolation from one side and not an interpolation across the discontinuity. This is a key point in this work.

5 Results

Fig. 3 shows the results of processing two synthetic scenes. In this case the problem size is small with **x** having just 2500 elements (each element of **x** corresponds to a vertex in the mesh). With regard to the "three plane" case note how using just a few laser points in each distinct region of the image results in three distinct planes being generated in the reconstructed scene. The strong edges in the images prohibit information flow between planes. For the nodes at the very edge of a plane the extrapolation and interpolation weights have become such that the node is only influenced by (coupled to) other in-plane nodes. The 1st order method alone is unable to reconstruct the planes correctly as it tries to make all nodes have similar ranges.

In the case of the "dome" example note how while there is no range discontinuity there is a sharp discontinuity in surface gradient around the perimeter of the dome. Note also that the first order smoothness term is unable to reconstruct the curvature of the dome in the absence of laser measurements. In contrast, with a second order smoothness cost the curved shape of the dome is recovered well. This is an important result. The generated curved surface is the smoothest surface that can explain the existing measurements and minimize the bust in second order smoothness constraints implicit in **P**.

We now turn to processing some real data. We used a nodding SICK LMS200 laser scanner on a mobile robot to capture laser data. Images were captured by a camera mounted above the laser with a wide angle lens. The image used in this case was 518 by 259 pixels resulting in some 134,162 range nodes and is shown in Fig. 4 with laser measurements projected into it. For scale, the target is approx 1.7m wide. The reconstructed model is shown alongside. Using second-order smoothness alone provides reasonable results, but tends to introduce 'rippling' arartifactsround noisy measurements. A small amount of first-order smoothness is necessary to damp the oscillations. Fig. 5 shows points of interest in the reconstruction. We show an outdoor result of the same problem size in Fig. 6.

The algorithm is implemented in Matlab and the linear solve is performed with Matlab's backslash operator (though there is no reason not to use another method such as Conjugate Gradient). The Three Planes case and the Dome case in Fig. 3, with 2,500 nodes both took 0.021 seconds to solve in a single iteration. For the real data case in Fig. 4 with 134,162 nodes, the algorithm took around 30 seconds on a 2Ghz dual core laptop.

We now present some numerical analysis of the performance of our approach. It is a hard task to obtain a ground truth geometry for the complete real scene. Instead of comparing pixel ranges to ground truth we compare them to laser measurements taken of the scene over a long period of time and which are not used in the

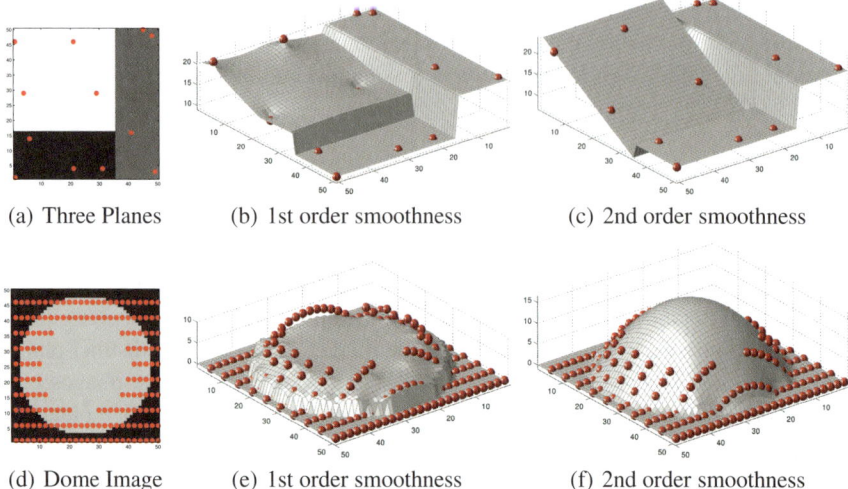

(a) Three Planes   (b) 1st order smoothness   (c) 2nd order smoothness

(d) Dome Image   (e) 1st order smoothness   (f) 2nd order smoothness

Fig. 3 Synthetic data examples which highlight important aspects of our approach. Each node in the mesh represents a single range node projected out from an image pixel. The images for each of the two cases are shown on the left. In all figures sparse laser measurements are shown in red. Note how the discontinuities in the image appear as discontinuities in the reconstructed surfaces. First-order smoothness alone tends to make surfaces have the same depth value whereas second-order smoothness is able to correctly reproduce both planar and curved surfaces.

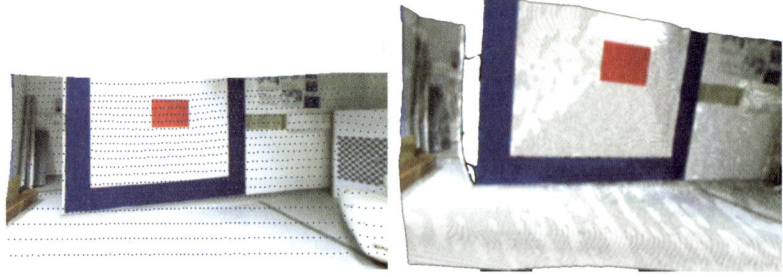

Fig. 4 Results from an indoor dataset. Image and laser measurements on the left, and the reconstructed model on the right.

Fig. 5 Details of a reconstructed scene from Fig. 4. Note the detail of the smooth floor and inferred sharp range discontinuity between two walls.

Fig. 6 Results from an outdoor dataset. On the left is the image with laser measurements overlaid. On the right is the reconstructed model.

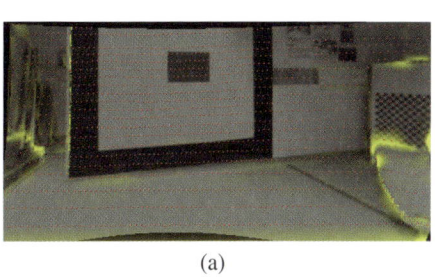
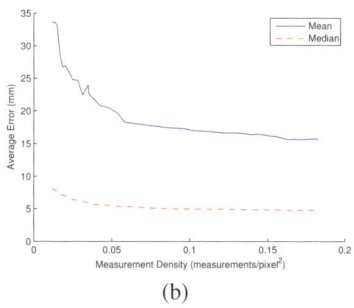

(a)   (b)

Fig. 7 The left image shows a comparison of range estimates to ground truth laser data for the indoor case. Areas in yellow show deviation from ground truth, with higher intensity representing larger errors. Laser measurements are shown in red. The graph shows average error of the estimate relative to the mean density of range measurements, when compared to laser measurements in the hold out set. The laser has a precision of 15mm.

opoptimizationConcretely, we collect a very dense cloud of laser data at the scene and draw from that a small sparse test set with which we reconstruct the scene shown in Fig. 4. The remaining laser data constitutes a dense hold out set, and for each unused laser measurement we can compare measured range to estimated range. Fig. 7(a) shows regions of the workspace which contain pixels with significant errors.

It is also instructive to consider how the accuracy of our approach depends on the density of laser measurements. Fig. 7(b) shows how the statistics (mean and median) of the pixel range errors change as a function of measurement density. Note that as expected, as measurement density increases the precision tends to that of the laser itself around 15mm. The results given in Figs. 4 and 6 are operating in the 0.01 measurements/pixel^2 region.

6 Conclusion

This paper has introduced a novel technique for fusing sparse laser data and images to enable a dense 3D scene reconstruction. Above and beyond existing prior work this technique uses a second order smoothness term which allows it to extrapolate both planar and curved surfaces. The problem is formulated as the solution of a sparse linear system, which allows the use of fast optimization techniques. The technique was applied to both illustrative synthetic cases as well as real data recorded in indoor and outdoor scenes containing challenging geometry.

The qualitative and quantitative results presented here suggest that our system provides 3D reconstructions of reasonable quality. Nevertheless, there is room for improvement. In particular we must consider how we can increase robustness to erroneous laser measurements (away from image edges) and how we might fuse multiple scenes in a principled way. The flip side of this problem is handling bonafide discontinuities in range when there is no change in image appearance and vice versa.

Acknowledgements. This work described here has been supported by the UK EPSRC (CNA and Platform Grant EP/D037077/1), the Office of Naval Research (grant N000140810337) and the European commission (grant agreement FP7-231888-EUROPA). The authors would also like to thank Ingmar Posner for his invaluable contributions.

References

1. Hoiem, D., Efros, A.A., Hebert, M.: Putting objects in perspective. In: CVPR, pp. 2137–2144 (2006)
2. Saxena, A., Chung, S.H., Ng, A.Y.: 3-D depth reconstruction from a single still image. Int. J. Comput. Vision 76(1), 53–69 (2008)
3. Hoppe, H., DeRose, T., Duchamp, T., McDonald, J., Stuetzle, W.: Surface reconstruction from unorganized points. In: SIGGRAPH 1992: Proceedings of the 19th annual conference on Computer graphics and interactive techniques, pp. 71–78. ACM Press, New York (1992)
4. Ohtake, Y., Belyaev, A., Alexa, M., Turk, G., Seidel, H.-P.: Multi-level partition of unity implicits. ACM Trans. Graph. 22(3), 463–470 (2003)
5. Scharstein, D., Szeliski, R.: A taxonomy and evaluation of dense two-frame stereo correspondence algorithms. Int. J. Comput. Vision 47(1-3), 7–42 (2002)
6. Hartley, R.I., Zisserman, A.: Multiple View Geometry in Computer Vision, 2nd edn. Cambridge University Press, Cambridge (2004)
7. Yang, Q., Yang, R., Davis, J., Nister, D.: Spatial-depth super resolution for range images. In: IEEE Computer Society Conference on Computer Vision and Pattern Recognition, vol. 0, pp. 1–8 (2007)
8. Torres-Méndez, L.A., Dudek, G.: Statistics of visual and partial depth data for mobile robot environment modeling. In: Gelbukh, A., Reyes-Garcia, C.A. (eds.) MICAI 2006. LNCS (LNAI), vol. 4293, pp. 715–725. Springer, Heidelberg (2006)
9. Diebel, J., Thrun, S.: An application of markov random fields to range sensing. In: Proceedings of Conference on Neural Information Processing Systems (NIPS). MIT Press, Cambridge (2005)
10. Andreasson, H., Triebel, R., Lilienthal, A.: Vision-based interpolation of 3D laser scans. In: Proc. International Conference on Autonomous Robots and Agents, ICARA (2006)

Relative Motion Threshold for Rejection in ICP Registration

François Pomerleau, Francis Colas, François Ferland, and François Michaud

Abstract. Simultaneous Localization and Mapping (SLAM) iteratively builds a map of the environment by putting each new observation in relation with the current map. This relation is usually done by scan matching algorithms such as Iterative Closest Point (ICP) where two sets of features are paired. However as ICP is sensitive to outliers, methods have been proposed to reject them. In this article, we present a new rejection technique called Relative Motion Threshold (RMT). In combination with multiple pairing rejection, RMT identifies outliers based on error produced by paired points instead of a distance measurement, which makes it more applicable to point-to-plane error. The rejection threshold is calculated with a simulated annealing ratio which follows the convergence rate of the algorithm. Experiments demonstrate that RMT performs better than former techniques with outliers created by dynamical obstacles. Those results were achieved without reducing convergence speed of the overall ICP algorithm.

Keywords: ICP, registration, scan matching, rejection, SLAM.

1 Introduction

Simultaneous Localization And Mapping (SLAM) algorithms use motion and observation probabilistic models to incrementally correct positioning problems. The mechanism used to transform different observation models into the same coordinate system is called registration (also known as data association or scan matching). Proposed SLAM solutions based on Maximum Likelihood

François Pomerleau, François Ferland, and François Michaud
Dept. of Elec. Eng. and Computer Eng., Université de Sherbrooke, Québec, Canada

François Pomerleau and Francis Colas
Autonomous Systems Laboratory, ETH Zürich, Switzerland
e-mail: `francois.{pomerleau,ferland,michaud}@usherbrooke.ca`,
`francis.colas@mavt.ethz.ch`

A. Howard et al. (Eds.): Field and Service Robotics 7, STAR 62, pp. 229–238.

(ML) present fast capabilities to minimize global positioning errors [7], but they still need to rely on efficient and robust registration algorithms to be stable in real robotic applications.

Registration can be done using landmarks (e.g., lines, circles, arcs, corners) [1]. When applied to the registration processes, landmarks confer the advantage of accelerating calculation by summing up information. However, landmark registrations can be sensitive to unstructured environments where landmarks are difficult to detect. A second type of registration is called Normal Distribution Transform (NDT). NDT segments spatial information and works on the first and second statistical moments to reduce the computational cost [9] while avoiding to define specific landmarks. However, NDT is still very sensitive to segmentation because large spatial cells filter out relevant details, whereas small cells augment the computational cost.

Another strategy is to directly use point clouds derived from exteroceptive data. One technique to find such matches is known as Iterative Closest Point (ICP) [2]. This method pairs points of both scans by finding for each point of the first scan the nearest point in the second one. From these pairs a motion vector is estimated to cope for their misalignment. This process is iterated until convergence. ICP variants were first developed for applications involving 3D model reconstructions [4], [6], [8]. When used in SLAM by an autonomous robot, these algorithms need to be adapted in several ways: 1) they must work in real-time [3]; 2) they must be adapted to the sensors to be able to use 2D and 3D spatial information [11], and to cope with sensor fusion (range, laser reflectivity [14], color [13], etc.); 3) they must be able to deal with occlusion and partially overlapping scans that frequently arise in a dynamic environment explored by a mobile platform.

This paper addresses the issue of occlusions and partially overlapping scans by using a new adaptive rejection technique called Relative Motion Threshold (RMT). For SLAM, changes in the environment and occlusions caused by the motion of the mobile robot are sources of outliers (i.e., points with no match). Fig. 1 presents an example for which an optimal rotation and translation of the blue point cloud (p_i) must be applied to align it with the reference red point cloud (q_j), with i and j being point indexes. Even after a small displacement of the robot between t_1 and t_2, maps can largely differ due to sensor occlusion. For example, in Fig. 1(b), 50% of the blue points are outliers when compared to the red point cloud. Also, the disambiguation of points obtained from obstacle 1 and obstacle 2 must be resolved to make ICP robust to initial positioning errors and overlapping. Removing outliers from the paired points is done during the rejection step of ICP, where it minimizes misalignment errors for the determination of the motion vector between two point clouds. The rejection technique introduced in this paper uses an adaptive threshold based on simulated annealing ratio to augment the robustness of ICP against outliers.

The paper is organized as follows. Section 2 presents an overview of rejection techniques. Section 3 describes the RMT rejection technique we propose.

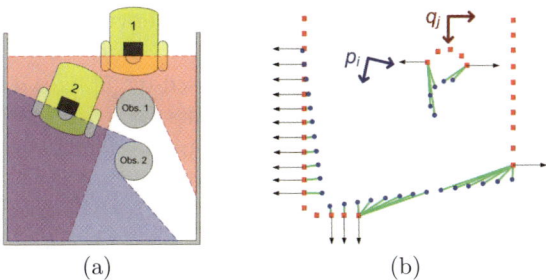

(a)   (b)

Fig. 1 Map registration example for SLAM, with misalignment caused by odometry error. The green lines represent the alignment error minimized by ICP algorithms. Obs.: Obstacle. (a) Two laser scans taken at time t_1 and t_2. (b) Point cloud p_i (blue) taken at t_2 is represented in the coordinate system of point cloud q_j (red) taken at t_1. Black arrows are surface orientations.

Section 4 presents experimental results that evaluate the performance of each rejection technique in terms of matching and convergence rate of ICP in simulated and real-world applications.

2 Rejection Techniques

Rejection techniques can be categorized as follows:

- *Fix*: Manual setting of the maximum distance d authorized between paired points. Then, all paired points with a distance higher than d are systematically rejected. This method is simple but does not adapt well to conditions that would require different thresholds.
- *Zhang*: Strategy based on statistical moments of the distribution of the distances between the paired points [15]. Four conditions are needed to adapt the threshold d:

$$d = \begin{cases} \mu + 3\sigma, \text{ if } 0 \leq \mu < \eta \\ \mu + 2\sigma, \text{ if } \eta < \mu \leq 3\eta \\ \mu + \sigma, \quad \text{if } 3\eta < \mu \leq 6\eta \\ \rho, \qquad\quad \text{otherwise} \end{cases} \tag{1}$$

where ρ is the median of the distance between paired points and η is a distance-based parameter set by the user. This method can be adjusted to different statistical distributions of the distances but still lacks generality.
- *Mean*: Technique proposed in [6] which sets d equal $\mu + \sigma$ for each iteration, and where μ and σ are respectively the mean and the standard deviation of the distances between paired points. This method has no parameter and is flexible to different type of outliers. However it filters outliers on the assumption that the distribution of distances is Gaussian, which is not

a good assumption in the case of a dynamic environment where several local minima can emerge.

- *Median*: An other statistical method proposed by [5] fixes d to 3 times the median of the distances between paired points. The method has no parameter, but calculating a median over a huge point cloud is computationally expensive.
- *Trim.*: Overlapping parameter ξ defined by [4] is used to reject a percentage of outliers:

$$N_{trimmed} = \xi N_{total} \qquad (2)$$

where N is the number of paired points. This approach is less dependent on the shape of the distribution. However, it requires to sort all paired points based on their distances at each iteration, which increases computation time. It can also misled by a large change in the overlap that may occur with a moving platform due to occlusions.

In addition to these techniques, it is also possible to reject multiple pairing to a single point [16], [3], as shown in the lower right part of Fig. 1(b) where multiple green lines connect to the same red points. Instead of authorizing all pairs, only the one with the smallest distance is kept. This criterion has been shown to improve the performance of all standard rejection techniques, and thus all results presented here use this additional criterion.

3   Relative Motion Threshold Technique for Rejection

Existing rejection techniques rely mostly on the Euclidean distance between paired points. While this distance has a direct impact on point-to-point error metric, ICP implementations commonly use a point-to-plane error metric to pair the points because of its faster convergence speed. This point-to-plane error metric[1] assumes that there is a local surface orientation vector estimated for each point q_j and projects the Euclidean distance between p_i and q_j on this vector. The point paired to q_j then minimizes this error and not the Euclidean distance (see Fig. 1(b)). We introduce a new, more general, rejection technique called Relative Motion Threshold (RMT). RMT is an adaptive rejection technique that progressively identifies outliers that create most of the error during the process of ICP. Adaptation is based directly on the error created by paired points instead of the Euclidean distance between those points in accordance to the matching process. We propose to reject the outliers with a maximum authorized error e_t at iteration t, evaluated by:

$$e_t = \begin{cases} \lambda e_{t-1}, & \text{if } \lambda < 1 \\ e_{t-1}, & \text{otherwise} \end{cases} \qquad (3)$$

---

[1] In the remaining of the text, the term error will refer to the point-to-plane error metric when there is no ambiguity.

with λ being a simulated annealing ratio defined by:

$$\lambda = \frac{||T_{t-1}||}{||T_{t-2}||} \qquad (4)$$

This ratio uses past motion information to determine if the point cloud is converging to a local minima, where $||T||$ is the Euclidean norm of the translation vector T which minimize the alignment error of p_i at iteration t. The translation vector T and the rotational vector Ω are calculated during the Error and Minimization step at the end of each iteration and are used to move p_i toward q_i. A ratio λ smaller than 1 means that the position of p_i is stabilizing. All points with a translation error larger than $e_t + \epsilon$ are identified as outliers and rejected during the iteration t. If the ratio λ is larger than 1, the motion of point cloud p_i is accelerating toward q_i due to new appearing constraints. The maximum authorized error e_t is then kept stable until the point cloud starts to converge again.

The minimum error ϵ is the only parameter needed for our rejection technique. It represents noises from sensor readings. A simple way to evaluate this parameter is to take two scans of a static environment and look at the distribution of translation errors created by the error metric used. This translation error should be centered on zero and can be estimated by a Gaussian distribution. The parameter ϵ can be estimated using the standard deviation of this distribution, making the parameter sensor-dependent instead of situation-dependent.

Fig. 2(a) presents an example of the relative motion threshold in function of iterations. During the two first iterations, only the rejection of multiple pairings is active to initialize a value for T_{t-1} and T_{t-2}. Then, e_2 is initialized to the maximum Euclidean norm of translation error at iteration 2. The simulated annealing ratio reduces this error until the point cloud p_i temporarily stops converging. The ratio λ is higher than 1 between iteration 6 and 7, forcing a constant error threshold until the point cloud p_i start converging again. The final threshold error is reduced until iteration 11 where it equals

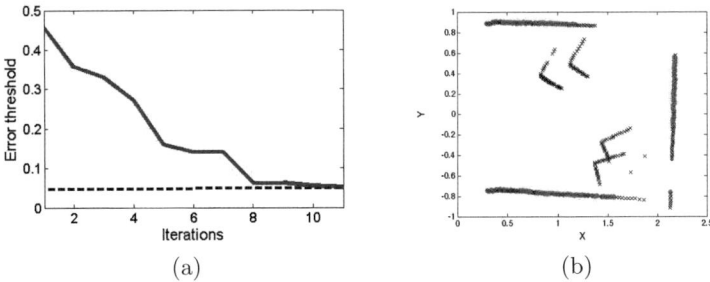

(a)   (b)

Fig. 2 (a) A generic example of maximal authorized error based on RMT in function of iterations. Dashed line represents minimum error ϵ. (b) Final position of the registered point clouds used in the generic example.

the minimum error ϵ represented by the dashed line. Fig. 2(b) shows the two point clouds used in this example at iteration 11. The point clouds were taken in a room where two boxes were moved in different directions. The thick points in light blue represent p_i which converges to the right position even with outliers caused by dynamic obstacles (i.e., boxes). Small black cross represent outliers detected by RMT.

4   Experimental Results

Experiments were conducted in simulation and real settings using the following ICP algorithm consisting in five steps [12]:

1. *Selection* reduces the number of points in p_i by selecting a representative subset of points p_s, where $s < i$. This step is a compromise between computation speed and robustness. Even if using a smaller number of points results in faster computation time, the result may diverge if not enough points are used, or if the selection process filters out necessary constraints.
2. *Matching* pairs each point of p_s in the point cloud q_j. This corresponds to a closest point search problem. One data structure often used to solve this problem is the k-d tree. It is a data structure that partitions the space into k dimensions, with the property of accelerating the nearest neighbor search. Recently, utilization of the approximate k-d tree [8] has shown to give faster results without altering ICP precision.
3. *Weighting* improves or reduces impacts of pairing point on the error matrix by using criteria such as distance, normal compatibilities and scanner noise. However, results suggest that weighting is data-dependent and does not increase convergence rate significantly [12].
4. *Rejection* uses techniques described in Section 2 and 3.
5. *Error and minimization* use all the remaining matched points to evaluate the misalignment error and a create a motion vector $m = [T, \Omega]'$ minimizing this error where T is the translation components and Ω the orientation components. This motion vector is applied to the point cloud p_i. Point-to-plane error function is shown to have a faster convergence rate than point-to-point error [12].

Steps 2 to 5 are repeated until any of the ending condition is reached. Several ending conditions have been proposed, e.g. number of iterations, error, relative motion between two iterations [4], [16], stabilization of mean and standard deviation of the distances between paired points, number of registered pairs [6]. Our complete ICP algorithm uses all of those.

4.1   Evaluation Method

To test the RMT rejection technique while dealing with outliers or occlusion caused by moving objects, we enriched the test protocol described in [10].

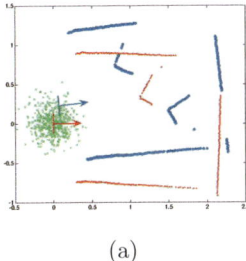 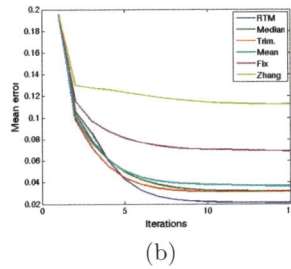

(a)   (b)

Fig. 3 (a) Two scans to be matched. The first scan is in red, whereas the second scan, after translation and rotation, is in thick blue. The green crosses show the distribution of displacement error used in our test with a standard deviation of 0.15 on x- and y-axis. (b) Comparison of the performance of several rejection methods in function of iterations. The performance is measured by the mean position error in respect to ground truth.

More specifically, we recorded data taken by a SICK LMS 200 laser range finder in a U-shaped room. Without moving the sensor, we added or moved boxes in its field of view. This way we generated 10 pairs of different scans with an overlapping ratio around 75%. For each trial, one of the two scans was transformed with a rotation and a translation vector drawn randomly according to a Gaussian distribution in order to fit the uncertainty of the localization of standard SLAM techniques. The standard deviations were $0.15\,m$ for each translation component and $0.15\,rad$ for the angle. Fig. 3(a) shows an example of two scans as well as the distribution of displacement of the second scan. We can see that in the second scan, one of the boxes moved while the other was removed. As the sensor is fixed between the scans, the result of the registration algorithm is exactly the inverse of this transformation. Fig. 3(b) shows the results for each rejection techniques. Curves represent the mean of the XY alignment error of p_i over 4000 trials. For rejection techniques that require the setting of parameters, optimal values were derived by sampling the parameter space and computing the percentage of good registration over a training set. The final performances were evaluated using the remaining configurations. Theses optimal parameters are presented in Table 1. RMT provides a large improvement over the other rejection techniques, since other methods tend to wrongly categorize points as outliers and converge towards local minima.

Table 1 Parameters used for each rejection techniques during comparison test.

	RMT	Median	Trim.	Mean	Fix	Zhang
Parameter	$\epsilon = 0.05$	none	$\xi = 76\%$	none	$d = 0.3\,m$	$\eta = 0.02\,m$

Table 2 Robustness of the rejection techniques with respect to initial error. The performance is measured in term of the mean final of XY alignment error of point clouds (in meter).

Std	RMT	Median	Trim.	Mean	Fix	Zhang
0.05	0.001	0.005	0.013	0.009	0.027	0.021
0.10	0.004	0.016	0.015	0.013	0.029	0.056
0.15	0.021	0.031	0.032	0.039	0.055	0.107
0.20	0.034	0.051	0.060	0.059	0.077	0.132
0.30	0.135	0.145	0.189	0.203	0.261	0.282
0.40	0.272	0.294	0.356	0.362	0.412	0.423

In terms of speed, ICP is an iterative algorithm known to converge in a small number of iterations. RMT rejection method does not impair the convergence rate of the matching algorithm. The mean and covariance on the number of iterations for the Median and RMT are respectively ($\mu = 9.5$, $\sigma = 2.4$) and ($\mu = 9.9$, $\sigma = 1.9$). Those results were obtained while keeping registration converging to the right value for 4000 trials. No significant difference were observed between all rejection techniques tested.

Looking at how rejection techniques perform with shifting initial positions, Table 2 presents the correct registration computed for 4000 trials of each rejection technique for various standard deviations on the initial error. RMT rejection technique performs better than the others for all conditions tested by having the lowest residual error. Moreover, a rise in the initial position variance decreases the performance for every methods, as expected. For this setup, it means that uncertainty on the position of the robot should be kept under 15 cm before applying ICP to achieve good registration. However, RMT rejection method is more robust than other techniques as performance loss occurs at a higher variance while being less computationally expensive than median technique.

4.2 Real-World Application

The last section described experiments with outliers mainly due to dynamical obstacles. Another main source of outliers can be created by low overlapping percentage of scans. The Canadian Space Agency (CSA) uses a rotating laser range finder installed on a robot to test Mars exploration algorithms which is showed in Fig. 4 (a). The robot typically moves few meters on a simulated Martian terrain, takes a 3D scan and decides where to go next. In this kind of application, overlapping between scans can vary between 50% and 90% and few 3D features are available, making the registration very sensible to outliers. The RMT was used on 3D point clouds extracted within this context of application. Fig. 4 presents the result of one registration with a distance of 15 m between the two scans. The grayscale surface correspond to the section used for the registration. The maximum height of the surface is about 1 m.

(a)   (b)

Fig. 4 (a) Robot and environment of the Canadian Space Agency for the Mars exploration project. (b) RMT applied to scans with low overlapping. In grayscale, the surface recovered from match points. In red and blue, the outliers of each scan.

Points on both side of the surface represent outliers removed during the registration. This demonstrates that the RMT can also deal with outliers created by low overlapping scans. Moreover, the algorithm is currently used by the Space Technologies Research Program of the CSA for complete mapping of the experimental Martian terrain.

5   Conclusion and Future Work

This paper presents a novel rejection technique called RMT in the context of ICP registration applied to SLAM in mobile robotics. Results show promising performance, making RMT a very interesting alternative to other rejection techniques. In particular, RMT allows better registration with point clouds containing dynamical obstacles. RMT also demonstrates its applicability in a Mars exploration context with low overlapping percentages. It also gives good results for identifying dynamical obstacles. In future work, we plan to characterize the stability of the approach in a complete SLAM algorithm, and to further extend the range of initial error that ICP can resolve.

Acknowledgements. The authors gratefully acknowledge the contribution of the Natural Sciences and Engineering Research Council of Canada (CRSNG), the Canada Research Chairs (CRC) and the Fondation UdeS for their financial support. We extend our thanks to David Gingras for his work realized under the Space Technologies Research Program at CSA. This work was also partially supported by Robots@home STREP EU Project IST-6-045350.

References

1. Altermatt, M., Martinelli, A., Tomatis, N., Siegwart, R.: Slam with corner features based on a relative map. In: Proceedings of the IEEE/RSJ International Conference on Intelligent Robots and Systems (IROS), vol. 2, pp. 1053–1058 (September-October 2004)

2. Bool, P., McKay, II.. A method for registration of 3-d shapes. IEEE Transactions on Pattern Analysis and Machine Intelligence 14(2), 239–256 (1992)
3. Censi, A.: An icp variant using a point-to-line metric. In: IEEE International Conference on Robotics and Automation (ICRA), pp. 19–25 (May 2008)
4. Chetverikov, D., Svirko, D., Stepanov, D., Krsek, P.: The trimmed iterative closest point algorithm. In: Proceedings of 16th International Conference on Pattern Recognition, vol. 3, pp. 545–548 (2002)
5. Diebel, J., Reutersward, K., Thrun, S., Davis, J., Gupta, R.: An icp variant using a point-to-line metric. In: IEEE/RSJ International Conference on Intelligent Robots and Systems (IROS), pp. 3436–3443 (September-October 2004)
6. Druon, S., Aldon, M., Crosnier, A.: Color constrained icp for registration of large unstructured 3d color data sets. In: IEEE International Conference on Information Acquisition, pp. 249–255 (August 2006)
7. Grisetti, G., Grzonka, S., Stachniss, C., Pfaff, P., Burgard, W.: Efficient estimation of accurate maximum likelihood maps in 3D. In: Proceedings of IEEE/RSJ International Conference on Intelligent Robots and Systems, pp. 3472–3478 (October 2007)
8. Ho, N., Jarvis, R.: Large scale 3d environmental modelling for stereoscopic walk-through visualisation. In: 3DTV Conference, pp. 1–4 (May 2007)
9. Huhle, B., Magnusson, M., Strasser, W., Lilienthal, A.: Registration of colored 3d point clouds with a kernel-based extension to the normal distributions transform. In: IEEE International Conference on Robotics and Automation (ICRA), pp. 4025–4030 (May 2008)
10. Minguez, J.: Metric-based scan matching algorithms for mobile robot displacement estimation. In: Int. Conf. on Robotics and Automation (2005)
11. Nuchter, A., Lingemann, K., Hertzberg, J., Surmann, H.: 6d slam with approximate data association. In: Proceedings of the 12th International Conference on Advanced Robotics (ICAR), pp. 242–249 (July 2005)
12. Rusinkiewicz, S., Levoy, M.: Efficient variants of the icp algorithm. In: Proceeding on 3DIM, pp. 145–152 (2001)
13. Strand, M., Erb, F., Dillmann, R.: Range image registration using an octree based matching strategy. In: International Conference on Mechatronics and Automation (ICMA), pp. 1622–1627 (August 2007)
14. Yoshitaka, H., Hirohiko, K., Akihisa, O., Shin'ichi, Y.: Mobile robot localization and mapping by scan matching using laser reflection intensity of the sokuiki sensor. In: 32nd Annual Conference on IEEE Industrial Electronics (IECON), pp. 3018–3023 (November 2006)
15. Zhang, Z.: Iterative point matching for registration of free-form curves and surfaces. International Journal of Computer Vision 13(2) (1994)
16. Zinsser, T., Schmidt, J., Niemann, H.: A refined icp algorithm for robust 3-d correspondence estimation. In: Proceedings of 2003 International Conference on Image Processing (ICIP), 2:II-695–8, vol. 3 (September 2003)

Bandit-Based Online Candidate Selection for Adjustable Autonomy

Boris Sofman, J. Andrew Bagnell, and Anthony Stentz

Abstract. In many robot navigation scenarios, the robot is able to choose between some number of operating modes. One such scenario is when a robot must decide how to trade-off online between human and tele-operation control. When little prior knowledge about the performance of each operator is known, the robot must learn online to model their abilities and be able to take advantage of the strengths of each. We present a bandit-based online candidate selection algorithm that operates in this adjustable autonomy setting and makes choices to optimize overall navigational performance. We justify this technique through such a scenario on logged data and demonstrate how the same technique can be used to optimize the use of high-resolution overhead data when its availability is limited.

1 Introduction

Autonomous UGVs have advanced to a point where they are competent and reliable a large portion of the time. However, even the most robust autonomous robotic systems will struggle with certain situations. Fortunately, in some domains it is reasonable to assume that a human operator may be available for periods of time to provide remote tele-operation support. Full tele-operation is prohibitively expensive for many applications due to the degree of required human attention and communications bandwidth, so a policy must determine under which conditions the robot or the human are to take control.

It is important for such a system to be well-suited for online use. Not only is it unpredictable in advance how well the autonomy system will perform in novel environments, but human operator performance can also vary depending on factors such as bandwidth limitations, operator handicaps such as limited skill or familiarity with the interface, fatigue and weather conditions. When little prior knowledge about the

Boris Sofman, J. Andrew Bagnell, and Anthony Stentz
Robotics Institute, Carnegie Mellon University, Pittsburgh, PA 15213
e-mail: {bsofman,dbagnell,axs}@ri.cmu.edu

A. Howard et al. (Eds.): Field and Service Robotics 7, STAR 62, pp. 239–248.

Fig. 1 For our experiments, we used logs from the Crusher unmanned ground vehicle. The robot operates in complex, natural environments where the goal is to navigate across large distances with the aid of on-board and overhead sensor data. Further information about the system can be found in [1]. The overhead processing capabilities used to plan prior routes and generate features for our experiments are described in [2].

operators' abilities is available, a learning system can observe the performance of the autonomous vehicle in particular situations and compare that to performance under remote human-control in similar situations. When the vehicle encounters similar situations in the future, it can then invoke whichever expert demonstrated better performance: the remote human or autonomous vehicle. Such a capability would enable a single operator to assist many UGVs, ensuring peak performance for the entire team with minimal human involvement.

We pursue this problem using an on-line, reinforcement learning approach and demonstrate its performance on logged data from the rugged, all-terrain UGV shown in Figure 1. The candidate selection system's goal is to learn to interpret available overhead sensor data in order to make decisions that maximize its overall long-term performance. This inevitably becomes a trade-off between exploring candidates' performance in situations that will allow it to learn more about the world and taking advantage of their learned models to maximize current performance.

We also show how this technique can be used to deal with scenarios where limited high-resolution overhead data is available to aid the robot in navigating through an environment. The Digital Terran Elevation Data (DTED) level of an overhead elevation data set specifies its density of coverage. Higher resolution overhead data can be used to produce more accurate traversal cost estimates that the UGV can use for better prior path computation but often require expensive and time-consuming aerial surveying and a large amount of bandwidth if remotely supplied to the vehicle. In scenarios where the availability of such data is limited, our algorithm can be extended to allow the robot learn to identify the situations where it will most benefit from high resolution data in order to allocate it to areas that maximize its impact.

The next section presents background on adjustable autonomy techniques and some example applications. Section 3 presents our online candidate selection algorithm, followed by experimental results in Section 4 and concluding remarks in Section 5.

2 Related Work

We deal with the scenario where a human can contribute limited attention to improve a mobile robot's performance. In this scenario, a robotic system operates somewhere on the spectrum between full autonomy, where there is no human involvement, and full tele-operation, where the human is in complete control at all times. Scenarios where the degree and methods of human interactions with robots within a system can be varied dynamically in order to optimize performance are often referred to as ones of *sliding autonomy* or *adjustable autonomy* [3, 4]. While most mobile robot systems tend to lie on one of the two extremes of this spectrum, effectively balancing autonomy with limited human involvement can lead to significant improvements in safety, efficiency and overall cost.

In some scenarios where the human is the primary operator, the autonomy system is intended to aid by request or when it detects a dangerous situation [5, 6, 7, 8]. Similar approaches have been applied to automating repetitive tasks in surgery to decrease surgeon fatigue [9].

In scenarios where the autonomy system is the default operator, the system must reason about whether and when to transfer control to a human [10, 11, 12]. Some have suggested relinquishing control when there is an expectation of high benefit [13, 14] or the degree of uncertainty is high [15].

Goodrich and Schultz have written an extensive survey article on the field of Human-Robot Interaction exploring many additional approaches and applications [16].

The key difference in our approach from the above-mentioned approaches is that we do not constrain the system by any pre-determined rules or models. Since in many scenarios prior performance information is unavailable, the ability to learn the capabilities of each potential expert online allows systems to better adapt to more diverse and challenging environments.

3 Approach

3.1 Contextual Multi-armed Bandit Setting

The candidate selection problem involves choosing an operator for each encountered situation from a set of candidate systems, in our case the autonomy system and the human tele-operator, whose performance we assume comes from some unknown distribution. It is therefore intuitive to frame this problem as an instance of the commonly studied *multi-armed bandit* problem [17, 18, 19].

In the k-armed bandit setting, at each time step the world chooses k losses (or rewards), $l_1, \ldots, l_k$, and the player makes a choice of an arm $i \in \{1, k\}$ without knowledge of the hidden losses. The player then observers only the loss l_i corresponding to the chosen arm. Since the loss distributions are unknown, there is an inevitable conflict between minimizing the immediate loss and gathering information that will be useful for long-term performance. This is often referred to as the *exploration-exploitation trade-off* since we must choose between *exploring* our unknown loss distributions and *exploiting* the arm we currently believe to be best.

We deal with a more suitable variation of this setting called the *contextual bandits* setting where at each time step t the player also observes some contextual information x_t which can be used to determine which arm to pull [20]. We compute these features from commonly available overhead imagery and DTED 3 elevation data for the given environment as described in [2] and convolve them with a Gaussian kernel in order to blur the data, in effect introducing an influence from surrounding areas into each location. This creates a more realistic modeling problem since the rate of progress at a given location is heavily influenced by factors from the surrounding area.

As is common with bandit problems, our goal is to minimize regret, the difference between the performance of the algorithm and that of the optimal algorithm in hindsight:

$$R = \sum_{t=1}^{T}(l_t - l_t^*) \tag{1}$$

where l_t^* is the loss incurred in round t by the optimal strategy.

3.2 Exploration-Exploitation Trade-Off

We choose to deal with the exploration-exploitation trade-off through the use of confidence bounds. With a model that is able to supply confidence bounds, the widths of the confidence bounds reflect the uncertainty of the algorithm's knowledge. By choosing the candidate with the highest upper confidence bound at each time step, the algorithm elegantly trades off between exploration and exploitation. When uncertainty is high, choosing that candidate will provide information that will quickly reduce uncertainty in that region of the model. As we gain knowledge about each candidate, confidence bounds will shrink and we will choose the candidate with the highest expected performance. This approach was well-justified for the bandits setting and shown to have small regret [21].

3.3 Formalization

We frame online candidate selection problems as follows. At each time step t, we get some contextual features x_t for our environment and must choose from one of k candidates to operate the robot for that time step[1]. The goal in such a setting is to minimize the loss at each time step, measured in the case of operator selection by the period of time it takes to enter and exit a 3 meter radius window around that location.

---

[1] In the case of choosing between a human and the autonomy system, $k = 2$. We discuss this problem in the more general case as it could also be applied to any candidate selection setting such as choosing between multiple autonomy systems, multiple human operators or multiple overhead data sources as shown later.

After each selection, the algorithm observes the noisy feedback l_t^i of only the chosen candidate i. We model the distribution for l_t^i as a Gaussian whose mean is a linear function of the contextual features x_t:

$$E(l_t^i | \mu^i, x_t) = \mu^i x_t \tag{2}$$

We assume the estimates have Gaussian noise and are therefore distributed:

$$\tilde{l}_t^i \sim Normal(l_t^i, \sigma^2) \tag{3}$$

We model this distribution online using a Bayesian linear regression model as described in [22]. This not only allows us to efficiently perform online updates of our model but also provides a variance estimate for each prediction. We therefore track k Bayesian linear regression instances in parallel, one for each candidate. At each time step we choose the candidate with the highest upper confidence bound prediction for that scenario.

4 Experimental Results

We validate this candidate selection algorithm offline through the following two applications relevant to mobile robot navigation[2].

4.1 Adjustable Autonomy

While we do not have the system infrastructure to be able to trade-off online between tele-operation and autonomous vehicle control, we simulated such an online scenario by using a pair of logged traversals of the same long course in western Pennsylvania by each candidate: a human tele-operator using a high-bandwidth camera system and the autonomy system. All locations where the path of the human driver and the autonomous driver were in sufficient proximity were used as a test point for the system. As the algorithm chose a candidate, the traversal time for only the specified candidate was revealed to the algorithm.

The course and estimated relative performance of each candidate using a trained model appear in Figure 2. Quantitative results comparing our algorithm to various alternatives appear in Figure 3 and Table 1.

4.2 Online Overhead Data Selection

We also show how this algorithm can be applied to direct the use of various-density overhead data where there is either limited time to gather that data or limited bandwidth for wireless transmission of the data to the vehicle during navigation. In such

---

[2] Many of the images in this paper are best viewed in color.

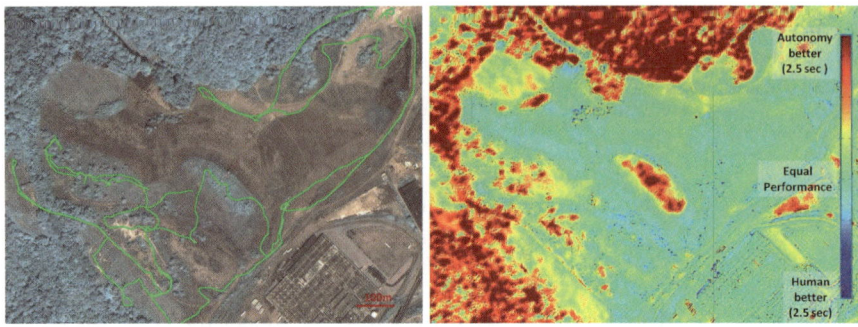

Fig. 2 Aerial image of test site with course driven using each operating mode (left) and the estimated differences in traversal time in seconds per meter for this site using the final models learned by the online candidate selection algorithm (right). The algorithm found that human performance tended to excel in open areas where the human was better able to interpret sparse obstacles and drive aggressively and at perimeters of heavy obstacles when the human's situational awareness allowed him to better interact with the environment.

Table 1 Online Operator Selection Performance

Algorithm	Cumulative Time (seconds)[a]	Percent Improvement over Always-Human
Online Algorithm	9551.7	9.41
Optimal	7809.3	25.94
Worst-Case	12791.0	-21.31
Always-Human	10544.4	0.00
Always-Autonomy	10055.9	4.63
Random Driver	10307.4	2.95

[a] Note that since 3 meter regions at example locations often overlapped with each other, these cumulative traversal times are greater than the total navigation time.

situations, the vehicle must decide online how to best utilize the availability of data for upcoming navigation.

We simulated this scenario by analyzing sets of multi-waypoint logged runs from a field test at Fort Carson in Colorado on sets of courses using DTED levels 3, 4 and 5 overhead data. The candidates for each waypoint in this case were the choice of density of aerial data for an area bounding that path segment. The candidate selection system therefore had the goal of learning a mapping from the average of feature values (computed as described earlier) within the segment's bounding box to the average traversal speed for the vehicle over that segment of the path using each candidate type of data. While DTED 5 data almost always resulted in the best performance, we simulated a scenario where high-density data is available for only a fraction of all segments: a maximum of 20% availability for DTED 5 and 30% availability for DTED 4.

At each step we used a linear program to optimize the allocations of remaining data availability using the predicted performance on all remaining segments from

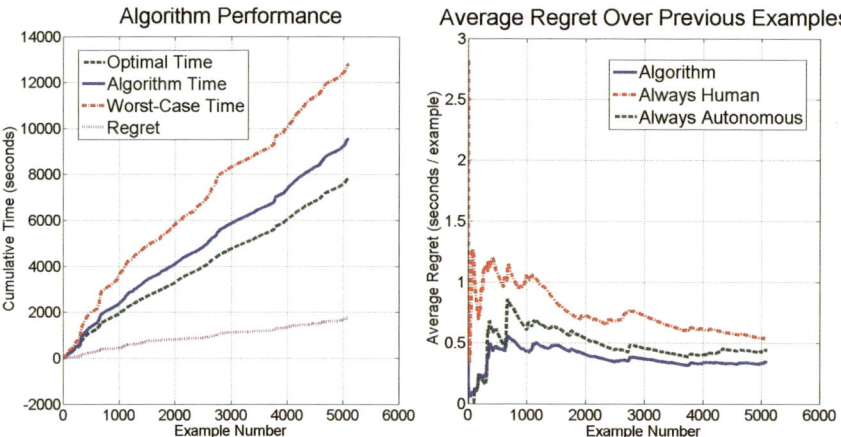

Fig. 3 Online operator selection performance: cumulative navigation time for our algorithm and various alternatives (left) and the average regret of our algorithm over previous examples compared to alternatives (right).

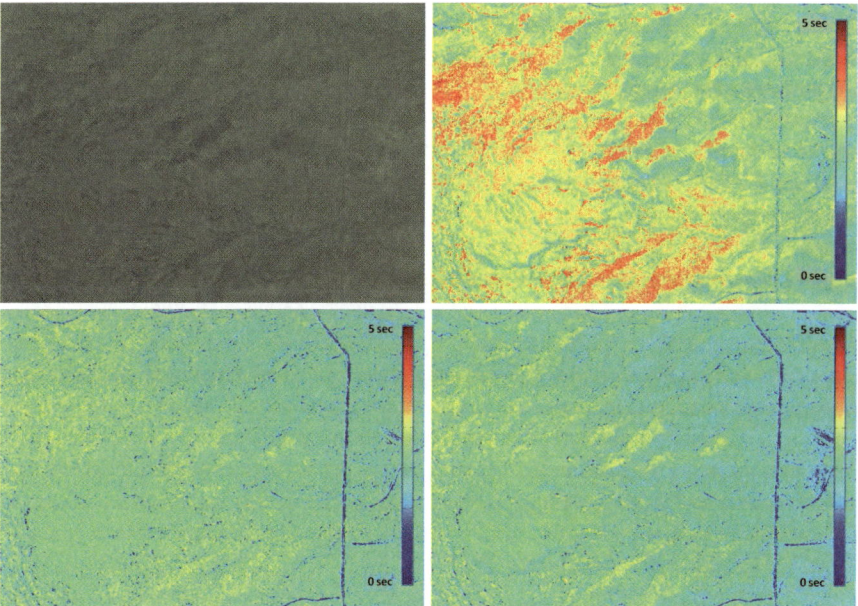

Fig. 4 Aerial image of sample terrain for data selection experiments is shown in top-left. Estimated traversal time in seconds per meter is shown for DTED 3, 4 and 5 data at top-right, bottom-left and bottom-right respectively. As expected, DTED 5 data shows large improvements in navigation speed for difficult terrain but does not provide much benefit on roads and open fields.

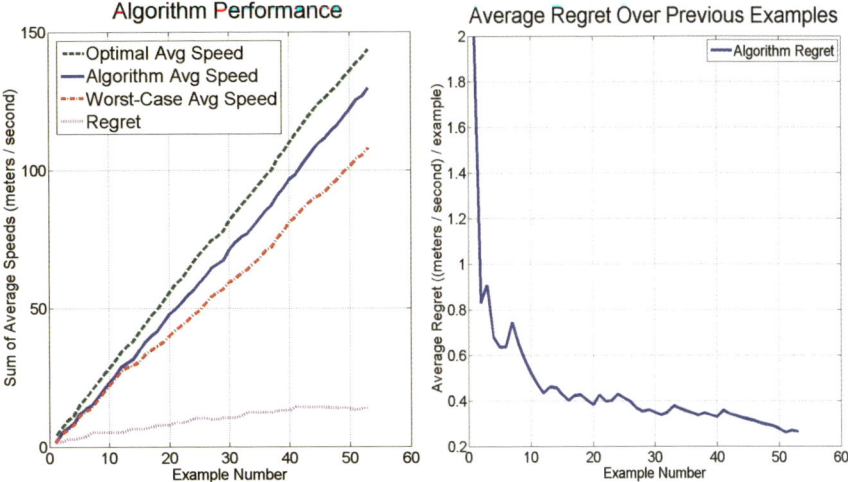

Fig. 5 Overhead data selection performance: sum of average navigation speed over each path segment for our algorithm and various alternatives (left) and the average regret of our algorithm over previous segments (right).

Table 2 Online Overhead Data Selection Performance

Algorithm	Average Speed (meters / second)	Percent Improvement over Random
Online Algorithm	2.45	5.60
Optimal	2.71	16.81
Worst-Case	2.04	-12.07
Random Data Source	2.32	0.00

the learned models for each candidate at that time. Selections at each step were based on the initial step of this locally computed optimal allocation. To avoid having to do integer programming, we chose the candidate with the highest allocation at the first step.

The course and estimated rate of progress using each data source predicted by the trained model appear in Figure 4. Quantitative results for this scenario appear in Figure 5 and Table 2.

Our algorithm shows a clear improvement over naive or random approaches for both scenarios with quickly-converging regret properties.

5   Conclusion

We have presented an online algorithm for dealing with scenarios where the robot must learn to trade-off between multiple operating modes. The proposed approach relies on a bandit-based framework and uses confidence bounds to deal with exploration-exploitation trade-offs. The algorithm was demonstrated on two scenarios relevant to the mobile robotics domain and showed improved performance over several alternatives. We hope that such techniques will increase the potential real-world applications of mobile robots by allowing them to adapt in real-time to changing environments and better allocate available resources.

Acknowledgements. This work was partially sponsored by DARPA under contract Unmanned Ground Combat Vehicle - PerceptOR Integration (contract number MDA972-01-9-0005) and by the U.S. Army Research Laboratory under contract Robotics Collaborative Technology Alliance (contract number DAAD19-01-2-0012). The views and conclusions contained in this document are those of the authors and should not be interpreted as representing the official policies, either expressed or implied, of the U.S. Government.

Boris Sofman is partially supported by a Sandia National Laboratories Excellence in Engineering Fellowship. The author gratefully acknowledges Sandia's Campus Executive Laboratory Directed Research and Development Program (LDRD) for this support.

References

1. Stentz, A., Bares, J., Pilarski, T., Stager, D.: The crusher system for autonomous navigation. In: AUVSIs Unmanned Systems North America (August 2007)
2. Silver, D., Bagnell, J.A., Stentz, A.: High performance outdoor navigation from overhead data using imitation learning. In: Robotics Science and Systems (June 2008)
3. Dias, M.B., Kannan, B., Browning, B., Jones, E., Argall, B., Dias, M.F., Zinck, M.B., Veloso, M.M., Stentz, A.: Sliding autonomy for peer-to-peer human-robot teams. In: 10th International Conference on Intelligent Autonomous Systems 2008 (July 2008)
4. Scerri, P., Pynadath, D.V., Tambe, M.: Towards adjustable autonomy for the real world. Journal of Artificial Intelligence Research 17, 2002 (2002)
5. Grace, R., Byrne, V., Bierman, D., Legrand, J.-M., Gricourt, D., Davis, B., Staszewski, J., Carnahan, B.: A drowsy driver detection system for heavy vehicles. In: Proceedings of the 17th Digital Avionics Systems Conference, vol. 2, pp. I36/1 – I36/8 (2001)
6. Vahidi, A., Eskandarian, A.: Research advances in intelligent collision avoidance and adaptive cruise control. IEEE Transactions on Intelligent Transportation Systems 4(3), 143–153 (2003)
7. Bishop, R.: Intelligent vehicle applications worldwide. IEEE Intelligent Systems 15(1), 78–81 (2000)
8. Krotkov, E., Simmons, R., Cozman, F., Koenig, S.: Safeguarded teleoperation for lunar rovers: From human factors to field trials. In: Proc. IEEE Planetary Rover Technology and Systems Workshop (1996)
9. Krupa, A., de Mathelin, M., Doignon, C., Gangloff, J., Morel, G., Soler, L., Marescaux, J.: Development of semi-autonomous control modes in laparoscopic surgery using automatic visual servoing. In: Niessen, W.J., Viergever, M.A. (eds.) MICCAI 2001. LNCS, vol. 2208, pp. 1306–1307. Springer, Heidelberg (2001)
10. Heger, F.W., Singh, S.: Sliding autonomy for complex coordinated multi-robot tasks: Analysis & experiments. In: Sukhatme, G.S., Schaal, S., Burgard, W., Fox, D. (eds.) Robotics: Science and Systems. The MIT Press, Cambridge (2006)
11. Fong, T.W., Thorpe, C., Baur, C.: Multi-robot remote driving with collaborative control. IEEE Transactions on Industrial Electronics (2003)
12. Stentz, A., Dima, C., Wellington, C., Herman, H., Stager, D.: A system for semi-autonomous tractor operations. Auton. Robots 13(1), 87–104 (2002)

13. Horvitz, E., Jacobs, A., Hovel, D.: Attention-sensitive alerting. In: Laskey, K.B., Prade, H., Cal, S.F. (eds.) Proceedings of the 15th Conference on Uncertainty in Artificial Intelligence (UAI 1999), July 30-August 1, pp. 305–313. Morgan Kaufmann Publishers, San Francisco (1999)
14. Hexmoor, H.: A cognitive model of situated autonomy. In: Kowalczyk, R., Loke, S.W., Reed, N.E., Graham, G. (eds.) PRICAI-WS 2000. LNCS (LNAI), vol. 2112, pp. 325–334. Springer, Heidelberg (2001)
15. Gunderson, J.P., Martin, W.N.: Effects of uncertainty on variable autonomy in maintenance robots. In: Workshop on Autonomy Control Software, pp. 26–34 (1999)
16. Goodrich, M.A., Schultz, A.C.: Human-robot interaction: A survey. Foundations and Trends in Human-Computer Interaction 1(3), 203–275 (2007)
17. Robbins, H.: Some aspects of the sequential design of experiments. Bull. Amer. Math. Soc. 58(5), 527–535 (1952)
18. Lai, T., Robbins, H.: Asymptotically efficient adaptive allocation rules. Advances in applied mathematics (Print) 6(1), 4–22 (1985)
19. Auer, P., Cesa-Bianchi, N., Fischer, P.: Finite-time analysis of the multiarmed bandit problem. Machine Learning 47 2(3), 235–256 (2002)
20. Wang, C., Kulkarni, S., Poor, H.: Bandit problems with side observations. IEEE Transactions on Automatic Control 50(3), 338–355 (2005)
21. Auer, P.: Using confidence bounds for exploitation-exploration trade-offs. The Journal of Machine Learning Research 3, 397–422 (2003)
22. Sofman, B., Ratliff, E.L., Bagnell, J.A., Cole, J., Vandapel, N., Stentz, A.: Improving robot navigation through self-supervised online learning. Journal of Field Robotics 23(1) (December 2006)

Applied Imitation Learning for Autonomous Navigation in Complex Natural Terrain

David Silver, J. Andrew Bagnell, and Anthony Stentz

Abstract. Rough terrain autonomous navigation continues to pose a challenge to the robotics community. Robust navigation by a mobile robot depends not only on the individual performance of perception and planning systems, but on how well these systems are coupled. When traversing rough terrain, this coupling (in the form of a cost function) has a large impact on robot performance, necessitating a robust design. This paper explores the application of *Imitation Learning* to this task for the Crusher autonomous navigation platform. Using expert examples of proper navigation behavior, mappings from both online and offline perceptual data to planning costs are learned. Challenges in adapting existing techniques to complex online planning systems are addressed, along with additional practical considerations. The benefits to autonomous performance of this approach are examined, as well as the decrease in necessary designer interaction. Experimental results are presented from autonomous traverses through complex natural terrains.

1 Introduction

The capability of autonomous robotic systems to successfully navigate through unstructured environments continues to advance. Ever improving high resolution sensors and perception algorithms allow a mobile robot to build a detailed model of its environment, and advances in planning systems allow for the generation of ever more complex routes and trajectories towards achieving a navigation goal. However, as perception and planning systems become more complex, so does the task of coupling these systems. This coupling often takes the form of a *Cost Function*. Using data from the perception system as input, the cost function maps to a scalar cost value, defined over the state space of the planning system (Figure 1). These costs are then used as the optimization metric by the planning system when determining the next action or sequence of actions.

David Silver, J. Andrew Bagnell, and Anthony Stentz
Robotics Institute, Carnegie Mellon University

A. Howard et al. (Eds.): Field and Service Robotics 7, STAR 62, pp. 249–259.
springerlink.com

Fig. 1 Crusher (Left) is capable of autonomous navigation through complex outdoor terrain. Perceptual data (Top Right) are converted to costs (Bottom Right) for use by the planning system.

In simple or structured environments, cost functions are often easily defined; for instance, in an indoor environment the cost of traversable freespace should be very low, and the cost of walls or other obstacles should be high. However, in rough or unstructured terrain, it is less intuitive how to define cost. A small obstacle should clearly have larger cost than flat ground, and smaller cost than a large obstacle. Explicitly defining these tradeoffs encodes the desired behavior of the robot and is quite challenging; defining a generalizeable function that maps from perceptual inputs to the proper cost is even more so.

This first step of defining the relative cost of various terrains requires a concrete definition of what metric a robot's performance will be measured against. Common metrics include maximizing safety or probability of success, minimizing distance traveled or time taken, minimizing net energy loss, minimizing observability or maximizing sensor coverage. Often, the actual desired robot behavior optimizes a combination of such metrics; for example, it may be desirable for a robot to approximately maximize safety but take certain risks to minimize distance traveled.

Previous work has focused on several differing approaches. Attempts to explicitly approximate traversability through simulation [3] or proprioception [5] limit the choice of metrics to maximizing safety; they also require a robot model capable of directly computing probabilities of mobility failure. Approaches focused on explicitly combining multiple metrics are limited to optimizing one metric subject to constraints on others [13, 15]. The most common general solution is to manually design and hand tune a cost function until the robot achieves the desired behavior. This can be an incredibly tedious process (as it requires a manual optimization in a potentially high dimensional space) and often results in systems that suffer from poor generalization and a lack of robustness to novel scenarios.

This paper explores the application of imitation learning to this challenge, specifically the LEARCH [10] algorithm described in the next section. The application of this approach to the Crusher autonomous navigation platform [14] (Figure 1) is reviewed, along with a discussion of practical considerations when applying this approach.

```
C_0 = Prior;
for i = 1...T do
    foreach P_e do
        foreach x ∈ getBoundingBox(P_e) do
            F[x] = getPerceptionFeatures(x);
            M[x] = C_{i-1}(F[x]) + L_e(x);
        P_* = planPath(s_e, g_e, M);
        P_e^* = replanExamplePath(P_e, β, M);
        {U_+^{P_e}, U_-^{P_e}} = computeVisitationCounts(P_e^*, P_*);
    R_i = trainBalancedRegressor(F, U_+, U_-);
    C_i = C_{i-1} * e^{η_i R_i};
```

Fig. 2 The LEARCH algorithm

2 Imitation Learning

Although explicitly defining the relative tradeoffs between different actions is a difficult task for a domain expert, indicating or demonstrating examples of correct behavior is often easier (otherwise, the task itself is not well defined). Therefore, the imitation learning framework seeks to learn the correct robot behavior from observation of expert behavior. Many previous applications of this framework to mobile robots [6, 7] sought to learn to predict what action to perform, based on the action performed by an expert at a certain state. In this way, action prediction essentially replaces the lower level motion planning operation on a mobile robot. However, this approach is inherently myopic and does not scale to longer range planning, as it requires all necessary information to be encoded in the current robot state.

Therefore, rather than learn a mapping from features of a state to actions, we seek to learn a mapping from features of a state to costs, such that the planning system will produce the correct behavior when provided with said costs. This approach has its roots in the concept of *Inverse Optimal Control*, and has recently been developed for use in robotic systems [1, 8]. By learning a cost function to reproduce expert behavior, the need for explicitly defining a metric or weighting between metrics is eliminated; the new metric is matching human performance, and it is left up to the human expert to define (through behavior) how to balance various options and considerations.

It is important to note that this approach learns the correct cost function for a specific planner or system of planners, and maps to cost from a specific perception system. The purpose is not to try and improve the separate performance of these systems; rather, it is to optimize the coupling of these modules to provide the best overall system performance. As the Crusher system operates with costs defined over a 2D grid, subsequent descriptions will deal specifically with this setting. Without loss of generality, it is easiest for now to consider the planning system as a basic A* planner, and a planned path as a sequence of 2D grid cells. Adaptation to more complex planning systems is covered in the next section.

Our imitation learning approach is based on the LEARCH algorithm, (Figure 2); for a full derivation, see [10]. The input to the algorithm is a set of example paths, each a sequence of 2D locations leading from a start s to a goal g and representing the correct path (according to the expert). LEARCH seeks to find a cost function C such that each example path P_e is the planner output under the cost function. The LEARCH inner loop iterates through each example. For each example path P_e with start and goal s_e and g_e a path P_* is planned under the current candidate cost function C_i (C_0 can be initialized to any prior). Since P_* is the output of an optimal planner, $C_i(P_*) \leq C_i(P_e)$. Since we desire a C such that $C(P_e) = C(P_*)$, we seek to minimize the difference in cost $C(P_e) - C(P_*)$. As the cost of a path is simply the sum of costs at states along it, P_e and P_* provide a list of states where the cost could be changed to lower this cost difference: states in P_e could have their cost lowered and states in P_* could have their cost raised (states in both simply cancel).

This list of candidate states (called the visitation counts) provides a local gradient in the space of cost functions. However, the cost at each state can not simply be raised or lowered, as the goal is not to identify the correct cost for each cell, but rather a function that maps perceptual features to an appropriate cost. If a function ΔC_i could be found that approximated this gradient (the output is positive or negative when provided with the appropriate features), adding it to C_i would lower the cost difference.

Finding a general function to match a list of input/output pairs can be solved through regression analysis. In this case, the inputs are well defined (perceptual features), but the outputs are not; for each input, the required cost delta is not known, just its sign. Therefore, outputs targets are specified as ± 1 depending on whether the costs need to be raised or lowered. In this way, the regressor R generalizes the local cost changes over the entire feature space. Each regression target can also be weighted to indicate that certain cost changes are more important relative to others. Determining these relative weights is discussed in Section 3.

The final learning procedure is summarized as follows: for each example path and the corresponding plan (under C_i), compute the set of visitation counts and the corresponding perceptual features. Next train a regressor R over these input/output pairs, and combine it with C_i, weighted by a learning rate parameter η. The cost of cells along an example path (which an expert specifically chose to encounter) will generally be lowered, while the cost of cells along a temporarily cheaper path (which the expert chose to avoid) will generally be increased. This loop is then iterated until convergence. Figure 3 provides a visual example of this procedure in action.

A few details remain. Rather than summation, we use an update rule of $C_{i+1}(f) = C_i(f)e^{\eta R_i(f)}$, resulting in exponentiated functional gradient descent [10] which makes better use of available dynamic range, as well as naturally enforcing a positivity constraint on costs. The choice of regressor (e.g. linear, neural net, etc.) is also unspecified. This decision helps define the balance between descriptiveness and generalization in the space of possible cost functions, and is discussed further in Section 3. Finally, by augmenting costs with a margin (determined by a loss function $L_e(x)$), trivial cost solutions can be eliminated and generalization improved.

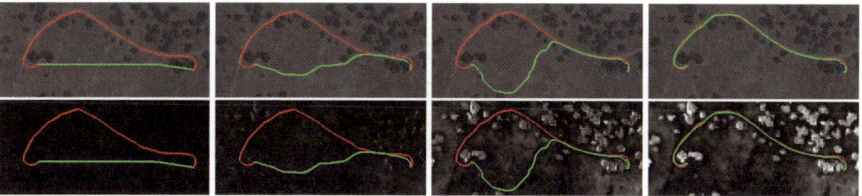

Fig. 3 An example of the LEARCH algorithm learning to interpret satellite imagery (Top) as costs (Bottom). As the cost function evolves (left to right), the current plan (green) recreates more and more of the example plan (red). Quickbird imagery courtesy of Digital Globe, Inc.

In addition to the algorithm as described, there are a few modifications that can increase robustness to noisy or imperfect expert demonstration [11, 12]. Human experts rarely demonstrate exact optimal behaviors, especially in large areas of similar terrain. By performing a balanced regression (normalizing relative weights such that positive and negative targets have equal total weight), the regressor can be forced to more strictly separate between terrains when changing cost. Another addition to the algorithm is to continually replan the example path. Human demonstration often contains a degree of high frequency noise; a smoothing operation is therefore beneficial. This smoothing can be performed by selecting a new example that is entirely contained within a corridor of width β around the original example. The new example is created by planning the optimal path using the current cost function within the corridor (and infinite cost elsewhere). This has the effect of adapting the example to the current cost hypothesis at a small scale (implicitly defined by β), while still adapting the hypothesis to the example at a large scale.

3 Application to Autonomous Navigation

Our imitation learning approach was applied to the task of interpreting perceptual data for the purpose of motion planning on the Crusher system. Crusher is provided with two main forms of perceptual data: static sources of prior data (overhead imagery and LiDAR), and dynamic sources of data collected in real time (onboard cameras and LiDAR). Once these data sources have been converted to 2D cost grids and fused together (at the cost level), Crusher's motion planning system is responsible for choosing vehicle actions. The motion planning system is similar to [4], and combines a global planner based on Field D* [2] and a local planner based on forward simulation of potential vehicle actions (specifically constant curvature commands) for a fixed horizon.

For the Crusher platform, imitation learning was first applied to the task of interpreting overhead data to create prior cost maps [11]. 2D feature maps are extracted from the input data sources, and then converted from maps of features to a map of costs. These maps are then used for global route planning offline, as well as online global planning when fused with current perceptual data. This context provides an ideal setting for the application of the LEARCH algorithm due to the static nature of the perceptual data, and the ease of collecting training examples: each example is simply a path that can be 'drawn' by an expert on top of imagery or other visualization of the underlying data.

Since overhead costs are only used for planning in regions that have not yet been directly observed by the robot, overhead costs are learned with respect to the Field D* global planner. As Field D* plans an interpolated path over a 2D grid, computing visitation counts is not as straightforward as simply marking which states each path traverses through. Instead, the distance traveled through each grid cell must be recorded. This results in visitation counts that are real valued instead of binary. During regression the output target for a real valued visitation count is still ± 1, but the target is now weighted relative to the visitation count. If a path passes through cell x_1 for twice as long as x_2, then x_1 has twice the impact on the cost of a path; moving the cost in the right direction is therefore twice as important.

Like many motion planning algorithms, Field D* also makes use of a configuration space expansion to account for the dimensions of the vehicle. A configuration space expansion also results in non-binary visitation counts; it is taken into account by incrementing the count of all states x_j relative to their contribution to the cost of state x_i when x_i is on a path. Crusher's planning system performs an expansion by averaging costs over a circular window. Therefore, for a path traveling distance d through cell x_i, all cells x_j within the expansion window have their counts incremented by d. More complex expansions can be accounted for in the same manner.

Imitation learning was next used to learn costs from features generated by Crusher's onboard perception system. Crusher's perception software processes raw sensor data into feature descriptions of voxels in real time; each column of voxels is then converted into a 2D cost. Therefore, unlike learning from overhead data, features are not static. While additional adaption of the LEARCH algorithm is required in order to deal with the dynamic and unknown nature of real time perceptual data, the approach remains conceptually similar and will be treated as such moving forward; for details see [12].

Instead of drawing a path on top of a visualization, collecting expert examples to train the perception system consists of the expert manually driving Crusher through an example behavior. Along with the path traversed, all raw sensor data is logged during this collection. Sensor data is then post-processed via playback through Crusher's perception software to generate the final features that will be converted into costs. By logging the raw sensor data, perception software does not need to remain static after training data collection. Whenever changes or improvements are made to perception software, features can simply be regenerated, and a cost function relearned. Therefore, training data in this form does not need to be recollected every time the system changes; the cost function is simply retrained offline.

Since costs from online perceptual data determine Crusher's actual motion commands, costs must be learned with respect to the local planning system. Figure 4(a) provides a simplified example to demonstrate why this is so. If costs were trained with respect to the global planner, LEARCH would be satisfied with the cost on the obstacle O when it is sufficiently high to make up for the extra distance $|P_g| - |P_c|$. However, since $|P_l|, |P_r| > |P_c| + O > |P_g|$, P_c remains the cheapest local planner

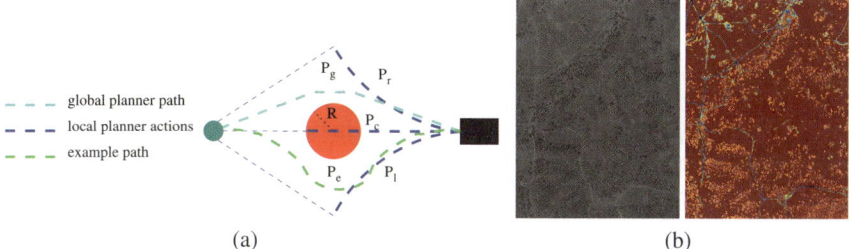

(a)   (b)

Fig. 4 **(a)** A simplified scenario where the local planner has only 3 possible actions.
(b) Example of a new feature (right) learned from panchromatic imagery (left).

option in this case. The result would be that the local planner would still choose to
drive over the obstacle. This result is observed empirically in Section 4.

Unfortunately, there are also problems with training directly for the local planner.
As the local planner only considers a discrete, kinematically feasible set of actions, it
is often the case that no series of actions will sufficiently match the expert example.
In this case, the LEARCH termination condition is undefined. Terminating when
the example path is lower cost than the planned path will not suffice; in Figure 4(a)
this could result in $C(P_e) < C(P_c) < C(P_l), C(P_r)$ (the colliding action would still be
preferred). Running until the cost function converges is therefore necessary, but has
its own side affects. Once $C(P_c) > C(P_l), C(P_r)$, LEARCH will start to try and raise
the cost along P_l or P_r. If the chosen regressor can differentiate between the terrain
under P_l or P_r and that under P_e, it will raise those costs without proper cause. The
end result is a potential addition of noise to the final costs, and lower generalization.
The degree of this noise depends on the resolution of the planner and the regressor.

If there were some way to know the 'right' planner action, then the selection
of that action could serve as a termination condition. Collecting this information
during demonstration by an expert would be extremely tedious, requiring an expert selection at every planning cycle. Instead, we propose the use of a heuristic
approach to approximate this decision. Essentially, we seek to 'project' the expert's
example behavior onto the space of possible planner actions. This is performed by
first learning a perception cost function for the global planner. As described above,
such a cost function will generally underestimate the cost necessary for the local
planner. Therefore, we score each local planner action by its *average* cost instead of
total cost[1]. An action with low average cost can not be said to be optimal, but it at
least traverses desirable(low cost) terrain. An additional distance penalty is added to
bias action scores towards those that make progress towards the goal[2]. After scoring
each action, that with the lowest score is used as the new example. The result of this
initial replanning step is to produce a new example behavior that is feasible to the
local planner.

---

[1] The global planner section of each action is still computed with respect to total cost.
[2] The weight of this penalty can be automatically tuned by optimizing performance on a
validation set, without any hand tuning.

When dealing with static overhead data a cost function, once learned, will generally only be applied once through a data set. In contrast, a perception cost function will be continually applied in real time on a robot when operating autonomously. Therefore, it is important that the cost function be computationally inexpensive. As combining multiple linear functions yields a single linear function, linear regressors have a significant computational advantage over nonlinear regressors (which would require a separate evaluation per regressor). Unfortunately, they also suffer from limited expressiveness. This can be dealt with by adding a feature learning phase, as described in [9, 11]. Such a phase automatically decides to occasionally learn simple non-linear combinations of the original input features that help differentiate terrains that are difficult for a linear cost function to discriminate. This approach is similar to the way in which cost functions are often hand-engineered: simple linear functions handle the general cases, with sets of rules to handle difficult special cases. Additionally, such new features can be used as a guide in the development of further engineered features. Figure 4(b) provides an example in the overhead context, demonstrating a new feature derived from only panchromatic satellite imagery. This new feature strongly disambiguates roads and trails from surrounding terrain, and could be taken as an indication that an explicit road extractor would be useful.

4   Field Results and Conclusions

The described imitation learning approach was implemented to learn mappings from both overhead and onboard perceptual data to cost. As described in [10], this approach is guaranteed to converge when using a properly chosen learning rate [3]. In practice, a decaying learning rate of the form $\eta/\sqrt{n}$ is used at the nth iteration. The parameter η affects only the rate of convergence; a well chosen η usually results in convergence after approximately 50 - 100 iterations. The computation required at each iteration for each example is dominated by the cost of applying the current cost function to local feature maps, and then planning through the resulting cost map. For the training sets used in this work, computation per iteration was approximately 5 minutes on a 2.4 Ghz processor[4].

The Crusher autonomy system originally made use of hand-tuned cost functions for converting overhead and perception features to costs. Comparing autonomous performance when using different cost functions can quantify the differences in performance brought about by these different approaches. Engineered prior cost maps were used for the first 4 Crusher field experiments. This process consisted of first performing a supervised classification of the raw feature maps, and then converting the classifier outputs into costs. This lossy compression of the feature space was performed to make designing a cost function easier and more intuitive. As different test sites provided differing data sources and resolutions, this process was repeated for each test site. Additionally, the desire to create cost maps from different subsets

---

[3] Use of a smoothing corridor removes the theoretical guarantee; however in practice this has not proven to affect convergence.

[4] If faster learning is required, LEARCH can be parallelized by example at each iteration.

and resolutions of prior data (in order to perform resolution comparisons), meant multiple cost functions were necessary. For each site, labeling training data and determining parameters for multiple cost functions would involve on average more than a day of a domain expert's time. Learned prior cost maps were then used for the remaining 6 field experiments. For each site, producing a series of example paths would take on average only 1-2 hours of an expert's time. These examples could then be used to train multiple cost maps using different data sources and resolutions.

In a timed experiment on a 2 km^2 test site, producing a supervised classification required 40 minutes of expert involvement, and tuning a cost function required an additional 20 minutes. In contrast, producing example paths required only 12 minutes. As Crusher has been tested on sites ranging up to 200 km^2, this time savings is magnified in importance. The final cost maps were also evaluated by comparing planned routes to an independent validation set of examples. The engineered map produced routes that matched 44% of states along validation paths on average. Using imitation learning to learn just the correct weights for the supervised classification produced a map that scored 48%. Imitation learning from the raw features scored 57%. This result demonstrates that the automated approach performs superior parameter tuning, and makes better use of all the available raw data. It has also been shown that Crusher navigates more efficiently when using learned prior maps online as opposed to engineered maps, driving safer routes at faster speeds [11].

During the more than 3 years of the Crusher program, an engineered perception cost function was continually redesigned and retuned, culminating in a high performance system [14]. However, this performance came at a high cost. Version control logs indicate that 145 changes were made to just the form of the cost function; additionally more than 300 parameter changes were checked in. As each committed change requires significant time to design, implement, and validate, easily hundreds of hours were spent on engineering the cost function. In contrast, the final training set used to learn a cost function consisted of examples collected in only a few hours.

The performance of different perception cost functions was compared through over 150 km of comparison trials. The final results comparing 4 different cost functions are presented in Table 1. In comparison to the engineered system, a cost function learned for the global planner resulted in overly aggressive performance. As discussed in Section 3, learning in this manner does not result in sufficiently high costs; the result is that Crusher drives faster and turns less while appearing to suffer from increased mobility risk. In contrast, the costs learned for the local planner

Table 1 Averages over 295 different waypoint to waypoint trials per perception system, totaling over 150km of traverse. Statistically significant differences (from Engineered) denoted by *

System	Avg. Distance Made Good (m)	Avg. Cmd. Vel. (m/s)	Avg. Cmd. Ang. Vel.(o/s)	Avg. Lat. Vel. (m/s)	Dir Switch Per m	Avg. Motor Current (A)	Avg. Roll(o)	Avg. Pitch(o)	Avg Vert. Accel (m/s^2)	Avg Lat. Accel (m/s^2)	Susp. MaxΔ (m)	Safety E-stops
Engineered	130.7	3.24	6.56	0.181	0.107	7.53	4.06	2.21	0.696	0.997	0.239	0.027
Global	123.8*	3.34*	4.96*	0.170*	0.081*	7.11*	4.02	2.22	0.710*	0.966*	0.237	0.054*
Local	127.3	3.28	5.93*	0.172*	0.100	7.35	4.06	2.22	0.699	0.969*	0.237	0.034
Local w/replan	124.3*	3.39*	5.08*	0.170*	0.082*	7.02*	3.90*	2.18	0.706*	0.966*	0.234*	0.030

performed very similarly to the high performance of the engineered system. Additionally, adding an initial replanning step further improved performance; by reducing cost noise, average speed increased, with a decrease in turns and direction switches, and no increase in mobility risk.

In conclusion, this work has demonstrated the applicability of imitation learning towards improving the robustness of autonomous navigation systems, while helping to minimize the necessary amount of expert interaction. Specifically, the parameter tuning problem that often results from the coupling of complex perception and planning systems can be automated through expert demonstration instead of expert intervention. In the future, we wish to expand this approach to also automate the selection of parameters internal to a planning system, further reducing the need for human tuning.

Acknowledgements. This work was sponsored by DARPA under contract "Unmanned Ground Combat Vehicle - PerceptOR Integration" (contract MDA972-01-9-0005) and by the U.S. Army Research Laboratory under contract "Robotics Collaborative Technology Alliance" (contract DAAD19-01-2-0012). The views and conclusions contained in this document are those of the authors and should not be interpreted as representing the official policies, either expressed or implied, of the U.S. Government.

References

1. Abbeel, P., Ng, A.: Apprenticeship learning via inverse reinforcement learning. In: International Conference on Machine Learning (2004)
2. Ferguson, D., Stentz, A.: Using interpolation to improve path planning: The field d* algorithm. Journal of Field Robotics 23(2), 79–101 (2006)
3. Green, A., Rye, D.: Sensible planning for vehicles operating over difficult unstructured terrains. In: IEEE Aerospace Conf. (2007)
4. Kelly, A., Stentz, A., Amidi, O., Bode, M., Bradley, D., Diaz-Calderon, A., Happold, M., Herman, H., Mandelbaum, R., Pilarski, T., Rander, P., Thayer, S., Vallidis, N., Warner, R.: Toward reliable off road autonomous vehicles operating in challenging environments. International Journal of Robotics Research 25(5-6), 449–483 (2006)
5. Kim, D., Sun, J., Oh, S.M., Rehg, J.M., Bobick, A.F.: Traversability classification using unsupervised on-line visual learning. In: IEEE International Conference on Robotics and Automation (2006)
6. LeCun, Y., Muller, U., Ben, J., Cosatto, E., Flepp, B.: Off-road obstacle avoidance through end-to-end learning. In: Advances in Neural Information Processing Systems, vol. 18 (2006)
7. Pomerleau, D.: Alvinn: an autonomous land vehicle in a neural network. In: Advances in neural information processing systems, vol. 1, pp. 305–313 (1989)
8. Ratliff, N., Bagnell, J., Zinkevich, M.: Maximum margin planning. In: International Conference on Machine Learning (2006)
9. Ratliff, N., Bradley, D., Bagnell, J., Chestnutt, J.: Boosting structured prediction for imitation learning. In: Advances in Neural Information Processing Systems, vol. 19. MIT Press, Cambridge (2007)
10. Ratliff, N.D., Bagnell, J.A., Silver, D.: Learning to search: Functional gradient techniques for imitation learning. Autonomous Robots (2009)

11. Silver, D., Bagnell, J.A., Stentz, A.: High performance outdoor navigation from overhead data using imitation learning. In: Proceedings of Robotics Science and Systems (2008)
12. Silver, D., Bagnell, J.A., Stentz, A.: Perceptual interpretation for autonomous navigation through dynamic imitation learning. In: ISRR (2009)
13. Stentz, A.: CD*: a real-time resolution optimal re-planner for globally constrained problems. In: Proceedings of AAAI National Conference on Artificial Intelligence (2002)
14. Stentz, A., Bares, J., Pilarski, T., Stager, D.: The crusher system for autonomous navigation. In: AUVSIs Unmanned Systems (2007)
15. Tompkins, P., Stentz, A., Whittaker, W.: Mission planning for the sun-synchronous navigation field experiment. In: IEEE International Conference on Robotics and Automation (2002)

Part VI
Underwater Localization and Mapping

Trajectory Design for Autonomous Underwater Vehicles Based on Ocean Model Predictions for Feature Tracking

Ryan N. Smith, Yi Chao, Burton H. Jones, David A. Caron,
Peggy P. Li, and Gaurav S. Sukhatme

Abstract. Trajectory design for Autonomous Underwater Vehicles (AUVs) is of great importance to the oceanographic research community. Intelligent planning is required to maneuver a vehicle to high-valued locations for data collection. We consider the use of ocean model predictions to determine the locations to be visited by an AUV, which then provides near-real time, *in situ* measurements back to the model to increase the skill of future predictions. The motion planning problem of steering the vehicle between the computed waypoints is not considered here. Our focus is on the algorithm to determine relevant points of interest for a chosen oceanographic feature. This represents a first approach to an end to end autonomous prediction and tasking system for aquatic, mobile sensor networks. We design a sampling plan and present experimental results with AUV retasking in the Southern California Bight (SCB) off the coast of Los Angeles.

1 Introduction

More than three-fourths of our earth is covered by water, yet we have explored less than 5% of the aquatic environment. Autonomous Underwater Vehicles (AUVs) play a major role in the collection of oceanographic data. To make new discoveries and improve our overall understanding of the ocean, scientists must make use of these platforms by implementing effective monitoring and sampling techniques to

Ryan N. Smith and Gaurav S. Sukhatme
Robotic Embedded Systems Laboratory, University of Southern California,
Los Angeles, CA 90089 USA
e-mail: {ryannsmi,gaurav}@usc.edu

Yi Chao and Peggy P. Li
Jet Propulsion Laboratory, California Institute of Technology,
4800 Oak Grove Drive, Pasadena, CA 91109 USA
e-mail: {yi.chao,p.p.li}@jpl.nasa.gov

Burto H. Jones and David A. Caron
Department of Biological Sciences, University of Southern California,
Los Angeles, CA 90089 USA
e-mail: bjones,dcaronusc.edu

A. Howard et al. (Eds.): Field and Service Robotics 7, STAR 62, pp. 263–273.

study ocean upwelling, tidal mixing or other ocean processes. One emerging example of innovative and intelligent ocean sampling is the automatic and coordinated control of autonomous and Lagrangian sensor platforms [4].

As complex and understudied as the ocean may be, we are able to model and predict certain behaviors moderately well over short time periods. Expanding our modeling capabilities, and general knowledge of the ocean, will help us better exploit the resources that it has to offer. Consistently comparing model predictions with actual events, and adjusting for discrepencies, will increase the range of validity of existing ocean models, both temporally and spatially.

The goal of this paper is to present an innovative ocean sampling method that utilizes model predictions and AUVs to collect interesting oceanographic data that can also increase model skill. Our motivation is to track and collect daily information about an ocean process or feature which has a lifespan on the order of a week. We use an ocean model to predict the behavior of an interesting artifact, *e.g.,* a fresh water plume, over a small time period, *e.g.,* one day. This prediction is then used as input to an algorithm that determines a sampling plan for the AUV(s). The AUV(s) are then retasked from a current mission or deployed. Afterward, the collected data is assimilated into the ocean model and an updated prediction is computed. A new sampling plan is created and the process repeats until the artifact is out of range or is no longer of interest.

We motivate the work from an oceanographic perspective and provide a realistic field application. Next, we briefly describe the ocean model and AUV used in this study. We discuss the waypoint selection algorithm and present results from a field implementation. We conclude with future research plans.

The work presented here serves as a proof of concept for the utilization of ocean model forecasts to design sampling missions for AUVs in particular, and aquatic mobile sensor platforms in general, to follow an ocean feature and collect data.

2   Oceanography Application and Ocean Model

Microscopic organisms are the base of the food chain: all aquatic life ultimately depends upon them for food. There are a few dozen species of phytoplankton and cyanobacteria that can create potent toxins when provided with the right conditions. Harmful algal blooms (HABs) can cause harm via toxin production, or by their accumulated biomass. Such blooms can cause severe illness and potential death to humans as well as to fish, birds and other mammals. The blooms generally occur near fresh water inlets, where large amounts of nutrient rich, fresh water is deposited into the ocean. This water provdes the excess food to support higher productivity and a *bloom* of microorganisms. It is of interest to predict when and where HABs may form, and which coastal areas they may affect. Harmful algal blooms are an active area of research along the western coast of the United States and are of large concern for coastal communities in southern California. The impact of HABs in this region can be seen in [7, 10]. With this motivation, we choose fresh water plumes as an ocean feature for which to design predictive tracking missions.

The predictive tool utilized in this study is the Regional Ocean Model System (ROMS) [9] - a split-explicit, free-surface, topography-following-coordinate oceanic model. We use ROMS because it is an open source ocean model that is widely accepted and supported throughout the oceanographic and modeling communities. Additionally, the model was developed to study ocean processes along the western U.S. coast which is our primary area of study.

Research is currently ongoing to update and improve ROMS for the Southern California Bight (SCB)[1] in an effort to characterize and understand the complex upwelling and current structure that exist and drive the local climate. The Jet Propulsion Laboratory (JPL) uses ROMS to provide nowcasts and hourly forecasts (up to 36 hours) for Monterey Bay, the SCB and Prince William Sound, see [6] for more information. The JPL version of ROMS assimilates HF radar surface current measurements, data from moorings, satellite data and any data available from AUVs operating in the area. Information regarding this specific version of ROMS and the data assimilation process can be found in [3].

3   Mobile Sensor Platform: AUV

The mobile sensor platform used in this study is a Webb Slocum autonomous underwater glider, as seen in Fig. 1. (http://www.webbresearch.com) The Slocum glider is a type of AUV designed for long-term ocean sampling and monitoring [8]. These gliders *fly* through the water by altering the position of their center of mass and changing their buoyancy. Due to this method of locomotion, gliders are not fast moving AUVs, and generally have operational velocities on the same order of magnitude as oceanic currents. The endurance and velocity characteristics of the glider make it a good candidate vehicle to track ocean features which have movements that are determined by currents, and that have a residence time on the order of weeks.

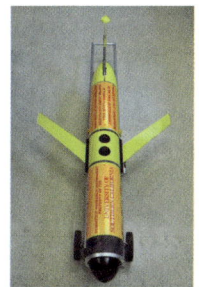

Fig. 1 *He Ha Pe*, one of two USC Slocum gliders, flying a mission off the coast of Catalina Island.

We utilize autonomous gliders because our collaborative research group owns two of them, and hence field experiments can be readily performed. We have upgraded the communication capabilities of our vehicles to take advantage of our local wireless network; details on this can be found in the concurrent article, [5].

Extensive research has been done on glider dynamics and controller design, *e.g.*, see [2] and the references therein. Thus, we do not discuss these details nor the trajectory along which the glider travels. We assume here that the glider can successfully navigate from one location to another.

---

[1] The SCB is the oceanic region contained within $32°$ N to $34.5°$ N and $-117°$ E to $-121°$ E.

4 Trajectory Design

We now present an algorithm which generates the locations for the AUV to visit to follow the general movements of a fresh water plume through the ocean.

Considerable study has been reported on adaptive control of single gliders and coordinated multi-glider systems, see for example [4] and the included references. In these papers, the trajectories given to the gliders were fixed patterns (rounded polygons) that were predetermined by a human operator. The adaptive control component was implemented to keep the gliders in an optimal position, relative to the other gliders following the same trajectory. The difference between the method used in [4] and the approach described here, is that here the sampling trajectory is determined by use of the output of ROMS, and thus is, at first glance, a seemingly random and irregular sampling pattern. Such an approach is a benefit to the model and scientist alike. Scientists can identify sampling locations based upon ocean measurements they are interested in following, rather than setting a predetermined trajectory and hoping the feature enters the transect while the AUV is sampling. Model skill is increased by the continuous assimilation of the collected data; which by choice, is not a continuous measurement at the same location.

For a fresh water plume, the low salinity and density imply that this feature will propagate through the ocean driven primarily by surface currents. A plume may dissipate rapidly, but can stay cohesive and detectable for up to weeks; we assume the later case. It is of interest to track these plumes based on the discussion in Sect. 2 as well as in [1]. In addition to tracking the plume, it is also important to accurately predict where a plume will travel on a daily basis. The ROMS prediction capabilities for a plume are good, but model skill can significantly increase from assimilation of *in situ* measurements.

A single Slocum glider is not optimal for the task at hand, as it is built for endurance missions and traveling at low velocities. Hence, we can not expect it to be able to collect samples over the entire area of a potentially large plume. Thus, we restrict ourselves to visiting (obtaining samples at) at most two locations for each hour of sampling. The primary location that we are interested in tracking is the centroid of the plume extent; analogous to the eye of the storm. Optimally, we would also like to gather a sample on the boundary of the plume. However, the glider may not be able to reach the plume centroid and a point on the boundary in a one hour time frame. With the given mission and the tools at hand, we present the following trajectory design algorithm.

4.1 *Trajectory Design Algorithm Based on Ocean Model Predictions*

We propose the following iterative algorithm for plume tracking utilizing ocen model predictions. This is the first known presentation of such a technology chain, and as such, is presented in a simplified manner. First, we assume the glider travels at a constant velocity v. Let d be the distance in kilometers that the vehicle can travel in a given time. We neither consider vehicle dynamics nor the effect of ocean currents upon the vehicle in this study; these are areas of ongoing research. Also, we only consider a 2-D planar problem as far as the waypoint computation is concerned.

The input to the trajectory design algorithm is a set of points, $\mathscr{D}$ (referred to as drifters) that determine the initial extent of the plume, and hourly predictions of the location of each point in $\mathscr{D}$ for a set duration. For the points in $\mathscr{D}$, we compute the convex hull as the minimum bounding ellipsoid, E_0. The centroid of this ellipsoid, C_0, is the start point of the survey. Next, we consider the predicted locations of $\mathscr{D}$ after one hour, $\mathscr{D}_1$. The centroid of $\mathscr{D}_1$ is C_1; the centroid of the minimum bounding ellipsoid E_1. The algorithm computes $d_g(C_0, C_1)$, the geographic distance from C_0 to C_1. Given upper and lower bounds d_u and d_l, resp., if $d_l < d_g(C_0, C_1) \leq d_u$, the trajectory is simply defined as the line $\overline{C_0 C_1}$. If $d_g(C_0, C_1) \leq d_l$, the algorithm first checks to see if there exists a point $p \in E_1 \cup \mathscr{D}_1$ such that

$$d_l \leq d_g(C_0, p) + d_g(C_1, p) \leq d_u. \tag{1}$$

If such a point exists, the trajectory is defined as the line $\overline{C_0 p}$ followed by the line $\overline{p C_1}$. If the set of points $p \in E_1 \cup \mathscr{D}_1$ which satisfy Eq. 1 is empty, then the algorithm computes the locus of points, $\mathscr{L} = \{p^* \in \mathscr{L} \mid d_g(C_0, p) + d_g(p, C_1) = d\}$. This locus $\mathscr{L}$, by definition, defines an ellipse with focii C_0 and C_1. We then choose a random point $p^* \in \mathscr{L}$ as another location for sampling. Here, the trajectory is the line $\overline{C_0 p^*}$ followed by the line $\overline{p^* C_1}$. If $d_g(C_0, C_1) > d_u$, the algorithm aborts as the plume is traveling too fast for the chosen vehicle. The algorithm then repeats this process for the defined duration of tracking. This selection process of waypoints for the AUV to visit to track the plume is presented in Algorithm 1. The overall iterative process to

---

Algorithm 1. Waypoint Selection Algorithm Based on Ocean Model Predictions

---

Require: Hourly forecasts, $\mathscr{D}_i$ for a set of points $\mathscr{D}$ defining the initial plume condition and its movement for a period of time, T.

  for $0 \leq i \leq T$ **do**

    Compute C_i, the centroid of the minimum bounding ellipsoid E_i of the points $\mathscr{D}_i$.

  end for

  while $0 \leq i \leq T - 1$ **do**

    if $d_l \leq d_g(C_i, C_{i+1}) \leq d_u$ **then**

      The trajectory is $\overline{C_i C_{i+1}}$.

    else if $d_g(C_i, C_{i+1}) \leq d_l$ and $\exists p \in E_i \cup \mathscr{D}_i$ such that $d_l \leq d_g(C_i, p) + d_g(p, C_{i+1}) \leq d_u$.

    then

      The trajectory is $\overline{C_i p}$ followed by $\overline{p C_{i+1}}$.

    else if $d_g(C_i, C_{i+1}) \leq d_l$ and $\{p \in E_i \cup \mathscr{D}_i \mid d_l \leq d_g(C_i, p) + d_g(p, C_{i+1}) \leq d_u\} = \emptyset$. **then**

      Compute $\mathscr{L} = \{p^* \in \mathscr{L} \mid d_g(C_0, p) + d_g(p, C_1) = d\}$, select a random $p^* \in \mathscr{L}$ and

      define the trajectory as $\overline{C_i p^*}$ followed by $\overline{p^* C_{i+1}}$

    else if $d_g(C_i, C_{i+1}) \geq d_u$ **then**

      Stop the algorithm. The plume is moving too fast for the selected AUV.

    end if

  end while

Algorithm 2. Ocean Plume Tracking Algorithm Based on Ocean Model Predictions

---

Require: A significant fresh water plume is detected via direct observation or remotely sensed data such as satellite imagery.

   repeat

      A set of points ($\mathscr{D}$) is chosen which determine the current extent of the plume.

      Input $\mathscr{D}$ to ROMS.

      ROMS produces an hourly forecast for all points in $\mathscr{D}$.

      Input hourly forcast for $\mathscr{D}$ into the trajectory design algorithm.

      Execute the trajectory design algorithm (see Alg. 1).

      Uploaded computed waypoints to the AUV.

      AUV executes mission.

      The AUV sends collected data to ROMS for assimilation into the model.

   until Plume dissipates, travels out of range or is no longer of interest.

---

design an implementable plume tracking strategy based on ocean model predictions is given in Algorithm 2.

Remark 1. In the SCB, a vertical velocity profile of ocean current is generally not constant. Since the plume propagates on the ocean surface $(1 - 3$ m$)$ and the glider operates at depths of $60 - 100$ m, it is not valid to assume that they are subjected to the same current regime, in both velocity and direction. Thus, it may be possible for a plume to *outrun* a slow-moving vehicle (*i.e.,* $d_g(C_i, C_{i+1}) \geq d_u$).

5   Implementation and Field Experiments in the SCB

The rainy season in southern California runs from November to March. During this time, storm events cause large runoff into local area rivers and streams, all of which empty into the Pacific Ocean. Two major rivers in the Los Angeles area, the Santa Ana and the Los Angeles River, input large fresh water plumes to the SCB. Such plumes have a high liklihood of producing HAB events. We deployed a Webb Slocum glider into the SCB on February 17, 2009 to conduct a month-long observation and sampling mission. For this deployment, the glider is programmed to execute a zig-zag pattern mission along the coastline, as depicted in Fig. 2, by navigating to each of the six waypoints depicted by the red and black bullseyes. Figure 2 also delineates the 20 m and 30 m isobaths, given by the green and red lines, respectively.

Unfortunately, weather and remote sensing devices did not cooperate to produce a rain event along with a detectable fresh water plume, so we were unable to retask the glider to track a real plume by use of Algorithm 2. Instead, we defined a pseudo-plume $\mathscr{D}$ with 15 initial drifter locations to demonstrate the proof of concept of this research. The pseudo plume is given by the blue line in Fig. 3.

The set $\mathscr{D}$ was sent to JPL and input to ROMS as the initial plume condition. The locations of the points in $\mathscr{D}$ were predicted for 15 hours. The initial time and location for the beginning of this retasking experiment coincided with predicted

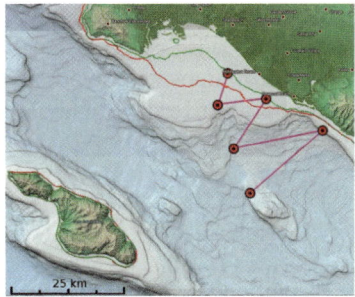

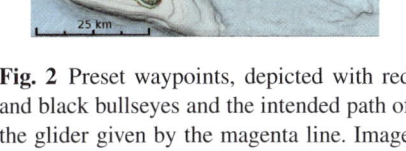

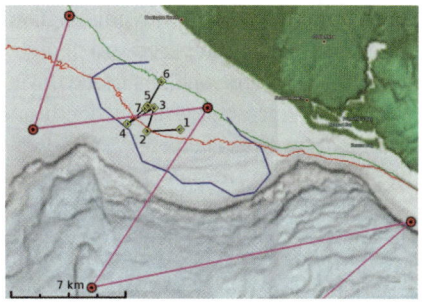

Fig. 2 Preset waypoints, depicted with red and black bullseyes and the intended path of the glider given by the magenta line. Image created by use of Google Earth.

Fig. 3 Plume (blue line), computed waypoints (yellow diamonds), and path connecting consecutive waypoints (black line). Image created by use of Google Earth.

coordinates of a future glider communication. The pseudo-plume was chosen such that C_0 was near this predicted glider surfacing location.

Based on observed behavior for our vehicle during this deployment, we take $v = 0.75$ km/h, and initially defined $d_l = 0.5$ and $d_u = 0.8$. The hourly predictions were input to the trajectory design algorithm and a tracking strategy was generated. Due to slow projected surface currents in the area of study, the relative movement of the plume was quite small. To keep the glider from surfacing too often and to generate a more implementable trajectory, we opted to omit visiting consecutive centroids. Instead, we chose to begin at the initial centroid, then visit the predicted centroid of the plume after five, ten and 15 hours, C_5, C_{10} and C_{15}, respectively. Between visiting these sites, the algorithm computed an additional waypoint for the glider to visit. These intermediate waypoints were chosen similarly to the p^* defined earlier, with $d = 3.75$; the distance the glider should travel in five hours. This design strategy produced seven waypoints for the AUV to visit during the 15 hour mission. The waypoints are presented in Table 1.

Note that we include the initial centroid as a waypoint, since the glider may not surface exactly at the predicted location. Upon visiting all of the waypoints in Table 1, the glider was instructed to continue the sampling mission shown in Fig. 2. Figure 3 presents a broad overview of the waypoints in Table 1, along with a path

Table 1 Waypoints generated by the plume tracking algorithm. Waypoint numbers $1, 3, 5$ and 7 are the predicted centroids of the pseudo-plume at hours $0, 5, 10$ and 15, respectively.

Number	Latitude (N)	Longitude (E)	Number	Latitude (N)	Longitude (E)
1	33.6062	-118.0137	5	33.6189	-118.0349
2	33.6054	-118.0356	6	33.6321	-118.0257
3	33.6180	-118.0306	7	33.6175	-118.0361
4	33.6092	-118.0487			

connecting consecutive waypoints. The plume is delineated by the blue line and the waypoints are numbered and depicted by yellow diamonds. Note that the glider did not travel on the ocean surface during this experiment. Between waypoints, the glider submerges below a set depth and performs consecutive dives and ascents creating a sawtooth-shaped trajectory as its glide path.

6 Results

In the study of path planning for field robots, planning the trajectory is usually less than half the battle, the real challenge comes in the implementation. This is exaggerated when dealing with underwater robots due to the complex environment. Next, we present results of an implementation of the designed sampling mission onto a Slocum glider operating in the SCB.

The waypoints given in Table 1 were computed under the assumption that the mission would be loaded onto the glider at a specific time and approximate geographic location. The glider arrived and communicated at the correct time and location, however, communication was aborted before the plume tracking mission could be uploaded. We were able to establish a connection two hours later at a different location, and successfully upload the mission file; this location is the red droplet labeled 1 in Fig. 4. We opted to not visit waypoint 1 based on the location of the glider and to get the glider back on schedule to track the plume. Figures 4 and 5 present magnified images of Fig. 3, where computed waypoints are the yellow diamonds and the red droplets are the actual locations visited by the glider.

We were able to sucessfully generate a plan and retask a deployed glider to follow an ocean feature for 15 hours. It is clear from the data that consideration has to be

Fig. 4 Computed waypoints (yellow diamonds) and actual glider locations (red droplets). Image created by use of Google Earth.

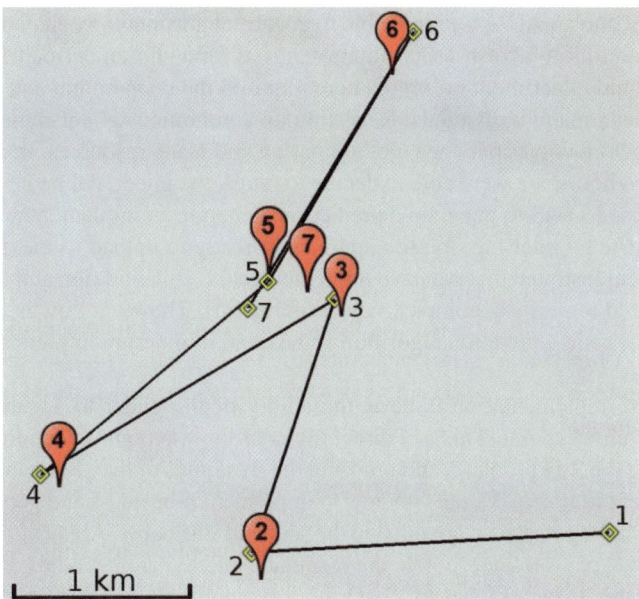

Fig. 5 Computed waypoints (yellow diamonds) and actual glider locations (red droplets). Image created by use of Google Earth.

made for glider dynamics and external forces from the ocean in the trajectory design algorithm. This is an area of active research. The motivation of this research is to follow plumes through the ocean via centroid tracking.

One element that we have neglected to discuss up to this point is that we have no metric for comparison. In particular, when we reach a predicted centroid, we do not have a method to check whether or not the plume centroid was actually at that location. We are planning experiments to deploy actual Lagrangian drifters to simulate a plume. This will give a comparison between the ROMS prediction and the actual movement of the drifters. This also provides a metric to determine the accuracy of the prediction and the precision of the AUV. Another component omitted from earlier discussion is time. When tracking a moving feature, a prediced waypoint contains time information as well as location. For this implementation, the glider began the mission at 0302Z and ended at 1835Z; a total time of 15.55 hours. Due to external influences, arrival at a few waypoints was not at the predicted time. Resolving this matter is contained within the addition of external forces, and is the subject of ongoing work.

7   Conclusions and Future Work

Designing effective sampling strategies to study ocean phenomena is a challenging task and can be approached from many different angles. Here, we presented a method to exploit multiple facets of technology to achieve our goal. Utilizing an ocean model and an AUV, we were able to construct a technology chain which outputs a path to follow a fresh water plume centroid for a chosen period of time. The successful field experiment presented here required the cooperation and communication between many individuals. Retasking an autonomous glider remotely while it is in the field involves patience, determination and many resources. In a period of less than two hours, we were able to decide to retask the glider, delineate a plume in the ocean, use ROMS to generate a prediction, generate an implementable tracking strategy, create a glider mission file and have it ready to upload to the glider. This paper has demonstrated that we have implemented the collaboration and technology chain required to perform complex field experiments. The work now is to improve upon the waypoint generation algorithm and extend it to design implementable 3-D trajectories.

The main implementation issue is the ability of the glider to accurately navigate to a given waypoint. This is a direct result of the waypoint selection algorithm only solving the 2-D problem, and ignoring the dynamics of the glider and the complex ocean environment. Details on how to implement robustness and generate more complex sampling missions are outside the scope of this paper. Areas of ongoing research include plans to incorporate the kinematic and dynamic models of the glider and extend this from a planar to a 3-D motion planning algorithm. Also, we plan to incorporate a 3-D current output of ROMS to plan a trajectory that exploits the currents to aid the locomotion of the glider. A more immediate step is to incorporate multiple AUVs, which leads to the development of an optimization criterion on which vehicle is best suited for a certain mission or to visit a chosen waypoint. A long-term goal is to facilitate autonomy for the entire system, leaving the human in the control loop as a fail-safe.

Acknowledgements. This work was supported in part by the NOAA MERHAB program under grant NA05NOS4781228, by NSF as part of the Center for Embedded Networked Sensing (CENS) under grant CCR-0120778, by NSF grants CNS-0520305 and CNS-0540420, by the ONR MURI program under grant N00014-08-1-0693, and a gift from the Okawa Foundation. The ROMS ocean modeling research described in this publication was carried out by the Jet Propulsion Laboratory (JPL), California Institute of Technology, under a contract with the National Aeronautics and Space Administration (NASA).

The authors acknowledge Carl Oberg for his work with glider hardware making field implementations possible and simple. We thank Ivona Cetinic for her efforts in the glider deployment and communications during the missions.

References

1. Cetinic, I., Jones, B.H., Moline, M.A., Schofield, O.: Resolving urban plumes using autonomous gliders in the coastal ocean. Journal of Geophysical Research/Americal Geophysical Union (2008)
2. Graver, J.G.: Underwater Gliders: Dynamics, Control and Design. PhD thesis, Princeton University, Princeton (2005)

3. Li, Z., Chao, Y., McWilliams, J.C., Ide, K.: A three-dimensional variational data assimilation scheme for the regional ocean modeling system. Journal of Atmospheric and Oceanic Technology 25, 2074–2090 (2008)
4. Paley, D., Zhang, F., Leonard, N.E.: Cooperative control for ocean sampling: The glider coordinated control system. IEEE Transactions on Control Systems Technology 16(4), 735–744 (2008)
5. Pereira, A., Heidarsson, H., Caron, D.A., Jones, B.H., Sukhatme, G.S.: An implementation of a communication framework for the cost-effective operation of slocum gliders in coastal regions. In: Proceedings of The 7th International Conference on Field and Service Robotics, Cambridge, MA (July 2009) (submitted)
6. Vu, Q.: JPL OurOcean portal (2008), http://ourocean.jpl.nasa.gov/ (viewed February 2009)
7. Schnetzer, A., Miller, P.E., Schaffner, R.A., Stauffer, B., Jones, B., Weisberg, S.B., DiGiacomo, P.M., Berelson, W.M., Caron, D.A.: Blooms of pseudo-nitzschia and domoic acid in the san pedro channela and los angeles harbor areas of the southern california bight. Harmful Algae/Elsevier 6(3), 372–387 (2003-2004)
8. Schofield, O., Kohut, J., Aragon, D., Creed, E.L., Graver, J.G., Haldman, C., Kerfoot, J., Roarty, H., Jones, C., Webb, D.C., Glenn, S.: Slocum gliders: Robust and ready. Journal of Field Robotics 24(6), 473–485 (2007)
9. Shchepetkin, A.F., McWilliams, J.C.: The regional oceanic modeling system (ROMS): a split-explicit, free-surface, topography-following-coordinate oceanic model. Ocean Modeling 9, 347–404 (2005), doi:10.1016/j.ocemod.2004.08.002
10. Wood, E.S., Schnetzer, A., Benitez-Nelson, C.R., Anderson, C., Berelson, W., Burns, J., Caron, D.A., Ferry, J., Fitzpatrick, E., Jones, B., Miller, P.E., Morton, S.L., Thunell, R.: The toxin express: Rapid downward transport of the neurotoxin domoic acid in coastal waters. Nature Geosciences (2009)

AUV Benthic Habitat Mapping in South Eastern Tasmania

Stefan B. Williams, Oscar Pizarro, Michael Jakuba, and Neville Barrett

Abstract. This paper describes a two week deployment of the Autonomous Under-water Vehicle (AUV) *Sirius* on the Tasman Peninsula in SE Tasmania and in the Huon Marine Protected Area (MPA) to the South West of Hobart. The objective of the deployments described in this work were to document biological assemblages associated with rocky reef systems in shelf waters beyond normal diving depths. At each location, multiple reefs were surveyed at a range of depths from approximately 50 m to 100 m depth. We illustrate how the AUV based imaging complements ben-thic habitat assessments to be made based on the ship-borne swath bathymetry. Over the course of the 10 days of operation, 19 dives were undertaken with the AUV covering in excess of 70 linear kilometers of survey and returning nearly 160,000 geo-referenced high resolution stereo image pairs. These are now being analysed to describe the distribution of benthic habitats in more detail.

1 Introduction

The Autonomous Underwater Vehicle (AUV) *Sirius* was part of a two week expe-dition in October, 2008, whose objective was to describe biological assemblages associated with rock reef systems in deep shelf waters on the Tasman Peninsula in SE Tasmania and in the Huon Marine Protected Area (MPA) to the South West of Hobart. Detailed multibeam sonar bathymetry data were previously collected by Geoscience Australia using a Simard EM3002 multibeam sonar system, Applannix motion sensor and C-Nav GPS to provide high-resolution Digital Elevation Maps (DEMs) of the study areas. The DEMs were used to determine suitable AUV survey

Stefan B. Williams, Oscar Pizarro, and Michael Jakuba
Australian Centre for Field Robotics, Uni. of Sydney, Sydney, NSW 2006, Australia
e-mail: {stefanw,o.pizarro,m.jakuba}@acfr.usyd.edu.au

Neville Barrett
Tasmanian Aquaculture and Fisheries Institute, Uni. of Tasmania, Hobart, Tas 7001, Australia
e-mail: Neville.Barrett@utas.edu.au

A. Howard et al. (Eds.): Field and Service Robotics 7, STAR 62, pp. 275–284.
springerlink.com   © Springer-Verlag Berlin Heidelberg 2010

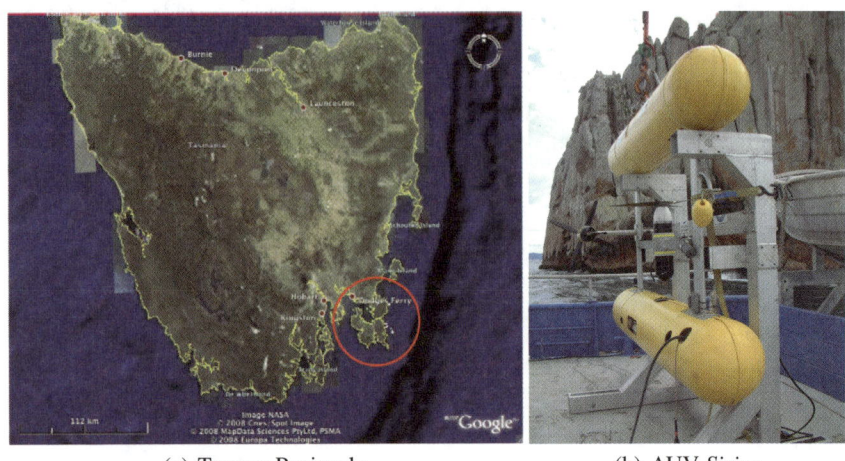

(a) Tasman Peninsula   (b) AUV *Sirius*

Fig. 1 (a) The site of survey work on the Tasman Peninsula in the South East of Tasmania. (b) The vehicle on-board the R/V Challenger prior to deployment. The dolerite cliffs of the peninsula can be seen in the background.

locations and to identify any hazards to operation. At each location, multiple reefs were surveyed at a range of depths from approximately 50 m to 100 m depth. Where distinct ectones (e.g. reef to sand) are present, transects were designed to cross transition zones and help determine the uniqueness of ectonal assemblages. Replication depended upon site logistics, however, dive profiles were designed to provide sufficient replication to quantitatively determine abundances of key species/features within depth strata, within reefs, between reefs (km to 100 km scale), and between differing levels of reef complexity.

2 Biodiversity Hub

While inshore reef systems are relatively easy to access and describe using methods such as dive surveys, offshore systems have remained relatively unknown because of the expense and complexity of available survey methods. A recent and very significant development contributing to our understanding of the physical environment of shelf habitats has been multibeam sonar and the interpretation of its associated backscatter. This has opened up opportunities for developing predictive capacity in this field where matching biological datasets are available. However, the effectiveness of this technique has yet to be fully tested as an appropriate surrogate for predicting patterns of biodiversity because of the lack of matching biological datasets collected at the same spatial scales and locations as fine scale acoustic surveys. In this context, the AUV deployments reported on here were part of a multi-disciplinary experimental program in eastern Tasmania, where the analysis of covariance is to be undertaken on co-located fine-resolution seabed habitat data, provided by the EM3002 multibeam sonar data, and biological datasets collected at similar spatial scales by Remotely Operated Vehicles (ROVs), Baited Underwater Video systems (BRUVs), towed video and the AUV. This research is being undertaken by the University of Tasmania and Geoscience Australia as part of the Marine Biodiversity Research Hub, which is a collaborative program funded under the Commonwealth Government's Commonwealth Environmental Research Facilities (CERF) Program. Within this Hub, several interrelated projects are designed to develop and test appropriate surrogates for biodiversity and incorporate these into an advanced predictive framework that covers a range of spatial scales [2].

The AUV data enables a finer-scale coupling of biological datasets with multibeam bathymetry than data collected through the use of ROVs, BRUVs and towed video alone, because of geo-referencing errors associated with USBL tracking systems. This additional data is expected to allow scale matching errors to be examined in more detail and allow surrogacy to be examined at the finest possible scale. Ultimately, CERF researchers expect to be able to compare the relative efficiency of using AUV, ROV and towed video systems for shelf habitat biological surveys. In addition, the high resolution images produced by the AUV are expected to significantly enhance the ability to identify taxa, adding finer taxonomic resolution, and hence value, to the data collection.

3   AUV-Based Benthic Habitat Mapping

One of the key features of the present cruise was the availability of a high resolution optical imaging AUV. The high spatial resolution and capacity to geo-reference the resulting imagery provides an invaluable mechanism for observing the extent and composition of particular benthic habitats. In this case, these data allow for post cruise analysis to validate habitat classification based on backscatter and slope data extracted from the ship-borne multibeam bathymetry.

AUVs are becoming significant contributors to modern oceanography, increasingly playing a role as a complement to traditional survey methods. Large, fast survey AUVs can provide high resolution acoustic multibeam and sub-bottom data by operating a few tens of meters off the bottom, even in deep water [6, 10]. High resolution optical imaging requires the ability to operate very close to potentially rugged terrain. The Autonomous Benthic Explorer (ABE) has helped increase our understanding of spreading ridges, hydrothermal vents and plume dynamics [16] both using acoustics and vision. The AUV *SeaBED* [14] is primarily an optical imaging AUV, used in a diverse range of oceanographic cruises including coral reef characterization [13] and surveys of ground fish populations [1]. Recently, the related AUVs Puma and Jaguar searched for hydrothermal vents under the artic ice [7]. Other AUV systems have been used to explore biophysical coupling, including mapping harmful algal blooms [11] and characterising up-welling around canyons [12].

The University of Sydney's Australian Centre for Field Robotics operates an ocean-going AUV called *Sirius* capable of undertaking high resolution, geo-referenced survey work [15]. This platform is a modified version of the WHOI

SeaBED vehicle. This class of AUV has been designed specifically for relatively low speed, high resolution imaging and is passively stable in pitch and roll. The submersible is equipped with a full suite of oceanographic sensors including a high resolution stereo camera pair and strobes, multibeam sonar, a depth sensor, Doppler Velocity Log (DVL) including a compass with integrated roll and pitch sensors, Ultra Short Baseline Acoustic Positioning System (USBL), forward-looking obstacle avoidance sonar, a conductivity/temperature sensor and combination fluorometer/scattering sensor to measure chlorophyll-a, turbidity and dissolved organic matter. The on-board computer logs sensor information and runs the vehicle's low-level control algorithms. *Sirius* is part of the Integrated Marine Observing System (IMOS) AUV Facility, with funding available on a competitive basis to support its deployment as part of marine studies in Australia.

Navigation underwater is challenging because electromagnetic signals attenuate strongly with distance. Absolute position estimates such as those provided by GPS are therefore not readily available. Simultaneous Localisation and Mapping (SLAM) is the process of concurrently building a feature based map of the environment and using this map to obtain estimates of the location of the vehicle. The SLAM algorithm has seen a considerable amount of interest from the mobile robotics community as a tool to enable fully autonomous navigation [3, 4]. Our current work has concentrated on efficient, stereo based Simultaneous Localisation and Mapping and dense scene reconstruction suitable for creating detailed maps of seafloor survey sites [8, 9]. These novel approaches, based on Visual Augmented Navigation (VAN) techniques [5], enable the complexity of recovering the state estimate and covariance matrix in a VAN framework to be managed. This has allowed these algorithms to run on significantly larger mapping problems than was previously feasible. These techniques have been used to renavigate the estimated vehicle trajectories using the data collected for this paper.

A typical dive will yield several thousand geo-referenced overlapping stereo pairs. While useful in themselves, single images make it difficult to appreciate spatial features and patterns at larger scales. It is possible to combine the SLAM trajectory estimates with the stereo image pairs to generate 3D meshes and place them in a common reference frame [15]. The resulting composite mesh allows a user to quickly and easily interact with the data while choosing the scale and viewpoint suitable for the investigation. Spatial relationships within the data are preserved and scientists can move from a high level view of the environment down to very detailed investigation of individual images and features of interest within them. This is a useful tool for the end user to develop an intuition of the scales and distributions of spatial patterns, even before any automated interpretation is attempted. Examples of the output of the 3D reconstructions for dives undertaken on this cruise are included below.

4 Deployments

The deployments undertaken over the course of the 10 day cruise in October 2008 were on shelf reef habitats at depths of between 50 and 100 m in eastern Tasmanian waters and in estuarine waters in 30 to 50 m deep in the Huon MPA and around Port Arthur, Tasmania. The vehicle was deployed on 19 dives over the 10 days of operation. During the course of these dives, the vehicle covered in excess of 70 linear kilometers of survey and collected nearly 160,000 high resolution stereo image pairs. Each dive ranged between 2.5 km and 6.5 km in total length with an average of 3.7 km covered per dive travelling at a speed of 0.5 m/s or approximately 1 knot. Table 1 shows summary statistics for the dives undertaken during the course of this cruise. Dives 1 through 4 were calibration runs undertaken prior to the scientific missions and are not show here.

Figure 2 shows the AUV dive profiles overlaid on the previously collected shipborne mutlibeam bathymetry, focusing on the Tasman Peninsula deployments. The AUV dive profiles were targeting particular rocky reef structures identified in the bathymetry derived from ship-borne multibeam surveys undertaken prior to the cruise. Figure 3 (a) shows two AUV dive profiles over the bathymetry at OHara reef (seen in the middle of Figure 2 (b)) with high resolution multibeam data collected by the vehicle from an altitude of 20 m over the eastern edge of the reef embedded in the figure. Figure 3 (b) shows details of one of the three dimensional seafloor reconstructions generated using the combined SLAM and stereo meshes for one of these dives.

Figure 4 shows dive profiles for six dives undertaken around the Hippolytes, a rocky island located group approximately 4 km offshore. These areas, in

Table 1 AUV Tasmania Dive Summary

Dive	Lat	Long	Max Depth [m]	Distance [m]	Avg. Alt. [m]	No. Stereo pairs
5	-43.0615	147.9648	67.0	5625	2.00	12262
6	-43.0631	147.9825	66.3	2947	2.00	7256
7	-43.084558	147.974086	77.9	4774	2.00	11278
8	-43.094119	148.024267	85.9	2817	2.10	7262
9	-43.120019	148.05373	90.9	2652	2.08	6336
10	-43.119798	148.047008	88.5	2717	2.06	6406
11	-43.119126	148.038401	83.3	2660	1.97	6727
12	-43.120581	148.03754	84.0	2875	2.00	6787
13	-43.123972	148.053974	96.3	2759	2.16	6870
14	-43.119887	148.045257	89.5	2995	2.06	7737
15	-43.040462	147.955845	57.8	4559	1.99	10563
16	-42.956996	148.005154	76.4	6542	2.05	15521
17	-43.165023	147.874164	52.7	4290	1.99	9768
18	-43.18674	147.88763	32.4	3889	2.02	8863
19	-42.913151	148.003898	60.0	6300	2.00	15162
20	-43.084558	147.974086	76.9	2780	2.04	6564
21	-43.29307	147.129389	45.9	4619	2.00	9655
22	-43.327701	147.166552	35.8	1153	1.51	2548
23	-43.270816	147.124965	33.0	3200	1.50	1623
TOTAL				70153		159188

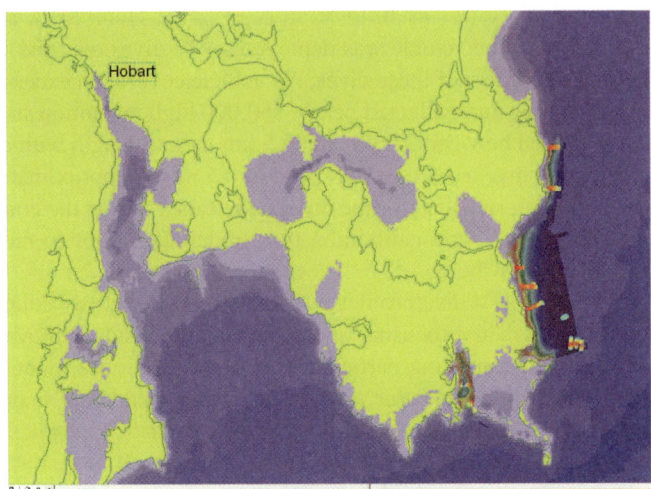

(a) Tasman peninsula deployments

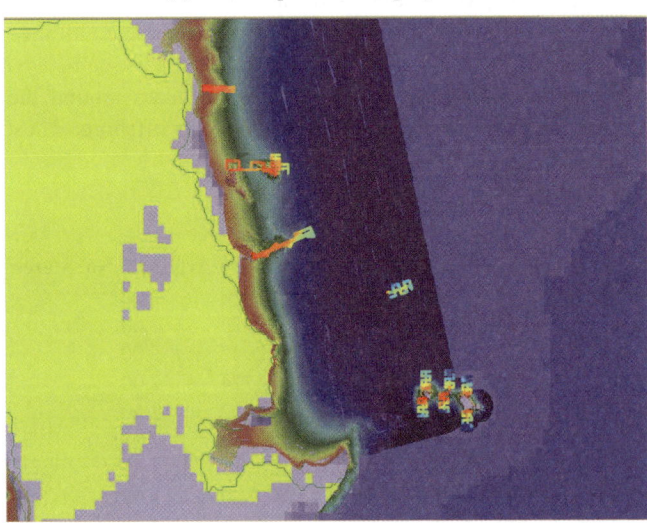

(b) Fortescue Bay

Fig. 2 Tasmanian Deployments 2008 illustrating the prior multibeam bathymetry available with AUV dive profiles overlaid (ship-borne multibeam sonar image courtesy of Geoscience Australia). The AUV profiles are colour coded by depth.

approximately 90 m of water, were expected to feature high levels of biodiversity. Finally, Figure 5 shows high resolution multibeam bathymetry collected by the vehicle in the Huon MPA showing a series of pockmarks mapped using the vehicle's on-board multibeam sonar at an altitude of 20 m during dive 23. These pockmarks had been identified in prior ship borne multibeam surveying but were targeted by the AUV to generate higher resolution bathymetry as well as undertaking a number of

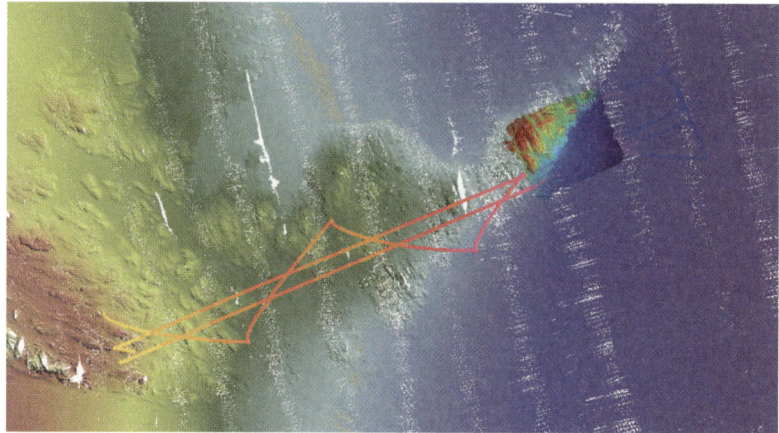

(a) Ohara reef bathymetry

(b) 3D reconstruction

Fig. 3 (a) Ohara reef bathymetry with two AUV imaging dive profiles (Dive 7 and 20) overlaid (ship-borne multibeam sonar image courtesy of Geoscience Australia). The long axis of these dives are 2 km in length. The AUV dive profiles are colour coded by measured depth. The colour coding between the AUV depth and the bathymetry are not consistent in order for the dive profile to stand out. Detailed bathymetry collected by the vehicle from a 20m altitude at the eastern edge of the reef is overlaid on the orginal ship-borne bathymetry showing the interface between the rocky reef and the deeper, sandy substrate. The AUV bathymetric data is gridded at 0.5m resolution. (b) Details of one of the SLAM loop closure points identified during Dive 7 at OHara reef. The vehicle path generated using SLAM has been used to place the stereo meshes into space and the resulting mesh has been texture mapped using the images. As can be seen, the loop closure has been successful, resolving the detailed structure of the scene in spite of the vehicle having travelled nearly 2 km between the two passes over this area.

visual passes at an imaging altitude of 2 m. The visual imagery revealed that these pockmarks, surrounded by muddy substrate, were filled with algal growth. Further investigation using divers is planned to determine the nature of this growth.

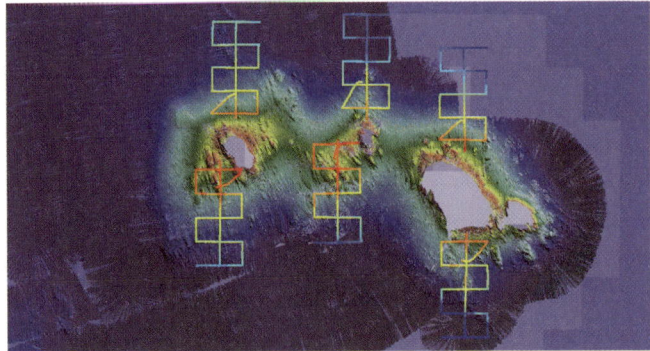

Fig. 4 Dive profiles for dives 8 through 13 around the Hippolytes, a series of islands located some 4 km offshore of the Tasman peninsula (ship-borne multibeam sonar image courtesy of Geoscience Australia).

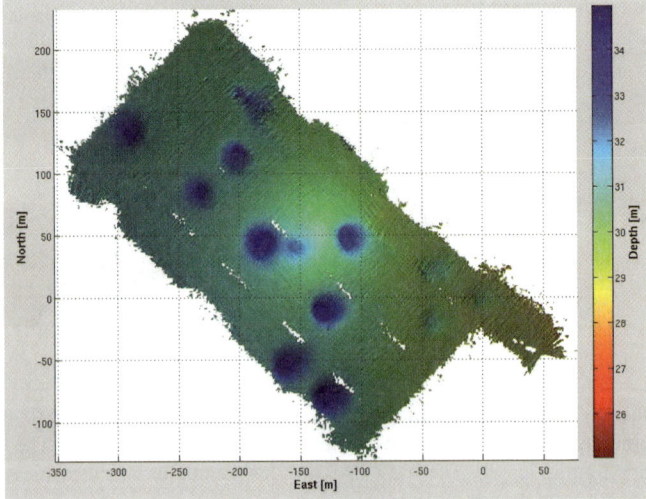

Fig. 5 A series of pockmarks mapped using the vehicle's on-board multibeam sonar at an altitude of 20 m during dive 23 in the Huon MPA.

The data gathered during the AUV dives will form a baseline at or near the time of protection for newly designated MPAs. This will allow changes related to protection, time and climate change to be assessed through time. This dataset will underpin proposals for ongoing research on the respective MPAs (Commonwealth and State) with respect to both ecosystem effects of fishing and climate change in an otherwise poorly understood ecosystem. The AUV, ROV and BRUV surveys will also provide, for the first time, a detailed quantitative inventory of the benthic faunal assemblages associated with temperate shelf reef systems in this region, and of the MPAs themselves. Shelf reef assemblages of this region remain relatively undescribed with respect to the quantitative composition of dominant species or taxonomic groupings, therefore providing that initial baseline description represents a significant contribution of this study. Like inshore systems, shelf reef assemblages in this region are influenced by fishing (e.g. scalefish and lobster), introduced species and climate change, yet nothing is currently known of the interactions occurring in these systems.

5   Conclusions and Future Work

This paper has reported on an expedition taken to survey biological assemblages associated with deep water, rocky reef systems off the coast of the Tasman peninsula and in the Huon MPA. Although the detailed analysis of the data and comparison against the benthic habitat classification based on sonar backscatter and slope is ongoing, we have illustrated how the data collected by the AUV is complementary to ship-borne multibeam.

We are currently preparing to return to Tasmania in early 2009 to undertake additional deployments associated with further multibeam swath mapping and to target urchin barrens and scallop fisheries. The impact of fishing and climate change on these poorly understood habitats stands to derive significant benefit from the detailed, optical surveying capabilities of tools such as the AUV.

Acknowledgements. This work is supported by the ARC Centre of Excellence programme, funded by the Australian Research Council (ARC) and the New South Wales State Government, and the Integrated Marine Observing System (IMOS) through the DIISR National Collaborative Research Infrastructure Scheme. The authors would like to thank the Captain and crew of the R/V Challenger. Their sustained efforts were instrumental in facilitating successful deployment and recovery of the AUV. Thanks to Justin Hulls and Jan Seiler for help and support on-board the ship. Thanks also to Dr. Vanessa Lucier of TAFI for providing the ship borne multibeam and to Dr. Tara Anderson from Geoscience Australia for her contribution to the AUV transect design for the surveys. The ship-borne multibeam sonar data were collected, processed and gridded to produce DEMs by Geoscience Australia. We also acknowledge the help of all those who have contributed to the development and operation of the AUV, including Duncan Mercer, George Powell, Ian Mahon, Matthew Johnson-Roberson, Stephen Barkby, Ritesh Lal, Paul Rigby, Jeremy Randle, Bruce Crundwell and the late Alan Trinder.

References

1. Clarke, M.E., Singh, H., Goldfinger, C., Andrews, K., Fleischer, G., Hufnagle, L., Pierce, S., Roman, C., Romsos, C., Tolimieri, N., Wakefield, W., York, K.: Integrated mapping of West Coast groundfish and their habitat using the Seabed AUV and the ROPOS ROV. EOS Trans AGU:Ocean Sci. Meet. Suppl., Abstract 87(36), OS46G–12 (2006)
2. Commonwealth Environmental Research Facilities (CERF) Marine Biodiversity Hub (2009), http://www.marinehub.org/index.php/site/home

3. Dissanayake, M., Newman, P., Clark, S., Durrant-Whyte, H., Csobra, M.: A solution to the simultaneous localization and map building (SLAM) problem. IEEE Transactions on Robotics and Automation 17(3), 229–241 (2001)
4. Durrant-whyte, H., Bailey, T.: Simultaneous localisation and mapping (SLAM): Part I the essential algorithms. Robotics and Automation Magazine 13, 99–110 (2006)
5. Eustice, R., Singh, H., Leonard, J., Walter, M.: Visually mapping the RMS Titanic: conservative covariance estimates for SLAM information filters. Intl. J. Robotics Research 25(12), 1223–1242 (2006)
6. Grasmueck, M., Eberli, G.P., Viggiano, D.A., Correa, T., Rathwell, G., Luo, J.: Autonomous underwater vehicle (AUV) mapping reveals coral mound distribution, morphology, and oceanography in deep water of the straits of Florida. Geophysical Research Letters 33(L23), 616 (2006)
7. Kunz, C., Murphy, C., Camilli, R., Singh, H., Eustice, R., Roman, C., Jakuba, M., Willis, C., Sato, T., Nakamura, K., Sohn, R., Bailey, J.: Deep sea underwater robotic exploration in the ice-covered arctic ocean with AUVs. In: IEEE/RSJ Intl. Workshop on Intelligent Robots and Systems, pp. 3654–3660 (2008)
8. Mahon, I.: Vision-based navigation for autonomous underwater vehicles. PhD thesis, University of Sydney (2008)
9. Mahon, I., Williams, S.B., Pizarro, O., Johnson-Roberson, M.: Efficient view-based SLAM using visual loop closures. IEEE Transactions on Robotics and Automation 24(5), 1002–1014 (2008)
10. Marthiniussen, R., Vestgard, K., Klepaker, R., Storkersen, N.: HUGIN-AUV concept and operational experiences to date. In: OCEANS 2004. MTTS/IEEE TECHNO-OCEAN 2004, vol. 2, pp. 846–850 (2004)
11. Robbins, I., Kirkpatrick, G., Blackwell, S., Hillier, J., Knight, C., Moline, M.: Improved monitoring of HABs using autonomous underwater vehicles (AUV). Harmful Algae 5(6), 749–761 (2006)
12. Ryan, J., Chavez, F., Bellingham, J.: Physicalbiological coupling in monterey bay, california: topographic influences on phytoplankton ecology. Marine Ecology Progress Series 287, 23–32 (2005)
13. Singh, H., Armstrong, R., Gilbes, F., Eustice, R., Roman, C., Pizarro, O., Torres, J.: Imaging Coral I: Imaging Coral Habitats with the SeaBED AUV. Subsurface Sensing Technologies and Applications 5, 25–42 (2004)
14. Singh, H., Can, A., Eustice, R., Lerner, S., McPhee, N., Pizarro, O., Roman, C.: SeaBED AUV offers new platform for high-resolution imaging. EOS, Transactions of the AGU 85(31) 289, 294–295 (2004)
15. Williams, S., Pizarro, O., Mahon, I., Johnson-Roberson, M.: Simultaneous localisation and mapping and dense stereoscopic seafloor reconstruction using an AUV. In: Proc. of the Int'l. Symposium on Experimental Robotics (2008)
16. Yoerger, D., Jakuba, M., Bradley, A., Bingham, B.: Techniques for deep sea near bottom survey using an autonomous underwater vehicle. The International Journal of Robotics Research 26, 41–54 (2007)

Sensor Network Based AUV Localisation

David Prasser and Matthew Dunbabin

Abstract. The operation of Autonomous Underwater Vehicles (AUVs) within underwater sensor network fields provides an opportunity to reuse the network infrastructure for long baseline localisation of the AUV. Computationally efficient localisation can be accomplished using off-the-shelf hardware that is comparatively inexpensive and which could already be deployed in the environment for monitoring purposes. This paper describes the development of a particle filter based localisation system which is implemented onboard an AUV in real-time using ranging information obtained from an ad-hoc underwater sensor network. An experimental demonstration of this approach was conducted in a lake with results presented illustrating network communication and localisation performance.

1 Introduction

Tracking an Autonomous Underwater Vehicle's (AUV's) position is essential for navigation and geo-referencing data gathered during survey tasks. Unlike some other types of field robots, AUVs suffer significant challenges in determining their location as GPS signals are only available whilst at the surface. There are three general approaches to maintaining an AUV's position estimate for navigation: (1) integration of motion estimates, (2) using acoustic transponders as a position reference, or (3) using the environment as a position reference [10, 12].

Path integration approaches accumulate motion estimates from either inertial measurement units, Doppler Velocity Logs (DVLs) or vision systems to determine position relative to a starting point [14]. These by themselves suffer integration drift with time and minimising this drift is highly dependent on hardware selection and algorithm choice.

Acoustic transponder systems provide absolute position estimates and do not suffer from error accumulation inherent in path integration approaches. These can

David Prasser and Matthew Dunbabin
Autonomous Systems Laboratory, CSIRO ICT Centre, PO Box 883, Kenmore QLD 4069
e-mail: David.Prasser@csiro.au, Matthew.Dunbabin@csiro.au

A. Howard et al. (Eds.): Field and Service Robotics 7, STAR 62, pp. 285–294.

either be characterised as long or short baseline systems. In long baseline systems, the AUV uses ranging information from acoustic transponders placed at known locations to determine its position. Short baseline systems typically localise the AUV relative to a support vessel. Although successfully used in many open ocean and near shore situations, key issues in this method arise from the propagation of sound underwater such as reflections, echoes, and changes in water density [12, 14].

An alternative to both odometry and acoustic transponders is to reference the vehicle's position against environmental features such as the seabed terrain or salient image features. This approach is advantageous as no other objects need to be introduced into the environment thus reducing equipment cost, deployment time and survey requirements. Many such approaches use Simultaneous Localisation And Mapping (SLAM) to produce an environment map for localisation which has the advantage of allowing exploration without a map and the side benefit of constructing a map of the environment [9]. These methods are often used in conjunction with an odometry system [8] for improved localisation in low feature environments and during decent to and ascent from the seafloor.

In this paper, we propose an intermediate approach to AUV localisation by utilising existing acoustic sensor network nodes (underwater modems) designed for ad-hoc data communication [2] instead of deploying a dedicated long or short baseline localisation system. Furthermore, this paper describes the development of a network based time-of-flight localisation algorithm which is implemented onboard the AUV in real-time to allow navigation within challenging and unstructured environments.

2   Sensor Network Based Localisation

Underwater wireless sensor networks provide a means to remotely monitor a set of parameters at multiple locations distributed throughout a marine environment. Many challenges exist to ensure reliable communications which include environmental effects as well as hardware limitations [1]. As such, most reported sensor nodes use either point-to-point communications between two nodes, or preset routing tables in multi-node systems. Previous work realised ad-hoc underwater communications by using commercially available acoustic modems that were controlled by low power processor boards attached to each modem to act as the sensing platform and implement a networking layer [2]. This forms the foundation for the network used in this investigation.

2.1   Scenario Description

This paper considers the use of acoustic sensor network nodes already deployed in the environment as a means of providing localisation information to an AUV in place of dedicated long baseline transponders. An AUV is equipped with an acoustic modem allowing it to communicate with the surrounding sensor nodes and from the round trip time determine its distance to the node. In a typical scenario, an AUV would operate in and around a sensor network, acting as an additional mobile sensor in a long term infrastructure or environmental monitoring project. In such a situation network relative localisation is necessary for both navigation and registering sensor data.

The system proposed here is based on the assumption that sensor nodes are placed arbitrarily within the environment, however, their position and depth below the surface is known to the AUV with a bounded uncertainty. It is also assumed that not all nodes are within communication range of the AUV and communication reliability between the AUV and a sensor node is not perfect. This is plausible in practice when the sensor network extends over great areas such as in lakes, rivers, or around coral reefs where line-of-sight communication is not available.

2.2 Localisation

In previous work [3, 4], a localisation system was developed and demonstrated which used custom built acoustic transponders utilising TDMA. This system allowed accurate localisation of the AUV and self-localisation of the transponders within the environment. The solution proposed here differs in that there is no clock synchronisation between the AUV and the sensor nodes, rather round trip times are measured. This also means that nodes are not consuming energy via transmission just to maintain network synchronisation when the AUV is not within range.

As the elapsed round trip time is being measured at the message (communication packet) level, it is influenced by factors such as the delay for modulating and demodulating the signals, packet length and other inherent hardware delays. Therefore, the estimate of range, R, can be determined by the measurement of the time-of-flight, t_{tof}, and systematic hardware and software delays by

$$
\begin{aligned}
t_{tof} &= \left(\frac{d_{A/N} + d_{N/A}}{v_w} + \Delta t_{Modem/Packet} + \Delta t_{node} \right) \\
R &\approx \left(t_{tof} - \Delta t_{Modem/Packet} - \Delta t_{node} \right) \tilde{v}_w / 2
\end{aligned}
\tag{1}
$$

where $d_{A/N}$ and $d_{N/A}$ are the actual distances from the AUV to the Node and from the Node to the AUV respectively, $\Delta t_{Modem/Packet}$ is the internal time delay of the modem for packet handling which is assumed deterministic, Δt_{node} is the time overhead of the node controller again assumed deterministic, and v_w and $\tilde{v}_w$ is the actual and approximated velocity of sound in water.

The system fuses time-of-flight data from the beacons (sensor nodes) with self motion and depth measurements made by the vehicle using a particle filter [13]. A Kalman filter could also achieve this, however, using a particle filter allows the possibility of multi-modal distributions which would be necessary in global localisation. The vehicle pose is represented as

$$
\mathbf{X} = \{x , y , z\}^T
\tag{2}
$$

where x and y are the vehicle's Cartesian coordinates referenced to a global frame, and z is the vehicle's depth. In practice the vehicle's depth sensor is sufficiently precise to not need filtering, however, a 3D filter is used to allow for potential calibration errors in the depth sensor and for differences between the depth zero point between the AUV and the beacons. In this study, an observational model on depth is applied as a normal distribution with $\sigma = 0.1$m.

The range estimation has two error conditions; (1) normally distributed measurement noise, and (2) echoes. Measurement noise is modelled as being normally distributed with a $\sigma = 3$m (See Section 4). The range measurements are modelled as having a 95% chance of being a true return as opposed to an echo ($P(E) = 0.05$), which removes the need for explicitly filtering outliers before updating the filter. Algebraically, the probability of a pose estimate, $\mathbf{X}$, given a range estimate, R, from modem i, is given by

$$P(\mathbf{X}|R) = (1 - P(E))N(R - |\mathbf{X} - \mathbf{m}_i|, \sigma) + \begin{cases} 0, & R \leq |\mathbf{X} - \mathbf{m}_i| \\ \frac{P(E)}{r_{max} - |\mathbf{X} - \mathbf{m}_i|}, & R > |\mathbf{X} - \mathbf{m}_i| \end{cases} \quad (3)$$

where r_{max} is the maximum permissible range, $N(x, \sigma)$ is a zero mean Gaussian and each of the modem locations, $\mathbf{m}_i$ is defined as

$$\mathbf{m}_i = \{x_{m_i}, y_{m_i}, depth_{m_i}\}^T \quad (4)$$

In this investigation, 1000 particles are used which can easily be processed in real-time on the AUV's onboard computer. Resampling occurs whenever range or depth measurements are made.

As the proposed AUV does not have Doppler or visual odometry (assumed turbid water in this investigation) the motion model for the vehicle is based on motor force and vehicle hydrodynamics. The dominant source of error in the motion model is assumed to be an unknown current acting on the vehicle. The unfortunate situation for AUVs without Doppler Velocity Logs or precision inertial measurement units is that there is no way to account for external disturbances. This places the burden on the localiser to detect and correct disturbances. This current is assumed to have a zero mean, normally distributed velocity with 0.5 to 1.2 m/s standard deviation.

3 Experimental Platforms

The acoustic sensor network nodes and CSIRO developed Starbug AUV used in this investigation are described below.

Acoustic Sensor Network Nodes

The sensor network nodes used in this investigation were developed for ad-hoc underwater communications and consist of an acoustic modem, a custom developed "node-controller", power supply and environment sensors as shown in Figure 1(a).

The modem used was a commercially available CDMA modem [1], chosen for its relatively low cost, in-built broadcast capabilities, and RS232 interface, although

---

[1] Aquacomm acoustic modem from DSPComm (www.dspcomm.com)

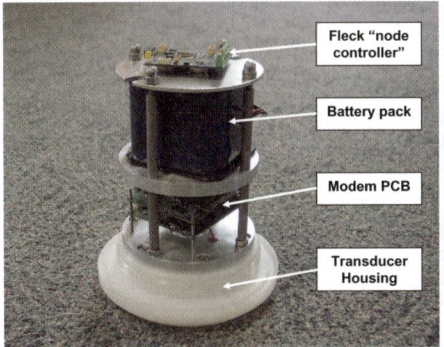

(a) Internal view of an acoustic underwater sensor network node.

(b) The Starbug MkIII AUV.

Fig. 1 Acoustic underwater sensor network node and the Starbug AUV used in network based localisation trials.

its maximum transmission rate is only 480bps. It should be noted that any acoustic modem that permits broadcast messaging at a sufficient bit rate could be used with only slight changes to the software that runs on the node controller.

The "node controller" is a custom system designed to provide the networking layer that most commercial modems lack [2]. The ad-hoc networking protocol is based on a modified Dynamic Source Routing (DSR) approach and can be configured for maximizing information throughput or minimising energy expenditure. The node controller software is implemented on CSIRO's "Fleck" wireless sensor network embedded hardware [11]. Apart from the need to model message handling delays, the localisation algorithm is not strongly dependent upon the details of the node controller. Range measurement messages are short (4 bytes of data) consisting of source and destination addresses, message type and a temporary message identifier.

Autonomous Underwater Vehicle

Figure 1(b) shows the Starbug AUV that was used in these experimental trials [6], which is a shallow water research vehicle with an operating speed of 0.5 to 1.0 m/s and a maximum depth of 40 m. Unlike many other AUVs, Starbug uses only visual odometry provided by a downward facing pair of stereo cameras which also record video data [7]. While this approach provides accurate odometry information it is only available when the AUV's altitude is low and degrades when operating in low-light or turbid water, or when descending through the water column. This capability is not used in these experiments so that the worst case localisation is presented.

One of the sensor nodes described above is mounted on the AUV allowing it to communicate with underwater nodes that are within range. The AUV interfaces with the onboard sensor node via a serial connection allowing it to interrogate surrounding nodes in turn and estimate time-of-flight. In general operation, one range measurement is made every two seconds.

4 Experimental Results

The proposed sensor network based localisation system described in Section 2 was tested using an underwater network at Lake Wivenhoe in Queensland, Australia. Experiments included characterisation of the modems with node-controllers for inclusion in the particle filter, as well as evaluation of the AUV localisation performance with varying number of sensor nodes and geometric distribution.

4.1 Acoustic Modem Characterisation

The first experiments consisted of assessing the communication throughput for a static network in which data packets are transmitted from a source to sink node in a multi-hop fashion. Figure 2(a) shows that packet throughput with increasing range illustrating that under static conditions data throughput between the deployed sensor nodes deteriorates significantly after approximately 1200m.

The modems used have an onboard active receive gain adjustment not accessible to our node-controller. During initial trials it was observed that as the AUV moved around the environment the range request packets from the AUV would not be reliably received. As such, each modem was sent two range requests in succession (two seconds apart) to allow the modem to adjust its receive gain if necessary. Figure 2(b) shows the results of successful range request returns as a function of range as the AUV moves throughout the sensor fields. As seen, on average, a greater number of second requests were successful. This assisted in receiving ranges to nodes, however, it slowed the rate at which all nodes in the sensor field could be interrogated.

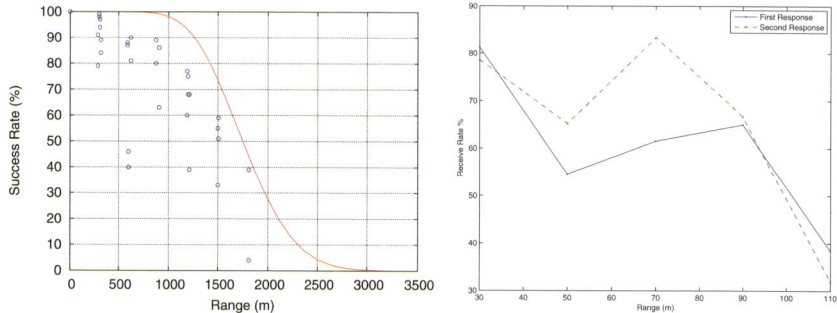

(a) Measured modem packet loss statistics with a calculated limit curve based on cylindrical spreading and a Gaussian channel.

(b) Rate of success from first and second range requests.

Fig. 2 Static network node performance for normal internode communication and the success rate of ranging requests from the AUV.

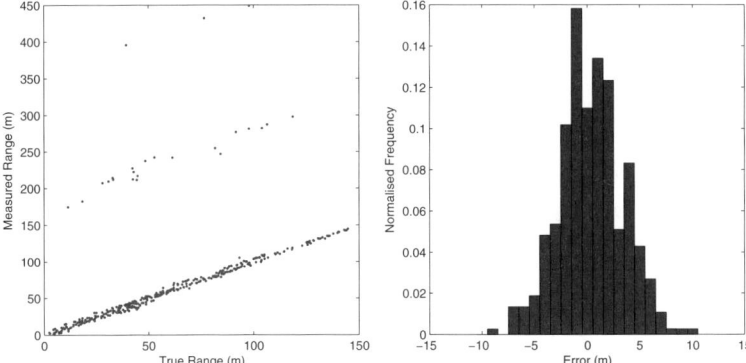

(a) Effect of echoes within environment and false range estimates.   (b) Normalised distribution of range error after the removal of multipath effects.

Fig. 3 Measured modem ranging performance with echoes and range estimate performance.

Figure 3(a) shows the range to a sensor node from the AUV as it moves throughout the sensor field from time-of-flight measurements against the range as calculated from GPS (AUV antenna was at the surface). As seen, a general linear relationship exists, however, a number of false readings are present as a result of echoes within the environment creating longer return path lengths (and hence time-of-flight).

Figure 3(b) shows the range estimation error for the data in Figure 3(a) after discarding range measurements that can be attributed to echoes. The remaining range errors are approximately normally distributed with a standard deviation of approximately 3 m. For this environment a worst case estimate of the probability of multi-pathing is 5%. Both of these effects are used in the sensor model described in Section 2.

4.2 Localisation Performance

The localisation performance was measured by sending the AUV on a trajectory consisting of a set of waypoints such that the GPS antenna was at the water's surface for ground truth. The experiments consisted of a different number of sensor nodes within communication range and in different locations of Lake Wivenhoe with water depth ranging from 6 to 28 m. The vehicle speed was set to 0.5 m/s and the particle filter was initialised to the GPS position at the start of each experiment.

Figure 4 shows the results of a trial with three sensor nodes within communication range of the AUV. The error in position between ground truth and localiser is bounded with an RMS value of 4.4m. Figure 5 shows a second trial with four sensor nodes within communication range of the AUV during a rectangular survey mission. As seen, the localisation error is bounded with an RMS value of 3.1m illustrating an improved localisation estimate with a different node arrangement.

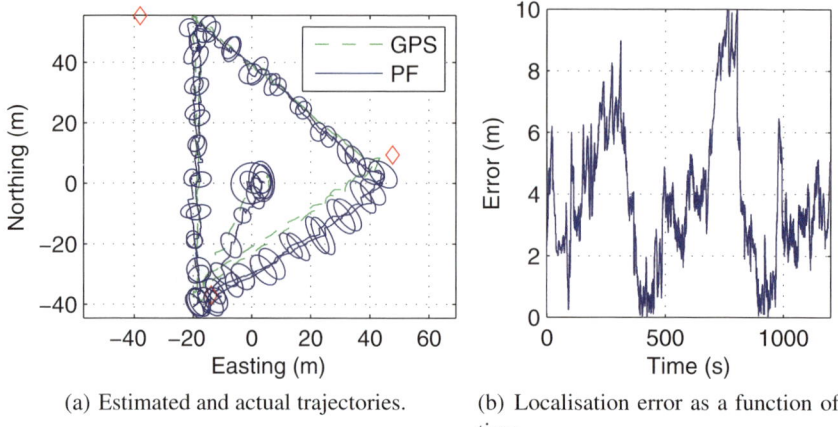

(a) Estimated and actual trajectories.

(b) Localisation error as a function of time.

Fig. 4 (a) The actual and estimated path using the localiser with three sensor nodes within communication range of the AUV. Also shown are the locations of the sensor nodes (diamonds) and the covariance of the localisation estimate. A total of 397 range updates (out of approximately 600 attempted) were received during the experiment. (b) The error in position between ground truth and localiser as a function of time with a bounded RMS value of 4.4m.

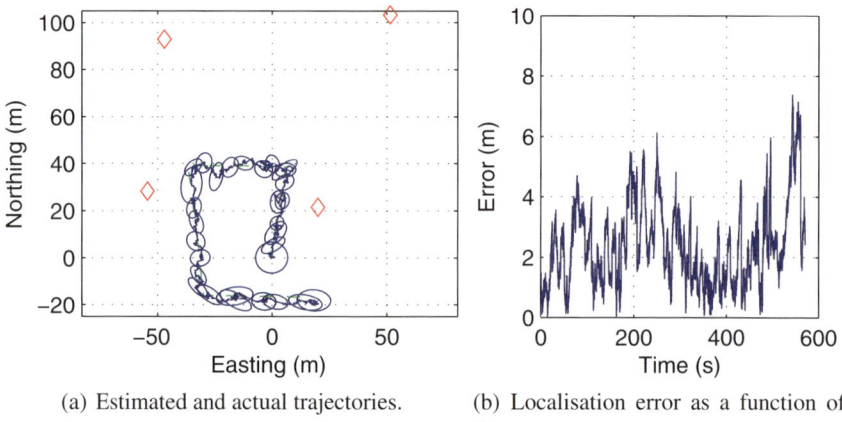

(a) Estimated and actual trajectories.

(b) Localisation error as a function of time.

Fig. 5 Localisation performance of particle filter over the course of a rectangular survey experiment with four sensor nodes within communication range. A total of 154 range updates were received (out of approximately 285 attempted) during the experiment and the RMS error was 3.1 m.

The performance of the system was evaluated in the limiting case where only one sensor node is within communication range (single beacon localisation). Using the data acquired from the three sensor node case, Figure 6 shows the performance of the system using only one of the sensor nodes. Compared to the first scenario the

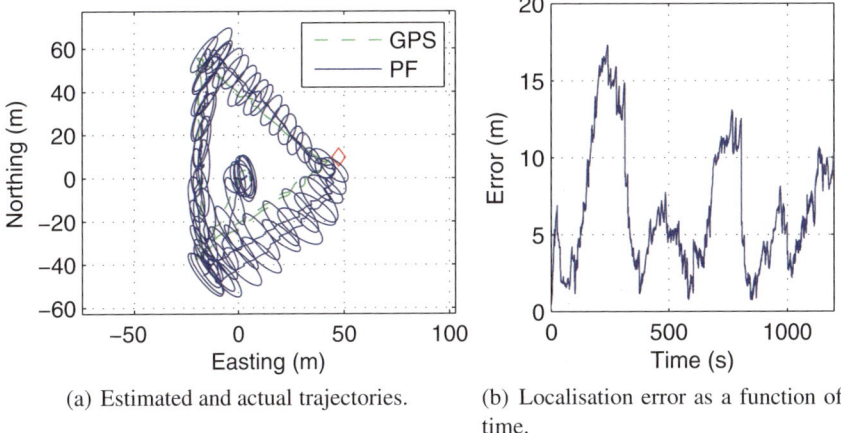

(a) Estimated and actual trajectories.

(b) Localisation error as a function of time.

Fig. 6 Localisation performance of particle filter over the course of a survey experiment with one sensor node within communication range.

number of range measurements has been reduced from 397 to 168. The RMS localisation error is 7.7 m, although greater than previous experiments it does illustrate that the AUV is still capable of estimating its position with an accuracy suitable for navigating in most underwater environments.

5 Conclusions

An approach to localisation of an AUV using an ad-hoc acoustic underwater network has been presented. The system utilises existing sensor network infrastructure with estimates of communication overhead to determine time-of-flight information between the AUV and individual sensor nodes. A particle filter based localisation system was developed and implemented in real-time onboard the AUV. Experimental field results have demonstrated that, despite the absence of precise self motion estimate and a slow rate of ranging updates, the system is able to provide localisation performance sufficiently accurate for navigation within and immediately outside the sensor network communication field.

   Current research is focused on improving localiser performance by incorporating more accurate vehicle motion estimates using the AUV's existing visual odometry system [7]. Additionally, the system is being expanded to perform network self localisation. This could be accomplished using network localisation techniques such as in Djusgash, Singh, et al. 2008[5]. In the scenario of long term underwater monitoring this would allow extra sensor nodes to be added to the network without requiring precise knowledge of the sensors location. Finally, the integrated AUV and sensor network system performance is currently being evaluated in different ocean environments including bays and coral reefs.

References

1. Akyildiz, I.F., Pompili, D., Melodia, T.: Underwater acoustic sensor networks: Research challenges. Journal of Ad Hoc Networks 3, 257–279 (2005)
2. Bengston, K.J., Dunbabin, M.D.: Design and performance of a networked ad-hoc acoustic communications system using inexpensive commercial CDMA modems. In: IEEE OCEANS 2007, Europe (June 2007)
3. Corke, P., Detweiler, C., Dunbabin, M., Hamilton, M., Vasilescu, I., Rus, D.: Experiments with underwater localization and tracking. In: IEEE ICRA, Rome, Italy, April 2007, pp. 4556–4561 (2007)
4. Detweiler, C., Leonard, J., Rus, D., Teller, S.: Passive mobile robot localization within a fixed beacon field. In: Proceedings of the International Workshop on the Algorithmic Foundations of Robotics, August 2006. Springer, New York (2006)
5. Djugash, J., Singh, S., Grocholsky, B.: Decentralized mapping of robot-aided sensor networks. Robotics and Automation. In: IEEE International Conference on ICRA 2008, May 2008, pp. 583–589 (2008)
6. Dunbabin, M., Roberts, J., Usher, K., Winstanley, G., Corke, P.: A hybrid AUV design for shallow water reef navigation. In: Proc. of the International Conference on Robotics and Automation, ICRA (2005)
7. Dunbabin, M., Usher, K., Corke, P.: Visual motion estimation for and autonomous underwater reef monitoring robot. In: Proc. of the International on Field & Service Robotics, pp. 57–68 (2005)
8. Eustice, R., Pizarro, O., Singh, H.: Visually augmented navigation in an unstructured environment using a delay state history. In: Proceedings of the 2004 IEEE International Conference on Robotics & Automation, pp. 25–32 (April 2004)
9. Eustice, R., Singh, H., Leonard, J., Walter, M., Ballard, R.: Visually navigating the rms titanic with slam information filters. In: Proc. of the International Symposium of Robotics Research, RSS (2005)
10. Leonard, J.J., Bennett, A.A., Smith, C.M., Feder, H.J.S.: Autonomous underwater vehicle navigation. Technical Report 98-1, MIT Marine Robotics Laboratory, Cambridge, MA, USA (1998)
11. Sikka, P., Corke, P., Overs, L.: Wireless sensor devices for animal tracking and control. In: Proc. First IEEE Workshop on Embedded Networked Sensors, Tampa, Florida, November 2004, pp. 446–454 (2004)
12. Stutters, L., Liu, H., Tiltman, C., Brown, D.J.: Navigation technologies for autonomous underwater vehicles. IEEE Transactions on Systems, Man, and Cybernetics, Part C: Applications and Reviews 38(4), 581–589 (2008)
13. Thrun, S., Burgard, W., Fox, D.: Probabilistic Robotics. MIT Press, Cambridge (2006)
14. Whitcomb, L., Yoerger, D., Singh, H.: Advances in doppler-based navigation of underwater robotic vehicles. In: Proc. of the International Conference on Robotics and Automation, ICRA (1999)

Experiments in Visual Localisation around Underwater Structures

Stephen Nuske, Jonathan Roberts, David Prasser, and Gordon Wyeth

Abstract. Localisation of an AUV is challenging and a range of inspection applications require relatively accurate positioning information with respect to submerged structures. We have developed a vision based localisation method that uses a 3D model of the structure to be inspected. The system comprises a monocular vision system, a spotlight and a low-cost IMU. Previous methods that attempt to solve the problem in a similar way try and factor out the effects of lighting. Effects, such as shading on curved surfaces or specular reflections, are heavily dependent on the light direction and are difficult to deal with when using existing techniques. The novelty of our method is that we explicitly model the light source. Results are shown of an implementation on a small AUV in clear water at night.

1 Introduction

We are interested in the localisation of underwater robots around fixed infrastructure. There are many applications of underwater robotics where it is critical for the robot to know where it is with respect to a structure, such as inspection tasks and welding. Assuming that most structures are passive, ie. they do not transmit any location information, then there are two viable sensing modalities that can be used to

Stephen Nuske
The Robotics Institute, Carnegie Mellon University, Pittsburgh, PA 15213, USA
e-mail: nuske@cmu.edu

Jonathan Roberts and David Prasser
Autonomous Systems Lab, CSIRO ICT Centre, PO Box 883, Kenmore,
Queensland 4069, Australia
e-mail: jonathan.roberts@csiro.au, david.prasser@csiro.au

Gordon Wyeth
School of Information Technology and Electrical Engineering, University of Queensland,
St Lucia, Queensland 4072, Australia
e-mail: wyeth@itee.uq.edu.au

A. Howard et al. (Eds.): Field and Service Robotics 7, STAR 62, pp. 295–304.
springerlink.com   © Springer-Verlag Berlin Heidelberg 2010

image a structure; sonar and computer vision. Of these, we have been investigating the use of vision in order to localise an Autonomous Underwater Vehicle (AUV) with respect to a known piece of underwater infrastructure - the leg of a surface platform.

Typically, the visual environment around such a structure is poor. Firstly, suspended particles in the water reduce visibility. Secondly, there is minimal or no natural lighting deep underwater, thus requiring an artificial light source to be mounted on the AUV. Thirdly, the visual appearance of the structure in this scenario is highly dependent on the incident angle of the light source. The light source is constantly moving (as it is on the AUV) and consequently the visual appearance of the structure varies dramatically over time. This is quite different from typical well lit environments, where the light source (typically the Sun) is far less dynamic and also where there is a significant level of ambient lighting. However, rather than this poor visual environment being a negative, we would argue that we can turn it to our advantage. By modeling the light source mounted on the AUV we can predict the appearance of the structure (the legs of a platform in our example) from different viewing poses. The process is to use an a priori 3D-surface model of the permanent structure being navigated with the light model to generate artificial images which are compared against the real camera image to localise the AUV.

2 Previous Work

Kondo et al. present two methods of navigating underwater structures in [10] and [9]. In [10], two laser beams are directed at the structure which are detected in the camera images to triangulate the relative distance and orientation of the vehicle. In [9], Kondo et al. use a light stripe to illuminate a 2D profile of the structure which is detected in the camera images. A common feature of the two systems developed by Kondo et al. is the use of active lighting. In our work an artificial light source is also used, but unlike the focused beams or light stripes of Kondo et al., the light source is unfocused.

Stolkin et al. [13] present work for a submarine localising from a leg of a platform and use an explicit 3D model of the structure, projecting the model onto the image plane to predict the shape of the structure. Model based tracking is an attractive approach for this application as the form of the structure is well known *a priori*. Fig. 1 shows the basic idea behind model based visual tracking. Synthetic images are generated for a large number of possible robot poses and each of these images is compared with the actual image captured by the robot. The comparator can take many forms. Examples include taking the pose that gives the best match, or using a multi-hypothesis framework such as a particle filter [8].

In the wider robotics field, model-based tracking has received much attention. The works of Gerard and Gagalowicz [4], Noyer et al. [12] and Ho and Jarvis [5] present pose estimation systems based on 3D-surface maps. They perform correspondences between real and synthetic images. Both Noyer et al. [12] and Ho and Jarvis [5] estimate pose with a probabilistic particle filter, which is an efficient

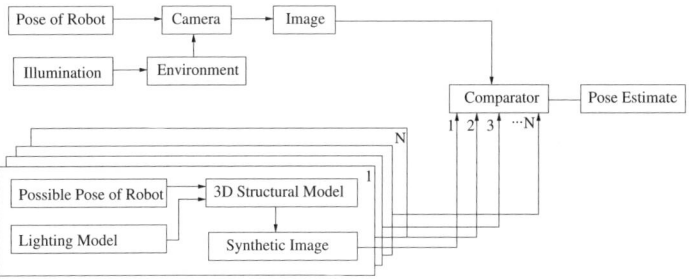

Fig. 1 An overview of the idea of model-based visual pose estimation. Multiple synthetic images (generated at different possible poses of the AUV) are compared with the real camera image of the scene.

means of sampling the solution-space, whereas Gerard and Gagalowicz [4] present a more brute force evaluation of the solution-space. None of these 3D-surface based methods consider reflectance and lighting properties in their work – they only use textured 3D models – which do not generalize to any lighting condition. A textured model would not suffice for the application presented in this paper, because the structure is made of one material and is therefore essentially texture-less. The images of the structure are also highly dependent on the light source, indicating that both reflectance properties of the structure and a light model should be known.

The work of Kee et al. [7] and Blicher et al. [1], in the domain of face identification, introduce the idea of using a 3D-surface model together with a light model. They show how to perform face identification in unknown lighting conditions by first estimating the current light source, then generating synthetic images of each face model using the estimated light source model. They used a database of many different 3D-surface face models. A single fixed pose of the faces with respect to the camera was assumed, then multiple synthetic face hypothesis images were matched to the real image. However, in our work there is a single 3D-surface map of the environment (the structure) and multiple pose hypotheses that are matched to a real image (taken from the AUV). The pose hypotheses with the best image match to the camera image from the robot will provide the pose estimate. This idea of estimating and incorporating a light model has not yet been applied to visual localisation.

3 Localisation Framework

Our framework uses a model of the structure and a model of light source together to generate synthetic images that are expectations of the real camera images. The synthetic images are compared to the real images to estimate the pose of the camera. The pose estimation is facilitated in a probabilistic multiple pose hypothesis framework – a particle filter – which uses a synthetic image of the structure from each pose hypothesis to derive a comparison score against the real image.

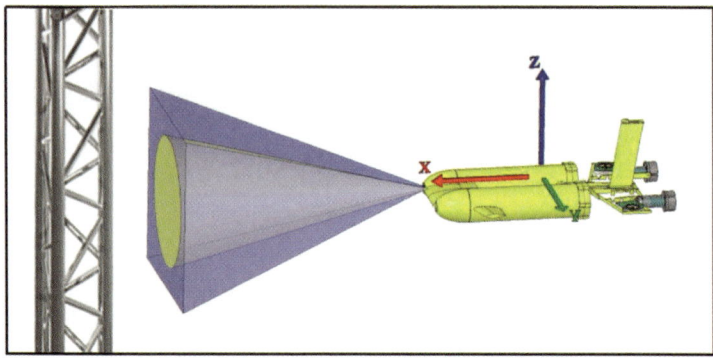

Fig. 2 The AUV has a forward facing camera with a field-of-view depicted in the figure by pyramid viewing volume. The spotlight is also facing forward and partially illuminates the field-of-view of the camera (depicted by the inner cone).

A forward looking spotlight and camera are mounted rigidly to the AUV[3] which is inspecting a structure. Fig. 2 shows the AUV, the coordinate system, the camera and spotlight-setup. A single forward-facing camera, from the AUV's stereo pair, is used. The field-of-view of the camera, is shown as pyramidal viewing volume in Fig. 2. The extrinsic pose of the camera is calculated with respect to the vehicle, and is facing along the vehicle's positive x axis. The camera's intrinsic parameters were calibrated using the OpenCV library[6]. The camera images are then undistorted by these parameters, and the model of the structure can then be projected directly onto the image plane.

3.1 Synthetic Image Generation

The synthetic images are generated from a polygon mesh of the structure. The mesh includes the surface normal, diffuse and specular reflectance properties. Meshes using such detailed surface properties have rarely been applied to visual localisation. These photometric properties are incorporated into the Blinn-Phong model [2] using the OpenGL library for generating synthetic images. There are other lighting models which could be also used, but the Blinn-Phong model is chosen for its simplicity, speed of computation and prevalent implementation on most graphics processors. In addition to the pose of the light, the model incorporates a number of other parameters to account for the attenuation by water and angular spread of the light source.

The structure to be localised against is comprised of three steel tubes, linked together by smaller rungs at approximately 45 degree angles. This structure is modelled as a polygon mesh of each of the tubes by defining the number of slices and stacks in the mesh of each tube. The rendering is performed with per-fragment lighting calculations, and shading interpolation to the pixels within the fragment, according to the light model defined above. A series of tests were performed to calculate the optimum number of polygons taking into account render times and model accuracy. For this model and using an Intel dual core 2.33GHz CPU and a NVIDIA Quadro FX 350M GPU with 320x240 images, it was found that optimum value of 3120 polygons equated to a time of 0.325ms to render a single synthetic image.

3.2   Particle Filter Localisation

The use of a particle filter is described in detail by Thrun et al. in [14]. The particle filter is a set of N pose hypotheses (particles) $X_t = x_t^{(1)}, x_t^{(2)}, x_t^{(3)} \dots, x_t^{(N)}$. The set is sampled from the previous set X_{t-1} using a propagation model m_t and a corresponding set of weights (probabilities), W. The weights are calculated from an observation of the environment, y, as follows:

$$W_k^{(n)} = p(y_k | x_k^{(n)}) \tag{1}$$

the observation of the environment is a camera image, y, which is compared with each pose particle x by rendering a synthetic image. The measurement of probability is provided from an image matching technique (discussed in Section 3.3). The concept is that a synthetic image generated from the particles nearest the correct pose will give the best match with the real image. These particles are then the most likely to be re-sampled for the next iteration. The current pose estimate of the AUV is extracted from the filter as the mean pose of the particles with the highest weights (top 5%). We use roll and pitch estimates from the AUV's Inertial Measurement Unit (IMU) as each particle's roll and pitch estimate. The remaining four degrees of freedom are propagated using a constant velocity model calculated from a set of previous pose estimates extracted from the particle filter.

3.3   Gradient-Domain Image Matching

The comparison process between the camera image and a synthetic image provides a likelihood measure for the set of particles in the filter. The works of [1, 4, 7, 12], all use image matching techniques which compare real and synthetic images. All of the techniques are variants of the Mean Absolute Difference (MAD).

The simple image matching techniques assume that it is possible to generate pixel intensities for the synthetic image that are equivalent to those in the real image. This is different from photo-realistic rendering, which is only interested in making the synthetic image appear real. Whereas, these image intensity matching techniques require the environment model and light model to be accurate estimates of the actual physical properties of the surrounding environment. The parameters of such models are difficult to estimate accurately. Furthermore, the Blinn-Phong image formation model [2], employed for real-time performance, does not incorporate the ability to model poor visibility. Suspended tiny particles that cause poor visibility have two effects on the lighting; absorbing light and reflecting light. These effects would need to be modelled before accurate intensity values could be generated in the synthetic image. It is difficult to generate accurate images of simulated poor visibility conditions at a rate quick enough for this framework. For this reason, intensity-based matching

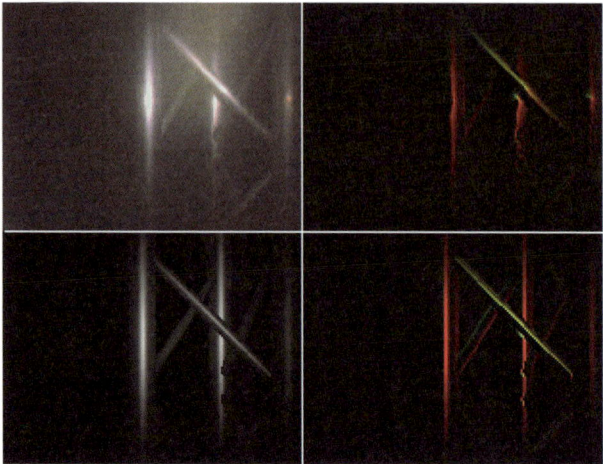

Fig. 3 Top Left: Real camera image. Top Right: Real Sobel image. Bottom Left: Synthetic image. Bottom Right: Synthetic Sobel Image. Horizontal gradient is shown in red and vertical in green; therefore pixels with diagonal gradients are yellow.

is not used and, instead, a gradient-domain image matching technique is developed. The gradient-domain removes the absolute intensity levels whilst capturing the subtle shading in the environment. This behaviour is different to an edge-image that identifies drastic boundaries of intensity.

The first step is to pass the real-image through a Gaussian filter, which removes the effects of noise. The synthetic and real images are then both passed through a horizontal Sobel operator to generate gradient images in both the x and y directions; G_x and G_y are the real Sobel images and g_x and g_y are the synthetic. Example synthetic and real images are shown in Fig. 3. The x direction is shown in red and y in green, therefore pixels with high x and y gradients are yellow. To compare the real and synthetic Sobel images, it would be possible to turn these two images into a gradient magnitude image and a gradient orientation image, which would enable a more logical means for comparison. But to avoid the expensive square root and arc tan computations, the images are compared directly in x and y gradients.

Firstly, a sum is taken of the gradient magnitude in the real, S_r, and synthetic, S_s, images:

$$S_r = \sum_{p=0}^{N} (|G_x(p)| + |G_y(p)|) \qquad S_s = \sum_{p=0}^{N} (|g_x(p)| + |g_y(p)|) \qquad (2)$$

where N is the number of pixels. Secondly, a sum of the difference in gradients between real and synthetic is calculated in each direction, D_x, D_y;

$$D_x = \sum_{p=0}^{N} (|G_x(p) - g_x(p)|) \qquad D_y = \sum_{p=0}^{N} (|G_y(p) - g_y(p)|) \qquad (3)$$

the final image matching score is derived by subtracting the sum of the gradient difference from the sum of the gradient magnitude and normalizing by the sum of the gradient magnitude:

$$D_\mu = \frac{(S_r + S_s) - (D_x + D_y)}{S_r + S_s} \qquad (4)$$

This result equates to the observation y and the particle $x^{(n)}$ from (1). The better the match between the images the larger value of D_μ. This score can be incorporated into the observation y and the particle $x^{(n)}$ from (1) as follows:

$$W_k^{(n)} = p(y_k | x_k^{(n)}) \propto e^{\rho D_\mu} \qquad (5)$$

Where ρ is a positive constant that adjusts the convergence of the particle filter.

4   Results

An experiment was conducted at night in clear water with the aim of determining if the visual localisation system in combination with the IMU could localise the AUV as it moved freely in all six degrees of freedom. The experiment began with the AUV approximately 1.5m away from the structure. The AUV approached the structure, strafed side to side, descended and rotated around the structure. Note that there was no ground truth data available during this experiment and hence the performance of the system could only be checked manually by inspecting the projected centre lines of the structure from the estimated pose, and confirming they align correctly in the raw camera images. Tracking images from the experiment are presented in Fig. 4, along with a movie of the results can be found in the video attachment located at:
    http://www.cat.csiro.au/ict/download/nuske/auv_pooltest1.mpg

The visual localisation system maintained accurate track of the structure for 440 frames where there were significant changes in scale, orientation and translation. The system then made a mistake when one of the columns of the structure disappeared behind another, and then reappeared on the other side. The system estimated the column reappearing on the same side, and did not recover from this error. The frames just before and just after the disappearing column can be seen as the bottom two images of Fig. 4. When the system was run again, and again over the same data, it did occasionally correctly estimate that the rear column appeared on the other side. However, it failed more times than it succeeded and a solution to this problem is currently being investigated.

Algorithmic speed and efficiency are critical to ensure a practical system for the target application. There is a demand to use low power processors on AUVs due to the on board power limitations of the vehicles. The results reported above were processed off-line. However, the algorithm has been implemented in a such a way as to

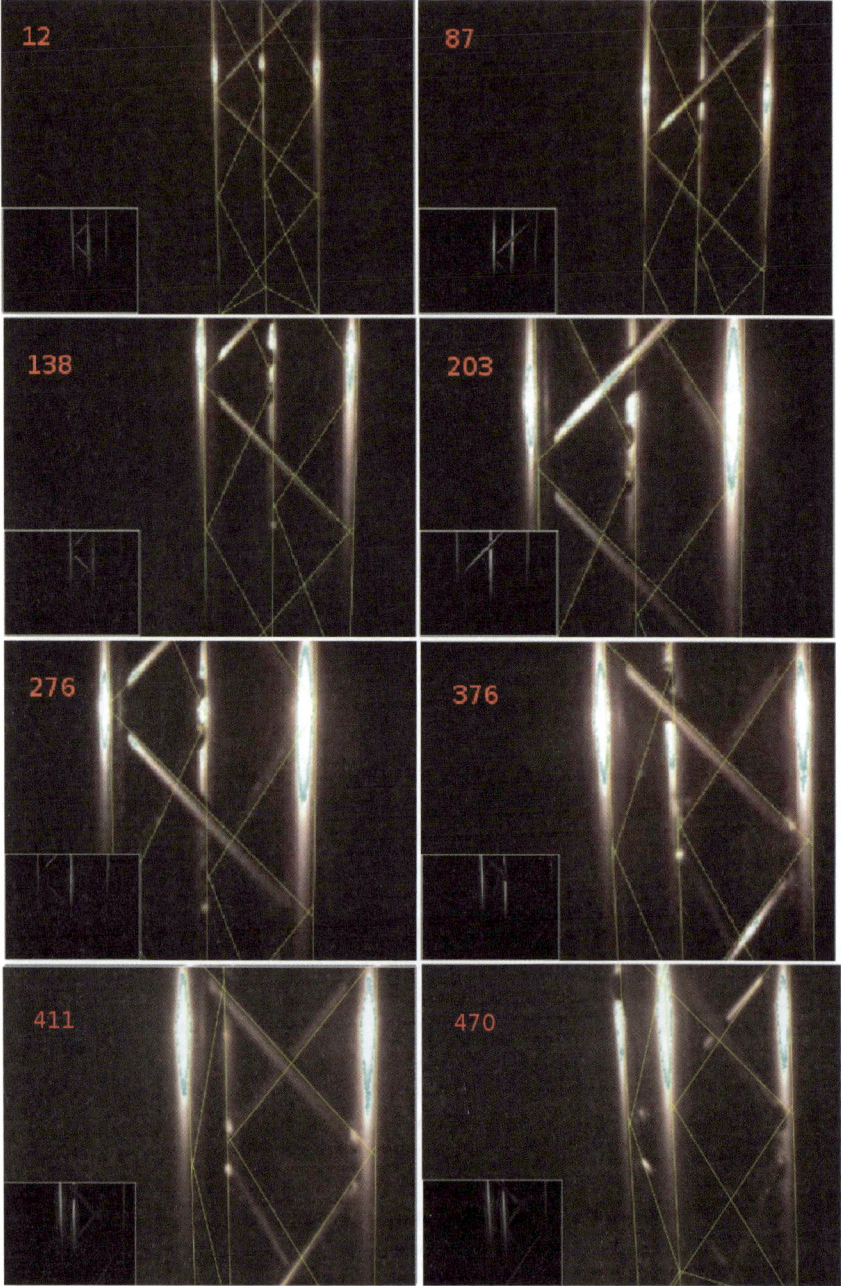

Fig. 4 Images showing the tracking of the oil rig structure. Real camera image is overlaid with the centre lines of the structure projected from the estimated pose. Bottom left corner of each image is a synthetic rendering of the structure. The frame number is located in the top left corner of each image. Data was collected at 15Hz.

make it amenable to on-board impementation. The resolution of the 3D-surface was minimised, and the image processing and matching algorithms were implemented on a GPU, which all brought the processing times much closer to real-time rates. The system was run on a laptop computer, which used small mobile processors and graphics hardware. It achieved a frame rate of 0.5Hz with 800 particles using 320x240 sized images and would therefore require the AUV to be moving slowly with respect to the structure to be able to track it reliably in real time.

The two potential methods to achieve higher efficiencies, would be to further reduce the render times of the synthetic images, and also to reduce the number of particles in the filter. Reducing the render time could involve further reductions in the polygon counts, only passing sections of the model that are in view to the graphics pipeline, optimising the lighting calculations or with improved/multiple GPUs. Future improvements to reduce the number of particles could include using a two-stage coarse-to-fine particle filter, such as used in the work of Klein and Murray [8], or to develop a better propagation model. The depth sensor and the magnetic compass are two sensors which could be included in the propagation model. However, it would need to be confirmed that these sensors are locally consistent in the desired environment (that is, if their inter-frame motion estimates are accurate). Another possible method of improving the propagation model is to use a visual odometry system. Marchand et al. [11] present such an approach.

5   Conclusion and Future Work

We have presented a visual localisation system that explicitly models the spotlight of an AUV navigating underwater structures. The light model is used in conjunction with a surface model of the structure to generate synthetic images that are accurate representations of the real camera image. A particle filter framework is employed where a synthetic image is rendered from each pose hypothesis and a observation function computes a probability through comparison with the real camera image. The observation function compares the real image with the synthetic images operates in the intensity gradient domain, avoiding the need to generate precise intensity values in the synthetic image and allows the system to operate in poor visibility conditions which are difficult to replicate in the synthetic image. The system was tested using a monocular vision system, spotlight, steel structure and low-cost IMU. Results show that the system can localise the vehicle in challenging image sequences where the light source is constantly moving and illuminating the scene non-uniformly.

In future work the system will continue to be developed with the goal of a fully functioning system in the targeted offshore environments. Localising from structures of other shapes and surface characteristics will be evaluated. The image processing algorithms presented here are essentially generic and are expected to be able to provide similar results from other structures. Improvements to the lighting model will also be investigated, such as modelling the spotlight as an area light source and also using a more accurate model of the surface reflectance properties to generate images with more precise representation of the shading. More accurate odometry information will also be employed which may come from a compass, a pressure (depth) sensor or potentially a visual odometry algorithm. This information is expected to greatly improve the accuracy and computational efficiency of the system by significantly reducing the area of the state space that must be evaluated. Ambiguous visual scenarios which have been the cause of divergence in the localisation filter will also be investigated in more detail.

References

1. Blicher, A.P., Roy, S., Penev, P.S.: Lightsphere: Fast lighting compensation for matching a 2d image to a 3d model. In: 17th International Conference on (ICPR 2004), pp. 157–162. IEEE Computer Society, Washington (2004)
2. Blinn, J.F.: Models of light reflection for computer synthesized pictures. SIGGRAPH Comput. Graph. 11(2), 192–198 (1977)
3. Dunbabin, M., Roberts, J., Usher, K., Winstanley, G., Corke, P.: A hybrid AUV design for shallow water reef navigation. In: Proceedings of the 2005 IEEE International Conference on Robotics and Automation, ICRA 2005, pp. 2105–2110 (2005)
4. Gerard, P., Gagalowicz, A.: Three dimensonal model-based tracking using texture learning and matching. Pattern Recognition Letters 21, 1095–1103 (2000)
5. Ho, N., Jarvis, R.: Global localisation in real and cyber worlds using vision. In: Australasian Conference on Robotics and Automation (2007)
6. Intel: Open Source Computer Vision Library: Reference Manual (2000), http://www.intel.com/technology/computing/opencv
7. Kee, S.C., Lee, K.M., Lee, S.U.: Illumination invariant face recognition using photometric stereo. IEICE Trans. on Information and Systems 7, 1466–1474 (2000)
8. Klein, G., Murray, D.: Full-3d edge tracking with a particle filter. In: British Machine Vision Conference, pp. 1119–1128 (2006)
9. Kondo, H., Maki, T., Ura, T., Nose, Y., Sakamaki, T., Inaishi, M.: Relative navigation of an autonomous underwater vehicle using a light-section profiling system. In: Proceedings of the 2005 IEEE International Conference on Intelligent Robots and Systems, IROS, pp. 1103–1108 (2004)
10. Kondo, H., Ura, T., Nose, Y., Akizono, J., Sakai, H.: Visual investigation of underwater structures by the AUV and sea trials. In: Proceedings of OCEANS 2003, vol. 1, pp. 340–345 (2003)
11. Marchand, E., Bouthemy, P., Chaumette, F.: A 2d-3d model-based approach to real-time visual tracking. Image and Vision Computing 19(13), 941–955 (2001)
12. Noyer, J., Lanvin, P., Benjelloun, M.: Model-based tracking of 3d objects based on a sequential monte-carlo method. In: Conference on Signals, Systems and Computers, vol. 2, pp. 1744–1748 (2004)
13. Stolkin, R., Hodgetts, M., Greig, A.: An em / e-mrf strategy for underwater navigation. In: Proceedings of the British Machine Vision Conference, pp. 715–724 (2000)
14. Thrun, S., Burgard, W., Fox, D.: Probabalistic Robotics. The MIT Press, Cambridge (2005)

Part VII
Multi-Robot Cooperation

Leap-Frog Path Design for Multi-Robot Cooperative Localization

Stephen Tully, George Kantor, and Howie Choset

Summary. We present a "leap-frog" path designed for a team of three robots performing *cooperative localization*. Two robots act as stationary measurement beacons while the third moves in a path that provides informative measurements. After completing the move, the roles of each robot are switched and the path is repeated. We demonstrate accurate localization using this path via a coverage experiment in which three robots successfully cover a 20m x 30m area. We report an approximate positional drift of 1.1m per robot over a travel distance of 140m. To our knowledge, this is one of the largest successful GPS-denied coverage experiments to date.

1 Introduction

Localization is critical for the navigational aspect of many robotic applications. Without accurate positioning, a mobile robot would get lost, wander away from its target workspace, and fail to complete its intended task. Additionally, there are many situations where an external positioning system, such as GPS, is unavailable to the robot, e.g. indoors, within dense vegetation, and underwater. To solve the localization problem, a team of robots can employ *cooperative localization* [1] to incorporate relative sensor measurements into a Kalman filter framework that estimates the pose of the robots.

It can be shown that the accuracy of such a filter is dependent upon the path the robots take. This is due to the fact that certain measurements are more informative than others, depending on the vantage point of the sensor. We believe a "leap-frog" path, as in [2, 3], is desirable because it temporarily grounds the increasing uncertainty of the system via stationary robots.

Stephen Tully, George Kantor, and Howie Choset
Carnegie Mellon University
e-mail: {stully@ece,kantor@ri,choset@cs}.cmu.edu

A. Howard et al. (Eds.): Field and Service Robotics 7, STAR 62, pp. 307–317.
springerlink.com © Springer-Verlag Berlin Heidelberg 2010

Fig. 1 Three robots used for experimental evaluation of the proposed leap-frog localization strategy.

The contribution of this work is the introduction of a new leap-frog path designed to produce informative measurements for three robots performing cooperative localization. We also report a 20m x 30m large-scale GPS-denied coverage experiment with three robots (see Fig.1) that was only possible after the gain in positioning accuracy provided by this new leap-frog strategy.

2 Related Work

The majority of recent work on cooperative localization ignores the path planning aspect of this topic and instead focuses on filtering. In [1], Roume-liotis et. al. define a Kalman filter framework, similar to our formulation in Sec. 3, that can be used for cooperative localization. In [4], they iterate on this work and present a distributed method for computing the Kalman filter. Finally, the work in [5, 6] studies the growth of uncertainty under different sensing modalities and with a varying number of robots.

Some recent work has addressed path planning and control for robots performing cooperative localization. Hidaka et. al. [7] present derivations to show that with any number of robots, the optimal formation for accurate localization is a "packed circles" configuration. For three robots, they claim that the optimal formation is an equilateral triangle. Trawny et. al. [8], on the other hand, perform optimization over possible multi-robot paths and demonstrate a performance improvement in simulation. Although the optimization is beneficial, we believe this method is susceptible to local minima.

Finally, some research has involved the investigation of leap-frog paths for cooperative localization. Navarro-Serment et. al. [2] use a group of small heterogeneous robots (Millibots) for localization and mapping. The authors use leap-frog paths to help maintain a better estimate over the occupancy grid map and their EKF localization. Kurazume [3] introduces leap-frog paths for cooperative localization as well, with a path that is designed to represent triangle chains of different configurations. Although these paths prove to be accurate solutions for localization, Kurazume does not analyze these paths from an information theoretic standpoint.

3   Cooperative Localization

To reduce motion error, a team of robots can employ *cooperative localization*, which improves the estimate of the state via relative sensor measurements between robots. This type of filtering method can be implemented with an extended Kalman filter, as in [1].

The state vector for the filter X_k is defined,

$$X_k = \begin{bmatrix} X_k^{0^T} & X_k^{1^T} & \cdots & X_k^{N-1^T} \end{bmatrix}^T \qquad X_k^i = \begin{bmatrix} x_k^i & y_k^i & \theta_k^i \end{bmatrix}^T,$$

where N is the number of robots and X_k^i represents the state of the i-th robot at time step k. In this formulation, (x_k^i, y_k^i) represents the position of the i-th robot and θ_k^i represents that robot's heading.

The state process equation $f(X_k^i, u_k^i)$ is based on a unicycle model,

$$f(X_k^i, u_k^i) = \begin{bmatrix} x_k^i + v_k^i \cos \theta_k^i \Delta t \\ y_k^i + v_k^i \sin \theta_k^i \Delta t \\ \theta_k^i + \omega_k^i \Delta t \end{bmatrix} \qquad u_k^i = \begin{bmatrix} v_k^i \\ \omega_k^i \end{bmatrix},$$

where u_k^i is a motion input for the i-th robot, which is composed of a translational velocity v_k^i and a rotational velocity ω_k^i.

The measurement equation for our state is a bearing-only measurement,

$$h_j^i(X_k) = \arctan\left(\frac{y_i - y_j}{x_i - x_j}\right) - \theta_j$$

which represents the relative bearing angle to the i-th robot as measured by the j-th robot. A typical sensor that provides bearing measurements in this form is a monocular camera.

The purpose of the extended Kalman filter (EKF) is to recursively estimate the state mean and covariance matrix with two stages: the prediction step, which produces the estimated mean and covariance, $\hat{X}_{k+1|k}$ and $P_{k+1|k}$ respectively, as well as the measurement update step, which produces an update to the estimated mean and covariance, $\hat{X}_{k|k}$ and $P_{k|k}$ respectively.

3.1   Prediction Step

The EKF prediction step is applied when processing the robot's internal velocities (usually from wheel encoders). The state mean and covariance matrix are computed as follows.

$$\hat{X}_{k+1|k} = \begin{bmatrix} f(\hat{X}_{k|k}^0, u_k^0)^T & f(\hat{X}_{k|k}^1, u_k^1)^T & \cdots & f(\hat{X}_{k|k}^{N-1}, u_k^{N-1})^T \end{bmatrix}^T$$

$$P_{k+1|k} = F_k P_{k|k} F_k^T + W_k U_k W_k^T$$

where $\hat{X}_{k|k}$ and $P_{k|k}$ define the estimate from the previous time step, and F_k and W_k are the Jacobians of the state process equation, as defined below.

$$
F_k = \begin{bmatrix} \frac{\partial f(\hat{X}^0_{k|k}, u^0_k)}{\partial X^i_k} & 0 & \cdots & 0 \\ 0 & \frac{\partial f(\hat{X}^1_{k|k}, u^1_k)}{\partial X^i_k} & & \vdots \\ \vdots & & \ddots & 0 \\ 0 & \cdots & 0 & \frac{\partial f(\hat{X}^{N-1}_{k|k}, u^{N-1}_k)}{\partial X^i_k} \end{bmatrix}
\quad
W_k = \begin{bmatrix} \frac{\partial f(\hat{X}^0_{k|k}, u^0_k)}{\partial u^i_k} & 0 & \cdots & 0 \\ 0 & \frac{\partial f(\hat{X}^1_{k|k}, u^1_k)}{\partial u^i_k} & & \vdots \\ \vdots & & \ddots & 0 \\ 0 & \cdots & 0 & \frac{\partial f(\hat{X}^{N-1}_{k|k}, u^{N-1}_k)}{\partial u^i_k} \end{bmatrix}
$$

The matrix U_k is a covariance matrix with 2x2 matrices along its diagonal (all U^i_k for $0 \leq i < N-1$ as defined below).

$$
U^i_k = \begin{bmatrix} \alpha \, |v^i_k| & 0 \\ 0 & \beta \, |v^i_k| + \gamma \, |\omega^i_k| \end{bmatrix}
\tag{1}
$$

Matrix U^i_k represents the covariance matrix for the additive white Gaussian noise that is expected to perturb robot i's motion input u^i_k. Conventional implementations use a static covariance for this purpose, but we believe a velocity dependent noise model is more accurate. The model in Eq. 1 accounts for the fact that wheel slippage is more pronounced at higher speeds and that zero additive noise should be expected when the robots are stationary.

3.2 Measurement Update Step

To properly incorporate the information provided by the bearing sensors, we perform a correction to the predicted state estimate,

$$
\begin{aligned}
H^i_j &= \frac{\partial h^i_j(\hat{X}_{k|k-1})}{\partial X_k} \\
K &= P_{k|k-1} H^T (H P_{k|k-1} H^T + R)^{-1} \\
\hat{X}_{k|k} &= \hat{X}_{k|k-1} + K(z_k - h(\hat{X}_{k|k-1})) \\
P_{k|k} &= P_{k|k-1} - K H P_{k|k-1}
\end{aligned}
\tag{2}
$$

where, for M bearing measurements, K is the Kalman gain, H is the measurement Jacobian, and R is an MxM matrix with diagonal elements σ_z^2 (the variance associated to a single measurement). The Jacobian H is constructed by appending together all row vectors H^i_j for each measurement between a robot i and another robot j. Likewise, h is constructed by appending together all h^i_j to form a column vector. z_k is the measurement vector to which h is associated.

It is important to realize that the effectiveness of this cooperative localization filter is dependent upon the path of the robots. This can be seen in Eq. 2 where a positive definite matrix $KHP_{k|k-1}$ is subtracted from the predicted covariance $P_{k|k-1}$. Since H is dependent upon the state estimate $\hat{X}_{k|k-1}$, the reduction in uncertainty via subtracting $KHP_{k|k-1}$ will vary in amount depending on the configuration of the robots.

4 Leap-Frog Path Design

In Sec. 3, we discuss how the effectiveness of the cooperative localization filter is dependent upon the path of the robots. This suggests that by careful path planning, we can achieve better position accuracy during experiments. As in [2, 3], we suggest the use of a "leap-frog" path for a team of robots, where at any given time, a subset of the robots temporarily act as stationary measurement beacons while the other robots are in motion.

A "leap-frog" path is intuitively beneficial for the Kalman filter because when robots are stationary, they will not gain any positioning noise, thus temporarily grounding the normally increasing uncertainty of the system. A moving robot can move around at will without concern for its added prediction noise because it can easily visit the nearby stationary robots to drive its position uncertainty down to their level via relative sensor measurements.

4.1 *Three-Robot Path Design via Information Gain*

The use of three robots for localization is a good fit for bearing-only measurements because the intersection of two bearing rays from two different robots will triangulate the location of a third robot, albeit with error due to noise. To investigate path design for a team of three robots, we consider the measurement update equation for the information filter, which is a dual to the Kalman filter and is commonly used in localization and mapping algorithms, such as [9]. The measurement update is as follows,

$$I_{k|k} = I_{k|k-1} + H^T R^{-1} H,$$

where the information matrix $I_{k|k} = P_{k|k}^{-1}$ is the inverse of the covariance matrix. In this work, we define the information gain $G(X_k)$ as a norm of the positive definite matrix that is added to the information matrix during a measurement update,

$$G(X_k) = \mathrm{tr}\left(H^T R^{-1} H\right).$$

The information gain depends on the state X_k through the measurement Jacobian H. We argue that states producing a larger information gain will offer measurements that are more informative to the Kalman filter.

To investigate the path optimization problem for three robots, we consider the situation in Fig. 2 where two robots (0 and 1) lie stationary on the y-axis

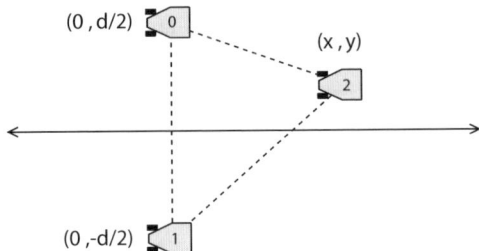

Fig. 2 Stationary robots 0 and 1 are a distance $d/2$ away from the x-axis. An analysis of the information gain is used to obtain the appropriate leap-frog path for robot 2.

an equal distance away from the x-axis (with a separation distance d). The third robot (robot 2) acts as the moving robot in this leap-frog strategy.

For any pose (x, y, θ) of robot 2 in Fig. 2, the filter will have the following information gain $G(X_k)$,

$$X_k = \left[0 \; \frac{d}{2} \; 0 \; 0 \; \frac{-d}{2} \; 0 \; x \; y \; \theta \right]^T$$

$$G(X_k) = \sigma_z^{-2} \left(6 + \frac{4}{d^2} + \frac{4}{(y - d/2)^2 + x^2} + \frac{4}{(y + d/2)^2 + x^2} \right)$$

To determine the optimal y value for any x position of robot 2, we can take the derivative of the information gain with respect to y, as follows,

$$\frac{\partial G(X_k)}{\partial y} = 8\sigma_z^2 \left(\frac{y - d/2}{\left((y - d/2)^2 + x^2\right)^2} + \frac{y + d/2}{\left((y + d/2)^2 + x^2\right)^2} \right)$$

By setting the derivative to zero, we can find the y that maximizes the information gain. The solution is $y = 0$, independent of the robot's x position. This implies that for a robot that is "leaping" past the two stationary robots along the direction of the x-axis, the optimal trajectory is for the robot to trace the x-axis itself, with position $y = 0$ throughout the path, and pass through the other two robots. This can be generalized for any position of robots 0 and 1 in the plane: the trajectory of robot 2 should move along the equidistant path between the two stationary robots to achieve maximum information gain.

We introduce a new three robot leap-frog path in Fig. 3 to build off this result. To our knowledge, this is the only path for which the moving robot, at every time step, will trace the equidistant path between the stationary robots. The implementation of this path involves the trailing robot of an equilateral triangle configuration to pass through the stationary robots, establishing a new position and a new equilateral triangle configuration on the other side. The robots then switch roles and repeat the sequence.

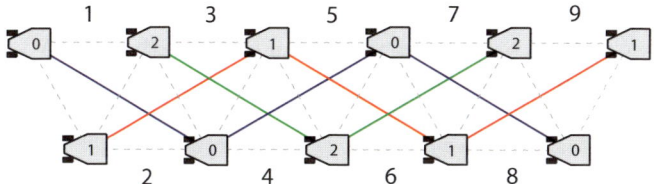

Fig. 3 This is the leap-frog path we use for simulations and experiments, which is based on an analysis of information gain for a team of three robots.

4.2 Empirical Results

To test the generated path displayed in Fig. 3, we have performed a series of Monte Carlo simulations for three mobile robots performing cooperative localization. The robots (when moving) are instructed to drive at a constant 0.5 m/s with $\alpha = 0.006$, $\beta = 0.02$, and $\gamma = 0.003$ for the motion noise model in Eq. 1. Relative bearing measurements are obtained at 10 Hz and are assumed to have additive Gaussian noise with a standard deviation of 1 degree.

In Fig. 4, we compare the results of the Monte Carlo simulations for three different paths. Path (a) is a smoothed version of the leap-frog path designed in Sec. 4, path (b) is a trajectory obtained when the robots move in an equilateral triangle formation, and path (c) is the same as (b) but omits the measurements.

Each path was simulated for 1000 different trials with randomly generated noise for measurements and motion. While the estimate of the state for each trial follows the intended path due to feedback control, the actual state for each trial is affected by the noise and drifts from the path. We measure the filter performance by observing the distribution of the robot state over all trials. A larger spread of data points implies worse tracking of the actual state. To quantify the performance, we compute the trace of the sample covariance

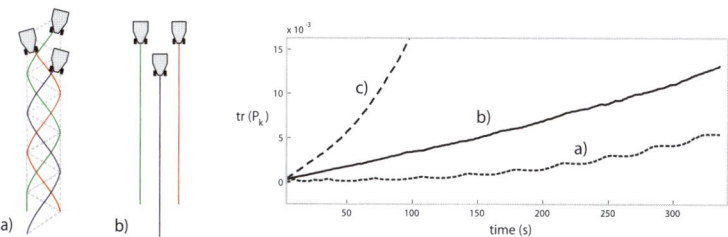

Fig. 4 Path (a) represents the leap-frog path presented in Sec. 4, path (b) represents the optimal formation for localization, and path (c) represents odometry only. The plot depicts the trace of the sample covariance matrix generated for a collection of 1000 Monte Carlo simulations.

matrix for the actual robot state computed over the 1000 simulated trials. Fig. 4 shows a graph of this metric for each of the path types. We note that our leap-frog path outperforms the optimal formation.

5 Experimental Evaluation

The motivating application for this work is GPS-denied autonomous coverage, for which accurate positioning is of critical importance. We apply a smoothed version of the path depicted in Fig. 3 to a team of real robots performing coverage.

5.1 Experimental Setup

We use three robots for outdoor localization experiments, each of which is based on the Learning Applied to Ground Vehicles (LAGR) platform [10]. Each mobile robot has three on board computers, wheel encoders for odometry, and a set of four stereo cameras mounted above the chassis. We choose to treat the 4 stereo pairs as 8 individual bearing sensors in order to reduce the computational load. The filter is implemented according to Sec. 3 and is centralized (meaning that only one of the robots is running the Kalman filter at any given time). The robots measure bearing to each other by detecting large red spheres in the camera images with a circle Hough transform [11]. See Fig. 1 for a photograph of the robots in their experimental configurations.

5.2 Coverage Experiments

The photos in Fig. 5 are from a video sequence recorded during one of our coverage experiments at Gesling stadium at Carnegie Mellon University. We are able to use this video sequence to post-process ground truth position data for each of the three robots. In order to do this, we compute a camera projection matrix based on known 3D points in the image (the markings on the football field). Then, after manually selecting a robot's location in the image plane, we can infer its 3D position via its projection onto the plane.

Fig. 5 An experiment on the football field at Gesling Stadium at Carnegie Mellon University. The three photos here are extracted from a video sequence used to record the ground truth position of the robots throughout the experiment.

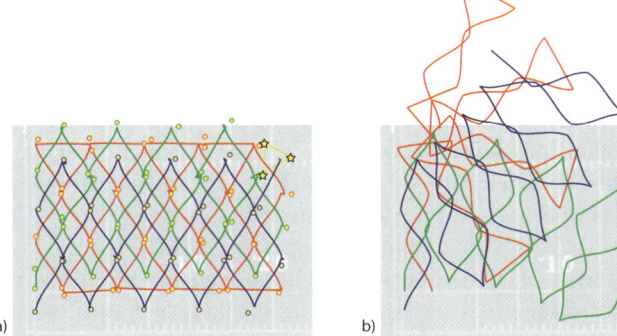

Fig. 6 (a) shows the estimated trajectory of the robots during a coverage experiment along with ground truth points (yellow circles). The final ground truth position is displayed with yellow stars. (b) shows the same experiment with a filter that ignores the bearing measurements (dead reckoning only).

The estimated path of the robots during the aforementioned coverage experiment is drawn in Fig. 6 (a). This estimated path follows an overall desired path that is composed of multiple smoothed versions of the leap-frog path displayed in Fig. 3 pasted together so as to sweep a region for coverage. The travel distance for each robot was approximately 140m during the experiment. The covered area was 20m x 30m. Ground truth points are shown in Fig. 6 for comparison and to help quantify the localization performance. Fig. 6 (b) shows the odometry-only estimate of the path. It is worth noting how erroneous this path is (most likely due to turning biases from unequally inflated tires) and how effective the filtering is in correcting the erroneous path to agree closely with the ground truth data.

The true final position of the three robots is also depicted in Fig. 6. The approximated error between the filtered estimate and the measured final ground truth pose is: 1.09 meters for robot 0 (the red trajectory in Fig. 6), 1.01 meters for robot 1 (the green trajectory in Fig. 6) and 1.15 meters for robot 2 (the blue trajectory in Fig. 6). We note that the accuracy of these ground truth measurements is subject to possible user error when manually selecting the image points that correspond to the robots in the video sequence.

The localization accuracy for this experiment is quite remarkable for this type of outdoor robot. The presence of wheel slippage coupled with a difficult terrain can cause severe drift in the odometry estimate over a path this long. Additionally, the measurements that we acquire with vision can be fairly noisy compared to more expensive laser range finders. But when an informative path, such as the one we present in Sec. 4, is used, the accuracy improves significantly, as shown in our experiment.

6 Conclusion

This work presents a leap-frog path designed to aid localization for a team of three robots. The path is designed such that the moving robot travels along a path that adds maximal information to the filter. The resulting path outperforms the optimal formation-based path. The experiment that we describe is, to our knowledge, one of the largest outdoor GPS-denied coverage results, successful in part because of precise localization.

Although we believe the absolute optimal path (in terms of localization accuracy) for a team of three robots would involve a leap-frog motion strategy, the path we introduce in this paper is most likely not optimal. Precisely defining the optimal path is still an open problem, which may require running an exhaustive simulation to optimize over all possible combinations of motion inputs: a task that would be computationally infeasible.

Also, this paper has focused on developing paths for a team of three robots. We believe that a three robot team is a good fit for applications that require accurate positioning, in part because three robots can provide proper triangulation. That said, it is always beneficial to add additional information to the Kalman filter, and a way to do this would be to add additional robots.

References

1. Roumeliotis, S., Bekey, G.: Collective Localization: A distributed Kalman filter approach to localization of groups of mobile robots. In: Proceedings of the International Conference on Robotics and Automation (April 2000)
2. Navarro-Serment, L.E., Paredis, C.J.J., Khosla, P.K.: A beacon system for the localization of distributed robotic teams. In: Proceedings of the International Conference on Field and Service Robotics (August 1999)
3. Kurazume, R., Nagata, S.: Cooperative positioning with multiple robots. In: Proceedings of the International Conference on Robotics and Automation, May 1994, pp. 1250–1257 (1994)
4. Roumeliotis, S.I., Bekey, G.A.: Distributed multirobot localization. IEEE Transactions on Robotics and Automation 18(5), 781–795 (2002)
5. Rekleitis, I.M., Dudek, G., Milios, E.: Multi-robot cooperative localization: A study of trade-offs between efficiency and accuracy. In: Proceedings of the IEEE/RSJ International Conference on Intelligent Robots and Systems (2002)
6. Roumeliotis, S.I., Rekleitis, I.M.: Propagation of uncertainty in cooperative multirobot localization: Analysis and experimental results. Autonomous Robots 17(1), 41–45 (2004)
7. Hidaka, Y.S., Mourikis, A.I., Roumeliotis, S.I.: Optimal formations for cooperative localization of mobile robots. In: Proceedings of the International Conference on Robotics and Automation, pp. 4137–4142 (2005)
8. Trawny, N., Barfoot, T.: Optimized motion strategies for cooperative localization of mobile robots. In: Proceedings of the International Conference on Robotics and Automation (April 2004)

9. Thrun, S., Liu, Y., Koller, D., Ng, A., Gharamani, Z., Durrant-Whyte, H.: Simultaneous localization and mapping with sparse extended information filters. Intl. Journal of Robotics Research 23(7-8), 693–716 (2004)

10. Jackel, L., Krotkov, E., Perschbacher, M., Pippine, J., Sullivan, C.: The darpa lagr program: Goals, challenges, methodology, and phase i results. Journal of Field Robotics 23, 945–973 (2006)

11. Davies, E.R.: A modified hough scheme for general circle location. Pattern Recognition Letters 7, 37–43 (1988)

A Location-Based Algorithm for Multi-Hopping State Estimates within a Distributed Robot Team

Brian J. Julian, Mac Schwager, Michael Angermann, and Daniela Rus

Abstract. Mutual knowledge of state information among robots is a crucial requirement for solving distributed control problems, such as coverage control of mobile sensing networks. This paper presents a strategy for exchanging state estimates within a robot team. We introduce a deterministic algorithm that broadcasts estimates of nearby robots more frequently than distant ones. We argue that this frequency should be exponentially proportional to an importance function that monotonically decreases with distance between robots. The resulting location-based algorithm increases propagation rates of state estimates in local neighborhoods when compared to simple flooding schemes.

1 Introduction

Robots in a team need to communicate state estimates to self-organize. Since many applications desire the team to spread over large-scale domains, resulting distances between robots can become larger than their capable peer-to-peer transmission ranges. These configurations require multi-hop networking to distribute state information over the entire system. To facilitate the

Brian J. Julian, Mac Schwager, and Daniela Rus
Computer Science and Artificial Intelligence Laboratory, Massachusetts Institute of Technology, 77 Massachusetts Avenue, Cambridge, MA 02139, USA
e-mail: bjulian@mit.edu, schwager@mit.edu, rus@csail.mit.edu

Brian J. Julian
Currently working at MIT Lincoln Laboratory, 244 Wood Street,
Lexington, MA 02420, USA

Michael Angermann
Institute of Communications and Navigation, German Aerospace Center (DLR),
P.O. Box 1116, D-82234 Wessling, Germany
e-mail: michael.angermann@dlr.de

A. Howard et al. (Eds.): Field and Service Robotics 7, STAR 62, pp. 319–329.
springerlink.com   © Springer-Verlag Berlin Heidelberg 2010

transportation of data packets in a multi-hop fashion, many mobile ad hoc networks implement sophisticated routing schemes. Due to the mobile nature of such networks, these schemes consume a significant amount of communication capacity for maintaining knowledge about network topology. While some routing strategies take spatial configurations into account, the robots are agnostic to the relevance of the actual data being transferred. There is no concept of data importance from the robots' point of view, often resulting in the suboptimal allocation of communication resources (e.g. time, bandwidth, power) to transfer packets.

The strategy in this paper allows robots to better manage communication resources for relaying state estimates. Since the collaboration of robots takes place in the physical world, spatial relationships between robot states can give insight into the importance of transferring each estimate. This location-based approach gives a quantitative answer to the question: how important is it for one robot to broadcast state information about another robot? We represent the importance of transmitting a state estimate as a function that is inversely proportional to the distance between robots.

From this importance function we develop a deterministic algorithm that ensures state estimates propagate throughout a robot network. The proposed location-based algorithm is efficient in terms of bandwidth and computational complexity; it does not require network topology information to be transmitted or computed. We used Monte Carlo simulations to show increased propagation rates of state estimates in local neighborhoods. Then with real control and wireless hardware, we simulated a nine robot team running a Voronoi coverage controller to show the algorithm's effectiveness in solving distributed control problems. Experimental results for the propagation of state estimates are also presented with five AscTec Hummingbird quad-rotor flying robots and four stationary robots.

A substantial body of work exists on location-based routing for mobile ad hoc networks. Haas proposed a zone-based routing protocol using a radius parameter to reduce the number of control messages [4]. Ni et al. developed a distance-based scheme to decide when a node should drop a rebroadcast [5], while Sun et al. adapted a similar scheme for setting defer times [7]. Cai et al. discussed how these ad hoc schemes influence flooding costs [1].

Our proposed algorithm is related to this body of work in that location is used to broadcast information through a mobile ad hoc network. However, instead of routing actual data packets to a predetermined receiver, we are deterministically transmitting state information to be used by the entire team of robots. This allows all transmissions to be treated as simple broadcasts, for which the sender uses the algorithm to select state estimates. This strategy is applicable for many distributed control problems, such as coverage control algorithms for mobile sensing networks.

Fig. 1 A simple example where robots i and j share a Voronoi boundary but cannot communicate their state estimates directly. This problem is easily resolved using a mobile ad-hoc network topology to route information through robot k.

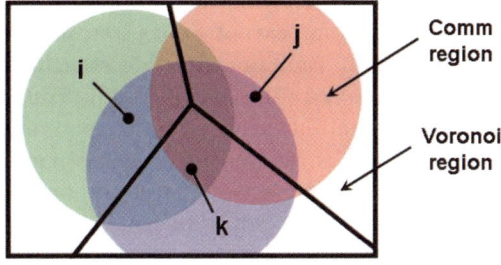

2   Importance of Broadcasting State Estimates

A common assumption for distributed control algorithms is that robots have access to state estimates of other nearby robots. This assumption is often translated into unrealistic requirements on communication range. The most common requirement is that estimates need to be directly shared between robots that are within a specified distance. Another common requirement is for information to be shared between robots of a defined spatial relationship (e.g. adjacent Voronoi regions [2] or overlapping fields of view [6]).

These communication requirements are too simplistic to be realized in practice. Actual network topologies depend on more than simple distance criteria, such as environment geometry, channel interference, or atmospheric conditions. Even if transmission ranges are ideal in the physical sense (e.g. the ideal disk model), spatial relationships for certain distributed controllers cannot guarantee peer-to-peer connectivity. Figure 1 shows a configuration where a direct communication link cannot be created between the Voronoi neighbors i and j. Moreover, robots that are spatially disconnected may decide not to route state estimates to one another. If they move to become spatially connected, the lack of shared data will prevent the robots from learning about their new neighbors. Thus, no new communication links will be established. We are motivated by these serious and unavoidable complications to develop an algorithm that ensures state estimates flow throughout a team of robots.

2.1   Broadcast Scheme

Consider n robots moving in a space[1], $\mathcal{P}$. Each robot, $i \in \{1, \ldots, n\}$, knows its current state, $p_i(t) \in \mathcal{P}$, by some means of measurement (e.g. GPS or visual localization). We propose that each robot maintains a list of state estimates, $[p_1(t_{i1}), \ldots, p_n(t_{in})]$, where t_{ij} denotes a time stamp at which robot

---

[1] Although it is easiest to think of the space being $\mathbb{R}^2$ or $\mathbb{R}^3$, the strategy we describe is equally useful with more detailed state estimates (e.g. velocity, acceleration, joint positions, state machine information, etc.)

i's estimate of robot j's state was valid. We have that $t_{ij} \leq t$ and $t_{ii} = t$. Each robot's state estimate is initialized to infinity to indicate that a valid estimate is lacking, except for its own state which is always current.

We desire to communicate state estimates throughout the robot network. For simplicity, we use Time Division Multiple Access (TDMA)[2] to divide the data stream into time slots of equal length, m. During a time slot, one assigned robot is allowed to broadcast over the shared frequency channel. Other robots broadcast one after the other in a predetermined order. One complete broadcast cycle is referred to as a frame.

To broadcast its own state estimate once per frame, the robot's time slot must be long enough to transmit the estimate and an associated time stamp. Such a time slot is considered to have length of $m = 1$. Clearly time slots of unit length are not sufficient to transmit information throughout the network; each robot would only be updated with the state estimate of its neighbors on the network. For multi-hop networking, the robots need longer time slots to broadcast the estimates of other robots.

One naive strategy is to assign a time slot length equal to the number of robots, $m = n$, so that each robot can broadcast its entire list of state estimates, thus creating a simple flooding scheme. Robots that are adjacent on the network use this information to update their own list, retaining only the most current state estimates. The process is repeated for each time slot, naturally propagating state estimates throughout the network without the need of a complicated routing protocol.

Although simple to implement, this strategy is not scalable for a large number of robots. Consider the rate a system can cycle through all time slots to complete one frame. This frame rate, r_f, gives insight into how quickly state estimates are being forwarded, and therefore how confident distributed controllers can be in using the estimates. For a network of fixed baud rate, r_b, the maximum frame rate[3] is given by $\max(r_f) = r_b/mnb$, where b is the data size of a state estimate and its associated time stamp. For $m = n$, increasing the number of robots in the system will decrease the frame rate *quadratically*. This inherent trade-off provides motivation to reduce the length of the time slot.

2.2 *Importance Function*

Many distributed controllers are dependent on spatial relationships between robots. When selecting which state estimate to broadcast, the selection process should also depend on these relationships. This makes sense because a

---

[2] In this paper we primarily discuss implementing the proposed strategy using TDMA; however, many other channel access methods are appropriate (e.g. FDMA or CDMA).

[3] We are ignoring overhead associated with TDMA (e.g. guard periods, checksums, etc.)

robot's state is more likely to be useful to controllers in proximity. However, it cannot be considered useless to controllers that are distant due to the mobile nature of the system. We propose that the importance of robot i broadcasting robot j's state estimate is inversely proportional to the distance between robot states.

Since the robots only have access to the state estimates they receive, a distance estimate is used to give the following importance function

$$f_{ij}(t) = d\left(p_i(t), p_j(t_{ij})\right)^{-\alpha} \tag{1}$$

where $d(\cdot, \cdot) \geq 0$ is a distance function and $\alpha \in (0, \infty)$ is a free parameter, both of which are selected for the given distributed controller. For example, a Voronoi coverage controller dependent on linear spatial separation may use a Euclidean distance function with $\alpha = 1$. This same distance function is appropriate for a sensor-based controller dependent on light intensity, although $\alpha = 2$ may be used since light intensity decays quadratically with distance from the source. Conversely, the distance function does not need to be Euclidean or even of continuous topology, such as for truss climbing robots with a finite configuration space. In any case, a robot should consider its own state estimate to be the most important to broadcast. This is reflected in the model since f_{ii} is infinite for any valid $d(\cdot, \cdot)$ and α.

3 Location-Based Algorithm for Broadcasting States

We use the importance function in Equation (1) to develop a deterministic algorithm. For a given time slot, this algorithm selects which state estimates a robot will broadcast. We first describe a probabilistic approach to help formulate the final algorithm.

3.1 *Probabilistic Approach*

Consider a robot that needs to select m state estimates to broadcast during its time slot. We provided motivation in Section 2.2 that some selections are more important than others. However, the robot should *not* systematically select the state estimates associated with the highest importance; doing so can prevent estimates from fully dispersing throughout the system. Instead, we propose that the probability of robot i selecting the state estimate of robot j is

$$P_{\mathcal{M}_i}^{ij}(t) = \frac{f_{ij}(t)}{\sum_{k \in \mathcal{M}_i} f_{ik}(t)}, \quad j \in \mathcal{M}_i \tag{2}$$

where $\mathcal{M}_i$ is the set of robot indices associated with selectable estimates.

Algorithm 1. Deterministic Method for Selecting State Estimates

---

n is the number of robots in the system and m is the time slot length.

Require: Robot i knows its state $p_i(t)$ and the state estimate of other robots $p_j(t_{ij})$.

Require: Robot i knows its running counter $[c_{i1}, \dots, c_{in}]$.

$\mathcal{M}_i \leftarrow \{1, \dots, n\}; \quad \mathcal{N}_i \leftarrow \varnothing;$

for 1 to m **do**

$\qquad P^{ij}_{\mathcal{M}_i}(t) \leftarrow \frac{f_{ij}(t)}{\sum_{k \in \mathcal{M}_i} f_{ik}(t)}, \quad \forall j \in \mathcal{M}_i; \quad c_{ij} \leftarrow c_{ij}[1 - P^{ij}_{\mathcal{M}_i}(t)], \quad \forall j \in \mathcal{M}_i;$

$\qquad k \leftarrow \arg\max_{k \in \mathcal{M}_i}(c_{ik}); \quad \mathcal{M}_i \leftarrow \mathcal{M}_i \backslash \{k\}; \quad \mathcal{N}_i \leftarrow \mathcal{N}_i \cup \{k\}; \quad c_{ik} \leftarrow 1;$

end for

return $\mathcal{N}_i$

Prior to the first selection for a given time slot, $\mathcal{M}_i$ is the set of all robot indices. From the full set the robot always selects its own state since it has infinite importance. The robot then removes its index from $\mathcal{M}_i$ to prevent wasting bandwidth. Since Equation (2) is a valid probability mass function, the robot can simply choose the next state estimate at random from the corresponding probability distribution, then remove the corresponding index from $\mathcal{M}_i$. This means estimates of closer robots are more likely to be chosen than ones that are farther away. By repeating this process, the entire time slot of length m can be filled in a straightforward, probabilistic manner.

3.2 Deterministically Selecting Estimates

It is not ideal in practice to probabilistically select which state estimates to broadcast. Consecutive selections of a particular robot index can be separated by an undesirably long period of time, especially concerning distant robots. By developing a location-based deterministic algorithm, we can increase the average rate at which all state estimates of a given time stamp will propagate throughout a team. In the deterministic case, propagation time is bounded above by the longest path taken among the estimates. No such bound exists in the probabilistic case, resulting in a positively skewed distribution of propagation times and a larger mean.

We propose that each robot maintains a list of counters, $[c_{i1}, \dots, c_{in}]$, which are initially set to a value of one. Using the probability mass function in Equation (2), each counter represents the probability that the corresponding index has *not* been selected. Consider a robot's first selection, which will always be its own index. The probability, $P^{ii}_{\mathcal{M}_i}(t)$, of selecting index i is equal to one, while all other probabilities, $P^{ij}_{\mathcal{M}_i}(t)$ subject to $j \neq i$, are equal to zero. This implies that the counter c_{ii} is multiplied by $[1 - P^{ii}_{\mathcal{M}_i}(t)] = 0$, or a zero probability of not being selected, while all other counters, c_{ij}, are multiplied $[1 - P^{ij}_{\mathcal{M}_i}(t)] = 1$, or a probability of one. By selecting the index with the lowest counter value, we are deterministically guiding our method to behave according to the probability distribution described by Equation (2).

Fig. 2 This figure shows the average propagation time for the location-based algorithm running on a 10×10 stationary robot grid. Averages were taken over 1000 Monte Carlo simulations. For small subgraphs (i.e. 2×2), update rates of state estimates increased with decreasing time slot lengths. For larger subgraphs, the optimal length was around $m = 7$.

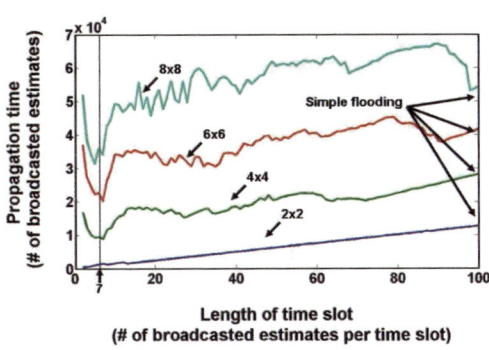

The selected index (in this case i) is removed from the set $\mathcal{M}_i$, and its corresponding counter (c_{ii}) is reset to a value of one. This process is iteratively applied to completely fill a time slot with m state estimates, with counters maintaining their values between frames. The complete deterministic strategy of $\mathcal{O}(mn)$ time is given in Algorithm 1.

4 Simulations and Experiments

We provide insight into the performance of the location-based algorithm in three ways: we conducted Monte Carlo simulations for 100 stationary robots, we used real control and wireless hardware to simulate nine robots running a distributed coverage algorithm, and we implemented this hardware on five flying and four stationary robots. We first describe the Monte Carlo simulations used to measure information propagation throughout the robot team. Propagation time is the main performance metric for the algorithm. This metric depends on the length of the time slot, or in other words, the number of state estimates communicated during one robot broadcast. We compare these results to the case when the time slot length equals the number of robots, since allowing robots to broadcast every state estimate is the simplest multi-hop scheme. This scheme is referred to as simple flooding.

In a MATLAB environment, we simulated a team of 100 stationary robots arranged in a 10×10 square grid. Each robot, initialized knowing only its own state estimate, was able to receive broadcasts from its adjacent neighbors along the vertical and horizontal directions. Each robot ran Algorithm 1 in distributed fashion. Over 1000 Monte Carlo simulations were executed for time slots of varying lengths, with each run having a random order for the time slot assignments. For the 2×2, 4×4, 6×6, and 8×8 subgraphs centered on the 10×10 graph, we measured the time needed for all subgraph members to exchange state estimates.

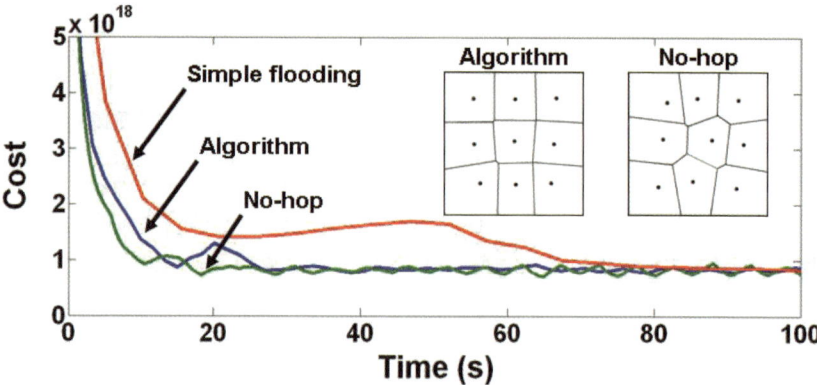

Fig. 3 Coverage costs are shown for a nine robot system simulated on real hardware running a Voronoi coverage controller. The system has a frame rate of 1.7 Hz when using a no-hop scheme ($m = 1$). The system initially performs well, but its inability to multi-hop state estimates resulted in a suboptimal final configuration. A simple flooding scheme ($m = 9$) improved steady state performance, however, the slow frame rate of 0.2 Hz caused the system to initially oscillate in a high cost configuration. The location-based algorithm with a time slot of length $m = 3$ performed the best overall by combining fast update rates with multi-hop capabilities. The final Voronoi configurations for the algorithm and no-hop simulations are also shown.

Figure 2 plots average propagation time for the Monte Carlo simulations. For the smallest subgraph (i.e. 2×2), state estimates propagated faster with smaller time slot lengths. This relationship makes sense since we are maximizing the frame rate, thus increasing update rates for the local state estimates of highest importance. As the subgraph size increases, very small time slot lengths become less effective at propagating estimates, especially between robots at opposite sides of the subgraph. By using a slightly larger time slot length, a significant improvement in performance over simple flooding is obtained; propagation times for all subgraphs decreased by more than 47% using a time slot length of $m = 7$. Analyzing such Monte Carlo plots provides a heuristic technique for selecting an acceptable time slot length for a given control problem.

We then tested the algorithm in a simulated robot scenario using real control and wireless hardware. We implemented a Voronoi coverage controller [2] on nine custom ARM microcontroller modules, each using a 900 MHz xBee module to wirelessly broadcast state estimates during its assigned time slot. Each control module simulated the dynamics of a flying robot, creating a virtual distributed robot team. In addition, a communication range was implemented such that packets from "out-of-range" robots were automatically dropped. We investigate the performance of the location-based algorithm in a simple scenario where nine virtual robots were tasked to cover a square area.

Fig. 4 An example mobile ad hoc network graph from the quad-rotor flying robot experiment is plotted in Google Earth. For this nine robot system, the location-based algorithm routes state estimates through the entire team. The bounded environment from the downward facing camera coverage problem is also shown.

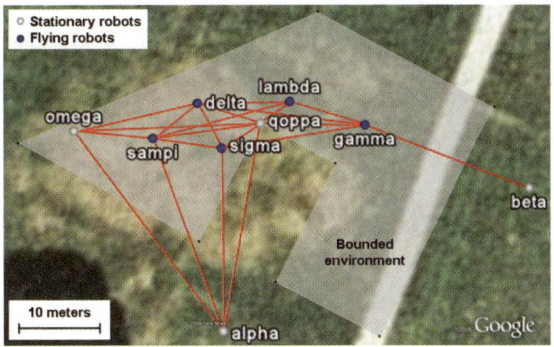

For this scenario the optimal configuration is for the robots to be arranged in a 3×3 square grid.

For the location-based algorithm, a time slot length of $m = 3$ was selected using the Monte Carlo technique previously discussed. We also selected the Euclidean distance function with $\alpha = 1$ given that the Voronoi coverage controller is linearly dependent on such distance. Each state estimate for the virtual flying robot is constructed of six 32-bit integers (robot identification, time stamp, latitude, longitude, altitude, and yaw), resulting in a data size of 192 bits. Given that the wireless hardware could reliably operate at 3000 baud, the resulting frame rate was about 0.6 Hz. For comparison, the simple flooding ($m = 9$) and no-hop ($m = 1$) schemes ran at about 0.2 Hz and 1.7 Hz, respectively. Figure 3 shows the resulting coverage cost profiles from these simulations. The location-based algorithm had better initial performance than the simple flooding scheme and better steady state performance than the no-hop scheme. The final Voronoi configurations for the algorithm and no-hop simulations are also shown.

Finally, we implemented the location-based algorithm on five AscTec Hummingbird quad-rotor flying robots [3] and four stationary robots, thus creating a nine robot team. Each flying robot was equipped with an AscTec AutoPilot board capable of capturing GPS, altitude, and yaw positions. The previously described control and wireless modules were installed on these AutoPilot boards. In addition, four separate modules were deployed at fixed locations to represent the stationary robots.

This experimental setup was designed to run a downward facing camera coverage controller for hovering robots [6]. Since this controller has a spatial dependence similar to the Voronoi coverage controller, the same time slot length, distance function, and α were used. Figure 4 shows the network topology of a random deployment configuration prior to starting the coverage controller. Here we limited the communication range to 30 meters; in previous experiments we were able to produce links in excess of 100 meters. Figure 5 plots the time stamp of the most current state estimates as received by the

Fig. 5 This plot shows the time stamp of the most current state estimates received by the stationary robot beta. Estimates of closer, more important robots are updated more frequently and tend to be more current, which validates the location-based algorithm.

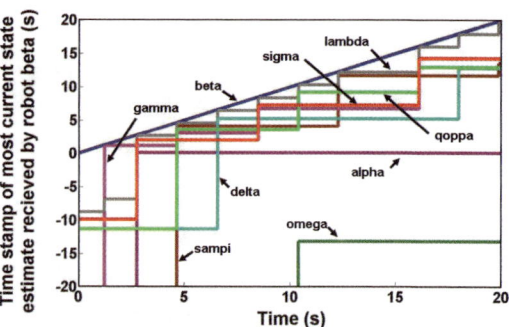

stationary robot beta, which can be considered the "worst case" receiver since it is the most remote robot in the team. As previously discussed, beta's own state estimate is always considered to be current. Estimates of other robots are updated as they are received by team broadcasts, whether directly from the originating robot or indirectly in a multi-hop fashion. Since closer robots are considered more important in the algorithm formulation, this results in their state estimates being more current with more frequent updates.

5 Conclusion

In this paper we presented a location-based strategy for exchanging state estimates in a distributed robot team. We developed a deterministic algorithm that, based on an importance function, broadcasts estimates of nearby robots more frequently than distant ones. Simulations using real control and wireless hardware show that the algorithm outperforms simple flooding schemes for large robot networks.

Our experiments consisting of five Asctec Hummingbird quad-rotor flying robots among four stationary robots showed the successful exchange of state estimates in a multi-hop fashion. Using the location-based algorithm, we successfully ran the coverage controller from [6] on five flying robots with downward facing cameras. Coverage results from this experiment will be presented in future publications.

We desire to further develop this work to exploit the spatial reuse of time slots for robots separated by multiple hops. This direction allows for virtually infinite team sizes and spatial coverage.

Acknowledgements. This work was done in the Distributed Robotics Laboratory at MIT and is supported in part by the MURI SWARMS project grant number W911NF-05-1-0219, NSF grant numbers IIS-0513755, IIS-0426838, CNS-0520305, CNS-0707601, EFRI-0735953, the MAST project, MIT Lincoln Laboratory, and the Boeing Corporation. This work is sponsored by the Department of the Air Force under Air Force contract number FA8721-05-C-0002. The opinions, interpretations, recommendations, and conclusions are those of the authors and are not necessarily endorsed by the United States Government.

References

1. Cai, Y., Hua, K., Phillips, A.: Leveraging 1-hop neighborhood knowledge for efficient flooding in wireless ad hoc networks. In: 24th IEEE International Performance, Computing, and Communications Conference, IPCCC 2005, pp. 347–354 (2005)
2. Cortes, J., Martinez, S., Karatas, T., Bullo, F.: Coverage control for mobile sensing networks. IEEE Transactions on Robotics and Automation 20(2), 243–255 (2004)
3. Gurdan, D., Stumpf, J., Achtelik, M., Doth, K.M., Hirzinger, G., Rus, D.: Energy-efficient autonomous four-rotor flying robot controlled at 1 khz. In: 2007 IEEE International Conference on Robotics and Automation, pp. 361–366 (2007)
4. Haas, Z.: A new routing protocol for the reconfigurable wireless networks. In: 1997 IEEE 6th International Conference on Universal Personal Communications Record, Conference Record, vol. 2, pp. 562–566 (1997)
5. Ni, S.Y., Tseng, Y.C., Chen, Y.S., Sheu, J.P.: The broadcast storm problem in a mobile ad hoc network. In: MobiCom 1999: Proceedings of the 5th annual ACM/IEEE international conference on Mobile computing and networking, pp. 151–162. ACM, New York (1999)
6. Schwager, M., Julian, B., Rus, D.: Optimal coverage for multiple hovering robots with downward facing cameras. In: Proc. of International Conference on Robotics and Automation (ICRA 2009), Kobe, Japan (2009)
7. Sun, M.T., Lai, T.H.: Location aided broadcast in wireless ad hoc network systems. In: 2002 IEEE Wireless Communications and Networking Conference, WCNC 2002, vol. 2, pp. 597–602 (2002)

Cooperative AUV Navigation Using a Single Surface Craft

Maurice F. Fallon, Georgios Papadopoulos, and John J. Leonard

Abstract. Maintaining accurate localization of an autonomous underwater vehicle (AUV) is difficult because electronic signals such as GPS are highly attenuated by water making established land-based localization systems, such as GPS, useless underwater. Instead we propose an alternative approach which integrates position information of other vehicles to reduce the error and uncertainty of the on-board position estimates of the AUV. This approach uses the WHOI Acoustic Modem to exchange vehicle localization estimates — albeit at low transmission rates — while simultaneously estimating inter-vehicle range. The performance capabilities of the system were tested using Oceanserver's Iver2 and the MIT Scout kayaks.

1 Introduction

Localization or navigation of vehicles using only onboard local sensors, such as a Doppler Velocity Logger (DVL) or Inertial Measurement Unit (IMU), are certain to experience accumulated positioning error. One can, of course, utilize more precise sensors to reduce the rate of accumulated error — DVL units with error accumulation rates as low as 0.2% are commercially available. However this approach may not be satisfactory due to practical, power or financial limitations.

Regardless of the platform used, the accumulation of error and uncertainty is simply slowed, rather than bounded. The result of this is that an AUV surveying the ocean floor or a land robot building a street map must be halted on occasion so as to reset the position uncertainty — either by surfacing for a GPS fix or by repositioning at a known location. This procedure wastes both energy and time, requires a human interface and may be unacceptable in many operating environments.

The standard approach for bounding error underwater is Long Baseline (LBL). Two or more beacons are deployed at known locations — either as buoys on the

Maurice F. Fallon, Georgios Papadopoulos, and John J. Leonard
Massachusetts Institute of Technology, 77 Massachusetts Avenue, Cambridge, MA 02139
e-mail: {mfallon,gpapado,jleonard}@mit.edu

A. Howard et al. (Eds.): Field and Service Robotics 7, STAR 62, pp. 331–340.
springerlink.com   © Springer-Verlag Berlin Heidelberg 2010

water surface or moored on the seabed. The AUV transmits an acoustic query to the beacons which reply in a manner which allows the AUV to estimate the beacon/AUV range and to then improve its own position estimate. Recent improvements to this system have removed the need for round-trip timing (Synced LBL) and also allowed for estimation of both range and angle using an array of receiving sensors (USBL).

While these technologies are now all commercially available, the mobility of the AUV is restricted as typical coverage is limited to an area within a few kilometers of the beacon. To relax this restriction an alternative approach considers a system in which a surface vehicle (with access to GPS) or a submerged vehicle (with accurate dead reckoning instrumentation) communicates with a fleet of much less accurately localized vehicles so as to improve the positioning of the latter. One example of this approach is the Moving Long Base Line (MLBL) navigation proposed by Vaganay *et al.* [9], in which typically two surface vehicles serve as mobile beacons for one or more AUVs. Other related recent research has been performed by Bahr *et al.* [1], Eustice *et al.* [4] and Maczka *et al.* [5]. It should also be recognized that multi-AUV navigation falls within the wider problem of multi-robot cooperative localization, see [6] for a more general introduction to the field.

In this paper, we describe experiments that extend the MLBL approach to situations in which a single surface vehicle is used to estimate the position of a submerged AUV using range-only measurements. In Section 2 the basic framework of this technique is discussed. Our algorithm is outlined in Section 3 followed by a number of modifications which improve performance. Section 4 presents the results of a combination of simulation and realistic experiments to illustrate the concept. Finally conclusions drawn from the experiments and the directions of future work are presented in Section 5.

2 Cooperative Localization under Water

This paper retains the framework for underwater localization previously introduced in [1] and also used in [4]. We shall assume there to be one surface vehicle providing the submerged fleet of vehicles with position information while perhaps operating as a communications moderator — in the dual role of a Communications and Navigation Aid (CNA). Each of the autonomous underwater vehicles maintains a dead reckoning filter, drawing upon measurements of velocity, heading and depth. Finally, communication through the water channel is possible using the WHOI Acoustic Modem — at transmission rates of the order of 32 bytes per 10 seconds — in a process which also yields a time-of-flight measurement which can be used to estimate the inter-vehicle range.

There are a number of methods which could be used to integrate the received position information. Our earlier work, [1], proposed an algorithm which utilized the on-board dead reckoning estimate of the AUV and a pair of CNA range estimates to form a complete estimate of the AUV state vector.

The seabed, the water surface and deep sea thermoclines within the water body have the ability to cause significant multi-path signal interference and the receipt of a substantial amount of infeasible outlier measurements. A typical dataset was illustrated for a regular Long Baseline systems in [8]. For these reasons it would be reasonable to assume that the received measurement set obtained from the WHOI modem would contain substantial multi-modality, thereby motivating this approach.

However the advanced processing within the WHOI modem decoder has the ability to suppress the bulk of these effects, such that the received measurements decoded by the modem contain only a moderate amount of noise. For this reason the proposed approach instead uses an implementation of the Extended Kalman Filter.

A particle filtering approach [3] could also have been considered as this would have more accurately incorporated the non-linearity of the correction step, however because we will maintain full control of the CNA's motion this issue can broadly be avoided.

Previous *proof-of-concept* experiments illustrated that the range variance is broadly independent of range itself, however detailed examination of this was not carried out [2]. The modem transducer was then directly clamped to the underside of the kayak. Our more recent experiments have instead hung the transducer 2-3 meters below the kayak hull. We expect less noise interference from the kayak motor and less reflections from the water surface in this configuration.

Figure 1 illustrates WHOI modem range data plotted versus GPS-derived 'ground truth', as measured in the Charles River adjacent to MIT recently. Because the ground truth distance between the two vehicles was determined using imprecise GPS measurements, it is difficult to precisely estimate the distribution of the range measurements. Other issues, such as the position of the GPS sensor relative to the modem on the kayak must also be recognized. In the absence of precise ground truth, we estimate the range variance to be between 4–8m.

3   Single Surface Craft Cooperative Navigation

The configuration we will consider in this work will be of a single CNA supporting N underwater vehicles[1]. Each AUV will maintain an estimate of its own position and uncertainty. This estimate will be propagated using the usual Kalman prediction step so as to integrate heading, forward and starboard velocity measurements.

As mentioned above, this estimate will be corrected using range and position information relative to an CNA using the WHOI acoustic modem. At present the 32 byte packet transmitted from the CNA shall contain latitude, longitude, depth and heading as well as a UNIX time-stamp. Transmission of a packet consists of two stages: first a *mini packet* is transmitted to initiate the communication sequence. The inter-vehicle range can be estimated using this mini packet. Following this, the information packet is transmitted in a process which lasts approximately 5-6 seconds. In all, it is prudent to reserve 10 seconds per transmission. Simularly the AUV will transmit a message containing its own position estimate as well the associated

---

[1] Subsequent research will aim to relax the necessity of a dedicated surface vehicle

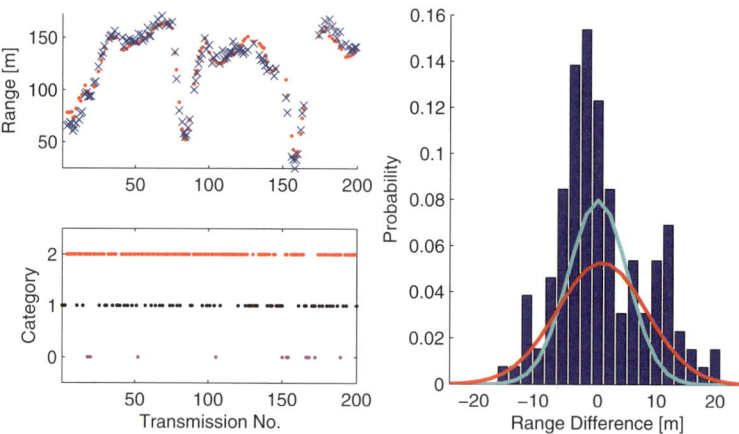

Fig. 1 Analysis of range estimates derived from the WHOI Modem. Upper Left: Comparison of modem range estimate (red dots) and range derived from GPS 'ground truth' (blue crosses) for each *fully successful* 10 second transmission period. Lower Left: Illustration of the frequency of successful transmissions. Category 0 represents an entirely failed transmission; Category 1: successful range transmission; Category 2: successful range and packet transmission. Category 2 corresponds to the modem ranges in upper left plot. Right: Histogram of range error (using estimated range versus GPS 'ground truth' range), also illustrated is a normal distribution fitted to the data (red, $\bar{r} = 0.66\text{m}, \sigma_r = 7.5\text{m}$) and the normal distribution used in the experiments in Section 4 with (cyan, $\bar{r} = 0\text{m}, \sigma_r = 5\text{m}$). This range data corresponds to Experiment 1.

covariance matrix which can be used to help the CNA plan its own supporting motion — also requiring 10 seconds per transmission.

It is envisaged that the MLBL will be integrated within a multi-AUV setup in which use of the communication channel is shared between many communicating processes. As a result the transmission rate of a position/range pair is likely to be substantially below one measurement per 10 seconds. Furthermore only a portion of transmitted messages will actually be received. For these reasons it is prudent to optimize the location from which the ASC transmits so as to maximize the benefit achieved from the correction step. Although a basic zig-zag motion plan was adopted in this work, future work will consider more elaborate motion planning for the CNA.

3.1 Utilizing Partial Messages

As illustrated in Figure 1, a significant proportion of the (range) mini packets are received without the information packet — meaning that the usual correction step cannot be made[2]

---

[2] For a typical mission in the open ocean inter-vehicle ranges of the order of 1km are expected, making this an even more significant issue.

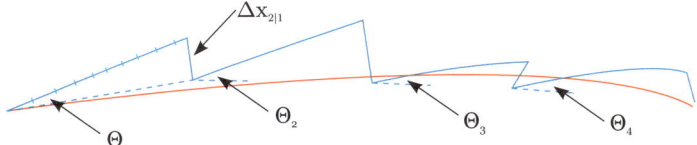

Fig. 2 Compass Bias Correction Example: MLBL position estimate (blue) is corrected towards the ground truth (red) in a consistent direction. The angular correction of the 4 correction steps, $\theta_{1:4}$, can be used to form an estimate of the bias angle, which is then removed. Note that multiple iterations of the prediction step take place between each correction step.

By linearly predicting the CNA position using previous position estimates, an estimate of the CNA at this time can be formed. This estimate can then be used with the previously orphaned range measurement to allow another correction step to occur. While post processing of the data from the experiments presented in Section 4 in this manner reduced the average error by approximately one meter, in future we propose to introduce redundancy into the transmitted messages so as to avoid this scenario. See Section 5 for more discussion.

3.2 *Online Compass Bias Correction*

A Bayesian filter - such as a Kalman filter or particle filter - assumes that measurements are formed using unbiased estimators. Heading is however particularly difficult measurement to estimate properly. Compass accuracy can be effected by the characteristics of the local region, the magnetism of the vehicle itself and magnetic declination. It is particularly severe for imprecise sensors used aboard the CNA platform. As a result, the compass used in the experiments presented in Section 4 is a dominant source of navigation error. Typically compass bias is corrected using a calibration process which can be both complex and time consuming. In this scenario, the EKF corrections garnered using the CNA range and position can be used to estimate the compass bias and to remove its effect.

Between successive corrections of the EKF, the filter will be predicted according to the dynamical model. The frequency of the prediction step will be much higher than the correction step. The distance between the posterior estimate of a correction step at time k_1 and the predicted position at time k_2 is the estimated relative distance traveled in that time

$$\triangle \mathbf{x}_{\bar{k}_2|k_1} = \bar{\mathbf{x}}_{k_2} - \mathbf{x}_{k_1}. \tag{1}$$

where $\mathbf{x}_{k_1} = [x_{k_1}, y_{k_1}]$ represents the state vector at time $k-1$. The CNA position and range measurement are then integrated to correct the posterior position estimate

$$\triangle \mathbf{x}_{k_2|k_1} = \mathbf{x}_{k_2} - \mathbf{x}_{k_1} \tag{2}$$

If the sensors contributing to the measurement, $\mathbf{z}_{k_2}$, are unbiased the expected value of the update will be zero. However if there exists a compass bias, the EKF will act to correct the filter in the direction opposite to the bias

$$\theta_{k_2|k_1} = \arccos\left(\frac{\triangle \mathbf{x}_{\bar{k_2}|k_1} \cdot \triangle \mathbf{x}_{k_2|k-1}}{|\triangle \mathbf{x}_{\bar{k_2}|k_1}||\triangle \mathbf{x}_{k_2|k-1}|}\right) \tag{3}$$

Figure 2 illustrates the issue for a sequence of MLBL corrections for a biased compass. It can be seen that the angle of the correction is consistently in the hypothesized bias direction. However as the CNA consistently maneuvers relative to the AUV, a closed form expression for the bias angle cannot be formed.

Instead we will propose to successively estimate the bias until this effect is removed. Consider the net angular correction set of N successive corrections, $(\theta_{k-N+1}, \dots \theta_k)$. We assume that the median of this set, given by $\tilde{\theta}_k$, will be in the direction of, but less than, the bias angle, i.e. $[0 \leqslant \tilde{\theta}_k < \theta_{\text{bias}}]$.

This value is assumed to be an initial estimate of the bias and used to correct the heading estimate subsequently. After the next N corrections, any remaining bias is again estimated and added to the running bias estimate. Eventually the bias will be assumed to be known and can be removed.

4 Experiments

A number of experiments were carried out in the Charles River, adjacent to MIT, to demonstrate the concept of Moving Long Baseline using the Surface Crafts for Oceanographic and Undersea Testing (SCOUT) kayaks designed in MIT and the low-cost Iver2 from Oceanserver (see Figure 4). Each of the kayaks was equipped with a WHOI modem, a compass and a GPS sensor while the Iver's basic sensor suite consisted of only a compass and a WHOI modem. The Iver2's only velocity estimate was a **constant** value of 1.028 m/s (2 knots) specified by the mission plan.

Each vehicle's onboard computer ran an implementation of the MOOS software platform [7]. Maintaining an accurately synchronized clock is essential for the estimation of inter vehicle ranges; to do so the Iver2 utilized a precisely synchronized

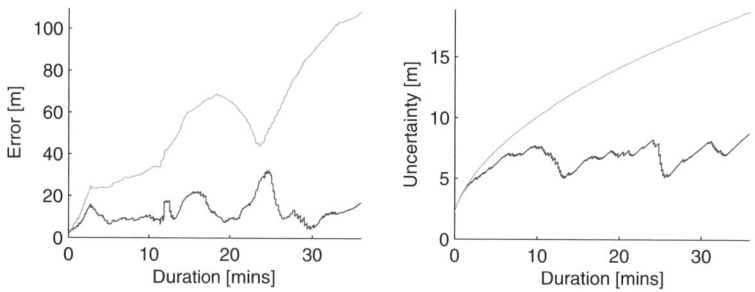

Fig. 3 Error (left) and 95% confidence (right) for the MLBL algorithm (blue) and the dead reckoning alone (green) for Experiment 1 where we have defined %90 confidence in terms of the largest eigenvalue of the covariance matrix.

timing board developed by Eustice et al. [4] while the SCOUT kayaks used the Plus-Per-Second (PPS) contained within its received GPS data messages.

Experiment 1: A single SCOUT kayak designated as the 'AUV' completed a survey-type mission while another kayak maintained a zig-zag pattern behind the 'AUV' — taking on the CNA role. The onboard GPS sensor was used to determine the ground truth position as well as to simulate forward and starboard velocities. Measurements drawn from the CNA transmissions were used by the 'AUV' to reduce its uncertainty. The designated 'AUV' carried out 1.5 circuits of a rectangle, covering approximately 1800 metres in total over a period of 37 minutes.

Note the large increase in the error of the position measurement between 22–26 minutes. This was caused by a combination of poor CNA position estimation (caused by visibility of just 4 GPS satellites) and the CNA moving close, yet parallel, to the AUV. It is envisaged that this could have been avoided with the use of a more accurate GPS unit or by forbidding the CNA from taking such a trajectory.

Fig. 4 Vehicles Used: OceanServer Iver2 (left) and the MIT Scout kayak (right)

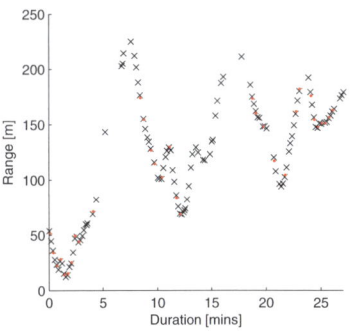 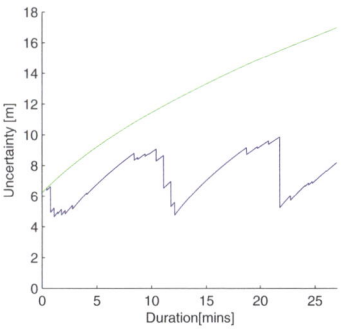

Fig. 5 Results for Experiment 2. Left: Modem range estimates with successful packet transmission (red dots) and modem range estimates but failed packet transmission (black crosses). Right: 95% confidence for the MLBL algoritm (blue) and the dead reckoning along (green). Note the two long portions of the run in which ranges were determined but no packet was successfully transmitted and the resultant growth in position uncertainty.

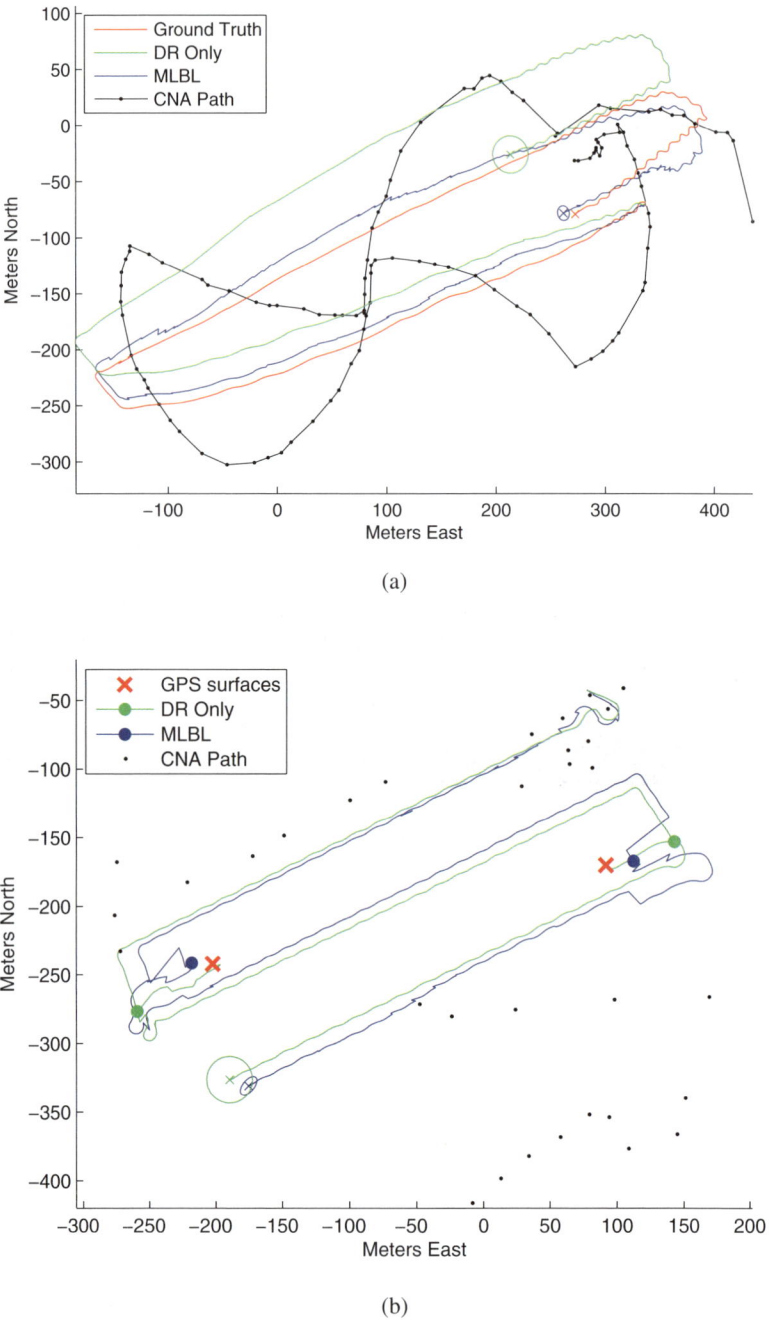

(a)

(b)

Fig. 6 Paths taken by the AUV and CNA during Experiment 1 (upper) and 2 (lower), see Section 4 for more details. CNA measurements were transmitted from the black dots. Note that the final 500m of Experiment 1 has been omitted as it overlaps with what is shown.

The following are a number of metrics for this test: mean error 12.5m, mean 'AUV' velocity 0.82m/s, mean CNA velocity 1.08m/s. There were 205 transmissions of which 130 were fully successful, 63 resulted in a failed packet transmission but a successful range estimate while 12 resulted in complete transmission failure. The algorithm can be seen to bound the error of the position estimate to approximately 20m.

Experiment 2: In a second fully realistic experiment, the Iver2 carried out a predefined 'lawnmower' pattern running at a depth of 2.4m while again the SCOUT kayak supported by transmitting its GPS position to the AUV via the WHOI modem. In addition the Iver2 transmitted its own MLBL position estimate, which was received by the CNA and used to plan locations from which to transmit.

Figure 6(b) illustrates the path taken by the vehicles. The test lasted 28 minutes and in total the Iver2 travelled 2 km. The AUV surfaced twice as a safety precaution. After 9 minutes the AUV first surfaced and received a GPS fix at (-201.6 -242.0) as shown as a red cross, at that time the front seat filter estimated a position of (-258.7,-276.5) while the MLBL filter estimate (-208.9,-238.1) giving an error of 66.7m and 8.3m error respectively. When the Iver surfaced for the second time (after 19 minutes), the corresponding errors were 53.7m and 14.1m. As well as estimating the AUV position with error, both of these MLBL filter estimates were within a 95% confidence interval upon surface. Note that after each surface the AUV transited from the GPS location back to its planned location on the mission path before diving and continuing the mission.

It should be mentioned that between 4–8 and 12–18 minutes no packets were successfully received by the AUV and as a result no MLBL corrections were possible (See Figure 5). This can be attributed to a number of factors

- The CNA was positioned behind the AUV and as a result churned water from the AUV propeller is likely to have reduced communication capabilities.
- With each failed transmission the AUV/CNA range grew until about 225m which is considered long for this experimental river environment [3].
- The presence of a tourist cruise ship nearby.

In future tests, precautions will be taken to avoid these issues.

5   Future Work and Conclusions

The concept of a single surface vehicle supporting the localization of an AUV has been outlined. Full experimental results with a single CNA supporting an Iver2 were presented. The resultant position estimate was shown to be substantially more accurate than the vehicle's own onboard navigation filter. Future work will focus on extending this framework for testing with three Iver2 vehicles and eventually towards the scenario in which a set of heterogeneous vehicles are continuously submerged with only a single vehicle occasionally surfacing to access the GPS.

---

[3] Note that the maximum range of the WHOI modem in the open ocean is estimated to be of the order of 4–5 times greater than the river environment.

Secondly, the performance of the algorithm is directly determined by the quality and frequency of received measurements. We will consider the optimization of the transmitted messages (and the re-transmission of failed data) so as to reduce the proportion of useless or partial messages received by the AUV. In this work the path taken by the CNA was an arbitary zig-zag behind the AUV. Motion planning of the CNA's path — so as to transmit messages from the most adventagous location — will also be carried out in future.

References

1. Bahr, A., Leonard, J.J.: Cooperative localization for autonomous underwater vehicles. In: Proceedings of the 10th International Symposium on Experimental Robotics (ISER), Rio de Janeiro, Brasil (July 2006)
2. Curcio, J., Leonard, J.J., Vaganay, J., Patrikalakis, A., Bahr, A., Battle, D., Schmidt, H., Grund, M.: Experiments in moving baseline navigation using autonomous surface craft. In: Proceedings of MTS/IEEE Oceans, September 2005, vol. 1, pp. 730–735 (2005)
3. Doucet, A., de Freitas, N., Gordon, N. (eds.): Sequential Monte Carlo methods in practice. Springer, Heidelberg (2000)
4. Eustice, R.M., Whitcomb, L.L., Singh, H., Grund, M.: Experimental results in synchronous-clock one-way-travel-time acoustic navigation for autonomous underwater vehicles. In: IEEE International Conference on Robotics and Automation (ICRA), Rome, Italy (April 2007)
5. Maczka, D.K., Gadre, A.S., Stilwell, D.J.: Implementation of a cooperative navigation algorithm on a platoon of autonomous underwater vehicles. In: Oceans 2007, pp. 1–6 (2007)
6. Mourikis, A.I., Roumeliotis, S.I.: Performance analysis of multirobot cooperative localization. IEEE Transactions on Robotics 22(4), 666–681 (2006)
7. Newman, P.M.: MOOS - a mission oriented operating suite. Technical Report Tech. Rep. OE2003-07 (2003)
8. Olson, E., Leonard, J., Teller, S.: Robust range-only beacon localization. IEEE Journal of Oceanic Engineering 31(4), 949–958 (2006)
9. Vaganay, J., Leonard, J.J., Curcio, J.A., Willcox, J.S.: Experimental validation of the moving long base line navigation concept, June 2004, pp. 59–65 (2004)

Multi-Robot Fire Searching in Unknown Environment*

Ali Marjovi, João Gonçalo Nunes, Lino Marques, and Aníbal de Almeida

Abstract. Exploration of an unknown environment is a fundamental concern in mobile robotics. This paper presents an approach for cooperative multi-robot exploration, fire searching and mapping in an unknown environment. The proposed approach aims to minimize the overall exploration time, making it possible to locate fire sources in an efficient way. In order to achieve this goal, the robots cooperate in order to individually and simultaneously, explore different areas of the environment while they identify fire sources. The proposed approach employs a decentralized frontier based exploration method which evaluates the cost/gain ratio to navigate to target way-points. The target way-points are obtained by an A* search variant algorithm. The potential field method is used to control the robots' motion while avoiding obstacles. When a robot detects a fire, it estimates the flame's position by triangulation. The communication between the robots is done in a decentralized control manner where they share the necessary data to generate a map of the environment and to perform cooperative actions in a behavioral decision making way. This paper presents simulated and experimental results of the proposed exploration and fire search method and concludes with a discussion of the obtained results and future improvements.

1 Introduction

Search operations inside buildings, caves, tunnels and mines are sometimes extremely dangerous activities. The use of autonomous robots to perform such tasks in complex environments will reduce the risk of these missions. In unknown environments, search operations are frequently complemented with the environment exploration.

Ali Marjovi, João Gonçalo Nunes, Lino Marques, and Aníbal de Almeida
e-mail: {ali,jgnunes,lino,adealmeida}@isr.uc.pt

* This work was partially supported by European project GUARDIANS contract FP6-IST-045269 as well as by the Portuguese Foundation for Science and Technology contract SFRH/BD/45740/2008.

A. Howard et al. (Eds.): Field and Service Robotics 7, STAR 62, pp. 341–351.
springerlink.com   © Springer-Verlag Berlin Heidelberg 2010

Autonomous environment exploration is a very fundamental issue in mobile robotics. This problem, complemented with map-building, is becoming increasingly solved in a robust way for single robot systems. Using multiple robot systems may potentially provide several advantages over single robot systems, namely higher speed, accuracy, and fault tolerance [1], [2], [3] and [4] . Nowadays, swarm based exploration and mapping where the robots can be smoothly added or removed to the operation is an area with increasing interests to the robotics community [5].

This study is integrated in a European project named Guardians[1]. The Guardians are a swarm of autonomous robots applied to navigate and search an urban environment. The project's central example is search and rescue in an industrial warehouse in smoke, as proposed by the Fire and Rescue Service of South Yorkshire. The job is time consuming and dangerous; toxins may be released and human senses can be severely impaired. They get disoriented and may get lost. The robots warn for toxic chemicals, provide and maintain mobile communication links, infer localization information and assist in searching. Map exploration and fire source detection are the topics in this paper.

The problem of coordination and control of multiple robots for mapping and exploration has been already addressed through several research approaches. Most approaches rely on centralized control to direct each vehicle. This centralized approach has been popular in the robotics community, because it allows near optimal behaviors in well understood environments. However, its performance decreases in new unidentified environments. Yamauchi [6] proposed a distributed method for multi-robot exploration, yielding a robust solution even with the loss of one or more vehicles. A key aspect of this approach involves sharing map information among the robotic agents so they execute their own exploration strategy, independently of all other agents. While this technique effectively decentralizes control, exchange of map information is not enough to prevent inefficient cooperative behaviors. This approach also required known starting positions and failed to provide a robust mechanism for map merging.

Simultaneous localization and mapping (SLAM) has been a topic of much interest because it provides an autonomous vehicle with the ability to discern and represent its location in a feature rich environment [11]. Some of the statistical techniques used in SLAM include extended Kalman filters, particle filters (Monte Carlo methods) and scan matching of range data. But if there is a local or global localization system where robots know their relative positions, SLAM techniques are not required.

Several researchers have suggested stigmergy methods [7] and [8]. Scheidt et. al. [8] uses stigmergy to achieve effects-based control of cooperating unmanned vehicles. They accomplished stigmergy through the use of locally executed control policies based upon potential field formulas. Nevertheless, this method is mainly useful when there are a lot of small robots working together.

Most of the existing approaches to coordinate multi-robot exploration assume that all agents know their locations in a shared (partial) map of the environment.

---

[1] http://www.guardians-project.eu

Effective coordination can be achieved by extracting exploration frontiers from the partial map and assigning robots to frontiers based on a global measure of performance [1], [2], [3] and [9]. Frontiers are the borders of the partial map, between explored free space and unexplored area [2]. These borders, thus, represent locations that are reachable from within the partial map and provide opportunities for exploring unknown terrain, thereby allowing the robots to greedily maximize information gain [10]. Compared to the problems occurring in single robot exploration, the extension to multiple robots poses new challenges, including:

Coordination and cooperation: Since there are several robots working in the same environment, they must have some kind of cooperation with each other in order to prevent collisions and share tasks. Effective cooperation can be achieved by having the robots into different non-overlapping areas [2], [3], [11]. The idea is that at a given time each robot should be dedicated to exploring one and only one frontier.

Integration of information collected by different robots into a single map: The main goal of exploration is to build a general map representing the environment. The robots should integrate all the data into a single map. Map merging is a big challenge in this field that has been address in several studies [12].

Uncertainty in localization and sensing: The effect of sensor errors ("noise") and errors in sensing the gradient of a "resource profile" (e.g., a nutrient profile) should be considered. Several researchers have illustrated that the agents can forage in noisy environments more efficiently as a group than individually [5], [13].

Decision making, reasoning, task sharing and navigation: Decision making for each robot in an unknown environment is a very complex problem. Since nobody knows what lies beyond the frontier of an unexplored area, there is no unique optimum algorithm that is completely reliable. In each situation, a robot should make a decision to progress exploration task based on a partial existing map and also the other robots' positions and objectives.

Most of the studies in multi-robot exploration do not address unknown environments. Moreover, most of the research in this field is based on centralized control of the robots. For example, in [14] and [6], the robots share a common map which is built during the exploration. Singh and Fujimura [14] presented a decentralized online approach for heterogeneous robots. Most of the time, the robots work independently. When a robot finds a situation that is difficult to solve by itself, it will send the problem to another robot which is likely to be able to solve the situation. The candidate robot is chosen by trading off the number of areas to be explored, the size of the robot and the straight-line distance between the robot and the target region. This technique generates a grid geometric map; therefore, the accuracy of the map depends on the grid size. Moreover, all the robots need to have a huge memory to keep the entire map. In the approach of Yamauchi [6], the robots move to the closest frontier according to the current map. However, there is no coordination component which chooses different frontiers for the individual robots.

Our approach, in contrast, is specifically designed to coordinate the robots so that they automatically do not choose the same frontier, so multiple robots can try to explore the same area. Additionally, our approach employs a topological map, so the robots only exchange environmental features. Topological map need much less memory capacity. As a result, this method needs significantly less time to accomplish the task.

The objective in this research is to generate the map of an unknown environment and also localize all the fire sources in the area. In fact, the final future goal is to create a fire risk map of an unknown environment with multiple robots, but this problem is not addressed here. A centralized global map is a requirement of Guardians project, but ideally the robots should be able to explore even with lack of communication, and in the case of nun-updated map.

During the exploration process, if there is a fire source, robots should detect it. The authors have addressed this issue in previous papers [15], [16], [17] and [18]. The last achievement of that research is kheNose. The kheNose is a device developed by the authors to sense olfactory information through the use of gas sensors, anemometers, a temperature and humidity sensor [19]. In the current study, the last version of kheNose has been used to detect the fire sources.

Collision avoidance between the robots during the exploration is a considerable issue that has not been addressed pragmatically in the previous studies. In this study, we propose a new practical method for multi-robot unknown environment exploration with fire source detection which takes "collision avoidance" and "task sharing" into consideration. This method has been tested in the real world and also in simulation. The effect of complexity of the environment and also the numbers of robots are the main parameters that have been studied in this paper.

2 The Proposed Method

This section explains the concept of the proposed multi-robot cooperation technique. This method is illustrated in the schematic diagram of Fig. 1. As shown in the diagram, the method includes three main tasks: navigation and exploration, decision making, and fire source detection. These tasks are briefly described below.

2.1 Decision Making

The main goal of the exploration process is to cover the whole environment in the minimum possible time. Therefore, it is essential that the robots share their tasks and individually achieve the objectives through optimal paths. In an unknown environment, the immediate goals are the frontiers. Most of the time, when the robots are exploring an area, there are several unexplored regions, which poses a problem of how to assign specific frontiers to the individual robots. We want to avoid sending several robots to the same frontier, which may result in collision concerns. Another issue is that we do have a base station, but the robots should be able to

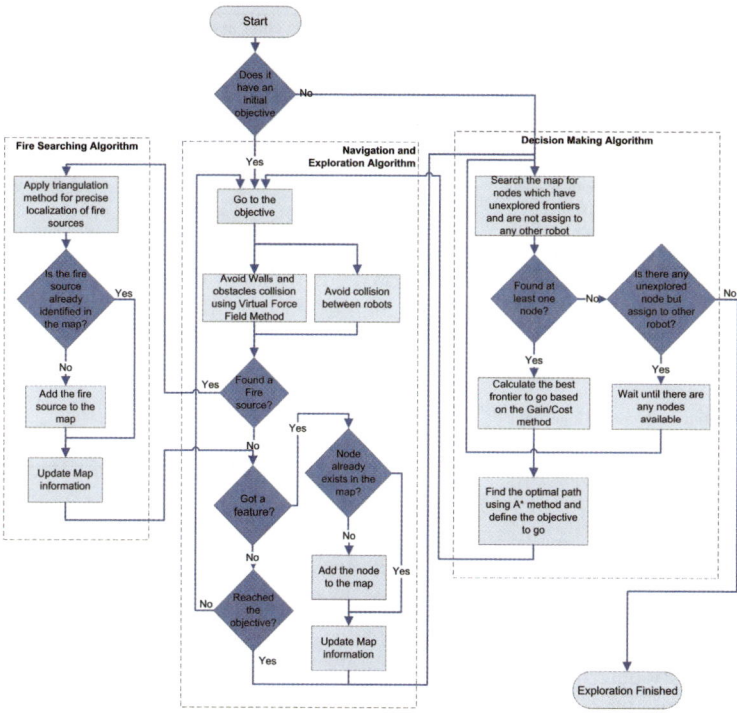

Fig. 1 Exploration and fire searching fluxogram

Algorithm 1. Map exploration algorithm - Decision making method

```
1   Receive_map_from_server()
2   while there is at least one unexplored link in the map do
3       do Follow the potential field algorithm until (getting a different feature in the environment)
4       Receive_map_from_server()
5       if the new node exists in the map then
6           Update the map's data with new information.
7           Send_map_to_the_server()
8       else
9           Add_new_node_to_map()
10          Send_map_to_the_server()

11      if the current node has any unexplored link then
12          Calculate direction gain of taking each unexplored link, based on the position of the other robots and
            get the higher gain direction to follow
13      else
14          Determine the best not-assigned unexplored frontier, based on their gain / cost.
15          Assign the frontier to the robot.
16          Calculate the best path to take, based on the A* algorithm and get that direction to follow.

17  End of algorithm
```

Algorithm 2. Collision avoidance between the robots (and increasing efficiency of collaborating)

```
1   Calculate a confined circular area around the robot
2   if any other robot is inside the circle then
3   │   Determine if the robots are in a collision pattern
4   │   if there is a possibility of being in a collision pattern then
5   │   │   //Follow the rules of engagement:
6   │   │   if they are in a direct collision path then
7   │   │   │   reevaluate their goals.
8   │   │   else
9   │   │   │   if they are both currently exploring frontiers OR they are both moving inside explored
        │   │   │   area then
10  │   │   │   │   give priority to the one which has lowest ID.
11  │   │   else
12  │   │   │   Give priority to the one that is exploring a frontier.

13  else
14  │   Continue exploration algorithm.
15  End of algorithm
```

---

explore autonomously. To address these problems, the proposed method is based on a decision-theoretic exploration strategy.

The frontier is selected based on the cost of reaching it and the utility it can provide to the exploration. The cost is calculated through the A* method which simultaneously determines the optimal path to reach the frontier and its distance. The utility depends on the number of the robots and their proximity to the frontier, which means that if there are several frontiers at similar distances, a given robot will go to the one that has higher utility. This procedure will make the robots disperse and explore the environment in a efficient way.

2.2 Task Sharing and Map Generation

The cooperation between the robots is based on the exchange of data allowing for task sharing and, consequently, an efficient distributed exploration. During the exploration, there is only one global shared map in the system. This map is in a base station that sends and receives the map to the robots whenever they request it. Within this map, besides having some information regarding the kind of nodes and their position, it also has data describing the location of the robots and their frontier target, as can be seen in Fig. 2. Through this data, a robot can see which frontiers are unexplored, their position and if any robot has targeted them as its objective, thus allowing a distributed efficient exploration (see Algorithm 1 and Fig. 1).

While dealing with multiple robots in one environment, collision between robots is a very important aspect. For instance, two robots might be in a narrow corridor with different directions and they may want to pass but cannot because they are facing each other or they may even treat each other as a dead end. This type of problems is avoided with a set of rules that prevents the robots to follow by the same corridor in facing directions (see Algorithm 2).

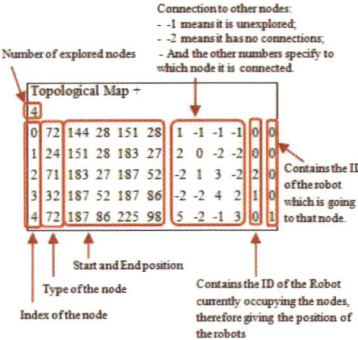

Fig. 2 Example of topological map data   **Fig. 3** Khepera III and kheNose

2.3 Fire Source Detection

During exploration and navigation, the robots are simultaneously acquiring information from the environment (see Fig. 1). All the robots are equipped with a set of sensors developed by this research group, which integrates temperature and chemical sensors named kheNose (Fig. 3).

When the robots are mapping the environment, they are constructing the map and verifying if the current node they have acquired is not already on the map, thus, assuring the coherence of the map and making the merging process simple, where most of the time it is only necessary to add new nodes to the global shared map.

An eight element thermopile array sensor is used in order to measure the absolute temperature as well as the ambient temperature on the robot to be able to distinguish the heat values. When the sensor detects hot-spots or areas with a temperature above a defined threshold, a heat source is identified and a pattern of motions is implemented in order to localize the position of that heat source.

3 Experiments

The algorithm has been tested in real world and also in a simulation world. For optimizing the exploration algorithm and measuring its performance, the Player/Stage simulator was used [20]. In the real world, there are several constraints that do not allow for testing the proposed method very easily. It is not effortless to build various test plans with different scales for testing and developing the method. Since there is no reliable simulator for fire and smoke in Player/Stage, the whole system has been tested in the real world.

3.1 The Real World Experiments

The proposed method was tested in different maze-like environments, like the one shown in Fig. 4, using three Khepera III robots equipped with KheNose sensing

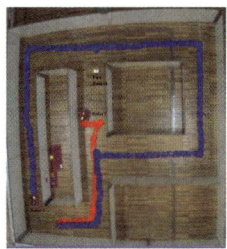

Fig. 4 Real maze experiment

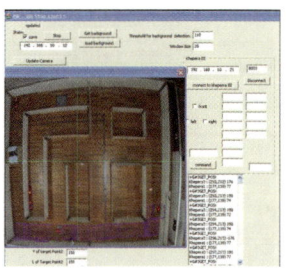

Fig. 5 Visual positioning application screen-shot

boards. The KheNose has multiple sensing capabilities, namely: an electronic nose, a gas sensing array, an anemometer array, and a thermal radiation array [19].

The localization of the robots is out of the scope of this work. This problem is solved with a ceiling camera. A network camera is mounted on the top of the environment and an image processing computer program is able to track and locate each robot. Each robot has two colored labels on the top that can be seen by the camera. The camera is connected to the network and an image processing program tracks the robots' position and provides the absolute position of each robot via wireless network. Image processing program is an object tracking application developed by the authors. By recognizing the center of each colored label and calculating the line crossing from these two centers, the orientation of the robot can be computed. The program is written in C++. Fig. 5 shows a screen shot of this program.

In terms of feature extraction, based on values measured by sonar and infrared sensors, the robot recognizes the features and should take an action and modify the shared map; it will save this data in the map structure as a new node, and will also update the data related to the previous feature. For each feature, the robot saves the data in the topological map, including the area of influence of that node and some other information (Fig. 2).

The system has been tested with different start positions for the robots in different maze structures. There is a small candle acting as a heat source in the environment which robots try to locate.

Fig. 4 shows two robots exploring a small maze and finding a fire source. Both robots started from the same point but not at the same time. We intentionally ran one of the robots a few seconds after the first one. The darker footprint shows the first robot's path and the lighter footprint is related to the second robot. As shown, the first robot found the fire source. For an example of the coordination algorithm, when the second robot reached the junction it figured out that the path in the front was already explored and it chose the right path.

Another parameter for evaluation of the method is the exploration time. The proposed method has been tested with a different number of robots in different mazes. The environment shown in Fig. 5 that is a 3.5 x 4 meters maze is tested by one, two and three Khepera robots separately. One robot could explore the environment in 412 seconds. This environment has been explored by two robots in 254 seconds.

 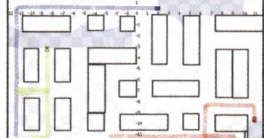 

Fig. 6 A maze with 34 nodes   **Fig. 7** 82 nodes   **Fig. 8** 135 nodes

The exploration time for the same maze with three Kheperas was 212 seconds. Each result is the average of five similar tests. Different tests with constant conditions had similar results with about seven percent variance. The maximum speed of the Kheperas is kept constant in all the tests.

In the real experiment the robots could locate the fire sources during the exploration. The performance of fire source detection has been addressed in previous studies [17], [19].

3.2 Simulation

Since there is no accepted standard benchmark, measuring the performance of a behavioral based multi-robot unknown area exploration algorithm is a very difficult job. One of the possible ways to do that is to compare the proposed method with a optimal method. But the issue is that there is no optimal method for exploring an unknown world. However, there is an optimal solution for minimizing the travelling path if the world (maze) is completely known before exploration.

The algorithm has been tested with different number of robots in specific mazes. The models of those mazes are also given to the optimal method and then we compared the results of the proposed algorithm with the optimal results. Since the optimal method has the world's model but the proposed method is exploring the unknown world, it is obvious that the results of the proposed method are always worse than the optimal but this can be a good criteria for evaluating the method.

The number of repeated nodes during travel can be another good parameter for measuring the performance of the method. A repeated node is a node that robots pass more than once. Fig. 9 shows the number of nodes that have been repeated more than once in the optimal method as well as in the proposed algorithm for the maze shown in Fig. 6. A good conclusion from the graph in Fig. 9 is that there is a trade-off between the number of robots and the size of the world. It shows that the proposed approach is acceptably comparable with the optimal method.

The mazes shown in Fig. 6, Fig. 7 and Fig. 8 have been tested separately with one, two, three and four robots and the results are shown in Fig. 10. The graph shows the average of five tests for each data. The variance was less than one percent. It is obvious that the exploration time improves with higher number of robots. Another conclusion from the graph is that having more robots is more advantageous in a complex maze than in a simple maze. This also proves that the cooperation algorithm in this approach is efficiently functional.

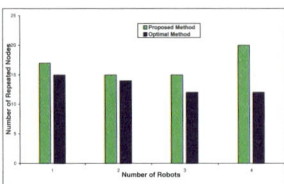
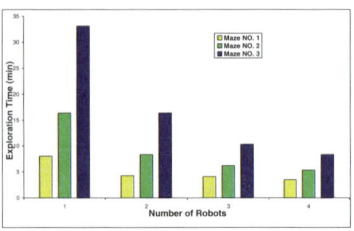

Fig. 9 Number of repeated nodes, Comparing the results of the proposed method with optimal method

Fig. 10 Test of various numbers of robots against complexity of the environment, 1: maze in Fig. 6, 2: maze in Fig. 7, 3: maze in Fig. 8

4 Conclusions and Future Works

A proposed method for multi-robot unknown environment exploration has been implemented and experimented in realistic reduced scale scenarios. The robots are able to cooperate and create a shared topological map of the unknown environment. Cooperation between the robots is done by sharing information in the shared map. The algorithm has been tested against a large variety of configurations in Player/Stage simulation program. The exploration algorithm is merged with fire source detection algorithm and has been tested in the real world. The effect of the number of the robots on exploration in different type of environment has been analyzed and discussed. The results show a high efficiency and reliability of this method.

In terms of implementation, more accurate sonar sensors should be installed on the robots.

Optimizing the searching process by integrating sensing cues in the frontiers selection can be another improvement of this research.

References

1. Fox, D., Ko, J., Konolige, K., Limketkai, B., Stewart, B.: Distributed multi-robot exploration and mapping. Proc. of the IEEE 94(7), 1325–1339 (2006), special Issue on Multi-Robot Systems
2. Burgard, W., Moors, M., Fox, D., Simmons, R., Thrun, S.: Collaborative multi-robot exploration. In: Proc. IEEE Int. Conf. on Robotics and Automation, pp. 476–481 (2000)
3. Burgard, W., Moors, M., Stachniss, C., Schneider, F.: Coordinated multi-robot exploration. IEEE Trans. on Robotics 21(3), 376–386 (2005)
4. Dedeoglu, G., Sukhatme, G.: Landmark-based matching algorithm for cooperative mapping by autonomous robots. In: Proc. of 5th Int. Symp. on Distributed Autonomous Robotic Systems, DARS 2000 (2000)
5. Liu, Y., Passino, K.: Stable Social Foraging Swarms in a Noisy Environment. IEEE Trans. on Automatic Control 49(1), 30–44 (2004)
6. Yamauchi, B.: Frontier-based exploration using multiple robots. In: Proc. of 2nd Int. Conf. on Autonomous Agents (1998)

7. Stipes, J., Hawthorne, R., Scheidt, D., Pacifico, D.: Cooperative Localization and Mapping. In: Proc. of the IEEE on Networking, Sensing and Control (2006)
8. Scheidt, D., Stipes, J., Neighoff, T.: Cooperating Unmanned Vehicles. In: Proc. of the IEEE on Networking, Sensing and Control (2005)
9. Koenig, S., Tovey, C., Halliburton, W.: Greedy mapping of terrain. In: Proc. IEEE Int. Conf. on Robotics and Automation (2001)
10. Zlot, R., Stentz, A., Bernardine Dias, M., Thayer, S.: Multi-robot exploration controlled by a market economy. In: Proc. IEEE Int. Conf. on Robotics and Automation (2002)
11. Dissanayake, M., Newman, P., Clark, S., Durrant-Whyte, H., Csorba, M.: A solution to the simultaneous localization and map building (SLAM) problem. IEEE Transactions on Robotics and Automation 17(3), 229–241 (2001)
12. Pfingsthorn, M., Birk, A.: Efficiently communicating map updates with the pose graph. In: IEEE/RSJ Int. Conf. on Intelligent Robots and Systems, pp. 2519–2524 (2008)
13. Liu, Y., Passino, K.: Biomimicry of social foraging behavior for distributed optimization: Models, Principles and emergent behaviors. Journal of Optimization Theory and Applications 115(3), 603–628 (2002)
14. Singh, K., Fujimura, K.: Map making by cooperating mobile robots. In: Proc. IEEE Int. Conf. on Robotics and Automation (1993)
15. Marques, L., Almeida, N., de Almeida, A.: Olfactory sensory system for odour-plume tracking and localization. In: IEEE Int. Conf. on Sensors, Toronto, Canada (2003)
16. Marques, L., Almeida, A.: Electronic nose-based odour source localization. In: 6th Int. Workshop on Advanced Motion Control (2000)
17. Marques, L., Nunes, U., de Almeida, A.: Particle swarm-based olfactory guided search. Autonomous Robots 20(3), 277–287 (2006), special Issue on Mobile Robot Olfaction
18. Marques, L., Nunes, U., Almeida, A.: Olfaction-based mobile robot navigation. Thin Solid Films 418(1), 51–58 (2002)
19. Pascoal, J., Sousa, P., Marques, L.: Khenose - a smart transducer for gas sensing. In: Proc. of the 11th Int. Conf. on Climbing and Walking Robots and the Support Technologies for Mobile Machines (CLAWAR 2008), Coimbra, Portugal (2008)
20. Gerkey, B., Vaughan, R., Howard, A.: The player/stage project: Tools for multi-robot and distributed sensor systems. In: Proceedings of the 11th International Conference on Advanced Robotics (ICAR 2003), Coimbra, Portugal, pp. 317–323 (2003)

Part VIII
Human Robot Interaction

Using Virtual Articulations to Operate High-DoF Inspection and Manipulation Motions

Marsette Vona, David Mittman, Jeffrey S. Norris, and Daniela Rus

Abstract. We have developed a new operator interface system for high-DoF articulated robots based on the idea of allowing the operator to extend the robot's actual kinematics with *virtual articulations*. These virtual links and joints can model both primary task DoF and constraints on whole-robot coordinated motion. Unlike other methods, our approach can be applied to robots and tasks of arbitrary kinematic topology, and allows specifying motion with a scalable level of detail. We present hardware results where NASA/JPL's All-Terrain Hex-Legged Extra-Terrestrial Explorer (ATHLETE) executes previously challenging inspection and manipulation motions involving coordinated motion of all 36 of the robot's joints.

1 Introduction

Due to their application flexibility, robots with large numbers of joints are increasingly common: humanoids with 20 or more DoF are now available off-the-shelf, many-link serpentine robots have been demonstrated with a wide range of locomotion modalities, and assemblies of modular and self-reconfiguring hardware have been constructed with many 10s of concurrently active joints. This flexibility is especially attractive for interplanetary and Lunar exploration contexts, where the extreme costs of transportation from Earth are balanced by maximizing versatility, reusability, and redundancy in the delivered surface system hardware. Such considerations have been a prime motivation for NASA/JPL's development of the 36-DoF All-Terrain Hex-Legged Extra-Terrestrial Explorer (ATHLETE) [13], with which astronauts will collaborate in our planned return to explore the Moon (figure 1).

Mission cost also dictates that we need operator interface systems that can rapidly and efficiently expose the maximum hardware capability to the humans that direct these robots, whether they are on-site astronauts or ground-based operators. This is a challenging problem in the high-DoF case: there are usually many ways the robot

Marsette Vona and Daniela Rus
Massachusetts Institute of Technology, 77 Massachusetts Ave, Cambridge MA 02139

David Mittman and Jeffrey S. Norris
Jet Propulsion Laboratory, California Institute of Technology, 4800 Oak Grove Drive, Pasadena CA 91109

A. Howard et al. (Eds.): Field and Service Robotics 7, STAR 62, pp. 355–364.
springerlink.com © Springer-Verlag Berlin Heidelberg 2010

Fig. 1 NASA/JPL's All-Terrain Hex-Legged Extra-Terrestrial Explorer (ATHLETE).

could move to achieve the task, and some may be better than others due to secondary goals. Sometimes a human operator can quickly visualize the desired motion, but till now the expression of this motion to the operations system has often been a tedious bottleneck. In this paper we present the design, implementation, and experimental results for a new operations system for high-DoF robots which employs *virtual articulations* to address this issue.

In our system, which we call the *mixed real/virtual operator interface*, the operator is presented with a graphical model of the robot and a palette of available joint types (figure 2, left). To constrain motion for a particular task, the operator instantiates virtual joints from this palette and interconnects them to the links of the actual robot and/or to new virtual links, constructing arbitrary virtual extensions to the actual robot kinematics. Virtual joints can be erected to parametrize specific task DoF; for example the long prismatic virtual joint in figure 4 parametrizes the length of a trenching motion. By closing kinematic chains, virtual articulations can also constrain whole-robot motion, thus narrowing the space of possible motions for a redundant task to those that satisfy the operator's intentions. The virtual *Cartesian-3* joint in figure 4, which allows three axes of translation but no rotation, constrains ATHLETE's deck to remain flat, even while moving to extend reach for the primary trenching task. Virtual links can serve as interconnection points for more complex constructions of virtual joints—the chain of two prismatic and two revolute virtual joints in figure 4 is interspersed with three virtual links—and can also model task-related coordinate frames or world objects (figure 3).

Once virtual articulations are constructed for a task, the operator can move any joint or link (e.g. with the mouse), and the system interactively responds in real-time with a compatible motion for all joints which best satisfies all constraints. For example, in the trenching task, the operator can effectively command "trench from -0.9m to +0.4m" by operating the corresponding virtual prismatic joint, or they may simply drag the constrained end effector with the mouse. We validate these motions in simulation and then execute them on the hardware.

Our system is generally applicable to kinematic operations in articulated robots of any topology, handling both open- and closed-chain constructions as well as both over- and under-constraint. In this paper we focus on our recent results operating new ATHLETE motions at JPL, but in other work we have also begun to demonstrate the usefulness of our approach in specifying motions for modular reconfigurable robots which can be assembled in arbitrary topologies. We expect that applications to other high-DoF kinematic motions, including in humanoids and in serpentine robots, will also be both useful and direct.

We describe related work next. Then we explain the architecture of our mixed real/virtual operator interface and detail the handling of under- and over-constrained cases, a key aspect of our system. Next we show several inspection and manipulation tasks on the ATHLETE hardware that would have been challenging with prior operator interfaces, including an experiment where we combine our virtual articulation interface with a direct-manipulation input device that mimics one ATHLETE limb. We developed this device, the Tele-Robotic ATHLETE Controller for Kinematics (TRACK), in prior work [6]. We conclude by summarizing known limitations of our approach and possible future directions.

2 Related Work

We see our new method of operating high-DoF robots using virtual articulations as filling a gap between existing low-level methods, including forward and inverse kinematic control, and existing high-level methods such as goal-based motion planning and programing-by-demonstration.

Bare kinematic control without higher-level goals or constraints is potentially tedious in the high-DoF case given the high dimension of the joint space. Task priority and task space augmentation approaches [7] can support high-DoF motion using holonomic constraints, but do not themselves offer any particular way to specify those constraints. Our virtual articulation approach addresses this with a concrete framework in which holonomic constraints can be constructed by an operator.

Goal-based motion planning, e.g. the classic "piano moving" problem of achieving a target configuration among obstacles, is typically not directly applicable in cases where the operator would also like to specify more detailed or continuous aspects of the motion. If we want such *scalable motion specification*, to constrain motion "on the way" to a primary goal configuration, we need something more. Virtual articulations are one language that does permit such scaling: the operator can constrain motion as much or as little as desired.

Programming-by-demonstration allows more specific motion specification, but is hard to apply when the robot topology diverges from preexisting systems and biology. Thus it has been used with some success for humanoids , or when mimicking hardware is available, as in our prior work with the TRACK direct-manipulation hardware interface. But, short of building a full 36-DoF scale model, how to apply the technique to the whole ATHLETE mechanism, or in general, for arbitrary topology robots? Virtual articulations are not tied to any particular topology. Further, in section 4 we show that an integration of TRACK with our virtual articulations system can have some of the advantages of both.

Though we don't find any prior authors using virtual articulations to build a general-purpose operations interface as we have done, there have been some related ideas. Virtual reality operator interfaces (e.g. [3]) have been explored where a model of the robot and its surroundings is provided to the operator for pose manipulation; we go beyond this by allowing the operator to virtually change and augment the kinematic structure. Our approach was motivated in part by past work with geometric constraints in graphics and animation [8, 12]; we show that a homogeneous model of only links and joints is sufficient in some practical applications.

Finally, we note that CAD constraint solvers [5] and physics simulators [10] have similar capabilities to our system. CAD solvers usually don't permit over-constraint, and typically use heuristics to infer constraints in under-constrained cases, which may or may not be what the operator intended; our system usefully handles both under- and over-constraint without heuristics. Physics simulators can also be problematic in over-constrained cases, and the need to specify physics parameters such as mass and friction properties could make the process of building virtual articulations much more tedious. Our current approach is purely kinematic, so constructing virtual articulations only requires posing them in space and connecting them. Pratt et al explored a dynamic correlate to our virtual articulation approach which they called virtual model control [9] for some applications in legged locomotion.

3   The Mixed Real/Virtual Operator Interface

The key advance that differentiates our system from prior approaches is that we permit the operator to interactively construct virtual links and joints both to constrain and parametrize the primary task and also to constrain coordinated whole-robot motion. In this section we give an overview of the architecture of our system and explain how we address handling of under- and over-constrained cases, which are both common and important. Due to space constraints we omit our approaches to a number of other issues which do need to be considered in a full implementation, including: joint pose representation, handling of joint motion limits, efficient and accurate Jacobian computation, joint inversions and re-grounding, model complexity management, automatic decomposition of the model into independently solvable pieces, adaptive step size and damping, and graphics/UI features.

Figure 2 shows our system's architecture. There are three categories of inputs: (1) robot models are loaded into the system from standard file formats such as VRML; (2) the operator may add and remove virtual articulations on-line with a variety of topological mutators; and (3) the operator may move joints and links, either virtual or actual, by a variety of means including mouse and keyboard actions, waypoint sequencing and interpolation, or by using special-purpose input hardware.

These inputs determine the evolution of both topological and numeric models that include the actual robot kinematics plus any virtual articulations. The topological model is a kinematic graph where the edges correspond to joints and the vertices to

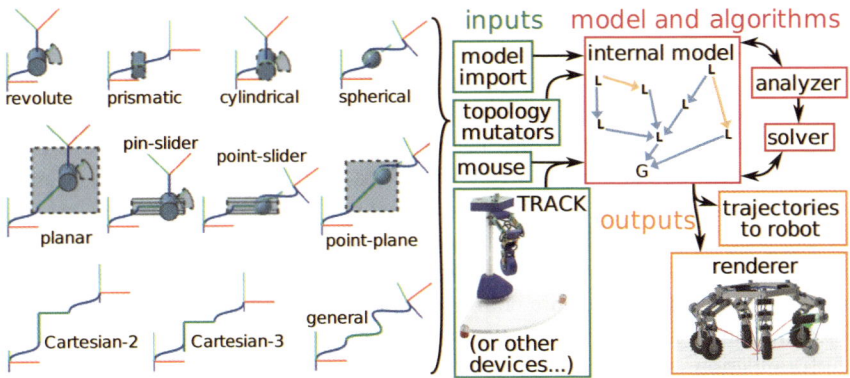

Fig. 2 Joint catalog and architecture of the *mixed real/virtual operator interface*. Arbitrary topology open- and closed-chain articulated robots are modeled with the set of 11 joints at left. Virtual articulations, such as adding or suppressing joints, are applied on-line with a set of topology mutation operations. Finally, the resulting mixed real/virtual structure is kinematically operated by the mouse or other input devices.

links. The numeric model consists of the current pose of each joint as well as the joint constraints, motion limits, and several kinds of goals, as described below.

The main action of the system is to compute feasible whole-robot motions in response to the operator's requests to move joints and links. Our approach is iterative local linear optimization based on the damped least squares Jacobian pseudoinverse and nullspace projection. This well-known approach has long been used in robotics; we apply a multi-priority formulation recently presented for a graphics application in [1]. The per-iteration time complexity of this formulation, typically dominated by a nullspace projection step, is quadratic in the number of joints. Nevertheless our implementation achieves real-time (several 10s of ms) response to operator motion gestures for the 50+ joints comprising the ATHLETE model plus various added virtual articulations. As with any local optimization approach, local optima must be avoided by higher-level means—the system is more of a controller than a planner. In our work thus far these global planning tasks are handled by the operator.

As feasible motions are computed they drive the two main outputs of the system: an interactive 3D graphical display of the robot plus any virtual articulations, and trajectories of the robot's joints that can be sent to the hardware for execution.

In general the space of feasible motions may be continuous (under-constrained case), discrete (well-constrained), or empty (over-constrained). Since the well-constrained case requires an exact balance of freedom and constraint, the under- and over-constrained cases are more common, and we give them special attention.

Handling Under-Constraint. When the operator adds a virtual joint closing a kinematic chain the dimension of the feasible configuration space can be reduced. This is the first of three ways that we address under-constrained (aka redundant) problems:

the operator may intentionally construct virtual articulations to express specific motion constraints and thus reduce redundancy.

The second way we handle redundancy is by exposing two levels of joint pose goals to the operator: *targets* and *postures*. A *target* in our system is the pose to which the operator has specifically manipulated (e.g. by mouse or keyboard interaction, or by waypoint interpolation) a joint or link.[1] A *posture* models a default pose; for ATHLETE operations we typically set joint postures according to the "standard driving pose" (figure 1). The system solves first for motions which best attain all targets, and within the set of motions which do, the system second tries to attain postures. Target and posture are both optional for each DoF of each joint.

Goal attainment is prioritized in our system by structuring the solvers according the formulation presented in [1], which we call *prioritized damped least squares* (PDLS). In this formulation there are an arbitrary number of priority levels, each containing an arbitrary set of constraints. The constraints at the highest priority level are solved first, and the solution for each subsequent level is projected onto the nullspace of the levels above it.

The least squares aspect of PDLS provides the third and ultimate means of handling under-constraint.[2] The least squares solution to an under-constrained problem will select a shortest step in joint space at each iteration, resulting in incrementally minimal motion: at a fine scale, the system will produce direct straight-line moves from one configuration to the next. In the under-constrained case a roundabout trajectory might also satisfy the constraints and maximize goal attainment, but would doubtless be surprising to the operator.

Priority Levels and Handling Over-Constraint. The least-squares nature of PDLS also means that within a priority level, over-constraint will result in a solution which minimizes the squared error across all constraints in the level. This is useful and can produce intuitive behavior from the operator's perspective. Another important feature of PDLS in over-constrained cases is the prioritization: satisfaction of constraints at a lower priority level will not compromise satisfaction at higher levels, even when the constraints conflict.

There are four priority levels in our system:

1. Joint *invariant*s are solved at the highest priority level. For example, a spherical joint permits no translation, so when closing a kinematic chain it induces three invariant goals expressing that its three translation components must be zero.
2. *Lock* goals model joints that have been "frozen" by the operator: each DoF of such a joint must remain as it was when the joint was first locked.
3. *Target* goals model intended joint and link poses as described above.
4. *Posture* goals model default poses, also described above.

---

[1] To model pose goals on a link *l* we transparently introduce a virtual general (unconstrained 6-DoF) joint *j* connecting *l* to the world frame, and set the goals on *j*.

[2] And to complete the terminology, damping refers to the well-known technique of numeric stabilization at near-singular configurations by introducing a damping factor.

It would also be possible to insert other (differentiable) optimality criteria, such as manipulability maximization or joint limit avoidance, as new priority levels.

To see how priority levels help in cases of over-constraint, consider the spherical object inspection task in figure 3. In this case we use TRACK to pose the limb holding the inspection camera. But there is also a virtual spherical joint constraining the camera, and TRACK has no haptic feedback. So, while the operator will generally try to pose it near to a feasible configuration, invariably this will diverge from the strict spherical constraint surface, over-constraining the limb. The spherical joint constraint is modeled at the invariant level, and TRACK's pose is modeled at the target level, so the system will automatically sacrifice the latter for the former. The overall effect is as if the virtual spherical joint was physically present and rigidly constraining the motion, and as if there were an elastic connection between TRACK and the motion of the actual limb.

4 Operating ATHLETE with Virtual Articulations

The object inspection task is one of four hardware experiments we present. All show the ability of our mixed real/virtual interface system to help design specific motions which are rapid for human operators to conceptualize but difficult to express in prior operations interfaces, including several other software systems under development within NASA [4, 11] as well as our own TRACK device used alone [6].

For the object inspection task, the operator designs a motion where a limb-mounted camera inspects a roughly spherical object while maintaining a constant distance. The operator directly models this constraint using a virtual spherical joint connecting the object (itself represented as a virtual link) and the camera. A secondary goal is to extend the space of reachable viewpoints by using the five other limbs to lean the hexagonal deck, but because the deck often carries a payload,

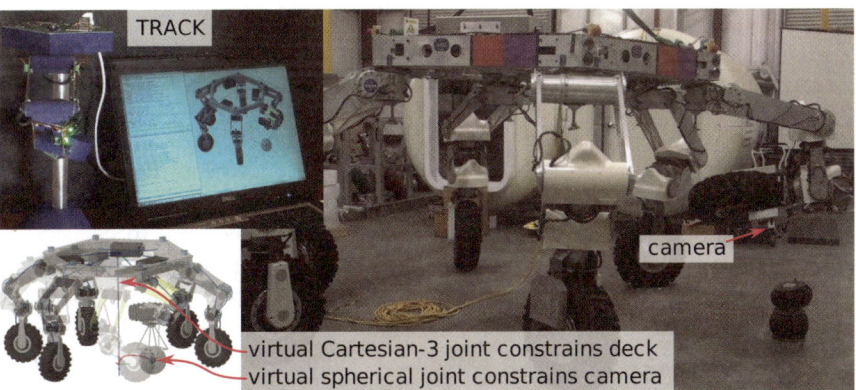

Fig. 3 ATHLETE inspecting an object using both the mixed real/virtual interface and TRACK, a special-purpose input device that mimics one limb.

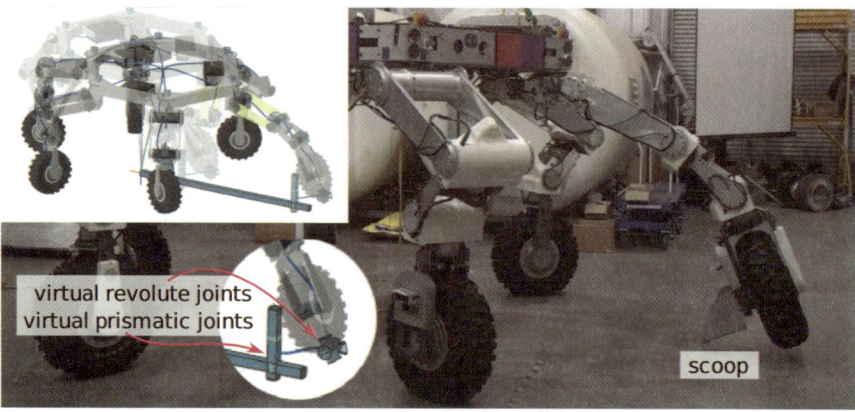

Fig. 4 ATHLETE performing a trenching motion (mixed real/virtual interface view inset).

Fig. 5 ATHLETE panning and tilting a fixed-mount camera with whole-robot motions.

we need to maintain its orientation. This is expressed by a virtual Cartesian-3 joint connected between the deck and the world frame.

After configuring the virtual articulations the operator can drag the camera with the mouse to scan the object. As described above, in this case we also integrated our TRACK hardware interface, which simplified motion specification. To save cost— total materials cost for TRACK was under $500 USD—we opted not to include haptic feedback in TRACK, potentially making it less applicable for constrained tasks. This example shows that constraint prioritization can mitigate the issue somewhat.

Figures 4, 5, and 6 give three additional examples: (1) a trench is inspected, with the support legs moving the deck to extend reachable trench length; (2) a rigidly mounted side-facing camera is made to pan and tilt with the motion both parametrized and constrained by virtual revolute joints; and (3) two limbs execute a pinching maneuver with the pinch distance and angles controlled by virtual prismatic

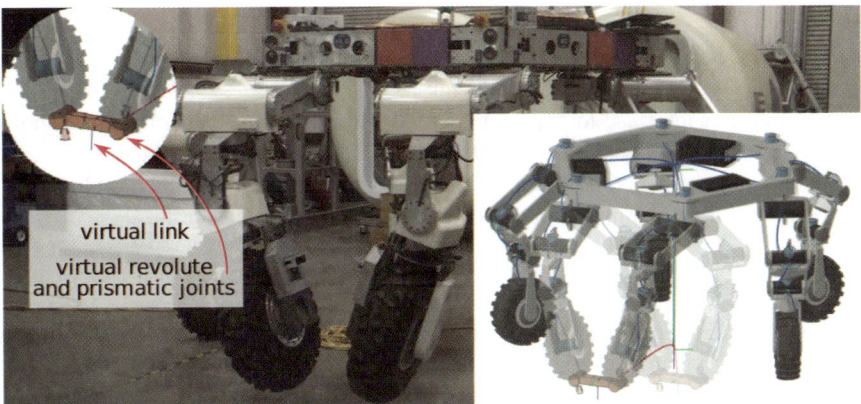

Fig. 6 ATHLETE performing a bimanual pinching motion.

and revolute joints. For the bimanual experiment the robot was partially supported by an overhead crane as simultaneously raising two limbs is not supported on the current hardware. The crane served as a safety-backup in the other experiments.

In our current implementation we design all virtual articulations and motions in simulation, export them as joint space waypoint sequences, typically generating a new waypoint whenever any joint moves at least $2°$. We then check the sequences in a previously validated simulator and execute them as position-controlled trajectories on the hardware. We performed each of the four experiments at least twice, though such repeatability is actually a property of the hardware, not our interface system.

5   Limitations and Future Work

A next step for this work will be to perform measured usability experiments; for those to be meaningful we should implement a few additional critical usability features including snap-dragging [2] and undo. We will also implement a more self-documenting drag-and-drop UI for constructing virtual articulations. We envision measuring both the operator learning time for our system vs. existing systems at JPL, and also the time required to design a complex motion in each system. The comparison may be indirect as our system is higher-level than the others.

Our purely kinematic implementation applies only to fully-actuated cases; we are developing quasi-static extensions for some underactuated tasks. Some constraints, for example helical motion, cannot be modeled with the current set of joints in the system. Possible extensions could increase the set of representable constraints.

6   Summary

Our *mixed real/virtual interface* implements virtual articulations as a rapid graphical operator interface for coordinated manipulation and inspection motions in high-DoF articulated robots. This new method fills a gap between existing lower- and higher-level interfaces. It is topology-independent, supports scalable motion specification, and usefully handles both under- and over-constraint.

We used our interface to experimentally demonstrate four new classes of coordinated motion for NASA/JPL's 36-DoF ATHLETE, all of which would have been difficult using prior methods, and we used constraint prioritization to combine our inexpensive direct manipulation device with virtual motion constraints.

Acknowledgements. ATHLETE VRML model provided by RSVP team, NASA/JPL/ Caltech. Work with ATHLETE hardware was carried out at the Jet Propulsion Laboratory, California Institute of Technology, under a contract with NASA and funded through the Director's Research and Development Fund. Additional funding came from the NSF EFRI program.

References

1. Baerlocher, P., Boulic, R.: An inverse kinematics architecture enforcing an arbitrary number of strict priority levels. The Visual Computer 20, 402–417 (2004)
2. Bier, E.A.: Snap-dragging: Interactive geometric design in two and three dimensions. PhD thesis, EECS Department, University of California, Berkeley (1988)
3. Flückiger, L.: A robot interface using virtual reality and automatic kinematics generator. In: International Symposium on Robotics, pp. 123–126 (1998)
4. Hauser, K., Bretl, T., Latombe, J.C., Wilcox, B.: Motion planning for a six-legged lunar robot. In: Proceedings of WAFR, pp. 301–316 (2006)
5. Hoffmann, C.M.: D-Cubed's Dimensional Constraint Manager. Journal of Computing and Information Science in Engineering 1, 100–101 (2001)
6. Mittman, D.S., Norris, J.S., Powell, M.W., Torres, R.J., McQuin, C., Vona, M.A.: Lessons Learned from All-Terrain Hex-Limbed Extra-Terrestrial Explorer Robot Field Test Operations at Moses Lake Sand Dunes, Washington. In: Proceedings of AIAA Space Conference (2008)
7. Chiacchio, P., Chiaverini, S., Sciavicco, L., Siciliano, B.: Closed-loop inverse kinematics schemes for constrained redundant manipulators with task space augmentation and task priority strategy. IJRR 10(4), 410–425 (1991)
8. Phillips, C.B., Zhao, J., Badler, N.I.: Interactive real-time articulated figure manipulation using multiple kinematic constraints. In: SIGGRAPH, pp. 245–250 (1990)
9. Pratt, J., Chew, C.M., Torres, A., Dilworth, P., Pratt, G.: Virtual model control an intuitive approach for bipedal locomotion. IJRR 20(2), 129–143 (2001)
10. Smith, R.: Open dynamics engine (2008), http://www.ode.org
11. SunSpiral, V., Chavez-Clemente, D., Broxton, M., Keely, L., Mihelich, P., Mittman, D., Collins, C.: FootFall: A ground based operations toolset enabling walking for the ATHLETE rover. In: Proceedings of AIAA Space Conference (2008)
12. Welman, C.: Inverse kinematics and geometric constraints for articulated figure manipulation. Master's thesis, Simon Fraser University (1993)
13. Wilcox, B.H., Litwin, T., Biesiadecki, J., Matthews, J., Heverly, M., Morrison, J., Townsend, J., Ahmad, N., Sirota, A., Cooper, B.: ATHLETE: A cargo handling and manipulation robot for the moon. Journal of Field Robotics 24(5), 421–434 (2007)

Field Experiment on Multiple Mobile Robots Conducted in an Underground Mall

Tomoaki Yoshida, Keiji Nagatani, Eiji Koyanagi, Yasushi Hada, Kazunori Ohno, Shoichi Maeyama, Hidehisa Akiyama, Kazuya Yoshida, and Satoshi Tadokoro

Abstract. Rapid information gathering during the initial stage of investigation is an important process in case of disasters. However this task could be very risky, or even impossible for human rescue crews, when the environment has contaminated by nuclear, biological, or chemical weapons. We developed the information gathering system using multiple mobile robots teleoperated from the safe place, to be deployed in such situation. In this paper, we described functions of the system and report the field experiment conducted in a real underground mall to validate its usability, limitation, and requirements for future developments.

Keywords: Search and Rescue, Teleoperation, Field Robotics, Mapping.

1 Introduction

Confined spaces such as underground cities, subways, buildings, and tunnels pose the maximum risk to first responders during urban search and rescue missions. Their advanced equipment and materials have the following objectives:

1. reduce the risk to personnel by using equipment instead of human for performing critical tasks;
2. perform tasks that humans can not execute; and
3. support personnel for rapid and sure execution of the task.

The responders will use robots and related technologies as advanced equipment to achieve these objectives.

Tomoaki Yoshida and Eiji Koyanagi
Chiba Institute of Technology

Satoshi Tadokoro, Kazuya Yoshida, Kazunori Ohno, and Keiji Nagatani
Tohoku University

Shoichi Maeyama
Okayama University

Hidehisa Akiyama
National Institute of Advanced Industrial Science and Technology

Yasushi Hada
National Institute of Information and Communications Technology

A. Howard et al. (Eds.): Field and Service Robotics 7, STAR 62, pp. 365–375.
springerlink.com   © Springer-Verlag Berlin Heidelberg 2010

The Ministry of Economy, Trade and Industry of Japan (METI) has investigated important issues that are required to be resolved in order to strongly promote robot applications, and they have designed a roadmap for the same.

In order to promote the development of disaster response robots, METI and New Energy and Industrial Technology Development Organization (NEDO) have set up "Project for Strategic Development of Advanced Robotics Elemental Technologies, Area of Special Environment Robots, RT System to Travel within Disaster-affected Buildings." The mission statement is as follows:

Fig. 1 Field experiment conducted in underground mall "Santica"

1. Gather information rapidly at the first stage of the disaster
2. Increase efficiency and accuracy of response by quick and distributed sensing.
3. Use RT (robot technology) in order to eliminate the risk of possible secondary disaster to human responders.

To meet the above demands, we have launched an industry-government-academia research project in collaboration with five universities, two national institutes, and three companies. The objective of the project is to develop an RT system for use in search and rescue missions; it consists of (1) highly maneuverable multiple robots, (2) a scalable communication system for long distance teleoperation of robots, (3) an intelligent remote control system for the robots used for assisting human operators, and (4) a 3-D mapping technology in no GPS environment and an environmental information management system for locating victims and aid rescue crews strategically. Disaster areas such as underground malls may be contaminated with nuclear, biological, or chemical weapons due to which they might be very dangerous for human responders during the initial stage of investigation. We have been developing the above RT system since 2005, which consist of multiple tracked vehicles, and have conducted a field experiment in an actual underground mall "Santica" located in Kobe, Japan. Fig.1 shows *Kenaf* moving toward simulated victims during the field experiment.

In this paper, we have described the RT system in brief. Then we have reported the results of the field experiment conducted to validate our RT system's usability and limitations, and identify requirements for future developments.

2 Fundamental Functions

RT systems used in search and rescue missions are required to consist of highly maneuverable robots, teleoperating system, a positioning system, and a

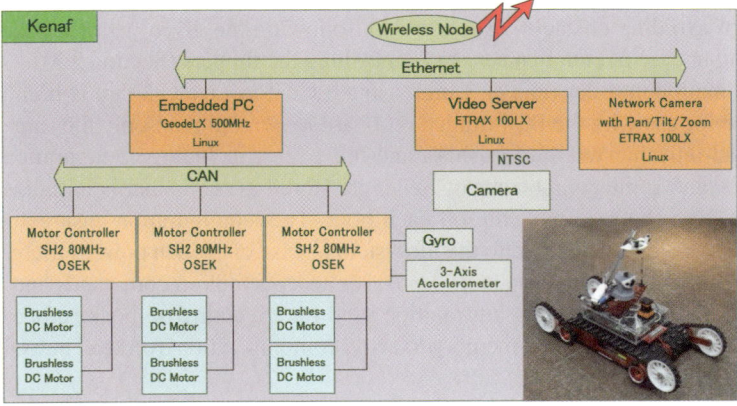

Fig. 2 Architecture of controller components

communication system. In order to integrate these systems and evaluate their performances, we have developed 10 tracked vehicles that serve as a research platform named "*Kenaf.*" The fundamental mechanism and functions of *Kenaf* are introduced in the following sections.

2.1 Highly Maneuverable Mobile Robots

Kenaf has a pair of full-body main tracks, and two pairs of sub-tracks (flippers) whose end pulley is larger than the hub pulley. Each flipper can change its orientation. It has a simple and tough ladder frame structure. The resulting total weight of the basic configuration is 20 [kg]. Heavy components such as batteries and motors are placed in the lower position of the robots to maintain a low center of gravity. This, in turn, ensures that *Kenaf* with the basic configuration, does not fall over until its roll angle exceeds 80[°], in theory. The main tracks are driven by a 90[W] brushless DC motor with a dedicated dual channel motor controller. The orientation axis of each flipper is driven by a 50[W] motor. The maximum running speed is approximately 0.8 [m/s] when a standard gear reduction ratio is employed and it increases to 2.5 [m/s] with a high speed configuration of gear reduction ratio.

2.2 Control Architecture

Kenaf has three Renesas SH2 embedded controllers as the motor controller and an AMD Geode-based low-power-consumption board computer as the main controller(Fig.2). Each motor controller is responsible for controlling the speed of the two motors. The controller used for the motor driving the main track employs 3D odometry (described in Section 2.4) and controls the trajectory of the robot so that it follows a given target line. The main controller coordinates with all the motor controllers by communicating over CAN. It has a certain degree of autonomy in terms of avoiding obstacles (Section 3.1), following the given path (Section 3.2), stopping in case of emergencies, and controlling the flippers (Section 3.3).

The main controller runs on Linux using a Gentoo Live CD that is highly customized for use with the PC installed on-board *Kenaf*. All read-only files are stored in a read-only filesystem (squashfs), and other files, including configuration files, log files, and some components of *Kenaf* are stored in tmpfs that is initialized and created from the read-only file on each boot. This configuration ensures that no permanent damage is caused to the filesystem in case of sudden power failure. A locomotion controller named *kenafLocoServer* accepts motion commands and status queries via CORBA. Status information used for executing on-board processes is also available on the shared memory to avoid communication overhead in CORBA.

2.3 Basic Operator Interface

As a baseline remote control function, we have developed a basic operator interface that can be used to control each 6DOF motion of the robot using a simple game pad and it can be used to monitor parameters such as battery voltage, pressure, and temperature. The operator console and *Kenaf* communicate over an IP network, which can be unreliable. To avoid a critical situation wherein the operator can not transmit a stop command to *Kenaf*, the operator console communicates with a dedicated remote control server via UDP instead of CORBA

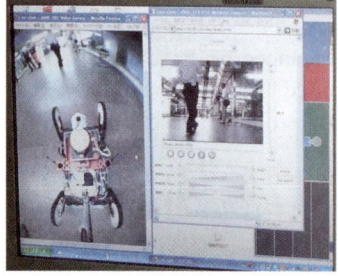

Fig. 3 Operator console with basic configuration

IIOP. If communication is not restored within a certain time period, the remote control server will disables all the actuators on *Kenaf* for safety.

The operator makes decisions on the basis of *Kenaf*'s camera view and its tilt status. Even though some various camera configurations are available, the primary camera configuration employed for basic teleoperation consist of a wide-view-angle (134[°] vertical, 103[°] horizontal) look down camera placed on top of a pole(Fig.1). The operator can perceive not only the surroundings but also the condition of *Kenaf* using this camera. Fig.3 shows images observed at the basic operator interface using the look down camera, and front camera.

2.4 3-D Odometry Using 3DOF Gyroscope

Generally, realizing an odometry system with tracked vehicles is difficult, because the turning motion of the vehicle generates positioning errors. Moreover, eliminating the positioning errors in our system is very important to transmit the position of the robot to the operator and to map the target environment. Therefore, we have proposed a novel odometry method for the position estimation of tracked vehicles with gyro sensors used in a 2-D environment tracking into account the slip characteristics of the tracked vehicle. The method is described in [7] in detail.

2-D odometry can not be used to provide accurate 3-D position information of our robots, and the temperature drift of the gyroscope is also a serious problem. Therefore, we have extended the above method for use in 3-D environments and have appended a drift cancellation function [8]. A preliminary experiment was conducted in an environment consist of standard stairs, and 60[cm] errors on an average were detected during 25[m] up-down navigation of the stairs. Other results were described in [8] in detail. These results are reasonable for our application, and this method was successfully implemented on all our robots.

2.5 Communication Network

Remote control of the communication system in an RT system is one of the major challenges. In Japan, the antenna power of wireless LANs is limited to 10 [mW]; therefore, its coverage area is in the range of 50 to 100 [m]. Moreover, increased traffic causes network congestion in communication infrastructure such as cellular networks at the time of disasters. Using cables or wireless mesh networks are inadequate in such situations. Cables ex-

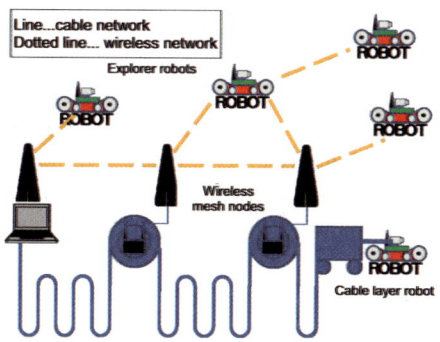

Fig. 4 Hybrid mesh network system

hibit better performance in terms of bandwidth and latency as compared to wireless networks, but they hinder robot motion owing to their weight and tendency to coil up.

Therefore, we have designed and developed a hybrid mesh network system consisting of a cable network and a wireless mesh network (Fig.4). The traffic between the wireless mesh nodes of this system is controlled using the Rokko Mesh Router designed by Thinktube Inc., which also serves as a 50 [m] network cable reel. The physical layer of the network complies with IEEE802.11g in the case of the wireless network and 100base-TX in the case of the cable network. The mesh network is based on the AODV routing protocol. A cable deployment robot is used to deploy the cables and wireless mesh nodes every 50 [m]. The other robots are then connected to the operator via the hybrid mesh network.

3 Operator Assistance Functions

3.1 Obstacle Avoidance Function

Using our RT system, operators manually control the robots on the basis of visual sensor data (described in Section 2.3). However, in situations where some evacuees obstruct the path of the robots, an autonomous obstacle avoidance function can prove advantageous for reducing the operator's work load. Therefore, we have appended a simple obstacle avoidance function on *Kenaf* which uses the sensor data acquired by a laser range finder (Top-URG UTM-30LX, Hokuyo Corp.). An important feature of the function is that it generates a path such that a certain distance is maintained from the moving obstacles for safety.

Fig.5 shows the actual momentary sensor data and the path generated in the case of a human standing in front of the robot at a distance of 4 [m]. First, the robot obtains 2-D range information (red dots in the figure). Then, possible paths for the robot are generated (green segments in the figure), and, finally, taking into account the boundary of the obstacles, the pink segment is selected as the robot's path. The autonomous obstacle avoidance function is run on a realtime basis by repeating the above procedure every 1/10 [s].

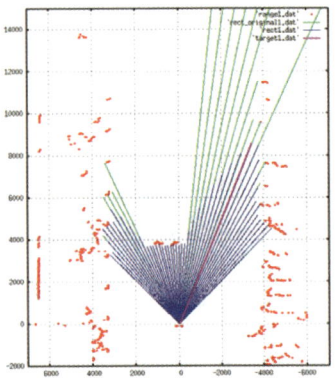

Fig. 5 Sensor data obtained by laser range finder and generated path

This obstacle avoidance function can anytime be replaced with conventional manual control.

3.2 Pointing Navigation Function

Teleoperating a mobile robot over long distances is a tough and tedious task for human operators. We have developed an operator interface for operating robots in flat and large areas, which reduces interaction between the robot and the operator. The interface accepts a target path expressed in terms of a sequence of waypoints, and the robot moves along the path autonomously. Thus, even in the case of long communication latency between the operator console and the robot, robot motion does not get affected.

We have setup two view modes for the operator. One is the local map mode that shows the sensor data obtained from a horizontal laser range finder(Fig.6, right), and the other is the camera video mode that shows

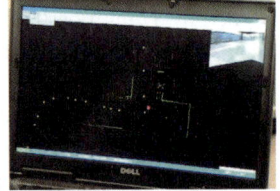

Fig. 6 Left: Robot's camera view. Right: Bird's-eye view.

images that are obtained from the on-board camera and that are superimposed by the sensor data(Fig.6, left). The operator can switch between these modes anytime during operation. Two cameras focused diagonally forward from left to right are used in addition to the standard look down camera for wide field of view. The operator clicks on these views to specify the waypoints.

The operator can interrupt the autonomous motion of the robot and switch to the manual control mode anytime during operation. Further, when obstacles are detected along the path, the operator can switch to the obstacle avoidance mode, which is described in the previous section.

3.3 Autonomous Flipper Control System for Operator'S Assistance

Flippers (sub-tracks) greatly assist robots in traversing large steps and rough terrains. However, it is challenging for an operator, particularly for one who does not possess necessary skills, to control such flippers remotely without a direct view of the actual environment.

To assist the operation of the tracked vehicle "Kenaf," we have been developing two autonomous flipper control systems based on different approaches. The common strategy of both the systems is to control each flipper angle on the basis of the sensor data to traverse bumps on the ground. The operator is required to indicate the direction to the robot for navigation.

One approach is on the basis of the contact detection of flippers to the ground, and the gap detection under main tracks. The contact of flippers is detected by measuring each flipper's motor torque, and the gaps under main tracks are obtained from PSD range sensors attached to the front and rear of *Kenaf*. Details are described in [1].

The other approach is to use two laser range finders to obtain the terrain shape information [6]. The two laser range finders are located on both side of *Kenaf*. Their sensing surfaces are perpendicular to the ground so that the ground shape in the vicinity of the two front flippers can be perceived. Fig.7 shows the locations of the sensors used to obtain the terrain information.

Both systems have been successfully implemented on *Kenaf*, and some preliminary experimental results have validated the usefulness of both systems. Fig.8 shows *Kenaf* traversing steps using the former autonomous flipper control system.

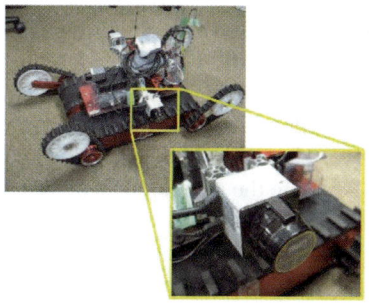

Fig. 7 Location of laser range sensors

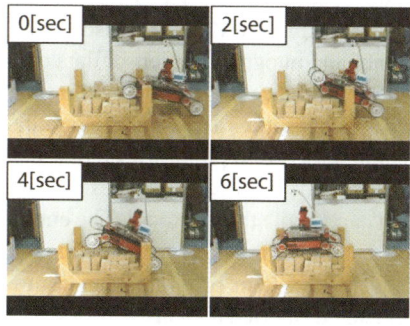

Fig. 8 Traversing random step fields under autonomous flipper control

4 Environmental Information Gathering Functions

4.1 3-D Mapping Using Laser Range Finder

3-D maps are very useful to rescue crews for strategizing. One of the features of a 3-D map is that it
can be viewed in multiple modes, such as a bird's-
eye view, which is a big advantage for strategizing
rescue plans.

The simplest method to obtain a 3-D map using
a robot is to fix a laser range finder on the upper
position of the robot. It directed upward to obtain
distance information leftward, upward, and right-
ward of its body. 3-D information is then obtained
by moving the robot in the forward or backward direction. This is a very simple and effective method
to obtain 3-D information. However, the quality of
the map depends largely on the accuracy of the estimated position of the robot.

Fig. 9 3-D scanner named TK-
scanner.

To obtain detailed 3-D environmental information, we have developed another
small-sized, wide-view, and lightweight 3-D scanner named TK-scanner [2] (Fig.9)
using a 2-D laser range finder and a pan-tilt mechanical base. The TK-scanner spun
the tilted 2-D laser range finder to obtain a set of 3-D information in 10 [s]. To
obtain a consistent 3-D information, the robot must keep still while TK-scanner is
scanning.

4.2 Geographic Information System

Information sharing is the most fundamental and important issue in managing rescue
operations in case of disasters. Because mobile rescue robots and devices provide
only fragments of information, we need a database system to store and integrate
them. Further, because this information is location and time-sensitive, the database
system should be similar to a geographic information system(GIS).

We used DaRuMa (DAtabase for Rescue Utility MAnagement) [5] as the GIS to
gather and integrate the sensor data obtained from our robots. DaRuMa is one of the
MISP (Mitigation Information Sharing Protocol) [3, 4] server implementation systems, and it serves as a database middleware. MISP provides functions to access and
maintain a geographic information database over networks. The protocol consists
of pure and simple XML representations; this facilitate the development of systems
that can handle this protocol. The entire system based on MISP forms a client-server
system wherein the server is the database and clients are data providers and/or data
requesters. Client programs can communicate with DaRuMa using MISP, and all
registered/queried data are transferred to/from a SQL server through DaRuMa. Our
robots can directly transmit their sensor data to DaRuMa through the hybrid mesh
networks.

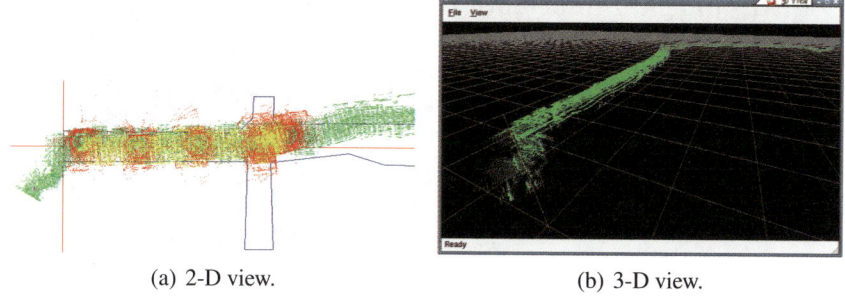

(a) 2-D view.　　　　　　　　　　　　(b) 3-D view.

Fig. 10 Integrated sensor data of laser range finders installed on several robots

5　System Integration and Field Experiment

The experiment was conducted on 11/6/2008 at 2:30 AM in "Santica" underground mall located in Kobe, Japan, to validate the integrated system in a real environment.

The starting position of three of the *Kenaf* robots was set close to the operator station in front of several blocks of random step field which simulate uneven terrain. The fourth robot was placed at a distance of 550 [m] from the station. All the four *Kenaf* robots were teleoperated from the operator station.

Because we only had a limited time to use the site, the network infrastructure for this experiment was set up by human in advance, instead of using cable deployment robot. It was 12 wireless nodes connected by 50 [m] LAN cables in serial.

The first robot with the basic configuration was teleoperated to travel as much distance as possible. The second robot employs intelligent operator assistance functions, and it was used to explore the environment in the vicinity of simulated victims. This robot was normally teleoperated using the pointing navigation interface, and the operator switched to the obstacle avoidance mode or the manual control mode as and when required. All the other robots were teleoperated using the basic operator interface with the video stream obtained from the look down camera installed on each robot. To verify the reliability of the operator assistance functions, a group of people was used to simulate evacuees in the real environment.

The third robot which was equipped with TK-Scanner (Fig.9), also was used to explore and obtain detailed 3-D information regarding the environment in the vicinity of the simulated victims simulated victims. Fig.11 shows a resulting 3-D map obtained using TK-Scanner. All the 3-D data measured by each *Kenaf* robot was input to the DaRuMa server.

The role of the fourth robot was to monitor the first robot, and to add more traffic load on the network for network performance test. The reliability of the network system was verified monitoring the network traffic from these four robots.

In general, the experimental results indicate that the integrated system shows good performance. The first *Kenaf* robot traversed a corridor and reached a dead end at a distance of 683 [m] from the starting point. The hybrid network system successfully provided coverage even when each robot used 4 [Mbps] of bandwidth

Fig. 11 The *Kenaf* with TK-Scanner in action and a resulting 3-D map.

for transmitting the video stream and control signal. An automatic cable deployment was one of the key technology to succeed in our scenario, but it was not used in this experiment. We will apply our developing cable deployment robot in our future field experiments for setting up the network infrastructure.

The GIS system worked successfully in the case of the four robots. The robots were operated for approximately 1 [h], and the GIS system could register and integrate more than five million data points during this time. Fig.10-(a) shows the example point data around starting point registered by two robots. Fig.10-(b) also shows 3-D viewer application, called DaRuMa Viewer, viewing all data points which was obtained by a fixed laser range finder on the first robot. Anyone can explore and interact collected data in the DaRuMa server with this viewer even when the operators operate robots at the same time.

On the basis of the experimental results, we have found that the current version of the DaRuMa server does not scale well because the data entry process could handle only one registration request at once. The function used for registering and handling the data is required to be modified in the next version. All the information is registered on a coordinate frame designed for each robot. The relationship between each coordinate frame was configured in advance on the basis of the initial position of each robot. However, the accumulated positioning error of each robot caused distortion in the resulting 3-D map. Since each map from each robot are generated independently from other map from other robot, there were some inconsistencies in resulting united 3D map. We propose to use the SLAM technique to refine the resulting map in our future experiment.

The intelligent operator assistance functions were very helpful in a particular situation. In case the simulated evacuees moved toward *Kenaf* in walking speed, the obstacle avoidance function successfully avoided the collision of the robot with the evacuees. Nevertheless, when evacuees moved toward *Kenaf* in running speed, it failed in avoiding collisions. In reality, a more intelligent function may be required. The pointing navigation function worked well when there were only a few obstacles in front of the robot. However, in other cases, manual navigation was effective in controlling the robot. This was because the paths could not always be determined in advance. In case of moving or large number of obstacles, the operator has to determine paths using imprecise, and incomplete information, which is not easy or even possible. This problem might be solved if an advanced obstacle avoidance function is integrated into the pointing navigation function. The two flipper control systems were successfully integrated in two of the *Kenaf* robots. Both systems drastically decreased operator interactions in the presence of a stairs and random step fields.

6 Conclusions

In this paper, we have presented the results of a field experiment conducted using a remote controlled multiple mobile robot system as an advanced tool for assisting first responders during an urban search and rescue mission. Four mobile robots were simultaneously successfully operated using a realtime video stream of the environment. Data such as 3-D information of the environment, location of the victims, and trajectory of each robot was input to GIS DB server, and this data could be used to design a unified map. Further, we are attempting to enhance the hybrid network system so that it can operate 10 robots by using IEEE 802.11n.

References

1. Ohno, K., Morimura, S., Tadokoro, S., Koyanagi, E., Yoshida, T.: Semi-autonomous Control System of Rescue Crawler Robot Having Flippers for Getting Over Unknown-Steps. In: Proc. of IEEE/RSJ International Conference on Intelligent Robots and Systems, pp. 3012–3018 (2007)
2. Ohno, K., Kawahara, T., Tadokoro, S.: Development of 3-D Laser Scanner for Measuring Uniform and Dense 3-D Shapes of Static Objects in Dynamic Environment. In: Proc. of the 2008 IEEE International Conference on Robotics and Biomimetics (2008)
3. NIED, AIST: Mitigation Information Sharing Protocol. rev.1.00.028s edn. (September 2006)
4. Noda, I., Hada, Y., Meguro, J., Shimora, H.: Information Sharing and Integration Among Rescue Robots and Information Systems. In: Tadokoro, S., Matsuno, F., Asama, H., Osuka, K., Onosato, M. (eds.) Proc. of IROS2007 Full-Day Workshop MW-3 (Rescue Robotics), IROS, October 2007, pp. 125–139 (2007)
5. DaRuMa: `http://sourceforge.jp/projects/daruma/`
6. Nagatani, K., Yamasaki, A., Yoshida, K., Yoshida, T., Eiji, K.: Semi-autonomous Traversal on Uneven Terrain for a Tracked Vehicle using Autonomous Control of Active Flippers. In: Proc. of IEEE/RSJ International Conference on Intellegent Robots and Systems, pp. 2667–2672 (2008)
7. Endo, D., Okada, Y., Nagatani, K., Yoshida, K.: Path Following Control for Tracked Vehicles based on Slip-compensating Odometry. In: Proceedings of the 2007 IEEE/RSJ International Conference on Intelligent Robots and Systems, pp. 2871–2876 (2007)
8. Nagatani, K., Tokunaga, T., Okada, Y., Yoshida, K.: Continuous Acquisition of Three-dimensional Environment Information for Tracked Vehicles on Uneven Terrain. In: Proceedings of the 2008 IEEE International Workshop on Safety, Security and Rescue Robotics, pp. 25–30 (2008)

Learning to Identify Users and Predict Their Destination in a Robotic Guidance Application

Xavier Perrin, Francis Colas, Cédric Pradalier, and Roland Siegwart

Abstract. User guidance systems are relevant to various applications of the service robotics field, among which: smart GPS navigator, robotic guides for museum or shopping malls or robotic wheel chairs for disabled persons. Such a system aims at helping its user to reach its destination in a fairly complex environment. If we assume the system is used in a fixed environment by multiple users for multiple navigation task over the course of days or weeks, then it is possible to take advantage of the user routine: from the initial navigational choice, users can be identified and their goal can be predicted. As a result of these prediction, the guidance system can bring its user to its destination while requiring less interaction. This property is particularly relevant for assisting disabled person for whom interaction is a long and complex task. In this paper, we implement a user guidance system using a dynamic Bayesian model and a topological representation of the environment. This model is evaluated with respect to the quality of its action prediction in a scenario involving 4 human users, and it is shown that in addition to the user identity, the goals and actions of the user are accurately predicted.

1 Introduction

Robots are more and more present in the daily life, not only in the industry but also at home as toys or as service robots such as vacuum cleaners. There is also a growing demand in the health-care domain for smart assistive device such as intelligent wheelchairs. The present paper is focused on an intelligent system designed to help the elderly or disabled people in their daily activities. For these people, moving in their houses or passing through doorways may represent challenging tasks. We developed a semi-autonomous robot for improving user mobility while minimizing the required input, i.e. having an interaction process adapted to low throughput devices such as single switches, sip and puff systems, brain machine interfaces, or simple voice recognition. More precisely, at each crossing, the robot proposes a direction of

Xavier Perrin, Francis Colas, Cédric Pradalier, and Roland Siegwart
ETHZ, Zurich, Switzerland
e-mail: xavier.perrin@mavt.ethz.ch

A. Howard et al. (Eds.): Field and Service Robotics 7, STAR 62, pp. 377–387.
springerlink.com   © Springer-Verlag Berlin Heidelberg 2010

travel to the user who will then either agree or disagree. The better the propositions are, the faster and easier the human-robot interaction is.

In this work, a dynamic Bayesian network (DBN) is used in order to learn the habits of multiple human users of a robotic helper. By *habits*, we mean the *succession of navigational tasks* one executes in a known environment: in a retirement home for instance, one resident wakes up in his bedroom, goes to the bathroom, then to the common room where the breakfast is served, and then goes to other rooms for his daily activities. Another resident will have other preferred locations for his own activities. All these accumulated additional information allow the robotic assistant to help its current human user from the first movements of a new travel until the destination. From the system's point of view, the actual identity of its user is not a relevant information. A user is merely defined by his activity pattern.

In the next section, we describe related works in recognition techniques for goal, user, or activity. In section 3, we detail the developed DBN, represented as a graphical model. Section 4 presents experiments in simulation as well as with a real robot and their results. The final discussion appears in section 5.

2 Related Works

Our aim is to ease the navigation of users, therefore we will consider activity recognition only from the point of view of navigation. In this case, inferring the user's intention requires techniques for plan recognition, which are used in a broad variety of domains, such as motion prediction, speech understanding, video surveillance, and so on. Uncertainty is inherent in plan inference, as the robot does not know in advance the intended destination of a user. Furthermore, many ways can lead to the same destination while one way can lead to several places. Probabilistic reasoning techniques are used in almost all works (review in [1]), as they help to express and maintain the beliefs in the possible goal destinations.

In the particular domain of *intention recognition* for navigation, two aspects were studied recently: local intention recognition (immediate action or location in the vicinity of the wheelchair from uncertain input) and global intention recognition (goal destination from local decision). Our work focuses on this latter issue, assuming that the recognition of the immediate intention is solved. In our test, this will be achieved by an interaction device with low uncertainty, such as a joystick or a reliable speech recognition. Inferring a global intention requires the ability to localize the robot on an available map of the environment. In a discrete environment represented by adjacent cells, Verma and Rao [12] described a Dynamic Bayesian Network (DBN) in the form of a graphical model composed of a Partially Observable Markov Decision Process (POMDP) enhanced with the notion of goal locations. Relying on reinforcement learning techniques, the agent explores how it should behave in order to reach three possible goals. Based on this acquired knowledge, the system could infer the most probable goal from a set of possibly noisy observations. Taha et al. [10] achieved similar results by fusing the robot position and the possible goals as the new state definition in a common POMDP. Vasquez [11] described a growing hidden Markov model algorithm dedicated to continuous learning, clustering, and making inference about car motions in a parking lot or people motions in a hallway.

Topological maps were successfully used in the works of Taha [10]. This environment representation is compact and matches the human's natural description of a path better than metric representations [7] (e.g. "go on the left at the second crossing" instead of "go straight for 100 meters, then turn left"). Many techniques exist for the map construction, e.g. the generalized Voronoi diagram, and applications based on imprecise human drawings have also been reported [9].

User models are often used in computer applications (e.g. intelligent help) or online search or e-commerce sites (e.g. book recommendation based on other customer's choices) [8]. Based on databases containing information from numerous users and some observations from the current user activity, reasoning techniques infer the next user action and try to help him. With the aging population, some research domains are focused on the *activity recognition* in so-called smart environments, i.e. environments where sensors have been installed in order to monitor the activity of human beings. The accumulated data are used to train algorithms which later serves to determine human activities, supervise the user's condition and medication, or detect anomalies [2, 4, 6, 13]. Some researches try to further determine the attributes (coffee drinker, smoker) of multiple users based on their location [3, 5].

As a summary, the studies on activity recognition perform well in capturing the user habits, but they can only monitor ongoing activities, not help the user to perform a task. On the other hand, the studies on intention recognition for facilitating the user's motion do not try to learn the typical user's daily habits. In this article, we propose a system that learns the habits of multiple human users controlling a robot and exploit this knowledge for the navigational control of the robot. In case of an unknown user, the same system first infers the identity of the user from the first movement in the environment before being able to exploit this knowledge for helping its user specifically. The robotic system is given a topological representation of the environment a priori. While incrementally learning the global intentions of a known user, it anticipates the user's destination and proposes better actions, as regular patterns of actions are performed day after day by the same user.

3 Model Description

The graphical model shown in figure 1 represents the dynamic Bayesian network which composes the core of the system. Our DBN is an extension of a Markov Decision Process (MDP), with the agent's *State* S_t and the *Action* A_t (lower part). The state S_t is the observed pose of the robot, in terms of position in a given topological map (a node) and orientation. [1] The action A_t is among the repertoire: forward, left, right, u-turn, and stop. An action always connects to nodes of the map. The MDP is enhanced by the notion of *Goal state* G_t, indicating the current global goal, and the

---

[1] In relatively small and not too dynamic environments as the ones considered for this work, localization can be considered as a solved issue.

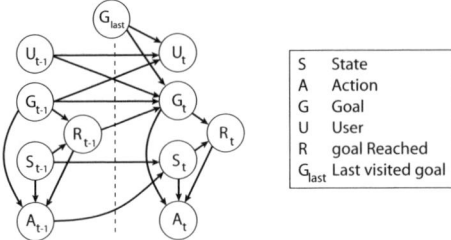

Fig. 1 Our dynamic Bayesian model for learning the habits of multiple users.

variable *Goal reached* R_t, a Boolean value indicating whether the agent's current state S_t is a goal state or not ($P(R_t|G_t\ S_t) = 1$ iff $G_t = S_t$).

The term $P(A_t|S_t\ G_t\ R_t)$ is specified separately according to whether or not the goal is reached. First, when $R_t = 0$, $P(A_t|S_t\ G_t)$ is the action model leading to each goal G_t from any state S_t as deduced from the topological map (i.e no reinforcement learning technique used). Then, when $R_t = 1$, the probability of staying still $P(A_t = stop)$ is higher than the other actions, so that the user can easily stop when it reaches a goal location. The last term of the MDP, the transition model $P(S_t|S_{t-1}\ A_{t-1})$, is also deduced from the topological map.

With the goal of learning the user habits for multiple users, we added two more variables. First, the user is symbolized with the variable U_t. Second, the variable G_{last}, representing the last visited goal, makes possible the learning of the succession of place visits, as will be shown later. Both the goal model $P(G_t|G_{t-1}\ R_{t-1}\ U_{t-1}\ G_{last})$ and the user model $P(U_t|U_{t-1}\ G_{t-1}\ G_{last})$ result from submodels, where predefined terms related to the persistence of a goal (keeping the same goal or switching to another one), resp. the persistence of a user, are combined with terms learned online representing the knowledge acquired by the system. These submodels are described in details in the appendix. When reaching a goal during an online supervised learning phase, the known user and the particular G_t and G_{last} are used for updating the histograms of the probability distributions of the goal and user models.

From the model shown in figure 1, we compute at each time step several probability distributions. After the learning phase, the user is not known any more. Nevertheless, given the known variables S_{t-1}, A_{t-1}, and G_{last}, we can infer his identity by computing $P(U_t|S_{t-1}\ A_{t-1}\ G_{last})$ using Bayes' rule and the law of total probability:

$$P(U_t|S_{t-1}\ A_{t-1}\ G_{last}) \qquad (1)$$

$$= \frac{1}{Z} \sum_{G_{t-1}} \left\{ \begin{array}{l} P(G_{t-1}) \sum_{U_{t-1}} [P(U_{t-1})\ P(U_t|U_{t-1}\ G_{t-1}\ G_{last})] \\ \times \sum_{R_{t-1}} [P(R_{t-1}|G_{t-1}\ S_{t-1})P(A_{t-1}|G_{t-1}\ S_{t-1}\ R_{t-1})] \end{array} \right\}$$

where the term $\frac{1}{Z}$ corresponds to a normalization factor. For a guidance robot, we also want to infer the goal the robot should be aiming to or the action it should propose to the unknown user. These two distributions can be computed as follows:

$$P(G_t|S_{t-1}\ A_{t-1}\ G_{last}) \tag{2}$$

$$= \frac{1}{Z}\sum_{G_{t-1}}\left\{P(G_{t-1})\sum_{R_{t-1}}\left[\begin{array}{l}P(R_{t-1}|G_{t-1}\ S_{t-1})P(A_{t-1}|G_{t-1}\ S_{t-1}\ R_{t-1}) \\ \times \sum_{U_{t-1}}\{P(U_{t-1})P(G_t|G_{t-1}\ R_{t-1}\ U_{t-1}\ G_{last})\}\end{array}\right]\right\}$$

$$P(A_t|S_{t-1}\ A_{t-1}\ G_{last}) \tag{3}$$

$$= \frac{1}{Z}\sum_{S_t}\left\{\begin{array}{l}P(S_t|A_{t-1}\ S_{t-1}) \\ \times \sum_{G_{t-1}}\left[\begin{array}{l}P(G_{t-1}) \\ \times \sum_{R_{t-1}}\left[\begin{array}{l}P(R_{t-1}|G_{t-1}\ S_{t-1})P(A_{t-1}|G_{t-1}\ S_{t-1}\ R_{t-1}) \\ \times \sum_{U_{t-1}}\left\{\begin{array}{l}P(U_{t-1}) \\ \times \sum_{G_t}\left\{\begin{array}{l}P(G_t|U_{t-1}\ G_{t-1}\ R_{t-1}\ G_{last}) \\ \times \sum_{R_t}\left[\begin{array}{l}P(R_t|G_t\ S_t) \\ \times P(A_t|G_t\ S_t\ R_t)\end{array}\right]\end{array}\right\}\end{array}\right\}\end{array}\right]\end{array}\right]\end{array}\right\}$$

If the user is known by the robot (e.g. through manual or visual identification), we can use a Dirac distribution for $P(U_{t-1})$ and recompute all the above equations with this additional information, the system giving back the learned user habits.

4 Experiments

We run experiments of our multi-user guidance robot in our laboratory environment, using a differential-drive robot (fig. 2a). Figure 3 shows that the topological decomposition of the environment is made of goal nodes and connecting nodes. The former ones represent either people's desks (D1-D6) or common rooms like the cafeteria (C), the printer room (P), the robot lab (RL), or the bathroom (B). Four users share the robot, each with a particular desk and typical sequence, all starting from the entrance E. User 1 executes the sequence D3-D4-P-C-D3-D2-D3, user 2 D4-RL-C-D3-D4-RL-D4, user 3 D2-C-D2-D3-D2-P-D2, and user 4 D6-RL-C-D4-D6-RL-D6 (fig. 4a–d). Additionally each user goes to the bathroom at a random point in his sequence. These sequences are repeated 20 times during the learning phase. In order to speed up this process, we used a simulation. However, for the tests on the 21th

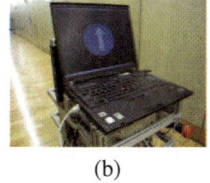

(a)   (b)

Fig. 2 (a) The differential drive robot used in our experiments. (b) Example of action proposed to the user, who then either agrees of disagrees orally (use of a speech recognition software).

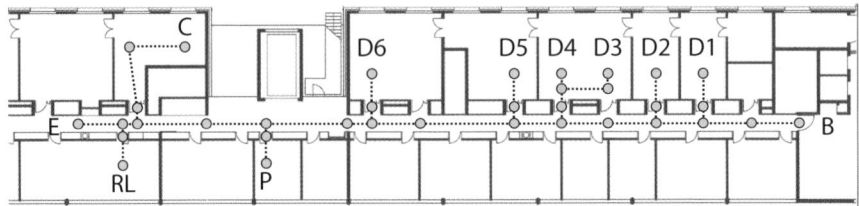

Fig. 3 Topological map of our laboratory used for the experiments with an entrance (E), desks (D1-D6), a printer room (P), a cafeteria (C), a robot lab (RL), and toilets (T).

day, a real robot was used. [2] In a second phase, the real robot was used by each of the user, the robot having no preliminary knowledge about the current user. At each time step, the system computes the probability distribution over the user, the goal, and the action with, respectively, equations 1, 2, and 3. This last distribution is used to propose an action to the user (see fig. 2b), who can confirm or not. In this precise experiment, this was achieved through a speech recognition system. [3]

Results of the user recognition are shown in figure 5, which present the evolution of the user probability distribution during a trial of the test phase. Starting from a uniform distribution, the actual user is correctly inferred as soon as a revealing action, mostly leading towards his desk, is executed. In the scenarios, the different user habits shared some common goals, or even some common sequences of goals. These similarities can be seen in figure 5, where the probability in the most probable user decreases and the one(s) for other user(s) increases (e.g. fig 5b, the sequence between goals D3 & D4 is shared by users 1 & 2). We can also notice that when a user goes to the bathroom, his inferred probability decreases slightly. As every user can go to the bathroom at any time, this is not a discriminant observation, the models tending thus slightly to a uniform due to the transition in the user submodel.

For quantifying our model, we compare the model prediction of the user, the goal, and the action with the real values. Based on their distributions computed at each time step, we can check if its maximum is the actual user, intended goal and proper action proposition. We thus introduce a measure μ_{max}, being the mean of the number of times the max of a distribution matches the real value. For example, the formula for the goal is $\mu_{max} = \frac{1}{N} \sum_{t=1}^{N} \delta(\hat{G}_t, \text{argmax}_{G_t} P(G_t|S_{t-1} A_{t-1} G_{last}))$, with N the number of steps, $\hat{G}_t$ the intended goal, and $\delta(a,b)$ the Kronecker function.

The results are displayed in table 1. As can be seen when comparing with figure 5, the user are properly recognized, the differences being explained by the amount of steps until a revealing action. Concerning the goals and the actions, their probabilities are related to the learned habits of the users and are also influenced by the geographical location of the visited goals. As an example, user 3 visited three

---

[2] In simulation, the robot moves instantaneously to the next node, the inputs to our DBN S_{t-1}, A_{t-1}, and G_{last} being identical to when using the real robot.

[3] The software performed well enough to assume a perfect recognition of the user agreement ("Yes") or disagreement ("No").

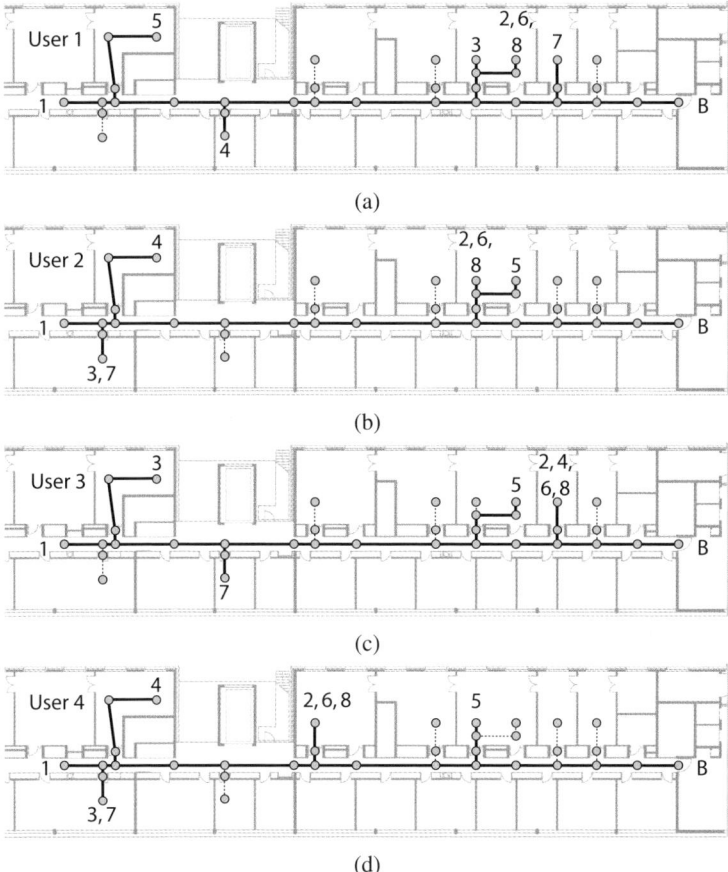

Fig. 4 Habit description of the four users. The succession of visited places is marked with increasing numbers. Each user visits the bathroom (B) at random among the sequence.

goals on the right when leaving his office, while the fourth one was on the left. Accordingly, the probability of going right, resp. of the three right goals were higher than going left to the bathroom. Furthermore, we can notice that the action inference is much better than the goal inference. Indeed, most of the time, the goal confusion is between two nearby goals that share a part of the sequence. As a consequence, the goal could be erroneously inferred during the first steps of the travel, while the action inference will be wrong only at a branching node.

Finally, we want to assess the robustness of our system. To this end, we run scenarios where two users took over the robot when user 2 reaches the cafeteria (C). First, the results with user 3 are displayed in figure 6a. As can be seen, while leaving C, the robot thought user 2 was going to D3, but also increased the probability in user 1, 3, and 4 as they all go in that direction after being in C. At each revealing action, the probability in specific users changes. But when reaching D2, user 3 starts

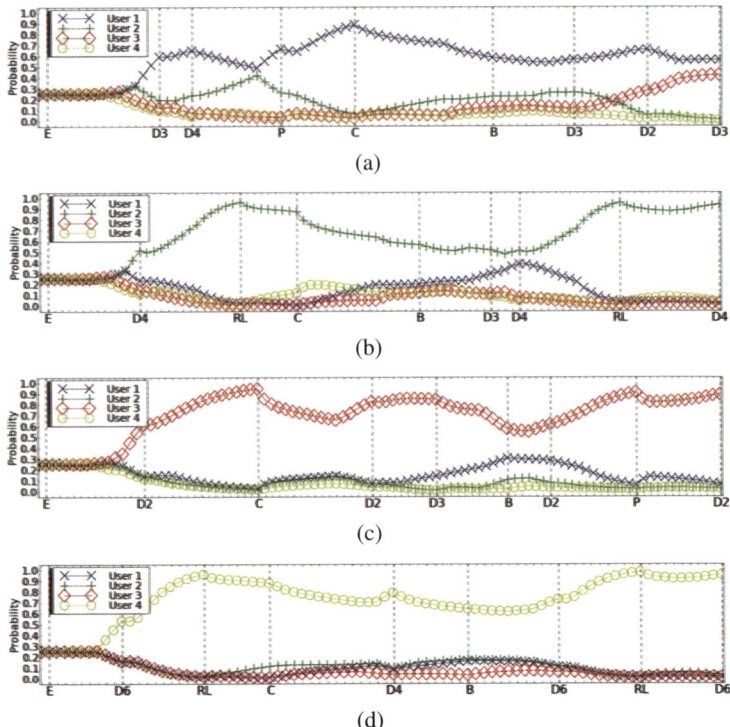

Fig. 5 User probability when (a) user 1, (b) user 2, (c) user 3, (d) user 4 was steering the robot. The dotted lines indicate the steps where a user stopped at a goal.

Table 1 Model performance for the user, goal, and action predictions of each of the four single user scenarios (left columns) and the two taking-over scenarios (right columns).

	Standard usage				Robot taken from user 2		Random
μ_{max}	User 1	User 2	User 3	User 4	User 3	User 4	
User	0.89	0.91	0.94	0.96	0.76	0.56	0.25
Goal	0.64	0.63	0.57	0.84	0.54	0.52	0.09
Action	0.94	0.93	0.94	0.96	0.89	0.92	0.20

being the most probable one. The increasing probability of user 1 is explained by the shared common goals with user 3, and starts to decline as soon as user 3 comes back to D2. In figure 6b, the results when user 4 takes the robot from user 2 are displayed. As user 4 first goes to the desk of user 2, this sequence is not revealing enough because only the former action differs from the habit of user 2. Then, going to the toilets does not bring any information about the user's identity. Finally, when user 4 reaches his desk, the remaining sequences are specific enough for the system to infer its user correctly. Overall, despite a performance loss in both goal and user recognition, table 1 shows that the actions proposed to the user are still relevant.

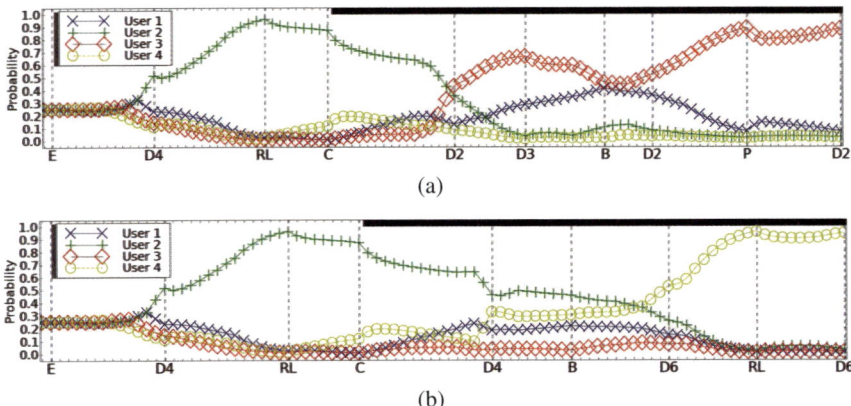

Fig. 6 Scenarios where (a) user 3 and (b) user 4 take the robot from user 2 in the cafeteria. The user probability is displayed. The dotted lines indicate the steps where a user stops at a goal. The black region indicates when a new user is steering the robot.

5   Conclusion

In this paper, we introduced a probabilistic model able to learn the daily habits of different users of a guidance robot. When the robot accumulated enough knowledge about a particular user, it is able to actively help its user to reach a goal destination. As shown in the experiments starting with an unknown user, the same model is able to infer the identity of the user after just a few revealing actions have been made, or specific goals have been reached. Having then a high confidence in a particular user, the system is again able to actively help its user. Furthermore, experiments where a different user takes over the guidance robot from an initial user have shown the ability of the system to recognize the switch between the users.

We tested our system with a reasonable environment size and user number, but it would be interesting to test its robustness to an increased user number or to complexer user patterns. If leading to pattern overlaps, the system should still be able to help the users as they would go to the same location. However, the extreme case would be if there is a uniform distribution on the goals when leaving a particular one (e.g. the bathroom). For these cases, discretizing the user sequences over different periods of the day or increasing the history length should help.

Acknowledgements. This work has been supported by the Swiss National Science Foundation NCCR 'IM2' and by the EC-contract number BACS FP6-IST-027140. This paper only reflects the authors' view and funding agencies are not liable for any use that may be made of the information contained herein.

Appendix: Submodels for the Goal and User Models

In this appendix, we first describe the submodel used for building the goal model in the main system, then the submodel for the user model. In the goal model $P(G_t|G_{t-1} R_{t-1} U_{t-1} G_{last})$, the goal at time t is dependent from a lot of variables issued from the previous time step. Some relations, e.g. between G_t and G_{last}, are user-dependent and some others, e.g. between G_t and R_t, can be predefined. A submodel is a Bayesian model on his own, with specific relation among the variables, which is then used in order to infer the probability distribution over some variables given some other known variables. In the case of the goal model, our submodel is defined as follows:

$$P(G_t\, G_{t-1}\, R_{t-1}\, U_{t-1}\, G_{last}) = P(G_{t-1}\, R_{t-1})\, P(G_t|G_{t-1}\, R_{t-1})\, P(U_{t-1}\, G_{last}|G_t)$$

We have $P(G_{t-1}\, R_{t-1})$ which is a uniform distribution. Then, $P(G_t|G_{t-1}\, R_{t-1})$ defines the persistence of a goal: if a goal is not reached, the probability of having the same goal in mind is much higher than changing, whereas when the user reaches a goal, the other places become equally probable with a preference of staying in the current place. This term is predefined. Finally, the $P(U_{t-1}\, G_{last}|G_t)$ distribution is derived from a histogram learned by the system when a user stops at a goal location. From this model, we compute the probability distribution $P(G_t|G_{t-1}\, R_{t-1}\, U_{t-1}\, G_{last})$, which is then used in the main model as the goal model, as follows:

$$P(G_t|G_{t-1}\, R_{t-1}\, U_{t-1}\, G_{last}) \propto P(G_t|G_{t-1}\, R_{t-1})P(U_{t-1}\, G_{last}|G_t)$$

The user model is build in a similar manner than the goal model and is the following:

$$P(U_t\, U_{t-1}\, G_{t-1}\, G_{last}) = P(U_{t-1})\, P(U_t|U_{t-1})\, P(G_{t-1}\, G_{last}|U_t)$$

Again, $P(U_{t-1})$ is a uniform distribution. $P(U_t|U_{t-1})$ describes the persistence of the human user, i.e. keeping the same user is more probable than switching between users. $P(G_{t-1}\, G_{last}|U_t)$ is also derived from a histogram, which is also learned when a user reaches a goal location. The probability distribution computed in order to build the user model is $P(U_t|U_{t-1}\, G_{t-1}\, G_{last})$:

$$P(U_t|U_{t-1}\, G_{t-1}\, G_{last}) \propto P(U_t|U_{t-1})P(G_{t-1}\, G_{last}|U_t)$$

References

1. Carberry, S.: Techniques for Plan Recognition. User Modeling and User-Adapted Interaction 11, 31–48 (2001)
2. Hristova, A., Bernardos, A.M., Casar, J.R.: Context-aware services for ambient assisted living: A case-study. In: ISABEL2008, vol. 1, pp. 1–5 (2008)
3. Liao, L., Fox, D., Kautz, H.: Location-Based Activity Recognition using Relational Markov Networks. In: IJCAI 2005, pp. 773–778 (2005)

4. Lu, C.-H., Ho, Y.-C., Fu, L.-C.: Creating Robust Activity Maps Using Wireless Sensor Network in a Smart Home. In: Conference on Automation Science and Engineering, pp. 741–746 (2007)
5. Matsuo, Y., Okazaki, N., Izumi, K., Nakamura, Y., Nishimura, T., Hasida, K., Nakashima, H.: Inferring Long-term User Property based on Users' Location History. In: IJCAI 2007, pp. 2159–2165 (2007)
6. Mori, T., Urushibata, R., Shimosaka, M., Noguchi, H., Sato, T.: Anomaly detection algorithm based on life pattern extraction from accumulated pyroelectric sensor data. In: Proceedings of the IEEE IROS, pp. 2545–2552 (2008)
7. Rawlinson, D., Jarvis, R.: Ways to Tell Robots Where to Go - Directing Autonomous Robots Using Topological Instructions. IEEE Robotics & Automation Magazine 15(2), 27–36 (2008)
8. Schickel-Zuber, V., Faltings, B.: Inferring User's Preferences using Ontologies. In: AAAI 2006, vol. 1, pp. 1413–1418 (2006)
9. Setalaphruk, V., Ueno, A., Kume, I., Kono, Y., Kidode, M.: Robot navigation in corridor environments using a sketch floor map. In: International Symposium on Computational Intelligence in Robotics and Automation, vol. 2, pp. 552–557 (2003)
10. Taha, T., Miró, J.V., Dissanayake, G.: POMDP-based Long-term User Intention Prediction for Wheelchair Navigation. In: Proceedings of the IEEE ICRA, pp. 3920–3925 (2008)
11. Vasquez Govea, A.D.: Incremental Learning for Motion Prediction of Pedestrians and Vehicles. PhD thesis, Institut National Polytechnique de Grenoble (2007)
12. Verma, D., Rao, R.: Goal-Based Imitation as Probabilistic Inference over Graphical Models. In: Weiss, Y., Schölkopf, B., Platt, J. (eds.) Advances in NIPS 18, pp. 1393–1400. MIT Press, Cambridge (2006)
13. Zheng, H., Wang, H., Black, N.: Human Activity Detection in Smart Home Environment with Self-Adaptive Neural Networks. In: International Conference on Networking, Sensing and Control (ICNSC 2008), pp. 1505–1510 (2008)

Long Term Learning and Online Robot Behavior Adaptation for Individuals with Physical and Cognitive Impairments

Adriana Tapus, Cristian Tapus, and Maja Matarić

Abstract. In this paper, we present an online adaptation approach and a long-term learning approach for socially assistive robotic (SAR) systems that aim to provide customized help protocols through motivation, encouragements, and companionship to users suffering from physical and/or cognitive changes related to stroke, aging and Alzheimer's disease.

1 Introduction

A recent trend in robotics is to develop a new generation of robots that are capable of moving and acting in human-centered environments, interacting with people, and participating in our daily lives. This has introduced the need for developing robotic systems able to learn how to use their bodies to communicate and to react to their users in a social and engaging way. Social robots that interact with humans have thus become an important focus of robotics research.

Research into Human-Robot Interaction (HRI) for socially assistive applications is in its infancy. Socially assistive robotics [4] is an interdisciplinary and increasingly popular research area that brings together insights from a broad spectrum of fields, including robotics, health, social and cognitive sciences, and neuroscience, among others.

Adriana Tapus
ENSTA-ParisTech, Paris, France
e-mail: adriana.tapus@ensta.fr

Cristian Tapus
Research Scientist at Google Inc., Mountain View, CA, USA
e-mail: crt@google.com

Maja Matarić
University of Southern California,
Computer Science Department, Los Angeles, USA
e-mail: mataric@usc.edu

A. Howard et al. (Eds.): Field and Service Robotics 7, STAR 62, pp. 389–398.
springerlink.com   © Springer-Verlag Berlin Heidelberg 2010

It is estimated that in 2050 there will be three times more people over the age 85 than there are today [1]. Most of the ageing population is expected to need physical and/or cognitive assistance. As the elderly population continues to grow, new research has been dedicated to developing assistive systems aimed at promoting ageing-in-place, facilitating living independently in one's own home as long as possible, and helping caregivers and doctors to provide long-term rehabilitation/ cognitive stimulation protocols. The first efforts towards having socially assistive robotic systems for the elderly have been focused towards constructing robot-pet companions aimed at reducing stress and depression [5], [12], [7], [6], and [8]. In addition to the growing elderly population, other large user populations represent ideal beneficiaries of socially interactive assistive robotics. Those include individuals with physical impairments and those in rehabilitation therapy, where socially assistive technology can serve to improve not only mobility [13], [2] [3] but also for outcomes in recovery. Finally, individuals with cognitive disabilities and developmental and social disorders (e.g., autism [9]) constitute another growing population that could benefit from assistive robotics in the context of special education, therapy, and training.

In order to be able to aid the target user populations, an effective socially interactive assistive robot must understand and interact with its environment, exhibit social behavior, and focus its attention and communication on the user in order to help the user achieve specific goals. Social behavior plays an important role in the assistance of people with special needs. An adaptive, reliable and user-friendly hands-off therapist robot can provide an engaging and motivating customized therapy protocol to participants in laboratory, clinic, and ultimately, home environments, and can establish a very complex and complete human-robot relationship. Therefore, such robots must be endowed with human-oriented interaction skills and capabilities to learn from us or to teach us, as well as to communicate with us and understand us. Hence, the work proposed here will focus on robot behavior adaptation to user's personality, preferences and disability level, aiming toward a long-term customized therapy protocol for stroke rehabilitation and other elderly specific application domains.

This paper presents two learning approaches, one based on on-line adaptation and the other based on long-term learning, for socially assistive robots designed for helping stroke patients and people suffering of age-related cognitive impairments (i.e., dementia). The rest of the paper is structured as follows. Section 2 illustrates the robotic test-bed. Section 3 describes the online learning and behavior adaptation approach and its validation in a rehabilitation-like context and Section 4 describes the long-term learning approach and its validation in a study with patients with dementia. Section 5 concludes the paper.

2   Experimental Platform

The experimental testbed used was a custom-designed humanoid torso robot mounted on a mobile robot base (Figure 1). The mobile base was an ActivMedia Pioneer 2DX robot equipped with a speaker, a Sony Pan-Tilt-Zoom (PTZ) color

Fig. 1 Robot test-bed: Bandit II humanoid torso mounted on the Pioneer mobile base

camera, and a SICK LMS200 eye-safe laser range finder. The anthropomorphic setup involved a humanoid Bandit II torso, consisting of 22 controllable degrees of freedom, which included: 6 DOF arms (x2), 1 DOF gripping hands (x2), 2 DOF pan/tilt neck, 2 DOF pan/tilt waist, 1 DOF expressive eyebrows, and a 3 DOF expressive mouth. All actuators were servos allowing for gradual control of the physical and facial expressions. We are interested in utilizing the humanoid's anthropomorphic but not highly realistic appearance as a means of establishing user engagement, and comparing its impact to our prior work with non-biomimetic robot test-beds [11].

3   Study 1

3.1   Robot Learning and Behavior Adaptation to User Personality and Preferences

The main goal of the first implemented methodology was to develop a robot behavior adaptation system that allows for dynamically optimizing three main interactional parameters (in our case: interaction distance/proxemics, speed, and vocal content) so as to adapt to the user's personality toward improving the user's task performance. These parameters defined the behavior (and thus personality) of the "therapist" robot. Task performance is measured as the number of exercises performed in a given period of time; the learning system changed the robot's personality, expressed through the robot's behavior, in an attempt to maximize the task performance metric.

A learning algorithm based on policy gradient reinforcement learning (PGRL) was developed. The n-dimensional policy gradient algorithm implemented for this work starts from an initial policy $\pi = \{\theta_1, \theta_2, \ldots, \theta_n\}$ (where n = 3 in our case). For each parameter θ_i we also defined a perturbation step ε_i to be used in the adaptation process. The perturbation step defined the amount by which the parameter may vary to provide a gradual migration towards the local optimum policy. The use of PGRL required the creation of a reward function to evaluate the behavior of the robot as parameters changed to guide it toward the optimum policy. The algorithm consisted of the following steps: (a) parametrization of the behavior initial policy

π; (b) approximation of the gradient of the reward function in the parameter space; (c) movement towards a local optimum.

The reward function was monitored to prevent it from falling under a given threshold, which would indicate that the robot's behavior at the time did not provide the user with an ideal exercise scenario. This triggered the activation of the PGRL adaptive algorithm phase to adapt the behavior of the robot to the continually-changing factors that determined the user's task performance. More details about this work can be found in [11].

3.2   Experimental Design for Learning in the Physical Exercise Context

We endeavored to develop an experimental design for a study involving stroke patients, and validate it first with non-patients, in a lab setting, in order to test the adaptation algorithm. In the experimental design, the participant stands or sits facing the robot. The experimental task is a common object transfer task used in post-stroke rehabilitation and consists of moving pencils from one bin on the left side of the participant to another bin on his/her right side. The bin on the right is on an electronic scale in order to measure the participant's task performance. The system monitors the number of exercises performed. The participants are asked to perform the task for a fixed amount of time (15 minutes for healthy adults, 6 minutes for stroke patients), but they can stop the experiments at any time. At the end of each experiment session, the experimenter presented a short debriefing. Before starting the experiments, the participants are asked to complete two questionnaires: (1) a general introductory questionnaire in which personal details such as gender, age, occupation, and educational background were determined and (2) a personality questionnaire based on the Eysenck Personality Inventory (EPI) for establishing the user's personality traits.

The learning algorithm is initialized with parameter values that are in the vicinity of what is thought to be acceptable for both extroverted and introverted individuals, based on the user-robot personality matching study described in [10]. The PGRL algorithm evaluates the performance of each policy over a period of 60 seconds. The reward function, which counts the number of exercises performed by the user in the past 15 seconds is computed every second and the results over the 60 seconds "steady" period are averaged to provide the final evaluation for each policy. The threshold for the reward function that triggers the adaptation phase of the algorithm is adjusted to account for the fatigue incurred by the participant. The threshold and the time ranges are all customizable parameters.

In the post-experiment survey, the participants are asked to provide their preferences related to the therapy styles or robot's vocal cues, interaction distances, and robot's speed from the values used in the experiments.

We designed four different scenarios for extroverted and introverted personality types; the therapy styles ranged from coach-like therapy to encouragement-based therapy for extroverted personality types and from supportive therapy to nurturing therapy for introverted personality types. We chose to use pre-recorded speech and selected words and phrases for each of these scenarios in concordance with encouragement language used by professional rehabilitation therapists. The challenge-based therapy script is composed of assertive language (e.g., "Keep going!" and "You can do more than that!"). Extroversion is also expressed with higher speech volume and faster speech rate. The aggressiveness of words, volume, and speech rate are adjusted to diminish along with the robot's movement towards the nurturing therapy style of the interaction spectrum. In contrast to the challenge-based script, the nurturing therapy script contains empathetic, gentle, and comforting language (e.g., "I'm glad you are working so well.", "I'm here for you.", "Please continue just like that", "I hope it's not too hard"). The speech uses lower volume and pitch. The transition from one personality-based therapy style to another is done smoothly (see algorithm above) in order to avoid any jarring influence on the human-robot interaction. We chose a set of three interaction distances and speeds for each introverted and extroverted personality type.

3.3 Experimental Results for the Physical Exercise Context

We performed the above-described experiment in a lab-like setting, with 12 non-patient, healthy adult participants (7 male, 5 female). The participants ranged in age between 19 and 35; 27% were from a non-technological field, while 73% worked in a technology-related area.

As shown in Figure 2, the robot adapted to match the preference of the participant in almost every case. The only exception was the interaction with participant 8. Despite the fact that the time spent in the preferred training style of that participant was smaller than the time spent in other training styles, the robot converged to it at the end of the exercise period. This was caused by the fact that the initial state of the robot was in a training style that was furthest from the participant's preference.

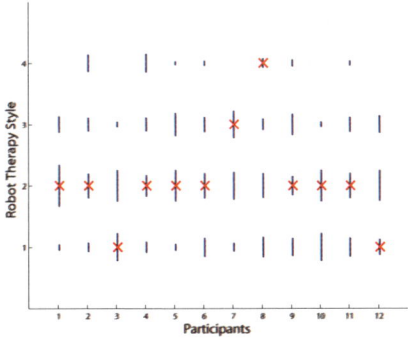

Fig. 2 The percentage of time that the 12 participants interacted with each of the four therapy styles of the robot. The crosses represent the participants' preferences.

The pilot results we obtained support our hypothesis that the robot could adapt its behavior to both introverted and extroverted participants. Further details about this work can be found in [11].

4 Study 2

4.1 Robot Learning and Behavior Adaptation to User Ability/Performance

The second learning and robot behaviour adaptation methodology was designed for the interaction between the robot and a user with dementia and/or Alzheimer's disease, with the main goal of helping users improve or maintain their cognitive attention through encouragements in a music-based cognitive stimulation game.

This approach consists of two parts: supervised learning and adaptation. The robot models the level of game challenge that can be: (a) Difficult: no hints; (b) Medium: when the song excerpt starts say "push the button" but do not indicate which button to push; and (c) Easy: when the song excerpt starts say which button to push. The supervised learning system learns the Accepted Variation Band (AVB) for each game level and for each disability bucket (mild, moderate, and severe), as a function of the user's task performance. The learning phase is followed by an adaptation phase, where the robot adapts its behavior so as to minimize the user's reaction time and maximize the correctness of the user's answers. If the user's task performance is below the Accepted Variation Band, the user is performing better than during the learning phase. The user is then promoted to the next level of game difficulty (if not already at the Difficult level). If the user's task performance is above the Accepted Variation Band, the user is not performing well enough. The user is then helped by having the game difficulty level decreased (if not already at the Easy level).

4.2 Experimental Design for Cognition Exercise Context

The experiment consists of repeated sessions, during which the user and the robot interact in the context of a cognitive game. The first session is the orientation, in which the participant is 'introduced' to the robot. The robot is brought into the room with the participant, but is not powered on. During this introduction period, the experimenter or the participant's nurse/physical/music therapist explains the robot's behavior, the overall goals and plans of the study, and what to expect in future sessions. The participant is also asked about his/her favorite songs from a variety popular tunes from the appropriate time period; those songs are later used in the subsequent sessions. At the end of the session, the Standardized Mini-Mental State Examination (SMMSE) cognitive test is administered so as to determine the participant's level of cognitive impairment and the stage of dementia. This test provides information about the cognitive (e.g., memory recall) level of impairment of the participant for use in initializing the game challenge level. The data determine the participant's initial mental state and level of cognitive impairment, and serve as a pre-test for subsequent end-of-study comparison with a post-test.

This experiment is designed to improve the participant's level of attention and consists of a cognitive game called Song Discovery or Name That Tune. The participant is asked to find the right button for the song, press it, say the name of the song, and sing along. The criteria for participation in the experiment (in addition to the Alzheimer's or dementia diagnosis) include the ability to read large print and to press a button. The participant sits in front of a vertical experimental board with 5 large buttons (e.g., the Staples EASY buttons). Four buttons correspond to the different song excerpts (chosen as a function of the user's preference) and the last button corresponds to the SILENCE or no song excerpt condition. Under each button, a label with the name of the song (or SILENCE) is printed. The robot describes to each participant the goal of the game before each session, based on the following transcript: "We will play a new music game. In it, we will play a music collection of 4 songs. The songs are separated by silence. You will have to listen to the music and push the button corresponding to the name of the song being played. Press the button marked "SILENCE" during the silence period between the songs. The robot will encourage you to find the correct song." Each participant is first asked by the music therapist or the robot to read aloud the titles of the songs and to press a button. Some additional directions are given. The participant is also directed to press the SILENCE button when there is no music playing. After a review of the directions, the participant is asked by the robot to begin the music game. The music compilation is composed of a random set of song excerpts out of the four different songs that form the selection and the silence condition. The entire music compilation lasts between 10 and 20 minutes, and is based on the user's level of cognitive impairment: the larger the impairment, the shorter the session. A song excerpt can be vocal, instrumental, or both. The order of song excerpts is random. The experiment was repeated once per week for a period of 8 months in order to capture longer-term effects of the robot therapist. A within-subject comparison was performed to track any improvement over multiple sessions. No between-subject analysis was done due to the small sample size and large differences in cognitive ability levels.

4.3   Experimental Results for the Cognitive Exercise Context

The initial pilot experimental group consisted of 9 participants (4 male, 5 female), from our partner Silverado Senior Living care facility. All the participants were seniors over 70 years old suffering of cognitive impairment and/or Alzheimer's disease. The cognitive scores assessed by the SMMSE test were as follows: 1 mild, 1 moderate, and 7 severe. Due to the total unresponsiveness of 6 of the severely affected participants, only 1 severely cognitively disabled participant was retained for the rest of the study, resulting in a final participant group composed of 3 participants (all female).

Fig. 3 Human-robot interacting during the music game: the robot gives hints related to the music game, the user answers, and the robot congratulates and applauds the correct answer

We constructed the training data and built a model for each cognitive disability level and for each game level. The participants played each game level 10 times (stages) in order to construct a robust training corpus.

The results obtained over 6 months of robot interaction (excluding the 2 months of learning) suggest that the elderly people suffering of dementia and/or Alzheimer's can sustain attention to music across a long period of time (i.e., on average 20 minutes for mildly impaired participants, 14 minutes for moderately impaired participants, and 10 minutes for severely impaired participants) of listening activity designed for the dementia and/or Alzheimer's population. Figures 4a, 4b, and 4c illustrate the evolution of the game difficulty over time, as well as response incorrectness and reaction time for user_id 1.

Outcomes are quantified by evaluating task performance and time on task. Based on the results we obtained, it can be concluded that the SAR system was able to adapt the challenge level of the game it was presenting to the user in order to encourage task improvement and attention training. Figure 4a shows the evolution in time of the game level for user_id 1. The participant started at the easy game level and remained there for several sessions. The participant then started to perform better and diminished the reaction time and reduced the number of incorrect answers, which, in turn, resulted in a game level evolution from the easy level to difficult. Starting from the 22nd trial, the participant consistently remained at the highest level of difficulty in the game (see Figure 4a). Figures 4b and 4c depict the evolution of the reaction time and the number of incorrect answers. The decrease of those metrics indicates improvement on the task. Similar improvement was observed for all participants.

The participants recognized the songs and identified the silence periods with the same probability. Hence, the analysis of the "no answer" situation among our elderly participants provides us with additional information. From our experiments, we noticed that the average rate of absence of response to silence was higher than the average rate of absence of response to songs, and that this phenomenon increased with the severity of the cognitive impairment. Our conjecture is that music stimulates the interest and responsiveness of the participants. Another interesting observation that deserves more study is the users' ability to participate simultaneously in different tasks (multitasking): the participants were able to sing and push

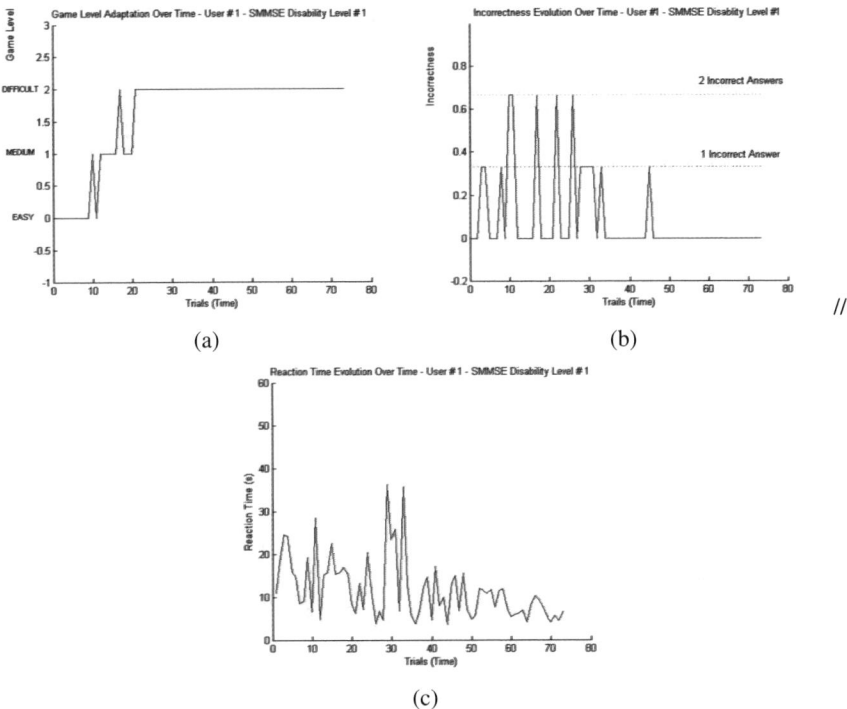

Fig. 4 Results: (a) Game Level Adaptation and Evolution Over Time (6 months) for User Id 1; (b) Incorrectness Evolution Over Time (6 months) for User Id 1; (c) Reaction Time Evolution Over Time (6 months) for User Id 1

the correct buttons at the same time. This is notable in particular for participants with cognitive disability, since multitasking requires dividing attention.

In summary, our social robot was able to improve or maintain the cognitive attention of users with Alzheimer's Disease in a specific music-based cognitive game. The robot's capability of adapting its behavior to the individual user's level of disability helped to improve the user's task performance in the cognitive game over time.

5   Conclusions

This research has aimed to develop adaptation and learning methods for socially assistive therapist robots that can provide customized physical rehabilitation and/or cognitive stimulation. We have presented results from two different adaptation and learning approaches, validated with healthy adults as well as with elderly users with Alzheimer's Disease. Our results are encouraging in light of our pursuit toward creating personalized socially assistive technologies that aim to improve human quality of life.

Acknowledgements. This work was supported in part by the National Academies Keck Futures Initiative (NAKFI), by the USC Alzheimer's Disease Research Center (ADRC), by the NSF IS-0713697 grant, and by the USC WiSE Program. The infrastructure for this research was supported by the NSF Computing Research Infrastructure grant CNS-0709296. We are also grateful to our partner: Silverado Senior Living - The Huntington, Alhambra, CA, USA.

References

1. American Alzheimer Association. About alzheimer's disease statistics. American Alzheimer Association (November 2007)
2. Brewer, B.R., Klatzky, R., Matsuoka, Y.: Feedback distortion to overcome learned nonuse: A system overview. IEEE Engineering in Medicine and Biology 3, 1613–1616 (2003)
3. Burgar, C.G., Lum, P.S., Shor, P.C., Vander Loos, M.: Development of robots for rehabilitation therapy: the palo alto va/stanford experience. Journal of Rehabilitation research and Development 37(6), 639–652 (2000)
4. Feil-Seifer, D., Matarić, M.J.: Defining socially assistive robotics. In: Proc. IEEE International Conference on Rehabilitation Robotics (ICORR 2005), Chicago, Il, USA, June 2005, pp. 465–468 (2005)
5. Nourbakhsh, I., Fong, T., Dautenhahn, K.: A survey of socially interactive robots. Robotics and Autonomous Systems 42(3-4), 143–166 (2003)
6. Kidd, C., Taggart, W., Turkle, S.: A sociable robot to encourage social interaction among the elderly. In: IEEE International Conference on Robotics and Automation (ICRA), Orlando, USA (May 2006)
7. Libin, A., Cohen-Mansfield, J.: Therapeutic robocat for nursing home residents with dementia: Preliminary inquiry. Am J Alzheimers Dis Other Demen 19(2), 111–116 (2004)
8. Marti, P., Giusti, L., Bacigalupo, M.: Dialogues beyond words. Interaction Studies (2008)
9. Michaud, F., Theberge-Turmel, C.: Mobile robotic toys and autism. In: Billard, A., Dautenhahn, K., Canamero, L., Edmonds, B. (eds.) Socially Intelligent Agents - Creating Relationships with Computers and Robots. Kluwer Academic Publishers, Dordrecht (2002)
10. Tapus, A., Matarić, M.J.: User personality matching with hands-off robot for poststroke rehabilitation therapy. In: Proc. International Symposium on Experimental Robotics(ISER 2006), Rio de Janeiro, Brazil (July 2006)
11. Tapus, A., Tapus, C., Matarić, M.J.: User-robot personality matching and robot behavior adaptation for post-stroke rehabilitation therapy. Intelligent Service Robotics 1(2), 169–183 (2008)
12. Walton, M.: Meet paro, the therapeutic robot seal. In: CNN (2003)
13. Yanco, H.: Evaluating the performance of assistive robotic systems. In: Proc. of the Workshop on Performance Metrics for Intelligent Systems, Gaithersburg, MD, USA (2002)

Part IX
Mining Robotics

Swing Trajectory Control for Large Excavators

A.W. Denman, P.R. McAree, M.P. Kearney, A.W. Reid, and K.J. Austin

Abstract. There is a strong push within the mining sector to automate equipment such as large excavators. A challenging problem is the control of motion on high inertia degrees of freedom where the actuators are constrained in the power they can deliver to and extract from the system and the machine's underlying control system sits between the automation system and the actuators. The swing motion of an electric mining shovel is a good example. This paper investigates the use of predictive models to achieve minimum time swing motions in order to address the question what level of performance is possible in terms of realizing minimum time motions and accurate positional control. Experiments are described that explore these questions. The work described is associated with a project to automate an electric mining shovel and whilst the control law discussed here is a much simplified form of that used in this work, the experimental study sheds considerable light on the problem.

1 Introduction

An electric mining shovel (EMS) is a large electro-mechanical excavator commonly used in open-cut mining to load haul trucks . They are critical production units at most open-cut mine sites and there is an ongoing need to improve their productivity through automation of the loading process. These machines are actuated by DC motors, have three primary degrees of freedom used for excavation and achieve mobility through crawler tracks. CRCMining and the CSIRO have been working together on the development of an automation system for mining shovels.

An automation system replacing the operator must necessarily implement a position servo capability. Our particular interest in this paper is in achieving position control for the swing motion. Control of this freedom presents challenges because (i) the rotational inertia of the machine house about the swing axis is very large relative

A.W. Denman

CRCMining, University of Queensland, Brisbane, Australia
e-mail: a.denman@uq.edu.au

A. Howard et al. (Eds.): Field and Service Robotics 7, STAR 62, pp. 401–410.

to the effective inertia of the swing motors (ii) the rates at which the swing motors can deliver/extract energy to the swing motion are sufficiently constrained that they become important influences on the control problem and (iii) in dealing with the issues that arise because of (i) and (ii), the existing control system for swing motion has a hybrid structure that must be accommodated for in the automation layer control system. The last of these points emphasizes the challenge of working with multi-layered control systems where various parties develop and support the different layers. In this problem the automation layer control system must be removable so that the machine can be operated by a human operator. To achieve this and for reasons of safety integrity, the automation layer produces outputs that feed in at the same point as the references provided by the operator joysticks.

The strategy we explore in this paper considers the swing motion in isolation from the other degrees of freedom and looks to develop an approximate minimum time controller that is inspired by the Pontryagin minimum principle [5]. The control law uses a model of the swing motion with the input being the joystick reference and the output being the swing angle in a receding-horizon framework to determine switching points for the joystick references that deliver near minimum-time trajectories. The model of the swing motion control system has been described previously in Ref. [6].

The paper has the following structure. Section 2 summarizes the characteristics of the control system and electric drive dynamics for the swing drive. Section 3 discusses the state-space model formulation and the application of this modeling technique to command the shovel to a desired swing angle. In Section 4 we provide a demonstration of the use of these models for trajectory control using the Pontryagin inspired framework control system.

2 Swing Drive Dynamics

The shovel used in this study has an ABB DCS/DCF600 Multi-Drive controller to regulate motor speed, armature current and field current in the swing DC motor. The controller is made up of four integral components; a PID or PI motor speed control loop, an armature current saturation limiter, a PI current control loop and an EMF-field current regulator.

The swing drive uses a combination of torque control and bang-bang speed control, whereby the swing joystick position generates a piecewise speed reference and an armature current saturation limit. A schematic of the swing drive model is shown in Figure 1. The difference between the reference and actual swing motor speed feeds the Proportional-Integral-Derivative (PID) speed controller incorporating derivative filtering. The output of the speed controller is scaled into a reference armature current that is the limited proportionally according to the amplitude of the swing joystick reference. The error between the limited current reference and the actual armature current feeds into a PI current controller that outputs an armature voltage to the swing motor. The swing motor has a constant field current with the DCF600 maintaining the field voltage at a steady level.

Table 1 Nomenclature

Category	Notation	Description
Model Inputs	f	coulomb friction (N)
	I_f	field current(A)
	j	joystick reference
	T	gravitational torque load (Nm)
	ω^d	desired motor speed (rad/s)
Model States	e_ω	motor speed error in the speed controller (rad/s)
	e_I	armature current error in the current controller (A)
	I	armature current
	I^d	reference armature current prior to saturation (A)
	Θ	motor position (rad)
	ω	motor speed (rad/s)
Controller Parameters	G	describing function gain
	K_ω	speed controller proportional gain
	K_I	current controller proportional gain
	K_{TI}	speed to current scaling
	T_ω^i	speed controller integration time constant (s)
	T_ω^d	speed controller differential time constant (s)
	T_ω^f	speed controller filter time constant (s)
	T_I^i	current controller integration time constant (s)
DC Drive Parameters	b	drive damping coefficient
	J	drive inertia resolved to the motor (Nm/s)
	K_T	motor torque constant (Nm/A)
	K_{emf}	motor back EMF constant
	L	motor armature inductance (Henries)
	R	motor armature resistance (Ohms)
Subscripts	s	swing drive

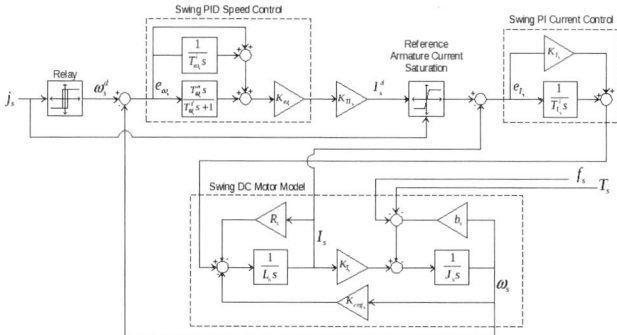

Fig. 1 Swing drive model schematic

Due to its hybrid nature, modelling the shovel swing drive effectively requires a means for incorporating the non-linear saturation effects seen in the motor armature currents. To include these effects into the prediction models a sinusoidal input describing function is used [6]. The describing function has been used for the study of limit cycles in non-linear dynamic systems [2, 3] and is used here for armature current saturation. The basic idea of the describing function approach is to replace each non-linear element in a dynamic system with a quasi-linear descriptor or describing function equivalent gain whose magnitude is a function of the input amplitude.

The drive prediction model is presented as continuous, linear state space systems with the form

$$\dot{x} = Ax + Bu \tag{1}$$

The input vector u, contains the reference motor speeds generated from the joystick signals, the static torque load on the motor due to gravitational effects and a coulomb friction disturbance input. The state vector x, contains armature current, the motor speed, the motor position, the integrals of the error in the speed and current controllers and the additional state of swing reference armature current prior to the saturation limit . This state arises from the derivative component in the swing motor speed controller. The full state space model for the swing drive is given in equation 2 [6]. The coulomb friction component is neglected in this work.

$$
\begin{pmatrix} \dot{i}_s \\ \dot{\omega}_s \\ \int e_{I_s} \\ \dot{I}_{s_{ref}} \\ \int e_{\omega_s} \\ \omega_s \end{pmatrix} =
\begin{pmatrix}
-\frac{R_s}{L_s} - \frac{K_{I_s}}{L_s} & -K_{emf_s} & \frac{1}{L_s T_{I_s}^i} & \frac{G_s K_{I_s}}{L_s} & 0 & 0 \\
\frac{K_{T_s}}{J_s} & \frac{b_s}{J_s} & 0 & 0 & 0 & 0 \\
-1 & 0 & 0 & G_s & 0 & 0 \\
A & B & 0 & -\frac{1}{T_{\omega_s}^f} & \frac{K_{\omega_s} K_{TI}}{T_{\omega_s}^i} T_{\omega_s}^f & 0 \\
0 & -1 & 0 & 0 & 0 & 0 \\
0 & 1 & 0 & 0 & 0 & 0
\end{pmatrix}
\begin{pmatrix} I_s \\ \omega_s \\ \int e_{I_s} \\ I_{s_{ref}} \\ \int e_{\omega_s} \\ \Theta_s \end{pmatrix} \tag{2}
$$

$$
+
\begin{pmatrix}
0 & 0 & 0 \\
0 & -\frac{1}{J_s} & -\frac{1}{J_s} \\
0 & 0 & 0 \\
\frac{K_{\omega_s} K_{TI}}{T_{\omega_s}^f}\left(1+\frac{T_{\omega_s}^f}{T_{\omega_s}^i}\right) & 0 & 0 \\
1 & 0 & 0 \\
0 & 0 & 0
\end{pmatrix}
\begin{pmatrix} \omega_s^d \\ T_s \\ f_s \end{pmatrix}
$$

$$A = -\frac{K_{T_s}}{J_s}\frac{K_{\omega_s} K_{TI}}{T_{\omega_s}^f}(T_{\omega_s}^f + T_{\omega_s}^d); B = -\frac{K_{\omega_s} K_{TI}}{T_{\omega_s}^f}\left(1+\frac{T_{\omega_s}^f}{T_{\omega_s}^i}\right) - \frac{b_s}{J_s}\frac{K_{\omega_s} K_{TI}}{T_{\omega_s}^f}(T_{\omega_s}^f + T_{\omega_s}^d)$$

3   Near-Minimum Time Control Law

The state space model for the swing drive is used to design a control loop to replace operator joystick input with the aim of minimum time swing motion between two points. To achieve as close as possible to minimum time, 'bang-bang' control action, using the Pontryagin minimum principle [5] was used to achieve time optimal

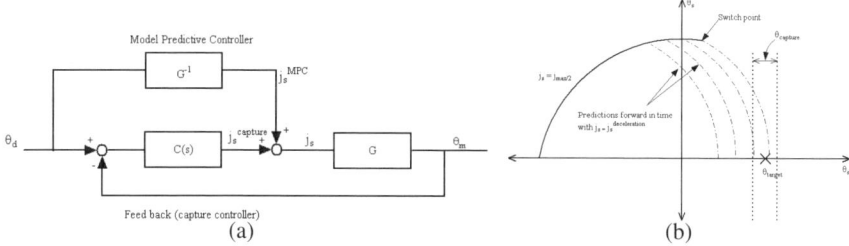

Fig. 2 (a) Swing drive controller. (b) Switch point determination.

motion via computation of an optimal switching point. This principle relies on the fact that solution, using on the maximum and minimum extremes of control input, must be time optimal. The integration of the state space model forwards in time includes the effect of the swing drive rate limiting. The drive is rate limited to 250%/s, such that transitioning between input extremes ($\pm100\%$) takes $0.8s$. This approach also required the addition of a terminal controller to achieve zero steady state error. The controller is presented in figure 2(a).

The switching point is defined as the moment where the input changes between maximum to minimum (or vice versa) such that the desired state is reached (zero position error and velocity). The state space model is used in two ways. Firstly, prior to reaching the switching point, the model is used to check that a desired decelerating swing joystick reference (to be time optimal, this reference would be 90 - 100%) would bring the swing drive to rest prior to reaching the target swing point (figure 2(b)).

Once the switching point is reached, the state space model is then used to compute what decelerating swing joystick reference is required to bring the swing drive to rest. The solution of this problem is obtained through the bisection method. A joystick reference is computed and applied at 10Hz to ensure real-time behavior of the controller.

As the capture region is entered (figure 2(b)), the state space model predictive controller is augmented by the feedback capture controller. The capture controller is added to account for any uncertainties that exist within the open loop state space model predictive controller. The capture controller contains terms for proportional, derivative and integral control with various anti-windup strategies. As the swing drive reached zero velocity, the contributions to the swing drive input from the model predictive controller are removed and the capture controller was left to maintain swing angle.

4 Demonstration of Swing Control

This section presents the results of the application of the prediction models for rope shovel trajectory control using model predictive controller alone. For the swing drive, we use the a model of the plant to be controlled to predict the future behaviour of the plants output. This feed-forward characteristic allows for future demands to be incorporated into the present control input, enabling control action to bring the shovel to rest to take place prior to attainment of the goal point. Feedback control alone would only allow for control action to occur once the goal point is attained while not at rest. The model is propagated forward in time using a stiff first order implicit integrator with a time interval of 0.05 seconds.

Here we present the results of requesting a 90° and 120° change in swing angle, beginning at rest. The swing angle of the shovel is sensed through a resolver attached to the swing drive transmission, with known gain, and the swing velocity is obtained through differentiation of the swing resolver output with time. Figure 3 shows change in swing angle against the change in swing velocity.

It can be seen that the desired changes in swing angle are faithfully reproduced, however, the swing drive is unable to come to a complete rest. It was expected that the open loop nature of this controller would struggle to bring the swing drive to rest. During the initial swing cycle, the swing drive accelerates under a constant swing reference of 50% (these trajectories will not be time optimal) until the switching point is reached. This choice of swing reference, however, still sees the machine reach close to it's maximum rotational speed of $14.4^o/s$ (see figures 3 and 6(a)). As the input switching point is reached, the prediction models are used to compute a reference that will bring the shovel to rest at the desired swing angle with the reference applied once it is computed. It can be clearly seen that as the switching point is reached, and a decelerating reference is applied, the shovel decelerates faster than expected. This conclusion is reached because after the initial deceleartion, the shovel speed decreases its deceleration rate. At the next time sample the model then predicts that a lesser decelerating reference is required to compensate. In the concluding stages of the swing cycle, the effects of mechanical backlash and the sensitivity of the predictive controller joystick reference output to small remaining swing angles results in the oscillatory behaviour seen in figure 3.

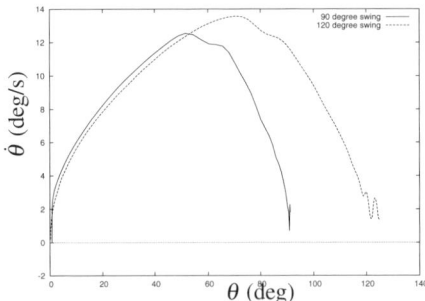

Fig. 3 Swing angle versus swing velocity for near-minimum time controller

4.1 Feedback Capture Control

The previous section demonstrated that the use of a predictive controller is not sufficient to bring the swing drive to rest. As indicated in the overall swing drive controller design in figure 2(a), a feedback capture controller is added. This controller becomes active once the measured swing angle enters the capture region (figure 2(b)). Proportional (P, K_p) , Proportional Derivative (PD, K_p, K_d, τ) and Proportional Integral (PI, K_p, K_i) controllers were examined for their suitability to this application.

A number of anti-windup strategies for the integrator within the capture controller were investigated due to the potential for the control input to saturate (and 'windup' the integrator) because of the limitiations in the existing control system and the swing drive. The discrete strategies investigated were resetting the integrator when saturation was detected (Reset), and holding the value of the integrator constant when saturation occurs (Hold-Reset). A feedback anti-windup strategy, where the difference between the unsaturated and saturated outputs of the controller (zero when there is no saturation) is fed back to the integrator to ensure that the integrator does not wind up was also used (Inner-Feedback).

The capture controller performance was evaluated on the research shovel from both a stationary and non-stationary initial condition. The resulting swing drive speeds and swing angle were measured. The results presented in figure 4 are for step inputs of positive 10° from a stationary position and a variable step input to bring the shovel to 0° from a non-stationary position. This replicates how the capture controller would have to behave as the shovel enters the capture region in the overall control strategy. Each capture test was done independently as the parameters governing the controllers were explored. Table 2 summarizes the parameters that were used.

Figure 4(a) plots the response of the controllers for a 10° change in angle from a stationary initial condition. The P and PD capture controllers perform best by minimizing the overshoot of the target angle, which is important when avoiding collisions with load devices, but are limited in the steady state error by the mechanical backlash present in the machine transmission. The backlash has been determined previously to be of the order of $\pm 4^o$. Adding an integral term added additional overshoot to the solution profile and the introducing different anti-windup strategies had

Table 2 Capture controller parameters

Condition	Type	K_p	K_i	K_d	τ	Anti-windup
1	P	200	-	-	-	-
2	PI	200	20	-	-	-
3	PI	200	20	-	-	Reset
4	PI	200	20	-	-	Hold-Reset
5	PI	200	20	-	-	Inner-Feedback
6	PD	200	-	50	0.1	-

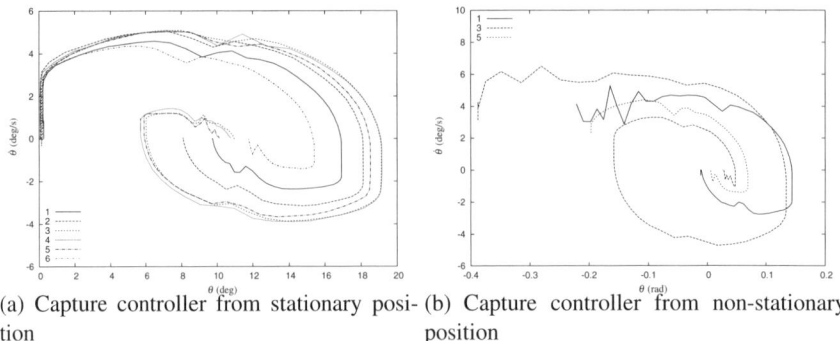

(a) Capture controller from stationary posi- (b) Capture controller from non-stationary
tion   position

Fig. 4 Capture controller test results on research shovel

negligible effect amongst themselves while adding more oscilation. Cases 1, 4 and 6
(table 2) were also investigated for a non-stationary initial condition to more closely
resemble how it would operate in the overall swing control. The results are given in
figure 4(b). Here, again, the P and PD capture controllers performed the best.

4.2 Near-Minimum Time Controller with Feedback Capture

This section presents the results of the application of the prediction models for
shovel swing trajectory control using the near-minimum time controller and its capture with feedback control. The capture controller used is the P controller (Case 1)
detailed in table 2. Figure 6(a) shows change in swing angle against the change in
swing velocity as a result of requesting a 90° and 180° change in swing angle, beginning and ending at rest. The initial behaviour of the swing speed versus swing
angle profile is highly dependent on whether or not the transmission was perfectly
meshed when started. If the transmission were not perfectly meshed it is possible
for swing speed to increase without any change in swing angle as the backlash is
taken up (see 90° swing angle of figure 6(a)).

As was observed in Section 4, the swing drive accelerates under a constant swing
reference of 50% until the switching point is reached. As the input switching point is
reached, the prediction models' joystick references are applied to bring the shovel to
rest at the desired swing angle. After the switching point the effects of unmodelled
friction are again observed. Unlike the results of Section 4, here, when combined
with the capture controller, the shovel swing drive is brought to rest at the desired
swing location within mechanical backlash tolerances.

Figure 5 plots the computed swing joystick reference profile against swing angle
for both cases presented in figure 6(a). Both profiles commence with an initial reference of −50% to move the swing angle towards the target. As the switching point
is determined, the reference become positive to decelerate the swing motion. From
this point on until the capture region is encountered (within 10° of the target), the
joystick reference is computed by the state space prediction model.

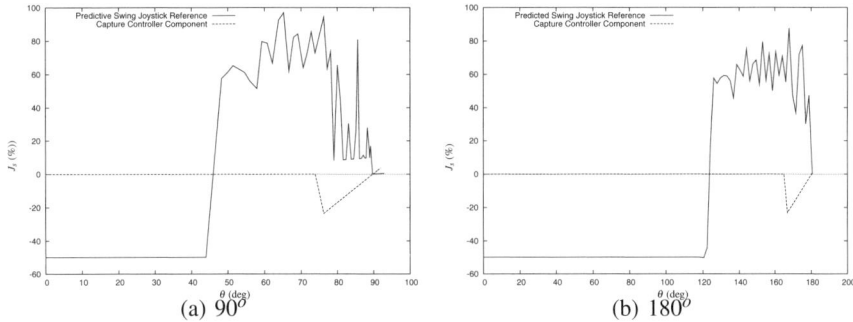

Fig. 5 Swing joystick reference contoller output

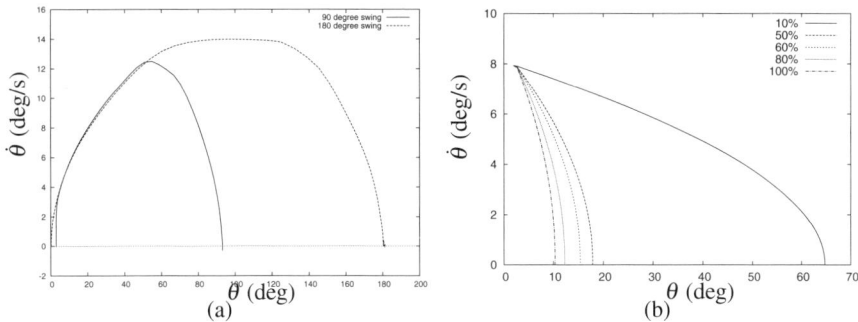

Fig. 6 (a) Swing angle versus swing velocity for near-minimum time controller and capture (b) Simulation of swing angle versus swing velocity for various decelerating references

There is clearly large variation in the applied joystick reference ($\pm 10\%$) during the decleration stage outside of the capture region. The reasons for this are twofold. Firstly, the sample rate of 10Hz, chosen to maintain realtime behavior while solving the bisection problem, appears to be too low. However, the behaviour of the shovel as seen in figure 6(a) indicates that the overall system is not susceptible to these variations in control input. The second reason lies in the sensitivity of the predicted swing angle at zero speed to the applied joystick reference. Figure 6(b) shows this behaviour. Clearly, when operating in the region where $J_s > 50\%$, a 20% change in joystick reference results in a comparitively small stationary swing angle value. This behavior can be understood conceptually by understanding that the joystick reference is essentially a torque with it having an inverse relationship to the total angular displacement over which it is applied.

As the capture region is encountered, the capture controller output is added to that of the predictive controller (figure 5). Initially, the capture controller acts to reduce the decelerating reference. This decrease is not incorporated in the predictive controllers calculations and it acts to increase the decelerating reference in the next prediction, leading to the initial oscillitary profile in joystick reference. Any other spikes in decelerating reference, particularly in figure 5(a), are believed to be due to model inaccuracies but this does not make it unfit-for-purpose. As the target swing angle is reached and the swing drive becomes stationary, the predictive controller input is stopped, leaving only the capture controller to regulate position.

5 Conclusions

This paper has presented an exploration of near-minimum time control on the swing drive of an electric mining shovel. While the state space model deals only with the dominant effects of the hybrid structure of the base machine controller, the main outcome of the paper has been to show that it is possible to achieve positional control within a predictive control framework. The single-input single-output structure of the controller limits its application for the broader automation problem which requires coordinated motion of all degrees of freedom. However, this work has established that a fit-for-purpose control structure can be implemented.

Acknowledgements. The authors would like to thank the Australian Coal Association Research Program (ACARP) and CRCMining for funding project under which the work presented in this paper was done and P&H Mining Equipment for the provision and support of the mining shovel used for this work.

References

1. ABB Automation Products GmbH.: DCS Thyristor Power Converters for DC Drive Systems 25 to 5150/10300 A, Software Description, DCS600 MultiDrive (2002)
2. Gelb, A., Vander Velde, W.E.: Multiple-Input Describing Functions and Nonlinear System Design. McGraw-Hill Book Company, New York (1968)
3. Graham, D., McRuer, D.: Analysis of Nonlinear Control Systems. John Wiley & Sons Inc., New York (1961)
4. Hendricks, P.A., Keran, C.M.: Automation and Safety of Mobile Mining Equipment. Engineering nad Mining Journal (February 1995)
5. Hocking, L.M.: Optimal Control; An Introduction to the Theory with Applications. Oxford applied mathdmatics and computing science series, pp. 82–98 (1991)
6. Siegrist, P.M., McAree, P.R., Wallis, D.L., Kearney, M.P., van Geenhuizen, J.A.: Prediction Models for Collision Avoidance on Electric Mining Shovels. In: 2006 Australian Mining Technology Conference, Australia (September 2006)
7. Singh, S.: State of the Art in Automation of Earthmoving. In: Proceedings of the Wrokshop on Advanced Geomechatronics (October 2002)
8. Wauge, D.H.: Payload Estimation for Electric Mining Shovels. PhD thesis, Mechanical Engineering, University of Queensland (2007)

The Development of a Telerobotic Rock Breaker

Elliot Duff, Con Caris, Adrian Bonchis, Ken Taylor, Chris Gunn, and Matt Adcock

Abstract. This paper describes the development of a tele-robotic rock breaker deployed at a mine over 1000kms from the remote operations centre. This distance introduces a number of technical and cognitive challenges to the design of the system, which have been addressed with the development of shared autonomy in the control system and a mixed reality user interface. A number of trials were conducted, culminating in a production field trial, which demonstrated that the system is safe, productive (sometimes faster) and integrates seamlessly with mine operations.

1 Introduction

A rockbreaker consists of a large serial link manipulator arm having 4 DOF that is fitted with a hydraulic hammer and used throughout the mining industry to break oversized rocks. CSIRO has been contracted by Rio Tinto Iron Ore to install a tele-robotic control system to the primary rockbreaker at the West Angelas mine, situated over 1000km north-east of Perth in Western Australia. Figure 1a shows the rockbreaker installation at the ROM (Run of Mine) bin. The bin is fitted with horizontal bars at the bottom that prevent oversized rocks from entering the crusher below (see Figure 1b). This arrangement is commonly referred to as a grizzly. Haul trucks carrying ore from a nearby quarry dump their load into the ROM bin. Any oversize rocks in the bin are crushed using the rockbreaker arm. Until now, an operator has had to step out of a control room adjacent to the bin and use a line-of-sight remote control to operate the arm. The rock breaking strategy in this context is determined from a quick visual examination of the rock (i.e. centre hit to shatter the rock, nibble the sides of the rock, or nudge the rock to let it fall through). The available time is

Elliot Duff, Con Caris, and Adrian Bonchis
CSIRO, PO Box 883, Kenmore QLD 4069, Australia
e-mail: first.last@csiro.au

Ken Taylor, Chris Gunn, and Matt Adcock
CSIRO, PO Box 664, Canberra ACT 2601, Canberra, Australia

A. Howard et al. (Eds.): Field and Service Robotics 7, STAR 62, pp. 411–420.
springerlink.com   © Springer-Verlag Berlin Heidelberg 2010

Fig. 1 Rock breaker (a) at rest over ROM bin (b) breaking a rock on grizzly.

limited by the level of ore in the hoper below the grizzly and the number of trucks queueing at the bin (typically less than a minute). If the rock cannot be broken in time, it must be pushed to the side and the arm returned to its rest position.

The objective of this project was to demonstrate the feasibility of effective and safe telerobotic control over long distances, as part of RioTinto's long term plan to tele-operate their mining operations. This distance introduces a number of technical (communications bandwidth and latency) and cognitive (lack of spatial and situational awareness) challenges that can be addressed by developing technologies at both the local and remote ends of the system. Improving the intelligence of the control system at the remote end (i.e. Cartesian motion, collision avoidance) can mitigate the effects of latency, whilst the development of mixed reality user interfaces (with a combination of live video and 3D computer visualization) can improve the situational awareness for the operator.

Mixed reality arises as a hybrid solution that attempts to overcome the weaknesses of the two extremes: **direct visualisation** of the environment by cameras and **synthetic visualisation** of a software model of the environment by computer graphics. Cameras provide a direct representation of the real world which includes all visible features, but typically only from a limited range of viewpoints. Virtual representations are very flexible with respect to viewing parameters and manipulation, but can only include information that is represented in the software model of the environment, and only to the limit of accuracy to which the dynamics of the real environment can be sensed. Thus, an interface that mixes the two paradigms for visualisation can take advantage of the best features of each, while overcoming some of the disadvantages. In practice, both paradigms provide situational awareness mediated by technology, and so are subject to failures of various sorts. Providing multiple streams of awareness incorporates a level of redundancy that protects against some failure modes.

This paper is divided as follows: Section 2 provides a brief background to tele-operation over the Internet and its applications to the mining domain; Section 3 describes the implementation of the tele-robotics system to a mining rock breaker in a production environmnet; Section 4 describes the results of the field trials; and Section 5, concludes with a discussion and proposal for future work.

2 Background

Tele-operation has been an active field of research and commercial activity for a number of years as it offers a means of isolating an operator from hazardous or un-inhabitable environments while retaining the reasoning powers of the human operator. It has a long history, dating back over sixty years to a "master-slave" system[5]. Many systems have subsequently been developed for underwater, radioactive, volcanic, and outer space environments.

During the late 1990s there was a great deal of interest in tele-robotics applications over the Internet [7]. One of the first Web-based tele-operation projects[6] involved a mock-up of an archaeological site situated in a radioactive area. Users could join a queue to take control of a 2.5DOF manipulator, searching for 'relics'. This work later evolved into a "tele-garden'" system in which users could remotely tend to a garden. To avoid latency induced instability both systems used supervisory control to specify a position in space[14]. Around the same time, researchers developed a Web-controllable, 6-axis robot that allowed operators to stack toy blocks by controlling the gripper and Cartesian position of the arm (again using supervisory control) [16]. This system was subsequently converted to a 'tele-laboratory' allowing students to perform parts of their coursework.

With respect to mining, Ballantyne et al. [1, 8] investigated the use of *virtual reality* displays for excavator tele-operation. The display enabled the operator to pre-set no-go areas for the excavator and to also mark areas of the excavation site in which the terrain was perhaps too dangerous for the excavator to navigate. Several Japanese groups have also been investigating tele-operation of mining and construction equipment [15, 10, 9]. Although research was conducted to develop a virtual reality system for the mining industry [2] this technology was never commercially realized. The reasons for this failure is unknown, but at this time, the technology was immature and probably did not provide the appropriate level of immersion and interaction necessary for control (i.e. due to latency and bandwidth issues).

Advances in our ability to develop autonomous systems have extended the possibilities for very high-level task specification, moving tele-operators from manual control to a role which is much more tactical or supervisory[12]. These layers of autonomy introduce different requirements for the human machine interface [4]. One of the main criticisms of tele-robotics is that it does not provide sufficient situational awareness [13] to the human operator to sustain the previous manual levels of production. This is being addressed by a number of research labs across the world (eg. NASA's Robonaut) and has become a significant research activity within CSIRO.

3 Implementation

The rockbreaker system architecture (see Figure 2) is based around two main components : the Remote Control System (RCS) located at the West Angelas mine and the Telerobotic User Interface (TUI) more than 1000kms from the Remote Operations Centre (ROC) in Perth, Western Australia. The RCS includes:

- CANBus tilt sensors fitted to the boom, jib and hammer and absolute encoder fitted to the slew axis (see Figure 3b);
- two analogue PTZ cameras and a fixed wide angle camera mounted diagonally across ROM bin, connected to three high speed video compression units;
- two pairs of Firewire megapixel stereo cameras, mounted 80cm apart and several metres above the ROM bin (see Figure 3a);
- industrial Ethernet I/O to generate voltages to drive solenoids and measure state of hydraulic control pack;
- a safety PLC that monitors the access in the rockbreaker workspace, and actuates safety relays that provide power to the control unit;
- a Linux PC to run control software and an XP machine to run stereo acquistion software. Both machine were connected to the mine site Intranet.

The control software is based upon our DDX (Dynamic Data eXchange) middleware [3]. It is split into a number of specialized modules. At the top is a communications layer that provides a simple web interface to the controller, and UDP communications for outgoing state information and incoming demands. The advantage of UDP is that it provides the lowest communications latency (which is an important consideration in this application) at the expense of reliability. At the next layer is the **trajectory planner**, which accepts the incoming demands and plans the arms trajectory (i.e. it is able to convert Cartesian demands into a sequence of joint space velocity demands). Below this is the **boom controller** which has PID loops on four joints and is able to detect and alert the operator to the joint limits. At the bottom is the **boom server** which sends the control signal to the Ethernet IO, which in turn generates the control voltages for the proportional directional control valves

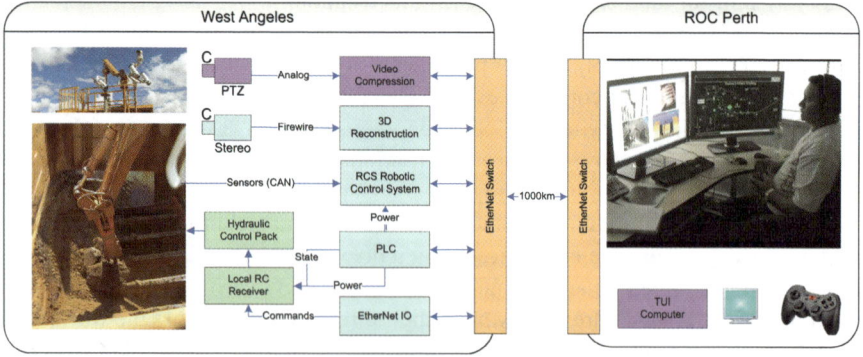

Fig. 2 System architecture showing components at the (a) remote and (b) local locations.

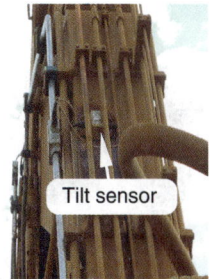

Fig. 3 Installed hardware: (a) cameras (b) tilt sensors and (c) calibration targets in ROM bin.

(solenoids) at the base of the rock breaker. These solenoids cannot be actuated by the computer without explicit control from the Safety PLC. Such control will not be given unless a number of fail-safe steps are taken, including; latching the access gates, a heartbeat from the control computer, and selecting "Computer Enable" in the control room. The Safety PLC also provides access into the site Citect system (which controls the crusher).

Particular care was taken to select hardware that could survive in the harsh mining environment. In summer, the ambient temperature can exceed 50 deg C, and drop below zero at night. The iron ore dust is particularly abrasive and can easily damage electronics. Since the arm dimensions were known, it was possible to use the estimated position of the hammer tip itself to measure the dimensions of the ROM bin (which were different from the mine plan). These dimensions were then used to place visual markers (see Figure 3c) in the ROM bin that were then used to calibrate the seven cameras. This meant that the arm, cameras and ROM bin were all measured in the same frame of reference. This frame of reference was used by the **collision detection** module (using openGL) to detect collisions between the model of the ROM bin and the arm.

Nodding lasers were initially proposed to acquire the 3D surface of the rocks in the ROM bin, but after discussions with the operators, we found that they rely heavily on the texture and colour of the rocks when deciding upon a breaking strategy. A second computer was used subsequently to acquire and process high resolution stereo images. A 3D surface was generated from advanced photogrammetry techniques (commercial product developed by CSIRO called Sirovison). To reduce the effects of stereo shadow a second pair of cameras was mounted diagonally across the ROM bin. To cope with the extreme lighting conditions (dark shadows across the bin in the morning and evening) contrast enhancement was achieved with exposures bracketing. In practice, the system is able to generate 3D surfaces with cm

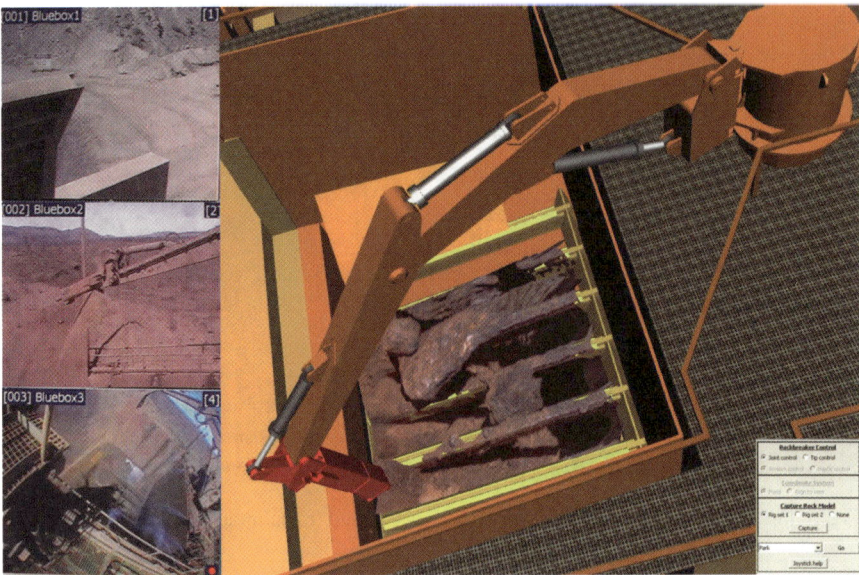

Fig. 4 Three video screens and augmented virtuality of rock breaker overlaid with 3D rock surface.

resolution. Once the 3D surface has been acquired it is converted to X3D and sent to the TUI for rendering onto the 3D virtual screen (see Figure 4). One future advantage of using this photogrammetry software is that it also has the ability to recognize joint sets and fracture surfaces - a valuable feature for future automation.

One of the shortcomings of the previous work has been the ad-hoc nature in which the user interfaces have been developed. Whilst components based robotics (such as Player/Stage, ORCA and Microsoft Robotic Studio) and Web-based toolboxes for LabView and Matlab have moved to the mainstream, the user interfaces have not provided the level of immersion necessary to provide sufficient situational awareness to control dangerous and expensive equipment in remote and unstructured environments. Some researchers have proposed a framework based upon gaming technology [11] which we have (i.e. using Second Life to control a simulation of the rock breaker) and will continue to look into (i.e. Unity). The proposed solution for this project has been to use AJAX3D which merges X3D and AJAX techniques. X3D is an ISO standard for real-time computer graphics that can be viewed with the appropriate viewer (eg. Flux from Mediamachines). Multimedia streams can be placed onto surfaces in the environment (e.g. video onto billboards). The X3D viewer can support audio, stereo displays and haptic devices. Being an open standard there are a number of CAD systems that are able to export drawings to X3D. With AJAX calls to the DOM (Document Object Module) or SAI (Scene Authoring Interface) it is possible to partially load parts of the scene (i.e. update the world) or reload features of the scene (i.e. a joint angle on the rock breaker).

A photograph of the TUI in operation at the ROC is shown in Figure 2b. In front of the operator there are two screens: on the right is the Citect system that is used to monitor the movement of the ore from the crusher down to the stacker/reclaimer; on the left is the user interface designed for the rock breaker control. It consists of the four windows (see Figure 4): three video screens and an augmented virtuality (a 3D computer graphics scene that includes some elements captured directly from the real world).[1] The operator is able to control the rock breaker with the mouse and the gaming joystick (Logitec RumblePad). Projected onto a screen above the two monitors are four video streams from various locations around the rock breaker: one to monitor approaching trucks and another to monitor the state of the secondary crusher. Speakers on either side of the screen reproduce the sounds made at the rock-breaker. This is a very important indicator of the state of the machine: the operator is able to hear the sirens that indicate that the machine is powered; and the sound of the hammer can also be used into indicate whether the hammer is making contact with the rocks (i.e. dry firing). Access to the mine communication systems (RF radio) is provided via a microphone/headset. With this the operator is able to inform the fleet management system of the availability of the crusher.

Once the user has established network communications with the RCS, most commands are accepted through the RumblePad. The right hand trigger button is used as a command validation switch (deadman). The movement of the arm being disabled when the switch is off. The left hand button over-rides the collision detection system to allow the operator to move the arm close to the ROM bin. This is typically used for the cleanup operations. When hydraulic pressure is requested, the RCS expects a heartbeat at 10Hz. The RCS will disable hydraulics after a specified number of heartbeats are missed (in this case, 20 heartbeats or 2 seconds).[2] In the TUI, the operator is able to select different modes of control. They can select in either velocity or position mode in joint, Cartesian or backhoe space (a combination of joint and Cartesian). The two joystick controls on either side of RumblePad are used to control the motion of the arm. The arm can also be sent to pre-configured set points (i.e. Home, Park etc.) or requested to move to a selected location on the 3D rock surface. The user interface is designed around the principal that there is only one primary view that determines the behaviour of the controls. For example, when the PTZ video is the primary display, then the arrow buttons (top left of the RumblePad) are able to pan and tilt the camera. However if the primary screen is the virtual screen, the arrow keys move you up and over the ROM bin (eg. jet-pack mode). The selection of primary screen is controlled by the numbered buttons.

---

[1] The system was designed so that the layout and size of the screens could be modified upon request. In the trial, rather than the original design of a primary screen with three thumbnails, the operator preferred four screens of equal size with minimal decoration.

[2] Measured over several days, the average round trip time was 56 milliseconds with a standard deviation of 3 milliseconds. On average, there were 3 pings per day that lasted more than 300 milliseconds. The video compression (IndigoVision) ranged in bandwidth from 339 Kbps to 1238Kbps at 25fps 4SIF. The state and demand UDP traffic consumed little bandwidth.

4   Field Trials

Field trials were conducted in mid December 2009. The trial consisted of three 12–hour shifts over 3 days during normal production runs. Two of the shifts started at 4am to allow for testing in night conditions. We used two operators, one at each end of the system. The operator based at the ROC in Perth had not been trained or introduced to the TUI prior to commencing the trials. A second operator was present in the control room at the mine site to supervise the rockbreaker operation and intervene in case of emergency.

During the field trials we were able to remotely replicate the work flow of the local human operator. When the operator is alerted to the presence of a large rock, the operator is presented with an overview of the rockbreaker from a wide-angle video stream and augmented virtuality (see Figure 4). The remote operator is able to "walk around the rockbreaker to inspect the rocks from different angles. Once they have established the appropriate breaking strategy, the operator is able to deploy the arm with the joystick. As the arm is commanded to move, the motion of the arm is replicated in the 3D scene. Simultaneously both PTZ cameras follow the tip of the hammer. When the operator is ready to break the rock, they can switch their attention to the live video stream, which they can use to monitor the breaking of the rock. Once complete, the arm can be automatically sent to the rest position.

Within half an hour of introducing him to the TUI, the operator based at the ROC in Perth, was breaking rocks. At first, the operator was unsure/sceptical of the game like controls, however after some experience with the new interface they were happy to accept the device. At the end of the trial the operator expressed the opinion that the deployment of the arm was faster for breaking "simple" rock configurations, but difficult to deal with complex rock configurations (where rocks are packed on top of one another). The operator made several useful suggestions for changes to the user interface that would address this problem (i.e faster 3D update of rock profile, and manual over ride of zoom control). Over the three days the operator was able to deal with all of the rocks without measurable disruption to production. However there were two safety incidents. In one case, there was a communications dropout, and in a second incident the operator moved the arm into the wall as a result of using a forth camera mounted on a hill over 200m away (this camera was not part of the rockbreaker system) and became confused with motion of the arm. At no time was any damage done to the arm or the ROM bin.

5   Discussion and Future Work

Perhaps the most significant comment made by the operator was when we asked what was the difference between local and remote operations. His reply was "*'In local operations, I can concentrate on the task at hand, and my peripheral senses deal with everything else. When I'm remote, I'm forced to redirect my attention to each screen/window that is front of me. This distracts me from what I am doing."* .

The current generation of control room used in the mining industry contains a number of custom built user interfaces: typically one for each mining process that needs to be monitored. To reduce the cognitive load of switching from one interface to another, we believe that the operator should be presented with a single interface. This interface should be highly immersive, interactive and reconfigurable.

The primary goal of our future work is the development and delivery of a **Mixed Reality Framework** that will provide a unified user interface for accessing and interacting with key areas of the mining operation. The intention of this work will not be to replace the existing remote control systems but to provide a portal to access various third-party systems via the Mixed Reality Framework. To use the web-browser metaphor, which is used to move from one web page to another, this system will provide a Mixed Reality Browser that will enable personnel to browse the state of various mining processes in a 3D context in conjunction with existing 2D interfaces.

In the context of teleoperation, "mixed reality" can be used to refer to interfaces that mix the different pathways to visualisation - direct visualisation via video and synthetic visualisation derived from a dynamic software model of the state of the world. Further, for this mixing to be effective, it must be based on information about the relationship between the pathways. In particular, the cameras themselves (and mobile camera platforms) must be modeled in some way, and the camera models may also be dynamically updated from sensor information. We refer to the combination of all these models, and the relationships between them, as the **composite situation model.** Several things need to be modelled in software as part of the process of situation representation. These models may exist on the same computer as the user interface, or they may be accessed from a centralised world model 'in the cloud'.

Situation awareness through visualisation is only one aspect of teleoperation. The actual operation and control of the remote machine must also be supported through the interface. A mixed reality interface creates opportunities for control paradigms based on direct selection and manipulation of objects within the interface. This includes real objects that have been visualised directly or synthetically, or virtual objects that have been added explicitly for the purpose of interaction. In each case, the same selection and manipulation techniques could be used. Just as a mixed reality interface incorporates a mixture of pathways for situation visualisation, it will also incorporate a mixture of operation pathways: direct operation (for example, the position of a joystick controls the degree of opening of a valve), and indirect or synthetic operation (for example, a virtual version of the arm is moved to a position, then commands are sent to move the real arm to that position). Mixed Reality Interaction represents a fertile area for investigation.

Acknowledgements. The authors would like to acknowledge the support of Rio Tinto Iron Ore Automation Group (Victor Schweikart, Iain Puddy, Roger Wainohu, Solomon Birnie, Robin Liu, Mark Miles, Graig Green), Campbell Nunn (from Transmin) and the remainder of the CSIRO team; Polly Alexander, Stephen Brosnan, Clinton Roy, Les Overs, Paul Flick, Dave Haddon, Kane Usher, Mathew Hutchins, Doug Palmer, Duncan Stevenson, Jock Cunningham, Eleonora Widzyk-Capehart, Peter Dean, George Poropat, Andrew Castleden.

References

1. Ballantyne, J., Wong, E.: A virtual environment display for teleoperated excavation. In: Proceedings of the IEEE International Conference on Intelligent Robots and Systems, Victoria, Canada, pp. 1894–1899 (1998)
2. Boulanger, P., Lapointe, J.F., Wong, W.: Virtualized reality: an application to open-pit mine monitoring. In: Archives of the 19th Congress of the International Society for Photogrammetry and Remote Sensing (ISPRS), Amsterdam, The Netherlands, vol. XXXIII, Part B5/1, pp. 92–98 (2000)
3. Corke, P., Sikka, P., Roberts, J., Duff, E.: DDX: A distributed architecture for robot control. In: Australian Conference on Robotics and Automation, Canberra, Australia (2004)
4. Duff, E., Usher, K., Taylor, K., Caris, C.: Web-based tele-robotics revisited. In: Proceedings of Australian Conference on Robotics and Automation, Brisbane, Australia (2007)
5. Goertz, R., Thompson, R.: Electronically controlled manipulator. Nucleonics 12(11), 46–47 (1954)
6. Goldberg, K., Mascha, M., Gentner, S., Rothenberg, N., Sutter, C., Wiegley, J.: Desktop teleoperation via the World Wide Web. In: Proceedings of the IEEE International Conference on Robotics and Automation, Japan, pp. 654–659 (1995)
7. Goldberg, K., Siegwart, R.: Beyond Webcams An Introduction to Online Robots. MIT Press, Cambridge (2001)
8. Greenspan, M., Ballantyne, J., Lipsett, M.: Sticky and slippery collision avoidance for tele-excavation. In: Proceedings of the IEEE International Conference on Intelligent Robots and Systems, Grenoble, France, pp. 1666–1671 (1997)
9. Kosuge, K., Takeo, K., Ishida, H.: Teleoperation system of (sic) power shovel for subterranean line work. In: Proceedings of the 23rd International Conference on Industrial Electronics, Control and Instrumentation, pp. 1421–1426 (1997)
10. Minamoto, M., Matsunaga, K.: Tele-presence information and remote-controlled task execution. In: Proceedings of the IEEE International Conference on Intelligent Robots and Systems, Victoria, Canada, pp. 1102–1106 (1998)
11. Richer, J., Drury, J.L.: A video game-based framework for analyzing human-robot interaction: characterizing interface design in real-time interactive multimedia applications. In: HRI 2006: Proceeding of the 1st ACM SIGCHI/SIGART conference on Human-robot interaction, pp. 266–273. ACM Press, New York (2006), http://doi.acm.org/10.1145/1121241.1121287
12. Roberts, J.M., Duff, E.S., Corke, P.I.: Reactive navigation and opportunistic localization for autonomous underground mining vehicles. International Journal of Information Sciences 145(1-2), 127–146 (2002)
13. Sellner, B.P., Hiatt, L.M., Simmons, R., Singh, S.: Attaining situational awareness for sliding autonomy. In: HRI 2006: Proceeding of the 1st ACM SIGCHI/SIGART conference on Human-robot interaction, pp. 80–87. ACM, New York (2006), http://doi.acm.org/10.1145/1121241.1121257
14. Seward, D., Margrave, F., Sommerville, I., Morrey, R.: LUCIE the Robot Excavator - Design for System Safety. In: Proceedings of the IEEE International Conference on Robotics and Automation, Minneapolis, USA, pp. 963–968 (1996)
15. Taketsugu, H., Takashi, Y., Junichi, A., Masaki, I., Hiroaki, Y.: Experimental land model of tele-operated underwater backhoe with AR technology. In: Proceedings of the International Symposium on Underwater Technology, pp. 339–344 (2004)
16. Taylor, K., Trevelyan, J.: Australia's telerobot on the web. In: 26th International Symposium on Industrial Robotics (1995), http://telerobot.mech.uwa.edu.au/Telerobot/

Camera and LIDAR Fusion for Mapping of Actively Illuminated Subterranean Voids

Uland Wong, Ben Garney, Warren Whittaker, and Red Whittaker

Abstract. A method is developed that improves the accuracy of super-resolution range maps over interpolation by fusing actively illuminated HDR camera imagery with LIDAR data in dark subterranean environments. The key approach is shape recovery from estimation of the illumination function and integration in a Markov Random Field (MRF) framework. A virtual reconstruction using data collected from the Bruceton Research Mine is presented.

1 Introduction

Mine accidents including those at Quecreek, Sago and Crandall Canyon highlight the urgency of estimating accurate 3D geometry in mines. Systems have been employed to map mines, from virtual reality systems for training rescue personnel [1] to automated survey robots and post accident investigation [2]. While many of these systems use state-of-the-art direct range measurement sensors, LIDAR sensors alone cannot meet the resolution, size, power or speed requirements to produce quality mine maps in a practical amount of time.

This research combines absolute range sensor data with high-resolution CCD imagery in a novel manner to achieve a quantitative increase in range data accuracy and density. In particular, the method targets application in artificial subterranean voids where assumptions can be used to constrain the image formulation problem. As both color and geometric information are of interest, cameras and range sensors commonly exist on modeling platforms [2]. Integration of the method presented here requires only calibration and low processing overhead.

The results from field experimentation in a working mine are discussed in detail. A dense visualization technique enabling mesh quality models to be displayed and updated in real-time on GPU hardware is explored. Lastly, a generalization of the method to similar domains in field robotics is made.

Uland Wong, Warren Whittaker, and Red Whittaker
Robotics Institute, Carnegie Mellon University
e-mail: uyw@andrew.cmu.edu, warrenw@andrew.cmu.edu, red@cmu.edu

Ben Garney
PushButton Labs
Ben.Garney@gmail.com

A. Howard et al. (Eds.): Field and Service Robotics 7, STAR 62, pp. 421–430.
springerlink.com   © Springer-Verlag Berlin Heidelberg 2010

2 Prior Work

The fusion of range and imaging sensors to improve 3D model quality has been studied in depth [3,4,5,6]. A general model for fusing raw LIDAR and image data into super-resolution range images using a Markov Random Field (MRF) was explored in Diebel and Thrun's seminal paper [4]. MRFs are undirected graphs that represent dependencies between random variables and have been used extensively in computer vision for noise removal, feature matching, segmentation and inpainting (see [3]). The popularity of the MRF stems from the ability to model complex processes using only a specification of local interactions, the regular grid nature of CCD images and the maximum *a posteriori* (MAP) solution requiring only direct convex optimization in many cases.

Diebel and Thrun surmised that higher resolution intensity (color) data could be used to texture range images and increase the range accuracy of interpolated points. The results in a uniformly and sufficiently illuminated regular office environment are quite compelling. Cameras are able to turn LIDAR scans into dense range images with very low computational overhead. However, the assumption that an image provides relative range information, even locally, is tenuous in unstructured environments. Generating 3D geometry from a general 2D projection is an ill-posed problem. The ability of Diebel's method to smooth point clouds using areas of flat image information was convincingly shown, but the converse of enhancing a point cloud using image texture was not. Recent research in range/camera fusion using MRFs include [5,6]; all of which also target indoor application.

This research extends MRF-based super-resolution to subterranean environments such as mines, caves, lava tubes and sanitary pipes. These environments have unknown but slowly varying albedos with a dominant diffuse reflectance term. These naturally-dark, enclosed spaces also require active illumination to image, enabling the use of calibrated lighting. With these assumptions we are able to provide a stronger depth estimate for texturing the interpolated LIDAR data.

3 Markov Random Field Framework

A range image is used as the common representation for fusion. The 3D range cloud data is registered to the pinhole of the camera, forming a range map (R) via projection of distances onto the $n \times m$ image plane at equivalent resolution. Many pixels in the range map will not contain range measurements; these holes are filled from nearby data through bilinear or nearest neighbor interpolation. The color image data can be then converted to intensity values or used as a raw RGB vector (I). A lattice MRF is formed where there is a single range and intensity measurement associated with each node. We propose an MRF fusion method similar to that documented in [4] that numerically integrates the image gradient.

The range map potential (3.1) promotes agreement between the estimated variables and the interpolated range data. The smoothness prior (3.2) regularizes large changes in the range estimate and like the image potential (3.3) connects potential transfer from a node to its neighbors.

$$\Psi = w_1 \sum_{i \in L} (R_i - x_i) \tag{3.1}$$

$$\Omega = \beta \sum_{i \in L} \sum_{j \in N(i)} (x_j - x_i)^2 \tag{3.2}$$

$$\Phi = \alpha \sum_{i \in L} \sum_{j \in N(i)} (x_j + \nabla I_{ij} - x_i)^2 \tag{3.3}$$

$$\alpha = w_2 \exp(-c \cdot \sigma)$$
$$\beta = w_3 (1 - \exp(-c \cdot \sigma)) \tag{3.4}$$

The image gradient is a reasonable predictor of depth change across neighboring pixels. However, integrating the gradient to produce depths over a large locality is prone to drastic shape distortions. The range estimate can be used to regularize numerical integration of the intensity gradient. The weights α and β are relatively scaled by an interpolation distance uncertainty (σ) for some weights w_2 and w_3 (3.4). σ can be generated from the range image during inpainting by using the Matlab command BWDIST, for example.

$$p(x \mid R, I, \sigma) = \tfrac{1}{Z} \exp\left(-\tfrac{1}{2}(\Psi + \Omega + \Phi)\right) \tag{3.5}$$

$$x_{MAP} = \arg\min_x f(\Psi + \Phi + \Omega) \tag{3.6}$$

Solving for the MAP of the distribution requires running a gradient descent algorithm on the target variables x in 3.5-3.6, where Z is the partition function [4].

4 Structure from Shading

The image gradient ∇I_{ij} in (3.3) can apply to either raw pixel data or better estimates of depth from the camera. As scene geometry cannot be ascertained from a single image without assumptions, often no better estimate exists. Definite reconstruction requires knowledge of image formation parameters like light field, surface reflectance (BRDF) and albedos. However, if assumptions like those commonly made in Shape-from-Shading are valid, the number of unknowns is greatly reduced.

The illumination and reflectance assumptions are appropriate for subterranean environments. Most dry underground mines and caves are located in Lambertian rock and many coal mine interiors are additionally covered with diffuse material like Shotcrete [7]. Low amounts of metallic meshing, industrial equipment, water and retro-reflectors are present, but the contribution of these specular surfaces can be reduced using the method documented below and in [10]. Robots in these naturally dark environments can be fitted to carry small area light sources for photography which produce simple light fields.

The MRF image observation (I) is estimated using Shape-from-Shading given the above assumptions. A lightness-based direct normal estimation method which uses range information is given below, but other techniques exist [8,9]. This method factors range information to allow varying albedos and trades accuracy for feature preservation. The effect of the light source's irradiance fall-off is first re-moved from the raw image data (E_0). We assume the following irradiance correction model for small area sources (4.1):

$$E_{unbiased} = \gamma(E_0) \cdot R^n \tag{4.1}$$

The radiometric function (γ) maps pixel values to irradiance, (R) is the interpolated depth estimate and (n) is the irradiance fall-off factor. For ideal point sources $n = 2.0$, while $n < 2.0$ for near-field area sources. The experimental setup described below exhibits an empirical decay of $n = 1.265$. The corrected image (E_c) is devoid of a near-field illumination intensity bias.

Converting RGB color into a single intensity value provides compactness and symmetry, and also minimizes chromaticity effects. Color space transformations such as CieLAB or YCbCr are often used to heuristically isolate the lightness component of an image, discarding chromaticity and albedo. Zickler's SUV transformation [10] describes a class of physics-based specular-invariant color spaces produced by rotating the RGB space such that a single channel is aligned with the illuminant color vectors. This method has produced excellent results with single-source images and enables many Lambertian algorithms to handle a large set of environments with specularities. The specular invariant image, as defined in eqs. 4.2-4.3, is used in experimentation:

$$[s, u, v]^T = R_r(\theta) \cdot \left[E_{unbiased}^{(r)}, E_{unbiased}^{(g)}, E_{unbiased}^{(b)} \right]^T \tag{4.2}$$

$$E_{inv} = \sqrt{u^2 + v^2} \tag{4.3}$$

$R_r(\theta)$ is defined as a (3×3) rotation matrix that aligns the red channel of an $\{r, g, b\}$ triple with the source color. The magnitude of the $\{u, v\}$ components is taken to be the diffuse image.

An albedo map is subsequently generated from the diffuse image using Blake's method for lightness computation [11]. Perceived intensity is a multiplicative relationship between surface slant angle and reflectance. The log image separates these components into additive terms. Scene albedos can be recovered from the gradient of the log diffuse image by thresholding to remove small changes and integrating. It is noted that the problem can be recast as finding the log albedo map (δ) that minimizes the following [11]:

$$\arg\min_\delta \left| \tfrac{\partial}{\partial x}\delta - T_\sigma\left(\tfrac{\partial}{\partial x}\log E_{inv}\right) \right|^2 + \left| \tfrac{\partial}{\partial y}\delta - T_\sigma\left(\tfrac{\partial}{\partial y}\log E_{inv}\right) \right|^2 \tag{4.4}$$

where (T_σ) is the threshold function. Exponentiating (δ) with the proper constant of integration produces the albedo values (4.5). The constant can be estimated from the range data to minimize depth discrepancy in the reconstruction.

$$\rho_{est} = \exp(\delta + c) \tag{4.5}$$

$$E_{inv} = \rho |n||l| \cos(\theta_{nl}) \tag{4.6}$$

$$\theta_{nl} = \arccos\left(\frac{E_{inv}}{\rho_{est}}\right) \tag{4.7}$$

The polar estimates (θ_{nl}) are combined with azimuth estimates (ϕ) from the range image and converted to gradients for integration in the MRF.

5 Experimental Results

The experimental setup uses both a continuously rotating planar LIDAR scanner and an 8 megapixel DSLR camera mounted to a mine robot. A small area light source is also mounted along the same axis to minimize cast shadows in the image. This replaces the normal flood lighting for the imager. The scanner has a practical throughput of ~40,000 points per second. The points are aligned along concentric rings with 0.5° angular separation in a 180° hemisphere in front of the unit. The camera takes hemispherical images using a constant angular resolution fisheye lens with a 182° field of view. The sensor mounting configuration and example data are shown in Fig. 1 below.

Fig. 1 (Left) Experimental setup with 1. LIDAR scanner. 2. Fisheye Camera, 3. Light Source. (Center) Raw fisheye imagery. (Right) Ground truth range image.

Thirty complete datasets consisting of LIDAR scans, High Dynamic Range (HDR) imagery and robot odometry were collected from the Bruceton Research Coal Mine in Pittsburgh, PA. LIDAR scans averaged 600,000 points. HDR images were each generated from a series of 5 images corresponding to exposures times of {¼, ½, 1, 2, 4} seconds using the method described in [12]. The 1.0 second exposure image was used as the Low Dynamic Range (LDR) reference image for analysis. An additional 16 datasets of LDR-only imagery were also collected.

A ground truth range map was generated for each LIDAR scan using the full point cloud. Multiple measurements mapping to the same pixel were averaged. The scans were subsequently down-sampled to 25,000 points and interpolated into a range image for testing the method. The datasets were further partitioned into 25 test sets and 5 training sets. Optimal weighting factors were learned using a simplex search on the training set, while validation occurred in the test set.

The proposed method was compared against Diebel's method and raw interpolation. The mean per-pixel L_1 norm (Manhattan distance) between the reconstructed range map and the ground truth map was used as a benchmark for

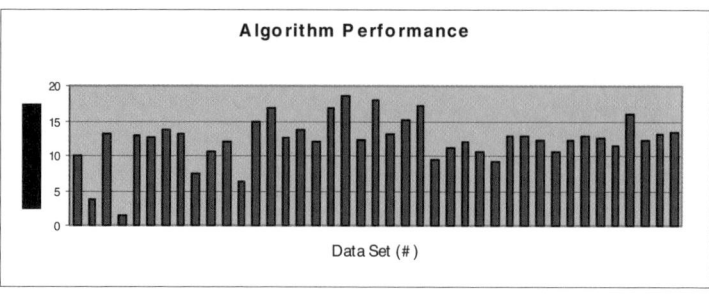

Fig. 2 Reconstruction Improvement vs. Raw Interpolation.

comparison. Ground truth data points outside the convex hull of LIDAR values in the interpolated map were discarded due to skew in scoring extrapolated points. The usable pixel area is determined for each scan by the number of saturated pixels, the range image convex hull and removal of high-gradient probable error values.

The results of the experiment are summarized in Table 1 and Fig. 2. Fig. 3 and Fig. 4 show an example reconstruction from a single view point. The scene features a yellow nylon mine curtain on the left side, wooden cribbing stacks on the right and aluminum meshing integrated into a mostly exposed ceiling.

Additional data of two corridors were also collected at the Bruceton Mine along evenly spaced intervals roughly 3 meters apart. Using robot odometry and

Table 1 Summary of Results

Quantity	Details
Total Test Datasets	41
{HDR, LDR-only} Datasets	{25, 16}
Interpolation Improvement	
Mean	12.2%
Max, Min	19.2%, 3%
Density Statistics	
LIDAR downsample	25,000 points
Ground Truth LIDAR	669,834 points
Mean Resultant	1,045,358 points
Mean Increase	41.8 x
Image Usability Information	
LDR Saturated	3.17% of total pixels
HDR Saturated	$4.20 \times 10^{-2}\%$ of pixels
HDR Accuracy Increase	20.5% over LDR-only
HDR Density Increase	51.5% over LDR-only

Fig. 3 Point Cloud of Cribbing. Low resolution cloud (left) and high resolution reconstruction from algorithm (right) showing stacked timbers supporting the roof.

Fig. 4 Colorized 3D Reconstruction. Full scene (left) and mine curtain detail (inset and right).

Iterative Closest Point (ICP) alignment, multiple scans were up-sampled using the proposed technique, fused together and color/illumination compensated. These models represent some of densest, most comprehensive mine reconstructions to date using a mobile robot. The results appear below:

Table 2 Corridor Modeling Statistics.

Model #	# of Scans	# of Images	# Points
1	4	16	5,543,451
2	8	32	9,680,105

The results are displayed using a hole-filling method similar to the multi-scale push-pull technique in [13]. This display system is adapted to benefit from high density clouds generated using super-resolution methods. Point clouds are rendered with push-pull interpolation in image space. A min-depth check and kernel density estimator are used to resolve edge discontinuities and remove occluded background measurements. The utilization of texture in-painting for both color interpolation and depth reconstruction provides the viewer with graphical continuity as well as proper occlusions, which standard point displays lack. In addition to fast rendering of huge datasets, the renderer allows the model to be updated in real

Fig. 5 Mine Corridor ICP model. (1) External view. (2) Internal view with rail tracks.

time as new data arrives without costly re-meshing operations. The system can generate real-time (>30Hz) imagery at 1080p HD resolution on commodity (Ge-Force GTX 260) hardware with point clouds of greater than 5 million points

6 Analysis

The results show that the method increases interpolation accuracy by up to 20% on the Bruceton Mine data, with an average improvement of 12%. The fisheye-spinner setup features density increases up to 70 fold, with an average of 40x increase in density (Table 1). Of note is that real resolution is created where LIDAR beam physics dictate a maximum angular resolution. This is apparent in 3D scanning mechanisms that actuate a planar sensor, where an increase in data collection time results in diminishing resolution returns.

To validate that true information is being stored in the interpolated values, a sliding-window 15x15 pixel Pearson correlation was performed. As shown in Fig. 6, the shaded image provides significant information about the ground truth that is not contained in interpolation. The fused range map correlates more than either source individually, concurring with the error estimation benchmark. While Diebel's method shows a numerical increase in accuracy, it is not statistically significant. This is corroborated by almost equal amounts of strongly negative and positive correlation in the raw image data.

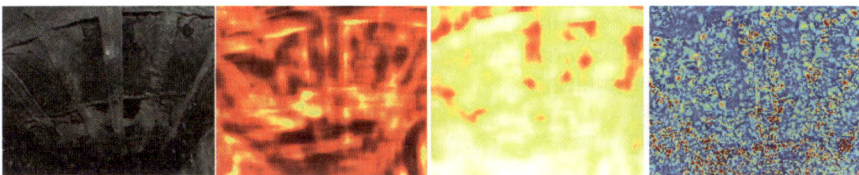

Fig. 6 (Left to right) Roof supports covered in Shotcrete, Image to ground truth correlation, Shaded image to ground truth correlation, and Reconstruction error reduction. Scale is brown to white over [-1, 1] for correlation and navy blue to red over [-0.025m, 0.05m] for error reduction.

The method encounters several drawbacks that prevent the fused result from achieving the same accuracy as LIDAR scans of equivalent density. Resulting range images are vulnerable to artifacts typical of raw interpolation, although to a lesser degree. Most reconstruction error occurs at occlusion edges where neighboring LIDAR points have large disparities. Regularization terms tend to over-smooth these edges and shading cues are ill-behaved due to cast shadows, among other reasons [9,11]. Attempting to isolate these specific edges in the image is difficult due to image noise, lighting and material specific effects and is not addressed in this research (see [5,14]).

7 Conclusion

A method was presented that fuses actively illuminated CCD imagery and LIDAR data. The method demonstrates increases in range accuracy of up to 20% on experimental data over interpolation and increases in measurement density of up to 70x using the experimental setup. The improvements are a result of calibrated imaging using additional knowledge of the image formulation model to reconstruct a 3D observation of the scene. This research demonstrated the efficacy of multi-sensor mapping systems as well as calibrated imaging for field robots.

Perhaps the greatest argument for range/image super-resolution is that it is easily bootstrapped to existing systems. Subterranean robots already require light sources for photography as well as range sensors for mapping and many high-throughput commercial scanners feature co-located cameras. The general use of illumination information for super-resolution is also applicable to other domains in field robotics. Planetary robots are likely to encounter highly diffuse environments (i.e. Mars) or characterizable reflectances on bodies lacking scattering atmospheres (i.e. moon, asteroids). Such development is likely to increase the safety of exploration and prospecting on the moon, where sensing is secondary to payload and comes at a premium cost.

In future work, agile and high accuracy applications will benefit from one or more actuated sensors. With an actuated camera, the technique can be used to zoom in on regions of interest and selectively up-sample a scan. With an actuated range scanner, such as a focal plane array LIDAR with variable optics, a static camera may be used to reconstruct rough low frequency data from a preliminary scan and detect areas of high frequency, while a second LIDAR pass focuses specifically on these areas. This setup can drastically compress the amount of data taken and reduce time required while producing an optimal reconstruction given certain throughput constraints.

References

1. Boulanger, P., Lapointe, J.-F.: Creation of a Live Virtual Reality Model of Mining Environments from Sensor Fusion. In: Proc. Minespace 2001 (2001)
2. Morris, A., Ferguson, D., et al.: Recent Developments in Subterranean Robotics. Journal of Field Robotics 23(1), 35–57 (2006)
3. Li, S.: Markov Random Field Modeling in Image Analysis (2001)
4. Diebel, J., Thrun, S.: An Application of Markov Random Fields to Range Sensing. In: NIPS 2005 (2005)
5. Torres-Mendez, L., Dudek, G.: Inter-Image Statistics for 3D Environment Modeling. IJCV 79, 137–158 (2008)
6. Gould, S., Baumstarck, P., Quigley, M., et al.: Integrating Visual and Range Data for Robotic Object Detection. In: ECCV 2008 (2008)
7. Clements, M.: Shotcreting in Australian Underground Mines: A Decade of Rapid Improvement. In: Shotcrete Spring 2003 (2003)
8. Zhang, R., Tsai, P., Cryer, J., Shah, M.: Shape from Shading: A Survey. IEEE PAMI 21(8) (August 1999)
9. Braquelaire, A., Kerautret, B.: Reconstruction of Lambertian Surfaces by Discrete Equal Height Contours and Regions Propagation. Image and Vision Computing 23(2), 177–189 (2005)
10. Mallick, S., Zickler, T., Kriegman, D., Belhumeur, P.: Beyond Lambert: Reconstructing Specular Surfaces Using Color. In: CVPR 2005 (2005)
11. Worthington, P.: Re-illuminating single images using Albedo estimation. In Pattern Recognition (2005)
12. Debevec, P., Malik, J.: Recovering High Dynamic Range Radiance Maps from Photographs. In: SIGGRAPH 1997 (1997)
13. Grossman, J., Dally, W.: Point Sample Rendering. In Rendering Techniques (1998)
14. Yang, Q., Yang, R., Davis, J., et al.: Spatial-Depth Super Resolution for Range images. In CVPR 2007 (2007)

Part X
Maritime Robotics

A Communication Framework for Cost-Effective Operation of AUVs in Coastal Regions

Arvind Pereira, Hordur Heidarsson, Carl Oberg, David A. Caron, Burton Jones, and Gaurav S. Sukhatme

Abstract. Autonomous Underwater Vehicles (AUVs) are revolutionizing oceanography. Most high-endurance and long-range AUVs rely on satellite phones as their primary communications interface during missions for data/command telemetry due to its global coverage. Satellite phone (*e.g.,* Iridium) expenses can make up a significant portion of an AUV's operating budget during long missions. Slocum gliders are a type of AUV that provide unprecedented longevity in scientific missions for data collection. Here we describe a minimally-intrusive modification to the existing hardware and an accompanying software system that provides an alternative robust disruption-tolerant communications framework enabling cost-effective glider operation in *coastal regions*. Our framework is specifically designed to address multiple-AUV operations in a region covered by multiple networked base-stations equipped with radio modems. We provide a system overview and preliminary evaluation results from three field deployments using a glider. We believe that this framework can be extended to reduce operational costs for other AUVs during coastal operations.

1 Introduction and Motivation

Autonomous Underwater Vehicles (AUVs) are revolutionizing oceanography. They have been widely used for *in-situ* measurements which would be difficult, expensive, and, in some cases, impossible to obtain by using traditional ship-based

Arvind Pereira, Carl Oberg, and Gaurav S. Sukhatme
Dept. of Computer Science, University of Southern California, Los Angeles
e-mail: {menezesp,gaurav}@usc.edu

Hordur Heidarsson
Dept. of Electrical Engineering, University of Southern California, Los Angeles
e-mail: heidarss@usc.edu

David A. Caron and Burton H. Jones
Dept. of Biological Sciences, University of Southern California, Los Angeles
e-mail: {dcaron,bjones}@usc.edu

A. Howard et al. (Eds.): Field and Service Robotics 7, STAR 62, pp. 433–442.
springerlink.com   © Springer-Verlag Berlin Heidelberg 2010

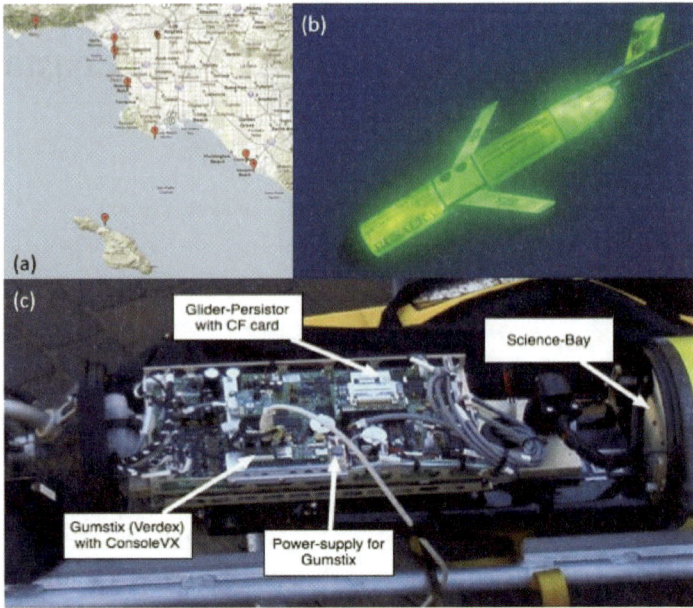

Fig. 1 (a) Basestation locations (current and planned) surrounding the Southern California Bight (SCB), (b) The Slocum glider, (c) Hardware modifications to the glider

sampling techniques [1]. AUVs typically use thrusters, rudders and fins as actuators [5]. *Gliders* [6, 3] are specialized AUVs that rely on buoyancy control and shifting center of mass for propulsion, to *fly* in the ocean - an energy-efficient technique that results in long mission times (3-4 weeks) at sea.

Table 1 shows several popular AUV platforms, and their primary modes of communication. The usual operation of AUVs involves the creation of a mission file during the mission planning stage (onshore). Most of the vehicles in Table 1 use a radio link (WiFi, radio modem) for operator-vehicle communications when the vehicle is near the operator. This typically occurs during the mission upload phase. Once deployed, AUVs typically communicate with a basestation onshore (or on a ship) using an acoustic modem or a satellite phone.

While satellite phones have the advantage of being usable at almost any ocean surface location, they are plagued by very low data-rates (*e.g.,* 2400 bps maximum for Iridium) and high costs for transmitted data or call time. Slow data rates imply longer times spent at the surface for data transfer. This is a safety concern in areas with high marine traffic such as the Southern California Bight (SCB), our region of interest and operation (the SCB is the oceanic region contained within 32°N to 34.5°N and 117°E to 121°E). Satellite phone communications are expensive. We estimate the nominal communication cost for Iridium usage to be approximately USD 2400 for a 3 week glider mission or approximately half of the total expendible cost of the deployment. Others [6] report their Iridium communications cost to be approximately USD 180/day which translates to approximately USD 3500 for a 3

Table 1 AUVs

Name	Manufacturer	Endurance	Radio	Acoustic	Satellite
Bluefin-12	Bluefin Robotics	10-23 h	Yes	Yes	Yes
HUGIN 1000	Kongsberg	17-24 h	WiFi	Yes	Yes
REMUS 600	Hydroid	20-45 h	WiFi	Yes	Yes
Gavia	Hafmynd	24 h	WiFi	Yes	Yes
SAUV II	Falmouth Scientific	Unlimited	Yes	Yes	Yes
Slocum Electric Glider	Webb	4 weeks	Yes	No	Yes
Spray Glider	UCSD	months	No	No	Yes
Seaglider	iRobot	months	No	No	Yes

week mission for a single glider. These limitations imply that during a surfacing, experimenters using satellite phones are often forced to transmit subsets of data from the AUV, instead of the entire dataset. The Iridium plan used in this work, is based on call time.

One strategy to mitigate the shortcomings of satellite phones is to use acoustic modems [9, 4, 2] or combined acoustic/optical strategies [8]. The obvious advantage is that AUVs need not surface to communicate if they are using acoustic modems. However, data rates on acoustic systems are typically low, they also have a high one-time cost and suffer from multi-path interference in shallower coastal regions. Optical techniques are typically shorter range and unsuitable for operations in deeper waters.

We remark that radio modems (*e.g.,* the Freewave^{TM}) used on AUVs (*e.g.,* the Webb Slocum glider (Table 1)) are rated for a range of 60 miles line-of-sight. Their use need not be restricted to dockside operations for mission upload; it could be extended to large near-coastal regions (*e.g.,* the SCB). A multi-AUV deployment over an extended time period in a region as large as the SCB could see significant cost reductions if the primary mode of communication with the AUV was a radio modem instead of a satellite phone. Our experimental platform, the Webb Slocum Glider, is primarily designed to communicate using a Iridium satellite modem during missions. When the operator is within Freewave range, the modem is typically used for launch, retrieval, data transfer, and maintenance of the vessel. In the course of a typical mission, the Freewave is used infrequently, and mostly at the dockside. This is because its effective range to the operator is rather small since it is rare to obtain line-of-sight connectivity between vehicle and operator during a mission due to occlusion.

Can the effective range of the radio modem be extended so that a region the size of the SCB would be effectively 'covered' thus rarely necessitating the use of a satellite phone for operator-AUV operations ? Here we report on the encouraging progress towards answering this question in the affirmative by 1. designing and augmenting coastal communication infrastructure (radio modems onshore at elevated sites for better line of sight connectivity to the vehicles), and 2. designing, implementing and testing protocols for data-transfer using radio modems.

To exploit the radio modem to its fullest potential we make the observation that the Southern California region has several HF-Radar (CODAR) sites at elevated locations. These provide accurate ocean surface current data, and are always instrumented with an internet connection. This infrastructure is a cost effective way to set up a network of radio modem shore stations to provide radio modem connectivity to vehicles on near-coastal missions. Elevated locations provide greater line-of-sight with the vehicle.

We contend that with a minimal modification of the vehicles and a small addition to existing shore locations it is possible to build a network of this kind that scales with multiple vehicles at limited cost. This paper describes the design and implementation of such a system. We report on communication tests using Webb Slocum gliders in the SCB, with the expectation that this strategy can pave the way for a similar use of a reduction in the communication costs for coastal operation of other AUVs.

2   The Webb Slocum Glider Communications System

The glider is a specialized robot driven by buoyancy which can fly in the ocean for extended periods of time at the expense of speed and maneuverability. Glider designers have devoted significant effort to power consumption minimization. The glider's navigation and communications are handled by a low-power microcontroller called the *Persistor*^{TM}. This computer performs standard navigational tasks and runs a modified version of PicoDOS^{TM} called GliderDOS^{TM} which contains glider-specific software.

In normal glider operations, the glider's Freewave modem is configured as a *slave* to connect to only a single *master* Freewave modem at the operator's end. The other side runs software from Webb Research called the DockServer^{TM}. The glider can also be operated via any terminal client since it provides a human-readable interface via ASCII strings. There is no inherent packetization of data being performed on the glider since it assumes that it is always connected to a single computer via either of its two links (Freewave or Iridium). This situation, coupled with the fact that the Freewave modems do not have a mode of operation which can independently handle hand-offs between modems, means that it is difficult to build a reliable end-to-end system to communicate with a glider without using packetization for the identification of sources and destinations. Any disruption in communication due to loss of a link or a reconnection of the glider via a new link, results in data corruption.

3   System Design

Our system consists of three main components. At the glider level, we have the communications module that handles the radio communications on the glider. At shore, we have the internet connected basestations which have the radio and antenna to communicate with gliders deployed in the ocean. Finally, we have the control server, a central data-aggregation and command/control server. This overall system

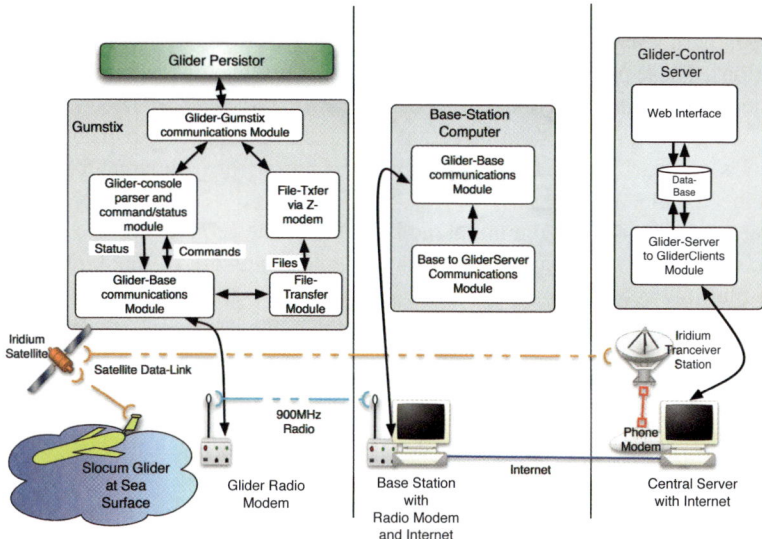

Fig. 2 The System Block Diagram

design, illustrated in Fig.2 adds more high-level control to the glider operations than currently possible which in turn facilitates autonomous re-tasking of the gliders on mission.

3.1 Communication Module and Protocol

The communications module is a combination of hardware and software to handle communications between the glider and shore. The hardware is specific to the robot platform (in this case a glider), and the software has some general building blocks as well as a platform-specific interface devoted to interaction between the communications code and the control software on the robot itself. We have implemented this module on a Webb Slocum glider, and this paper describes experiments with that particular AUV, but the module can easily be added to most other AUVs.

The basestations and gliders communicate through our own light-weight communication protocol. Each basestation can store and forward certain information between specified nodes (gliders in this case). We treat the Freewave modems as a serial link, and have incoming and outgoing packet queues which provide feedback to vary both the inter-packet delay as well as Freewave packet-sizes. Freewave modems make a "best-effort" delivery attempt on these packets, which can be fragmented into smaller pieces in the event of poor connections. This protocol (which we will describe in more detail in a future paper), also supports guaranteed delivery as well as a non-guaranteed mode of transmission. The packet structure contains 14 bytes without payload data, and allows several applications to multiplex data, such as file transfers, status packets, data packets, terminal commands and so on. We use

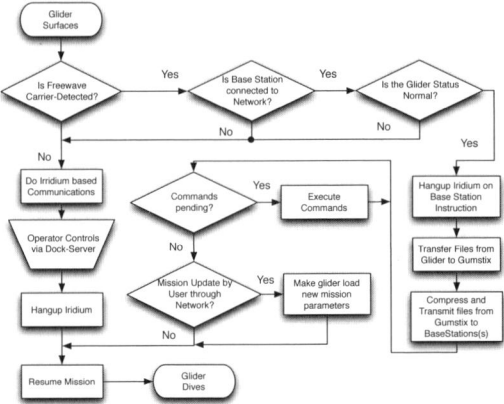

Fig. 3 Control Flow of tasks performed during glider surfacings

a *Selective-Acknowledgement* communications scheme with a fixed window which works well in regions with *persistent line-of-sight* between antennae.

On our Slocum gliders, the communication module hardware consists of a Gumstix computer and the aforementioned software. Adding the Gumstix between the glider's control computer and the Freewave is a minimally intrusive way of adding new communication capabilities to the glider. Using a separate computer to interface with a vehicle abstracts the interface between higher-level communications and lower level vehicle control cleanly and can also be used in the future to handle glider re-tasking. We chose a *Gumstix* because, besides its small size, it is a fully functional Linux-based computer, consumes very little power ($<$120mA @5V, 400MHz), and has good interfacing capabilities. Fig.1(c) shows the physical modifications to the glider due to the addition of the Gumstix. We need to make only 5 modifications to the glider to allow the glider to communicate with an external computer. Although the Gumstix consumes more power than the persistor alone, we have designed the system such that the Gumstix is powered up along with the Freewave modem - a feature that ensures that it gets automatically turned on at the water surface, while staying shut off when the glider is diving.

The platform-specific software on the glider intercepts all messages to and from the glider persistor and the Freewave modem, parses the ASCII strings from the glider and follows the basic control flow displayed in Fig. 3. It also sends necessary status information to shore using our packet protocol, gathers sensor data files from the persistor, compresses them and sends them to the shore station. If communication with the main server via Freewave is unavailable, the glider falls back to Iridium to call in.

3.2 Base Stations

The basestations are the shore stations that handle direct communication with the gliders. The hardware consists of an internet-connected computer, running Ubuntu

Linux Server, a Freewave radio and an antenna. The computer runs our communication software which essentially relays datagrams between the gliders and the control server. Communications between these basestations and the control server take place via TCP/IP. If the basestation loses connection to the control server it turns of its Freewave to ensure gliders connect to another basestation or fall back to Iridium for communication.

3.3 Control Server

The control server is the orchestrator of the overall system. It is written in C++ and runs on Ubuntu Linux Server and utilizes a MySQL database for storing data. It maintains a connection to each of the basestations, keeps track of the state of the system and is in charge of issuing commands to connected gliders. It also notifies the end users of events via email. All data it receives is logged to a database. On top of the control server, we have a web based user interface, written in Java^{TM}with Google Web Toolkit, to provide easy accessible control and visualization to the end user of the system. The server also runs 3rd party monitoring software to monitor the health of the basestation computers and software. If a problem is encountered system administrators are notified.

4 Experiments and Results

To test our design, we have performed 3 experimental deployments. The first was off the coast of Santa Catalina Island in November 2008. The other two were near Pt. Fermin in January and February 2009. The heat-map for File transfers (Fig. 4) is based on results extrapolated from glider surfacing and transferring files to shore from locations C, D and E. The heat-map for Carrier Detect (Fig.5(a)) shows a

Table 2 Results from Field and Laboratory Tests

Test	Maximum Distance [km]	Data Rate (KB/sec)	File Transfers	Freewave Switching	Quality of Link
A - Pt.Fermin *	2.6	Not Measured	Not Tested	No	Poor
B - Pt.Fermin *	5.2	Not Measured	Not Tested	No	Poor
C - Pt.Fermin †	12.3	Not Measured	Not Tested	Yes	Intermittent
D - Pt.Fermin +	9.2	0.1535	Yes	No	Slow
E - Pt.Fermin	3.5	1.46	Yes	No	V.Good
L - Lab Tests	N.A.	7.883	Yes	No	V.Good

* This test was conducted with a faulty antenna installation at Pt.Fermin

† Connection was made to both Pt. Fermin and Catalina Island. Distance to Catalina was 20 km.

+ Remote operator performed a glider re-tasking via network at this location through Pt.Fermin

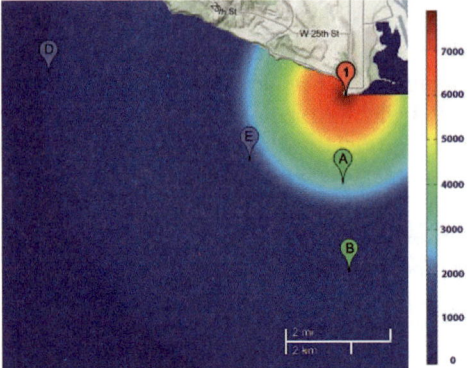

Fig. 4 File Transfer data rate off Pt. Fermin

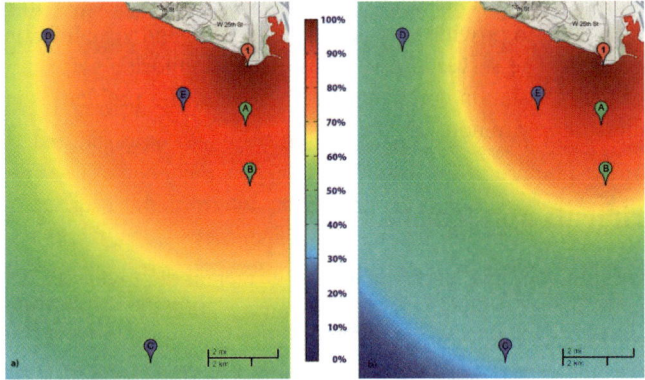

Fig. 5 a) Carrier-detect success percentage (CD on time/Total time) b) Link-Layer protocol's uptime percentage (Protocol time/CD on time)

percentage of time the base-stations had Carrier-detect with the gliders to the total time of a surfacing. This metric is based on 4 surfacings. Fig. 5(b) shows a heat-map based on the percentage of protocol-level Link States to the total Carrier-detect duration during a given surfacing. This metric gives us an idea of how well our protocol is performing in real conditions. Observations show that during intermittent communicatios, such as those at C and D, the protocol-link suffers. These heat-maps are based on simple linear-interpolation of very sparse data with no underlying communication model assumption. Although we have not analyzed the relationship between sea-state and the communication quality achieved, the average wave-height was approximately 1m with a dominant period of 6 seconds during the data-collection period.

5   Conclusion

This paper outlined the design behind a communication system using long-range RF-modems, to communicate with AUVs in a coastal area. Field trials using Slocum gliders indicate promising results and give us valuable insights into improvements in the design. On the gliders themselves, we add a small computer to interface between the original glider computer and its Freewave modem. Our communications proto-col supports datagrams with priority, can maintain in-sequence transmissions,and has both reliable and un-acknowledged datagram modes. Instead of transmitting all the data from the glider, we create status packets which contain a snapshot of the gliders state by parsing its lengthy ASCII transmission. We also utilize the Gum-stix's processing power to compress data files before transmission, which provides us with a typical space saving of approximately 4x. The high data rates our system achieves in the field (1.46KB/sec) is 6 times faster than Iridium while simultane-ously transmitting multiplexed glider console information and status packets. This combined speed increase translates to a 24x improvement, which allows us to send more data, while also reducing surface times and cutting down on Iridium data-transfer costs.

Fig. 4, a sparsely interpolated map based on only 3 averaged measurements, im-plies that fairly high throughputs are possible close to a base-station (<4km), with a fairly sharp bit-rate drop from 1.46KB/s to approximately 154bytes/sec. This sig-nificant drop is due to the links becoming more intermittent as carrier detect on the radio is only available 65% of the time at a distance of 9.2km, while it is more than 86.7% at 3.5km. The measurements of communication performance we present here are sparse and were collected at four surfacing locations, but they represent charac-teristic portions of a typical coastal belt that needs coverage. Field tests have shown that our system allows status packets to be reliably transmitted from distances upto 20km (Glider Surfacing C). We have successfully performed glider re-tasking via the network from a distance of 9.2km - a feature we will use in the future to enable mission re-planning based on data gathered by the glider. We have also developed a central server which allows us to easily collect and visualize data from the glider, or create and send new mission files through the network.

6   Future Work

Experimental results while promising, show that there is room for improvement. From Fig.5 (a) and (b) we make the observation that we can improve upon our protocol, such that its link spans all the time the radio has carrier. We understand that this is a consequence of protocol choices, which were tuned to obtain good results at the lab - which as observed, is significantly different from conditions in the field. We believe local conditions due to waves play a major role in causing communications disruptions, since the antenna of the Slocum glider is very close to the waters surface and local waves occlude line-of-sight between radios. We believe that by using better queue management, introducing variations of re-transmit time and packet sizes, based on the link state, can lead to significant improvements in ensuring we have a much better protocol-level link between the glider and base-stations. We are also in the process of mapping out the entire region of interest for link quality. Equipped with such a map, we can then design planners which incorporate the knowledge of communication link availability to bias the surfacings of AUVs such that they keep overall operation costs low. Concurrent work [7] in our lab, used Iridium to perform feature tracking based on ocean model predictions for the Southern California Bight region. By using our communication system to perform mission adaptations, we will get a more realistic comparison between cost-savings using it instead of Iridium.

Acknowledgements. The authors gratefully acknowledge the help provided by Ivona Cetinic, Matthew Ragan, Ryan Smith, Filippo Arrichiello, Ray Arntz, Kyaa Heller and Martin Getrich in the course of the experimental trials, and the support provided by the CINAPS team at USC. This work was supported in part by grants CCR-0120778, CNS-0520305, CNS-0540420 from the US NSF, ONR MURI N00014-08-1-0693, grant NA05NOS4781228 from the USNOAA MERHAB program, and a gift from the Okawa Foundation.

References

1. Creed, E.L., Kerfoot, J., Mudgal, C., Barrier, H.: Transition of slocum electric gliders to a sustained operational system. In: Oceans 2004. MTTS/IEEE Techno-Ocean 2004, November 2004, vol. 2, pp. 828–833 (2004)
2. Freitag, L., Grund, M., Singh, S., Partan, J., Koski, P., Ball, K.: The whoi micro-modem: An acoustic communications and navigation system for multiple platforms. In: Proceedings of the IEEE Oceans Conference (2005)
3. Griffiths, G., Jones, C., Ferguson, J., Bose, N.: Undersea gliders. Feeding and Healing Humans 2(2), 64–75 (2007)
4. Heidemann, J., Ye, W., Wills, J., Syed, A., Li, Y.: Research challenges and applications for underwater sensor networking. In: Proceedings of the IEEE Wireless Communications and Networking Conference, Las Vegas, Nevada, USA, pp. 228–235. IEEE, Los Alamitos (2006)
5. Erling, J., Refsnes, G.: Nonlinear Model-Based Control of Slender Body AUVs. PhD thesis, Norwegian University of Science and Technology, Department of Marine Technology, Trondheim, Norway (December 2007)
6. Schofield, O., Kohut, J., Aragon, D., Creed, E.L., Graver, J., Haldeman, C., Kerfoot, J., Roarty, H., Jones, C., Webb, D.C., Glenn, S.: Slocum gliders: Robust and ready. Journal of Field Robotics 24(6), 473–485 (2007)
7. Smith, R.N., Chao, Y., Jones, B.H., Caron, D.A., Sukhatme, G.S.: Trajectory design for autonomous underwater vehicles based on ocean model predictions for feature tracking. In: Proceedings of The 7th International Conference on Field and Service Robotics, Cambridge, MA (July 2009) (submitted)
8. Vasilescu, L., Kotay, K., Rus, D., Dunbabin, M., Corke, P.: Data collection, storage, and retrieval with an underwater sensor network. In: Proceedings of the 3rd international conference on Embedded networked sensor systems, November 2005, pp. 154–165. ACM, New York (2005)
9. Yu, X.: Wireline quality underwater wireless communication using high speed acoustic modems. In: Proceedings of the IEEE Oceans Conference (2000)

Multi-Robot Collaboration with Range-Limited Communication: Experiments with Two Underactuated ASVs*

Filippo Arrichiello, Jnaneshwar Das, Hordur Heidarsson, Arvind Pereira, Stefano Chiaverini, and Gaurav S. Sukhatme

Abstract. We present a collaborative team of two under-actuated autonomous surface vessels (ASVs) that performs a cooperative navigation task while satisfying a communication constraint. Our approach is based on the use of a hierarchical control structure where a supervisory module commands each vessel to perform prioritized elementary tasks, a behavior-based controller generates motion directives to achieve the assigned tasks, and a maneuvering controller generates the actuator commands to follow the motion directives. The control technique has been tested in a mission where a set of target locations spread across a planar environment has to be visited once by either of the two ASVs while maintaining a relative separation less than a given maximum distance (to guarantee inter-ASV wireless communication). Experiments were carried out in the field with a team of two ASVs visiting 22 locations on a lake surface (approximately $30000m^2$) with static obstacles. Results show a 30% improvement in mission time over the single-robot case.

1 Introduction

A significant body of literature deals with the motion control of aquatic vehicles for autonomous navigation [11]. Interest in the field is motivated

Filippo Arrichiello and Stefano Chiaverini
DAIEMI, Università degli Studi di Cassino, Via G. Di Biasio 43, 03043, Cassino (FR), Italy
e-mail: `f.arrichiello,chiaverini@unicas.it`

Jnaneshwar Das, Arvind Pereira, Hordu Heidarsson, and Gaura S. Sukhatme
Robotic Embedded Systems Laboratory, University of Southern California, Los Angeles, CA 90089, USA
e-mail: `jnaneshd,menezesp,heidarss,gaurav@usc.edu`

* This work was supported in part by the NOAA MERHAB program under grant NA05NOS4781228, by NSF as part of the Center for Embedded Networked Sensing (CENS) under grant CCR-0120778, by NSF grants CNS-0520305, CNS-0325875 and CNS-0540420, by the ONR MURI program under grant N00014-08-1-0693, and a gift from the Okawa Foundation.

A. Howard et al. (Eds.): Field and Service Robotics 7, STAR 62, pp. 443–453.

by different applications, e.g., naval system applications, harbor operations, defense and patrolling of coastal perimeters, and marine biology applications. Motion control of Autonomous Surface Vessels (ASVs) has been studied in the context of dynamic positioning for fully actuated [15] and under-actuated vessels [18], as well as trajectory tracking [14, 1]. Separately, collaboration in multi-robot teams [10] has been widely studied. In the intersection of these two areas - motion control of a fleet of ASVs - different approaches have been proposed, however most of them have only been validated by numerical simulations [8], or by experiments with a single real vessel and simulating the others [13]. Here, we focus on the cooperative control for ASVs and present the results of experiments with a team of two underactuated ASVs in the field performing a joint mission motivated by environmental monitoring.

We provide a list of target locations to the vessels. They must dynamically allocate locations among themselves such that each location is visited exactly once, and all obstacles are avoided. The approach proposed here is applicable to a scenario with dynamic obstacles and targets since it does not pre-plan a route. In the course of the mission the ASVs are required to maintain connectivity at all times. Unlike [16] where connectivity is maintained by measuring signal strength and the mobility controller is a spring-damper system, we use a layered hierarchical control decomposition which accommodates a dynamic environment (targets and obstacles may be added dynamically). When needed, a leader-follower configuration allows the team to allocate a target to the vessel closest to the next target, while the other vessel ensures the communication constraint is maintained. Our results from five field trials at a lake with a two-ASV team (communication constraint of 60m, in an exploration area of 300m x 100m) show a 30% improvement in exploration time compared to a single vessel.

We focus on navigation techniques for a realistic environment where several obstacles can be found. Thus, we make use of a behavior-based technique as a guidance control to take advantage of its reactivity to unpredicted conditions [9, 6]. In particular, we present the use of a behavior-based technique called the Null-Space based Behavioral (NSB) control as a guidance system for ASVs. The NSB approach has been extensively tested for the control of autonomous ground robots and results have shown robust control for formation and spread control for a team of robots [4], escorting an external agent with a team of robots [2], and the formation control of a fleet of ASVs [7].

2   Problem Description

We require a set of pre-specified locations to be visited by one (and only one) of the ASVs exactly once. During the execution of the mission, both ASVs must ensure that their relative distance does not exceed a preset bound. Needless to say, the vessels must avoid mutual collisions and collisions with obstacles in the environment. We decompose the overall mission into elementary tasks that can be then prioritized and implemented individually; in particular: 1. Avoid obstacles and inter-robot collisions, 2. Satisfy the communication constraint, and 3. Navigate to assigned target locations.

While we discuss the details in the next section, broadly speaking each ASV properly activates and combines these three modes. When the distance between the vessels is consistently lower than the communication bound, they are free to choose targets and navigate to them independently. When the distance is close to the communication range, a supervisory module on each ASV has to ensure the communication constraint is satisfied. This is done using a leader-follower policy where one of the vessels (the leader) continues with its mission, while the other (the follower) has to enforce the separation constraint adapting its motion to the leader. The navigation task moves each (single) vessel toward its assigned location. Since the vessels must avoid all collisions, the obstacle avoidance controller takes evasion action when obstacles or other vessels are within a safety margin.

The targets to be visited are dynamically chosen by the vessels on the basis of their locations. A communication sub-system ensures that a shared target and obstacle map are maintained both of which can be updated dynamically from shore, allowing dynamic missions.

3 Control Strategy

In order to achieve the proposed mission in a cooperative way, the control architecture has been organized into a three-level hierarchy, as shown in Figure 1: Supervisor, NSB, and Maneuvering control.

At the highest level, a supervisor is in charge of selecting the active tasks for the vessel and their reference values, i.e., it activates the obstacle avoidance behavior when the vessel is close to the other vessel or to a static obstacle, it defines the next target to be visited and it activates a leader-follower policy to avoid breaking the communication link. In order to make its decisions, the supervisor makes use of some information about the vessel position (read from the vessel's GPS), the map of the environment, the set of visited and un-visited locations, and information received from the other vessel's supervisor. An intermediate level implements a behavior-based technique, namely the Null-Space based Behavioral (NSB) control, to simultaneously achieve multiple tasks with different priority; the NSB, on the basis of the active tasks and their relative priority, defines motion directive for the vessel (e.g., the desired velocity and motion direction). Finally, the lowest-level controller is a maneuvering control that, taking into consideration the underactuated actuation system of the vessel, defines the reference commands for the actuators in order to follow the motion directives received by the NSB.

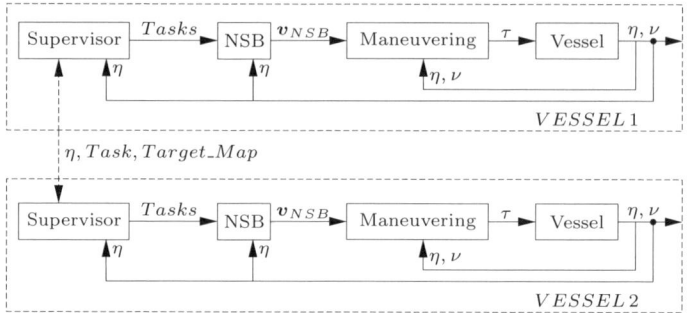

Fig. 1 Sketch of control architecture for a the team of two under-actuated vessels.

3.1 The Supervisor

The supervisor module on each vessel has the knowledge of the overall mission. In particular, it knows the target map and, in order to avoid revisiting the same targets during the course of the mission, it keeps track of targets visited by the vessel itself or by the other one. When a location is reached and a new target has to be chosen, the supervisor finds the target nearest to the current vessel position among the non-visited targets different from the one currently assigned to the other vessel. Once all the targets have been visited, the vessels are asked to reach and keep a final restoring configuration.

During the vessels' motion, the two supervisory modules are in charge of keeping the communication constraint. To fulfil this constraint, when needed, they activate a leader-follower policy to cause one vessel to take care of the communication constraint while the other continues its mission. The choice of which vessel is the leader and which the follower is negotiated between the supervisors on the basis of the distances between the vessels and their next targets. In particular, the vessel that is the closest to its next target becomes the leader while the other becomes the follower. The leader continues its mission ignoring the other vessel, while the follower has to control its distance from the leader while also trying to stay close to its target. The follower is also allowed to switch to a new target if this is closer than that previously assigned to it. If the communication link breaks or the distance between the vessels exceeds a maximum threshold, the leader stops to wait for the follower moving toward it. Moreover, the supervisory module decides when to activate the obstacle/collision avoidance. Finally, the supervisory module defines the priority order of the active tasks.

3.2 The Null-Space Based Behavioral Control

The Null-Space based Behavioral (NSB) control is a behavior-based approach aimed at controlling the motion of autonomous vehicles in dynamic scenarios.

In particular, the NSB, whose details can be found in [3, 2], uses a hierarchy-based structure to simultaneously achieve multiple tasks using a projection technique to delete the components of the lower priority tasks that would conflict with the highest ones. Here, the NSB is used as a guidance system for an ASV that, on the base of the active tasks and of their priority order, has to define the motion directives for the vessel.

Following the line of behavior-based approaches, the mission of the vessel is decomposed into elementary tasks. For each task a suitable task function is defined as $\boldsymbol{\sigma} = \boldsymbol{f}(\boldsymbol{p})$, where $\boldsymbol{\sigma} \in \mathbb{R}^m$ is the task variable to be controlled, m is the task function dimension, and $\boldsymbol{p} \in \mathbb{R}^n$ is the vessel position.

For each task, the velocity reference for the vessel is specified, starting from desired values $\boldsymbol{\sigma}_d(t)$ of the task function, solving the inverse kinematic problem at a differential level. Thus, the velocity reference of the generic i^{th}-task is calculated as $\boldsymbol{v}_i = \boldsymbol{J}_i^\dagger \left(\dot{\boldsymbol{\sigma}}_{i,d} + \boldsymbol{\Lambda}_i \tilde{\boldsymbol{\sigma}}_i \right)$, where $\boldsymbol{J}_i^\dagger$ is the pseudo-inverse of the task function Jacobian, $\boldsymbol{\Lambda}_i$ is a constant positive-definite matrix of gains and $\tilde{\boldsymbol{\sigma}}_i$ is the task error defined as $\tilde{\boldsymbol{\sigma}}_i = \boldsymbol{\sigma}_{i,d} - \boldsymbol{\sigma}_i$.

When the mission is composed of multiple tasks, the overall vessel velocity is obtained by properly merging the outputs of the individual tasks. A velocity vector for each task is computed as if it were acting alone; then, before adding the single contribution to the overall vehicle velocity, a lower-priority task is projected onto the null space of the immediately higher-priority task so as to remove those velocity components that would conflict with it. If the subscript i also denotes the priority of the task with, e.g., Task 1 being the highest-priority one, the overall vessel velocity is given by:

$$\boldsymbol{v}_{NSB} = \boldsymbol{v}_1 + \left(\boldsymbol{I} - \boldsymbol{J}_1^\dagger \boldsymbol{J}_1 \right) \left[\boldsymbol{v}_2 + \left(\boldsymbol{I} - \boldsymbol{J}_2^\dagger \boldsymbol{J}_2 \right) \boldsymbol{v}_3 \right], \tag{1}$$

where $\left(\boldsymbol{I} - \boldsymbol{J}_i^\dagger \boldsymbol{J}_i \right)$ represents the null-space projector of the ith-task, i.e., it filters the velocity components that would conflict with the ith-task.

To achieve the mission described in Sec. 2, three tasks have to be defined:

a) *Obstacle-avoidance*: This behavior, when active, is always the highest priority task because its goal is to preserve the integrity of the vessel. In the presence of an obstacle/vessel in the advancing direction, its aim is to keep the vessel at a safe distance from it. Thus, its implementation produces as output a velocity, in the vessel-obstacle direction, that keeps the vessel at a safe distance from the obstacle. Formally, the task function is $\sigma_o = \|\boldsymbol{p} - \boldsymbol{p}_o\| \in \mathbb{R}$ where $\boldsymbol{p}_o$ is the obstacle position, and $\boldsymbol{J}_o = \hat{\boldsymbol{r}}^{\mathrm{T}} \in \mathbb{R}^{1 \times 2}$ is the task Jacobian where $\hat{\boldsymbol{r}} = \frac{\boldsymbol{p} - \boldsymbol{p}_o}{\|\boldsymbol{p} - \boldsymbol{p}_o\|}$ is the unit vector aligned with the obstacle-to-vehicle direction. Defining as $\sigma_{o,d} = d$ desired distance, the task output is

$$\boldsymbol{v}_o = \boldsymbol{J}_o^\dagger \lambda_1 (d - \|\boldsymbol{p} - \boldsymbol{p}_o\|). \tag{2}$$

Possible motions in this task null-space are all the motions that do not change the distance from the obstacle. Thus, the null-space projector projects the

velocity commands of the lower-priority tasks along the tangential direction
of a circle centered in the obstacle and passing through the vessel itself.

b)*Move-to-target*: Defining the task function as $\boldsymbol{\sigma}_t = \boldsymbol{p} \in \mathbb{R}^2$, whose Ja-
cobian is $\boldsymbol{J}_t = \boldsymbol{I} \in \mathbb{R}^{2\times 2}$ and assigning the desired value as $\boldsymbol{\sigma}_{t,d} = \boldsymbol{p}_t$, then,
the output of the task is a velocity, in the target direction, proportional to
the distance from the target $\boldsymbol{p}_t$:

$$v_t = \Lambda_t \left(\boldsymbol{p}_t - \boldsymbol{p} \right) \tag{3}$$

c)*Keep the communication constraint*: To fulfill the communication constraint, a leader follower approach can be applied. The follow-the-leader task
is aimed at keeping the follower (whose position is $\boldsymbol{p}_f$) at a distance d from
the leader position $\boldsymbol{p}_l$. The task function mathematical definition is analogous to the obstacle avoidance task, while its output is a velocity, in the
leader-to-follower direction, proportional to the difference among desired and
measured distance; moreover, the desired velocity of the leader is added as a
feedforward term:

$$v_f = \Lambda_f \left(\|\boldsymbol{p}_l - \boldsymbol{p}_f\| - d \right) \frac{\boldsymbol{p}_l - \boldsymbol{p}_f}{\|\boldsymbol{p}_l - \boldsymbol{p}_f\|} + v_l. \tag{4}$$

3.3 Maneuvering Control

The maneuvering controller is an onboard controller aimed at steering the
vessel along a desired path and moving it with a desired velocity [11, 12].
Receiving motion reference commands from the NSB, the maneuvering controller has to generate the generalized forces applied by the actuators.

Basing on the model for ASV in [11] and considering the underactuated
propulsion system of the vessel (see Figure 2.a), a maneuvering controller is
designed following the approach proposed in [17]. From this approach, the
maneuvering control can be expressed as the sum of a heading autopilot and
a surge control aimed at causing the vessel to follow the velocity reference
commands. The heading autopilot is aimed at controlling the heading of the
vessel to make it move in the desired direction χ_{NSB}. In particular, it regu-
lates the propulsion torque and the rudder angle to correct the orientation of

Fig. 2 a) The ASV motion reference model; b) Two vessels during the experiment.

the vessel. The surge control has to make the norm velocity of the vessel to track the value generated by the NSB; however, the vessel has to move at full speed only when the orientation error is null. Thus, the surge control works as a PI controller regulating the advancing direction multiplied by a scaling factor depending on the orientation error. Following the control architecture of Figure 1, the output of the NSB, that is a velocity vector generated for a material point, can be geometrically represented through its norm U_{NSB} and its direction χ_{NSB} that are given to the maneuvering control as desired surge and heading/advancing direction.

4 Experimental Results

In this section the experimental results of the mission execution with the proposed control architecture are illustrated. The platform used for this experiment consists of two Autonomous Surface Vehicles (ASVs) designed by the University of Southern California's Robotic Embedded Systems Lab. Each ASV is an OceanScience QBoat-I hull with a length of 2.13 m and a width of 0.71 m at the widest section. Each ASV is equipped with an onboard computer, a wireless bridge, and a navigation package consisting of a GPS unit, a three-axis accelerometer, a compass, and a rate-gyro; the vessels weighs approximately 50 kg with instrumentation and batteries. The software for each vessel was written in C++ running under the linux operating system. Multiple processes manage mission planning, navigation, control and the communication between the vessels.

The mission for the team of two ASVs was to visit a set of 22 target locations spread in the Echo Park lake in Los Angeles (see Figure 3). Additionally, a communication constraint of a maximum distance of 60 m was to be maintained, and collisions between the vessels and with external static obstacles (small islands present in the lake) were to be avoided.

The GPS coordinates of the locations to be visited were provided to the ASVs at the beginning of the experiment. The two vessels had to autonomously navigate, communicate and cooperate to visit all the locations while avoiding multiple visits to any location and preserving the

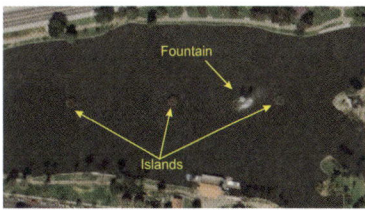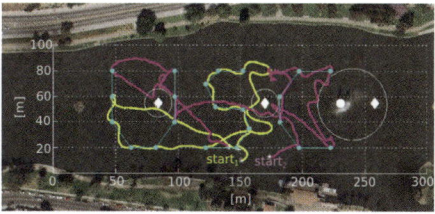

Fig. 3 a) Experiment site at Echo Park Lake, Los Angeles; b) Paths followed by the vessels during the first experiment, overlaid on the environmental map.

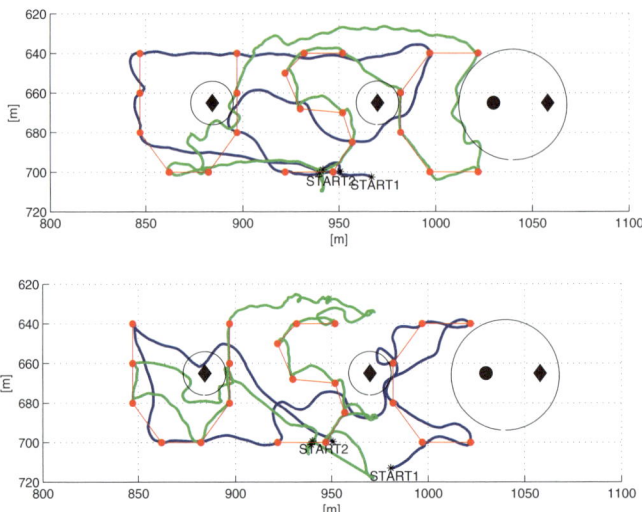

Fig. 4 Paths followed by the vessels during the second and third experimental trials at the Echo Park Lake, Los Angeles

communication constraint. Figures 3.b and 4 show the paths followed by the two vessels at the experimental site in the course of three different trials. In particular, Figure 3.b shows the obstacles' positions and the safety areas of the obstacle avoidance functions (activated only when inside these areas) overlaid on the environmental map. In this trial is clear that all the locations were visited while none of them was visited multiple times.

Figure 5.a shows the relative distance between the vessels during the experimental trial of Fig. 3.b. The leader-formation task was activated by the supervisory modules when the relative distance between the vessels was greater than 50 m. While the relative separation was below this value, the vessels moved independently. Since the leader-follower task is one-dimensional, the null-space projection of the task allows the system to attempt achieving lower-priority tasks, thus, the follower can use internal movements of the leader-follower task (i.e., movements that don't change the relative distance between the vessels) to try to achieve lower priority tasks (e.g., move toward the target). When the distance between the vessels exceeds 60 m, the leader stops to wait for the follower moving toward it. This situation arises when the follower encounters an obstacle along its path which activates the obstacle avoidance task with highest priority, thus causing the follower to lag.

A video of the experimental data (GPS and compass readings of the vessels) is available at http://webuser.unicas.it/arrichiello/video/collabASV.mpg. The video shows the dynamic behavior of the vessels while reaching the locations, avoiding collisions and obstacles. Moreover, it shows

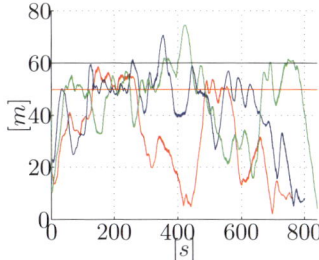 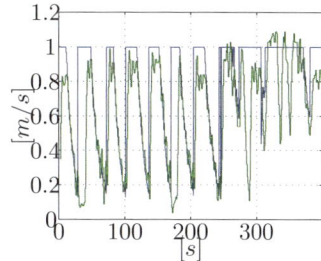

Fig. 5 a) Plot of the relative distance between the vessels during three field trials. The leader-follower strategy is activated when the distance is greater than 50 m. The leader has to stop and wait for the follower when their distance is greater than 60 m; b) Plot of one of the vessels' desired and measured velocity during the first 400 s of an experiment.

the activation of the leader-follower policy (represented by a change of color of the vessels) when out form the communication range.

Figure 5.b shows the desired and measured norm of velocity of one of the vessels during the first 400 s of the mission. It is worth noticing that the requested velocity is saturated to 1 m/s,moreover the assigned velocity decreases close to 0 when reaching the assigned location. The more irregular behavior after the first 300 s is due both to the leader-follower task and to the obstacle avoidance task.

We have performed several mission trials. The duration of each trial was between 13 and 14 minutes, during which time all targets were visited, no target was visited more than once, and the communication constraint was respected. The same mission has been also executed with a single vessel to provide a speedup benchmark. The mission with the single vessel (with the same target locations as in the two vessel case) takes approximately 20 minutes, thus the use of two vessel shows a 30% improvement in mission time over the single vessel case. While the present system works well, we believe that in the case of static targets, better results can be achieved by advance planning the target assignments in order to optimize the execution time. Our focus in the immediate future is to work with dynamic settings in which obstacles can be added or detected during the mission execution, and targets may appear or disappear during run-time. In principle, the control strategy proposed here will be effective in such settings.

5   Conclusion

In this paper, we presented a collaborative exploration technique for a team of two under-actuated ASVs designed for marine biology and oceanography experiments. The control approach, based on a behavior-based technique coupled with a maneuvering controller, was experimentally tested in a lake in

Los Angeles. The mission consisted of multiple targets that had to be visited while maintaining the communication link between the vessels, i.e., ensuring that the relative distance never exceeded a threshold. The robots successfully navigated to all the targets while avoiding collisions and obstacles, validating the presented control approach. In the future, we plan to use the presented techniques for large scale biological sampling experiments in marinas and lakes with obstacles with two or more ASVs. To extend the proposed technique to a larger team of ASVs we will start from the technique proposed in [5] to deal with Mobile Ad-hoc NETworks (MANETs) and ensure the global connectivity of the team while executing the assigned mission.

References

1. Aguiar, A.P., Hespanha, J.P.: Trajectory-tracking and path-following of underactuated autonomous vehicles with parametric modeling uncertainty. IEEE Transactions on Automatic Control 52(8), 1362–1379 (2007)
2. Antonelli, G., Arrichiello, F., Chiaverini, S.: The entrapment/escorting mission: An experimental study using a multirobot system. IEEE Robotics and Automation Magazine (RAM). Special Issues on Design, Control, and Applications of Real-World Multi-Robot Systems 15(1), 22–29 (2008)
3. Antonelli, G., Arrichiello, F., Chiaverini, S.: The Null-Space-based Behavioral control for autonomous robotic systems. Journal of Intelligent Service Robotics 1(1), 27–39 (online March 2007) (printed January 2008)
4. Antonelli, G., Arrichiello, F., Chiaverini, S.: Experiments of Formation Control With Multirobot Systems Using the Null-Space-Based Behavioral Control. IEEE Transactions on Control Systems Technology 17(5), 1173–1182 (2009)
5. Antonelli, G., Arrichiello, F., Chiaverini, S., Setola, R.: Coordinated control of mobile antennas for ad-hoc networks. International Journal of Modelling Identification and Control Special/Inaugural issue on Intelligent Robot Systems 1(1), 63–71 (2006)
6. Arkin, R.C.: Behavior-Based Robotics. The MIT Press, Cambridge (1998)
7. Arrichiello, F., Chiaverini, S., Fossen, T.I.: Formation control of marine surface vessels using the Null-Space-based Behavioral control. In: Pettersen, K.Y., Gravdahl, T., Nijmeijer, H. (eds.) Group Coordination and Cooperative Control. Lecture Notes in Control and Information Systems, pp. 1–19. Springer, Heidelberg (2006)
8. Borhaug, E., Pavlov, A., Ghabcheloo, R., Pettersen, K., Pascoal, A., Silvestre, C.: Formation control of underactuated marine vehicles with communication constraints. In: Lisbon, P. (ed.) Proceedings 7th IFAC Conference on Manoeuvring and Control of Marine Craft (2006)
9. Brooks, R.A.: A robust layered control system for a mobile robot. IEEE Journal of Robotics and Automation 2, 14–23 (1986)
10. Cao, Y.U., Fukunaga, A.S., Kahng, A.B.: Cooperative mobile robotics: Antecedents and directions. Autonomous Robots 4, 226–234 (1997)
11. Fossen, T.I.: Marine Control Systems: Guidance, Navigation and Control of Ships, Rigs and Underwater Vehicles. Marine Cybernetics, Trondheim, Norway (2002)

12. Fossen, T.I.: A nonlinear unified state-space model for ship maneuvering and control in a seaway. Journal of Bifurcation and Chaos (2005)
13. Ihle, I.A.F., Skjetne, R., Fossen, T.I.: Nonlinear formation control of marine craft with experimental results. In: Proceedings 43rd IEEE Conference on Decision and Control, Paradise Island, The Bahamas, Decenver 2004, vol. 1, pp. 680–685 (2004)
14. Lefeber, E.L., Pettersen, K.Y., Nijmeijer, H.: Tracking control of an underactuated ship. IEEE Transactions on Control Systems Technology 11(1), 52–61 (2003)
15. Loria, A., Fossen, T.I., Panteley, E.: A separation principle for dynamic positioning of ships: theoretical and experimental results. IEEE Transactions on Control Systems Technology 8(2), 332–343 (2000)
16. Mosteo, A.R., Montano, L., Lagoudakis, M.G.: Multi-robot routing under limited communication range. In: 2008 IEEE International Conference on Robotics and Automation, pp. 1531–1536 (2008)
17. Pereira, A., Das, J., Sukhatme, G.S.: An experimental study of station keeping on an underactuated ASV. In: 2008 IEEE/RSJ International Conference on Intelligent RObots and Systems, Nice, France (September 2008)
18. Pettersen, K.Y., Fossen, T.I.: Underactuated dynamic positioning of a ship-experimental results. IEEE Transactions on Control Systems Technology 8(5), 856–863 (2000)

A Simple Reactive Obstacle Avoidance Algorithm and Its Application in Singapore Harbor

Tirthankar Bandyophadyay, Lynn Sarcione, and Franz S. Hover

Abstract. Autonomous surface craft (ASC) are increasingly attractive as a means for performing harbor operations including monitoring and inspection. However, due to the presence of many fixed and moving structures such as pilings, moorings, and vessels, harbor environments are extremely dynamic and cluttered. In order to move autonomously in such conditions ASC's must be capable of detecting stationary and moving objects and plan their paths accordingly. We propose a simple and scalable online navigation scheme, wherein the relative motion of surrounding obstacles is estimated by the ASC, and the motion plan is modified accordingly at each time step. Since the approach is model-free and its decisions are made at a high frequency, the system is able to deal with highly dynamic scenarios. We deployed ASC's in the Selat Pauh region of Singapore Harbor to test the technique using a short-range 2-D laser sensor; detection in the rough waters we encountered was quite poor. Nonetheless, the ASC's were able to avoid both stationary as well as mobile obstacles, the motions of which were unknown *a priori*. The successful demonstration of obstacle avoidance in the field validates our fast online approach.

1 Introduction

The need for monitoring and securing harbor environments has grown in recent years, as a result of increased attention to pollution from runoff or other sources,

Tirthankar Bandyophadyay
Singapore-MIT SMART Program, Singapore
e-mail: tirtha@smart.mit.edu

Lynn Sarcione
Massachusetts Institute of Technology, Cambridge, MA USA
e-mail: sarcione@mit.edu

Franz S. Hover
Massachusetts Institute of Technology, Cambridge, MA USA
e-mail: hover@mit.edu

A. Howard et al. (Eds.): Field and Service Robotics 7, STAR 62, pp. 455–465.
springerlink.com   © Springer-Verlag Berlin Heidelberg 2010

natural processes such as sediment transport, water properties, and algal blooms, as well as security against threats. Harbors, with high density of goods, vessels, and people, are heavily utilized but fragile infrastructures. Among the world's harbors, Singapore Harbor is recognized as one of the largest in terms of total tonnage shipped [10], with several hundreds of large ships present at any given time. At the same time, the city of Singapore is intimately linked with the harbor. Any development on land directly affects the harbor. In many ways, Singapore represents the most important and difficult worldwide harbor environment for monitoring.

Autonomous systems are now at the level of maturity that they can be brought to bear on the overall needs of harbor observation. Autonomous surface craft (ASC) such as robotic kayaks are particularly well-suited due to their low unit cost and high loading capacity; such ASC's can be used in extremely shallow waters, where an autonomous underwater vehicle would be impractical physically and acoustic navigation would be difficult.

Several difficulties dominate autonomous agents in harbor environments. First, harbors have numerous structures and vessels both small and large which must be detected. The smaller vessels may not use the Automatic Identification System, or AIS, to broadcast the ship's data and are prone to unexpected maneuvers. Larger vessels, while presumably easier to detect at large distances, cannot realistically take actions to avoid hitting an ASC. Numerous underwater or near-surface obstacles such as shipwrecks are common in harbors. Above-water structures and vessels can also endanger communications between the vehicles and other parts of the system. Secondly, harbors can experience strong tides and tidal currents, which are often complicated by variable bathymetry. Currents can be predicted and made available to operators, but sometimes significant deviations occur, perhaps in the form of large eddies. Autonomous systems have to be able to develop optimized paths and adaptive actions that are robust against such disturbances. In this paper, we describe a series of tests that utilized autonomous kayaks in Singapore Harbor during January 2009, with a focus on the obstacle avoidance problem.

Prior Work on Obstacle Avoidance: Local reactive obstacle avoidance techniques [1, 2, 3] have been quite popular due to their simplicity and fast computation. Some works [2, 3], utilize the natural robot-centric polar frame to choose the best direction to move. While these algorithms plan in position space, others [11, 4] map the obstacles in the velocity space and choose suitable control parameters to satisfy kinematic constraints. Velocity obstacles (VO) [6], incorporate the dynamics of obstacle motion into the velocity space. A common way to handle obstacle motion in known environments is to augment the configuration space by a time axis [5, 7]. When exact motion of the obstacle are unknown, predictive techniques [8, 9] are used to identify the motion parameters.

Our approach is similar in principle to the VO approach except that it is formulated in position space. By planning in a relative frame, we avoid modeling the kayak and the obstacle motion individually in the rough sea. A simple linear prediction based on the immediate history is used to determine the relative obstacle

Fig. 1 (a) Selat Pauh operational area. (b) A range sensor on the ASC is used for data acquisition in real-time preparation for traffic avoidance

velocity. This keeps the computation load light as the algorithm runs at a high frequency and helps in bounding the uncertainty errors at each step.

2 Working in the Singapore Harbor Environment

Equipment: The ASC's utilized in Singapore Harbor are each equipped with a GPS receiver, a compass, and wireless communications gear in the base configuration. To support the aquisition of evironmental data a number of other sensors were added. A *Blueview* blazed array imaging sonar was used to image corals and shipwrecks. A *Velodyne* 3-D scanning laser imaged above-water structures including an oil platform. Also, an *RDI* doppler velocimeter measured the ASC's speed over ground. Each ASC has a full-thrust mission duration of about three hours.

For obstacle avoidance, a single *SICK* 2-D laser scanner was utilized. The range is 250m and the resolution is on the order of centimeters. We were able to use the full 10Hz scan rate of the sensor in our algorithm. The obstacle avoidance operation uses the onboard GPS and compass for waypoint navigation, and the 2-D laser for obstacle detection. In the remainder of this section, we describe several operational issues relevant to the Singapore Harbor environment: the effect of strong currents on navigation and the effect of waves on obstacle detection.

Effect of Ocean Currents on Navigation: It can be observed that ocean currents greatly affect navigation. The currents we encountered in Singapore Harbor reached 1.6m/s, whereas the maximum speed through water of the kayaks was about 2.4m/s on a fully charged battery. As seen in Figure 2, the closeness of these two values means that a simple waypoint-following controller can be unsatisfactory depending on the goal of the mission. Here the kayak was given four waypoints, effectively defining a square box to be traversed. The controller, developed for operation in low-current conditions, gives the kayak one of the desired waypoints to travel towards. Once the kayak reaches that waypoint, with some error tolerance, the next point of the square becomes the desired waypoint. In the test shown, significant currents to the southwest have deformed several of the vehicle paths up to fifteen percent of the

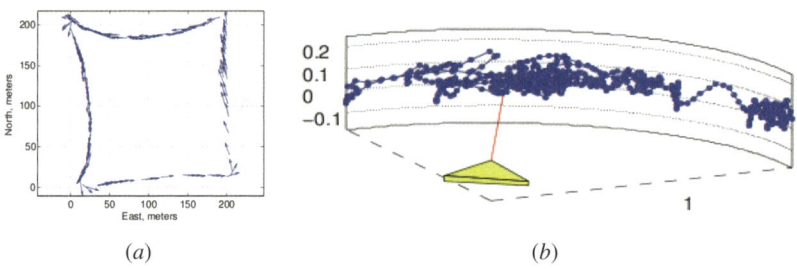

(a)   (b)

Fig. 2 (a) The GPS log data for an ASC being asked to navigate the perimeter of a square under waypoint control. The distortion of the paths are due to currents. (b) Projections of a single laser beam onto a constant-radius sheet showing the effects of large roll and pitch motions of the craft, at about 1Hz.

leg length. If the goal of the mission was to navigate a straight line for data sampling, this current effect would be unsatisfactory, and a controller with true cross-track error regulation would be needed. The results from this run also serve to motivate path and mission planning overall, because the current influences the amount of time and energy needed to complete each leg.

Effect of Waves on Object Detection Rate: Although utilizing the *SICK* 2-D laser has many benefits, it limitations are demonstrated when ocean waves are present. The unit is fixed on the vehicle, at a height of about 35cm above the waterline. Figure 2*b* shows the projected motion of a single laser beam fixed to the vehicle, and with the vehicle heading ranging from $[0 - 90]°$. The beam spends a fraction of time below the surface of the water, leading to no return. Other points are well above the water surface, perhaps yielding a return from the superstructure of a vessel instead of the desired hull. Figure 3*c* shows the overall performance of our cluster-based detection, as a function of range. In relatively calm waters, good hit rates can be found at about half the specified range of the sensor, but in waves virtually no hits are obtained outside 20m. The observations are based only on the data that was obtained during the runs, as we did not do a systematic study of the sensor clustering characteristics as a control experiment. We note that mounting the unit on a gimbal could take out some of the roll and pitch effects seen.

3   Online Navigation Approach

The purpose of this work is to devise a motion strategy that enables safe navigation along a desired direction for an ASC using only local 2-D range readings in the presence of unknown ocean currents and surface waves from nearby boats. At each step the ASC estimates the relative obstacle position and motion and subsequently chooses a direction that avoids collision. The approach is to follow the sense-plan-act paradigm at each step at a high frequency.

The basic model of the ASC is that of a point with controllable direction and a maximum powered velocity. A major difference between our application and

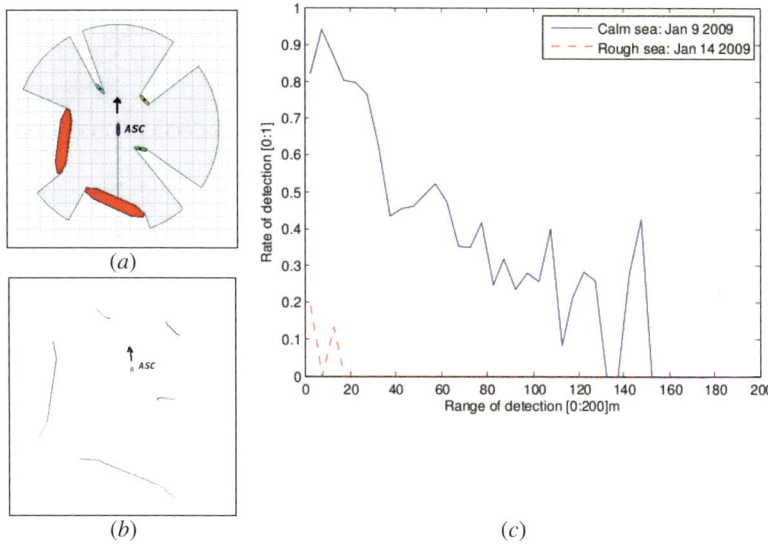

Fig. 3 (*a*) The typical schematic environment with the ASC facing north and surrounded by boats of varying size. (*b*) The local information available to the ASC using the 2-D laser. The ASC must plan its path while avoiding dynamic obstacles. (*c*) The fraction of boat detections per second vs. the distance. The higher curve is for a relatively calm day, while the lower curve is for a choppy day. In the second case, detection beyond 20m was impossible, and even below the 20m mark, it was less than 20%.

terrestrial robotics is the ability of the uncertain environmental factors, such as wind and currents, to move the ASC in any arbitrary direction. We model the velocity vector of the vehicle V_{asc} by a simple superposition of the velocity arising out of environmental factors, V_{drift}, and the velocity due to the ASC's own propulsion, V_{thrust}:

$$V_{asc} = V_{drift} + V_{thrust} \tag{1}$$

We represent the world in terms of clusters, which are determined from the scan data by a simple thresholding based range segmentation. As detected from the laser data range and angle data, any one obstacle is considered to be a single cluster, so that it has a starting point and an end point. Each such terminal point can either be an *occlusion point* that occludes the sensor's line of sight visibility, or a *range point* which is the limit of the cluster visible due to the sensor's range limit (Figure 4*a*). The range points are an artifact of the sensor limitations and do not reflect information about the obstacle.

In general, the occlusion points represent the shape characteristics of the silhouette and not of the actual object. Due to this, the motion of the occlusion points do not exactly represent the motion of the object, *i.e.* the rotational motion of the object can change the shape of the silhouette and give the occlusion points some

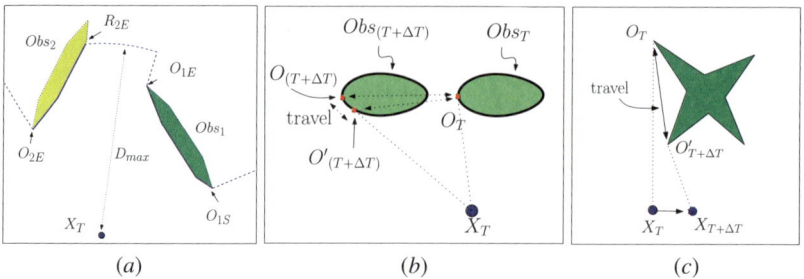

Fig. 4 (*a*) Both the end points O_{1S} and O_{1E} of the cluster for Obs_1 are occlusion points and can be used as reliable features in a short time duration. R_{2E} is a range point and does not provide any distinctive information about the obstacle; (*b*) Occlusion point travel due to curvature. $O_{(T+\Delta T)}$ is the actual point, while $O'_{(T+\Delta T)}$ is the detected point. (*c*) Occlusion point travel due to visual discontinuity.

velocity. However, the occlusion points can act as distinctive features of the obstacle, however, under the following conditions:

Low obstacle rotation rate: In open water, the translational velocity of many moving objects is quite high compared to their rotational speed. Due to this, the velocity of the occlusion point closely approximates the linear velocity of the moving objects:

$$V_{occ/asc} = V_{obs/asc} + \omega_{obs} \times r_{occ/obs} \approx V_{obs/asc}$$

Here $V_{occ/asc}$ is the relative velocity of the occlusion point with respect to the ASC, $V_{obs/asc}$ is the actual velocity of the obstacle with respect to the ASC, ω_{obs} is the rotation rate of the boat, and $r_{occ/obs}$ is the radius vector from a reference frame on the obstacle to the occlusion point.

Small radius of curvature: The occlusion points may travel along the physical object surface due to its curvature. The distance error in occlusion point, however, is usually much smaller than the actual travel if the radius of curvature is small compared to the distance from the sensor. In Figure 4*b* let the obstacle move as shown relative to the sensor. If the radius of curvature is small, the actual distance discrepancy is small, and $\overline{O_{T+\Delta T}O_T} \approx \overline{O'_{T+\Delta T}O_T}$

Limited sharp edges: Depending on the inherent shape of the obstacle, the occlusion points may jump a large distance. Figure 4*c* shows such a case. This sudden jump gives an erroneous measure of the motion of the obstacle. Running the detection at high frequency and maintaining a short motion history helps the algorithm recover from such an error which is unavoidable unless the obstacle is fully modeled.

Following these assumptions and ignoring the obstacle rotation, the obstacle motion is estimated simply as the average motion of the two occlusion points: $V_{obs} = (V_{occ,s} + V_{occ,e})/2.$

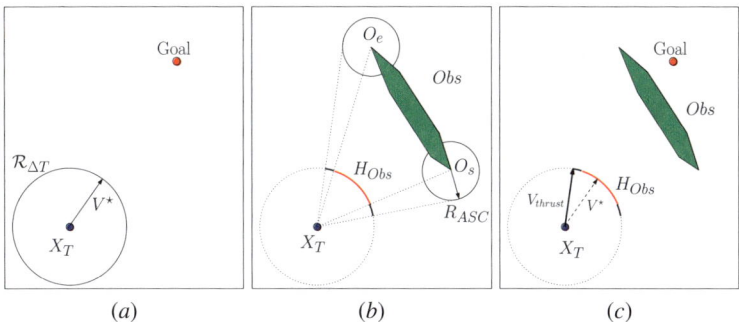

Fig. 5 In the stationary case, the dilation of the obstacle creates forbidden zones in the heading of the ASC.

3.1 Navigation Algorithm

In general, the ASC has an underlying objective such as waypoint navigation which generates a desired heading. Our algorithm modifies the heading command in light of nearby moving obstacles for collision avoidance. The higher level planning that generates the desired heading command is responsible for avoiding local minima, as the local reactive approach fails to address it.

We plan in the position space rather than velocity space due to the unreliability in velocity measurements of the ASC as well as the obstacles. Since the motion of the ASC and the environment are not modeled, we extrapolate the current velocity measurements in a simple linear model for a short duration ΔT. The position space in the planning horizon ΔT becomes the reachable set $R_{\Delta T}$ which is the set of all positions that the ASC can reach in time ΔT using this linear model. Using the simplified motion model in the previous section, we can determine directions that will cause collision with nearby obstacles. Each obstacle corresponds to one or two continuous sets of directions, termed as forbidden headings, that should be avoided. We denote the forbidden heading for a given obstacle Obs, by H_{Obs}.

Stationary case: In Figure 5, $R_{\Delta T}$ shows the reachable region. The optimal velocity V^* vector towards the goal position is shown in Figure 5a. The obstacle is represented by the occlusion points, O_e and O_s. Let the ASC have a bounding radius of R_{ASC}. To accommodate the size of the ASC, we extend the cluster by this measure. Without having to consider the whole obstacle, the Minkowski sum to represent the obstacle in the configuration space is reduced to dilating the occlusion points by R_{ASC}. The heading that the ASC must avoid in order to prevent collision is as shown in Figure 5b by the arc H_{Obs}. The decision of moving past O_e or O_s is made by choosing the shortest path to the goal. In general, the final choice of V_{thrust} once the H_{Obs} is established depends on the mission preferences. The *corrected* ASC heading is taken towards the corresponding endpoint of H_{Obs}. The same approach holds for multiple obstacles with the introduction of multiple forbidden regions.

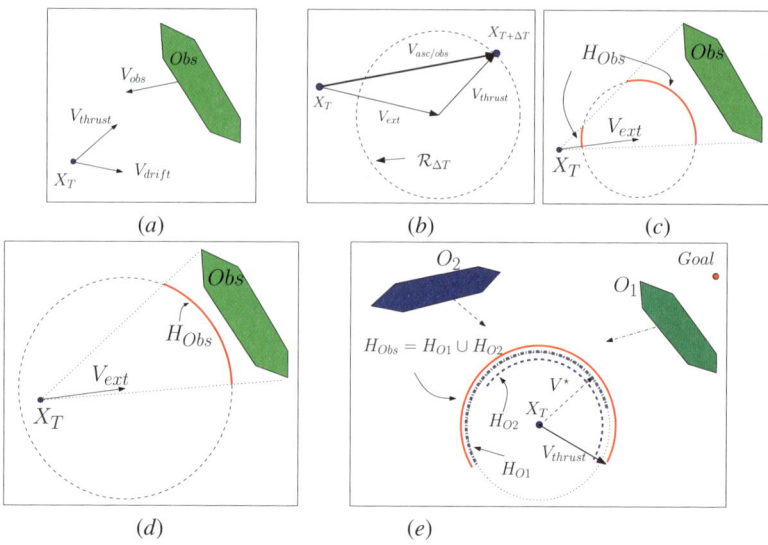

Fig. 6 (*a&b*) Velocity of the ASC with respect to the obstacle $V_{asc/obs}$ is the vector sum of uncontrollable velocity components affecting the obstacle and the ASC, and the controlled speed. The reachable state $R_{\Delta T}$ after ΔT is the circle shown; (*c-e*) The relative velocity of the ASC with respect to the obstacle is used to translate the reachable set. The forbidden heading regions are shown in red. (*c*) Scenario when $|V_{thrust}| < |V_{ext}|$; (*d*) Scenario when $|V_{thrust}| > |V_{ext}|$; (*e*) Forbidden regions for multiple moving obstacles.

Dynamic case: As discussed earlier in (Equation 1), environmental factors such as wind and current can introduce an additional velocity V_{drift} to the ASC. Also many of the obstacles in a harbor-like environment are mobile contributing to the dynamic environment seen by the craft. The velocities of these obstacles are unknown *a priori* and have to be deduced from the local range information. Let us take the case of a single obstacle moving with unknown velocity V_{obs}, while the ASC drifts with the velocity V_{drift}. As the sensing is done in the egocentric frame of the ASC, it is impossible to distinguish between these. Let V_{ext} represent the uncontrollable component of the ASC velocity towards the obstacle, *i.e.* the combined effect of the ASC drift and the obstacle motion V_{obs}, $V_{ext} = V_{drift} - V_{obs}$. Note that the ASC can only control V_{thrust}; using the onboard sensors to measure the obstacle velocity would give us $V_{ext} + V_{thrust}$. The net ASC velocity with respect to the obstacle and the reachable set $R_{\Delta T}$, is then given by:

$$V_{asc/obs} = V_{ext} + V_{thrust}$$
$$X_{T+\Delta T} = X_T + V_{ext}\Delta T + V_{thrust}\Delta T$$

For a constant estimate of V_{ext} in the duration ΔT, $R_{\Delta T}$ is shown in Figure 6.

The choice of the planning horizon ΔT depends on two factors: accuracy of velocity estimation, and the distance to the nearest obstacle. If the predicted motion is considered reliable, the ASC can plan for a much longer time step with confidence.

(a)   (b)

Fig. 7 (*a*) Selat Pauh Jan 14 ocean current forecast. The operational area is shown by the green polygon. The current direction is shown by the arrows, and the color shows the magnitude (red being of the order 1.6m/s). *Image courtesy: N. Patrikalakis (CENSAM)*. (*b*) SICK ld-1000 scanning laser being used in rough waters on Jan 14, for obstacle detection.

On the other hand, if the motion model is highy unpredictable or if the data is sporadic, it is advisable to plan for a shorter horizon. Given ΔT, choosing the maximum V_{thrust} is usually desirable from the point of view of the mission. However, in cases where the obstacle is too close, $R_{\Delta T}$ is further bounded by the minimum distance to the obstacle in consideration, i.e., $V_{thrust} = min(V_{thrust\_max}, dist(asc, obs)/\Delta T)$.

Estimating V_{ext}: From our velocity definitions, we have $V_{ext} = V_{asc} - V_{obs} - V_{thrust}$ where $V_{asc} - V_{obs} = V_{asc/obs} = -\dot{r}$. Here, r is the range vector from the ASC to the obstacle, and $\dot{r}$ is the vector of the time rates of change of r's components over time. V_{thrust} is estimated from the thrust command on the vehicle or a water velocity sensor, in union with the compass heading. In the absence of these estimates, the physical thrust can be briefly turned off, forcing $V_{thrust} = 0$. Note that this use of V_{thrust} should not be confused with the circle radius in defining the forbidden angles described in previously. Here it is used to describe an actual measurement or estimate of controlled velocity.

4 Experiments at Selat-Pauh

Selat Pauh was the test site utilized during the January 2009 experiments (Figure 1). Selat Pauh is located off the southern coast of Singapore where a significant amount of ship traffic is seen. In addition, the site has numerous stationary structures, such as buoys and oil rigs, and there are strong current fluctuations daily. Overall, this area is ideal for testing and observing the harbor environment.

Using the theory described, a number of obstacle avoidance tests were completed as summarized in (Figure 8). Current predictions provided for the experiment date, January 14, and a deployment photo are shown in Figure 7.

The weather conditions were challenging and tended towards strong winds and a choppy sea. Detection quality was on a par with that shown in the lower curve of Figure 3c, and this explains why the ASC only took avoidance action at a short range. However, as shown, the online approach was successful even under these dismal circumstances. The algorithm ran at 10Hz, on a Mini-ITX with 1GB RAM.

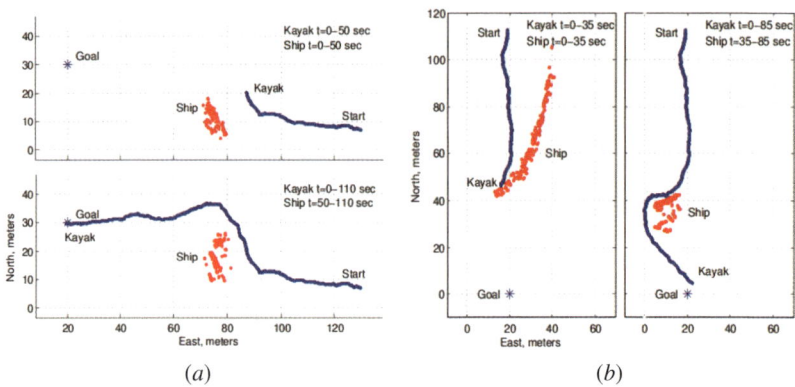

Fig. 8 (*a*) Avoiding a stationary boat (*b*) Avoiding a moving obstacle

In Figure 8 the GPS position logs (the blue points) of the ASC are broken into two time sets for clarity. The red points are the laser hits from the ASC plotted in the global frame. A simple waypoint based controller was used to navigate from the Start to Goal with known GPS locations. Since the percentage of detection is so low, 20%, we averaged the laser data over a moving window of 1sec before applying the obstacle detection algorithm, improving the detection rate significantly.

In Figure 8*a* the boat was kept stationary. We see that initially the ASC follows the V^* direction to go straight towards the goal before it detects an obstacle at a distance of about 20m. The ASC motion is then modified to go around the boat and as soon the the obstacle is safely cleared, it executes a new V^* direction. In the second run Figure 8*b*, the boat actively obstructs the path of the ASC, moving from top right corner of the plot to right in front of the kayak, from the side. The ASC detects the obstacle and modifies its motion accordingly. Such close range dynamic obstacle avoidance requires fast online algorithms like the one proposed.

5 Conclusion

In large-scale autonomous vehicle testing in Singapore Harbor, we have found that strong currents and heavy traffic are serious robustness concerns. Autonomous vehicles need to have more available speed and substantially increased energy storage in order to perform meaningful missions in these waters. Path planning for known current and robust control to reject unknown currents are also critical. We have made specific progress in obstacle avoidance which, as described here, is appropriate for day-to-day use to avoid fixed and slowly-moving obstacles. The main features of our algorithm are that it is neither probabilistic nor model-based, and that it is posed in position space; as a result, it scales seamlessly to situations with many objects, and with very low computational cost. In turn, our simple approach requires good confidence in the range data and obstacle detection, and for this we have successfully employed a clustering algorithm. The avoidance behavior is demonstrated for detection rates under 20%.

In future work, we plan to test avoidance of faster moving obstacles and to use vessel motions reported by AIS. The algorithm can be extended to formations, and including range information in the forbidden regions could lead to addtional trajectories that may be useful.

Acknowledgements. The research described in this project was funded in whole or in part by the Singapore National Research Foundation (NRF) through the Singapore-MIT Alliance for Research and Technology (SMART) Center for Environmental Sensing and Monitoring (CENSAM).

References

1. Khatib, O.: Real-time obstacle avoidance for manipulators and mobile robots. The Int. J. of Robotics Research 5(1), 90–98 (1986)
2. Minguez, J., Montano, L.: Nearness diagram navigation (nd): A new real time collision avoidance approach for holonomic and no holonomic mobile robots. In: Proc. of the IEEE/RSJ Int. Conf. on Intelligent Robots and Systems, Takamatsu, Japan (November 2000)
3. Ulrich, L., Borenstein, J.: VFH*: Local obstacle avoidance with look-ahead verification. In: IEEE Int. Conf. on Robotics and Automation, April 2000, pp. 2505–2511 (2000)
4. Fox, D., Burgard, W., Thrun, S.: Controlling synchro-drive robots with the dynamic window approach to collision avoidance. In: Proc. of the IEEE/RSJ Int. Conf. on Intelligent Robots and Systems (1996)
5. Hsu, D., Kindel, R., Latombe, J.C., Rock, S.: Randomized kinodynamic motion planning with moving obstacles. Int. J. Robotics Research 21(3), 233–255 (2002)
6. Fiorini, P., Shiller, Z.: Motion planning in dynamic environments using velocity obstacles. Int. J. of Robotics Research 17(7), 760–772 (1998)
7. Latombe, J.: Robot Motion Planning. Kluwer Academic Publishers, Boston (1991)
8. Chang, C.C., Song, K.-T.: Dynamic motion planning based on real-time obstacle prediction. In: Proceedings on IEEE Int. Conf. on Robotics and Automation, pp. 2402–2407 (1996)
9. Foka, A.F., Trahanias, P.E.: Predictive Autonomous Robot Navigation IROS. In: IEEE/RSJ Int. Conf. on Intelligent Robots and Systems (IROS) (2002)
10. Gordon, J.R.M., Lee, P.M., Lucas Jr., H.C.: A resource-based view of competitive advantage at the Port of Singapore. J. Strategic Information Systems 14(1), 69–86 (2005)
11. Simmons, R.: The Curvature-Velocity Method for Local Obstacle Avoidance. In: Int. Conf. on Robotics and Automation (April 1996)

Part XI
Planetary Robotics

Model Predictive Control for Mobile Robots with Actively Reconfigurable Chassis

P. Michael Furlong, Thomas M. Howard, and David Wettergreen

Abstract. Actively reconfigurable chassis enable planetary mobile robots to access more varieties of terrain. While typical approaches for exploiting such mechanisms reply on feedback control, it is beneficial to consider actively controlled elements at planning time rather than during motion execution. In this paper we present an approach for extending work in model-predictive trajectory generation to actively reconfigurable chassis. The motion planner uses a kinematic motion model and a terrain shape model to determine sequences of actions that minimize a cost function over vehicle attitude by modifying the shape of the velocity, curvature, and chassis configuration profiles. Simulation and field results are presented demonstrating the benefits of this technique on a prototype mobile robot for lunar excavation.

1 Introduction

As exploration of our universe continues, it becomes necessary to navigate in challenging, cluttered terrain to examine geologic records, search for records of microorganisms, and potentially discover life. In such environments, obstacles cannot entirely be avoided, they must be traversed in a manner that minimizes the risk to the platform. Adjustable chassis improve the mobile robots ability to overcome harsh terrain by shifting the center of gravity to maximize stability.

Most modern platforms with actively reconfigurable chassis adjust their configuration via feedback control, maximizing a stability criterion on the current state of the vehicle. A potentially better solution is to generate trajectories using a predictive model of vehicle motion or stability that includes the active chassis freedoms.

A potentially better, safer solution is to generate inputs and trajectories based on a prediction of vehicle motion and configuration articulation through

P. Michael Furlong, Thomas M. Howard, and David Wettergreen

Robotics Institute, Carnegie Mellon University, Pittsburgh, PA 15213-3980, USA

e-mail: {furlong,thoward,dsw}@ri.cmu.edu

A. Howard et al. (Eds.): Field and Service Robotics 7, STAR 62, pp. 469–478.

the environment. Deliberative planning allows a greater understanding of traversability and enables improved route selection decisions.

1.1 Related Work

Related work spans research on exploiting vehicles with reconfigurable chassis and research on the generation of trajectories for a robot that optimizes some cost function.

Iagnemma *et al.* present an optimization technique for externally reconfigurable four-wheeled vehicles that maximizes stability over rough terrain [5]. Schenker *et al.* incorporates that technique into a system for planning vehicle configurations at different points in the trajectory [10]. Ishigami *et al.* [6] modifies the technique in [4] to incorporate models of wheel slip to determine minimal slip paths that exploit the steering capabilities of the robot. Nakamura *et al.* [9] show that repositioning the center of mass of the vehicle improves the static stability while reducing energy consumption of the vehicle during traverse.

The work of Howard and Kelly [4] employs optimization procedures to produce trajectories that minimize terminal state position error and generates trajectories that account for predictive motion models of the vehicle as well as models of the terrain the vehicle has to cross. Farritor, Hacot and Dubowksy [2] use a genetic algorithm to combine pre-scripted action modules to produce plans that are then evaluated for feasibility and vehicle safety. While the approach does account for vehicle physics it deals with actions sampled from the configuration space of the vehicle and not the continuum of the command profile parameterization.

Mobile robot stability is addressed by Diaz-Calderon and Kelly [1], Schenker *et al.* [10], and Iagnemma *et al.* [5] by examining the relationship between the center of mass of the vehicle and the edges of the polygon of support. Messuri and Klein [8] as well as Hirose, Tsuagoshi, and Yoneda [3] determine the stability of the vehicle as a function of the energy require to induce a tip over.

1.2 Discriminators

While the results of this work are similar to the control approaches developed by [2, 6], it is distinct in two ways. First, the sidearm angle command profiles are produced through continuum optimization rather than the action-library based approach of [2]. As such the resulting trajectory is inherently feasible as the presented technique optimizes the inputs to the control system instead of sampled actions of the vehicle. Our approach differs from [6] in that it searches the local continuum and optimizes a criterion based on vehicle stability instead of minimizing slip.

2 Technical Approach

The trajectory optimization technique presented in this paper is an extension of model-predictive trajectory generation for mobile robots with actively reconfigurable chassis. This section presents an overview of the posture optimization trajectory generation technique, a discussion of the predictive motion model and the cost functions applied in this work.

2.1 Active Posturing

Actively reconfigurable chassis enable mobile robots to adjust their center of mass to improve traction and achieve a more stable configuration. Mobile robots such as Scarab [12] and the Sample Return Rover [7] are able to independently adjust the independent chassis wheel bases by setting the sidearm angle. Actively adjustable wheel bases enable posturing of the robot body in a ways that are not available to fixed chassis robots (Figure 1).

2.2 Model Predictive Trajectory Generation

Typical motion planning or navigation tasks simply require that a mobile robot move from one pose to another. Often there are no constraints imposed on the internal configuration of the vehicle over the course of the trajectory. Given a potentially infinite number of chassis configurations, the problem

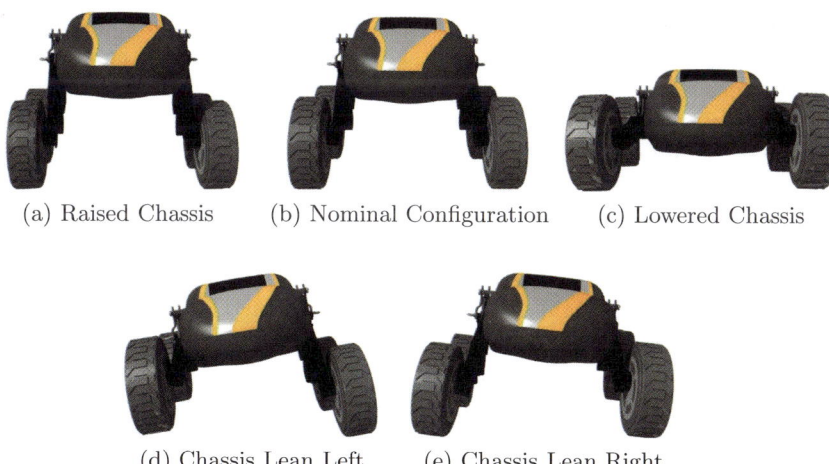

(a) Raised Chassis   (b) Nominal Configuration   (c) Lowered Chassis

(d) Chassis Lean Left   (e) Chassis Lean Right

Fig. 1 Articulation capabilities of the Scarab mobile robot. Symmetrical values for the sidearm chassis angles enables the body to raised (a) to avoid small obstacles or lowered (c) to reduce the risk of tip over. Asymmetrical sidearm chassis angles enables the body to be postured left (d) and right (e) to achieve more stable configurations on rocks and slopes.

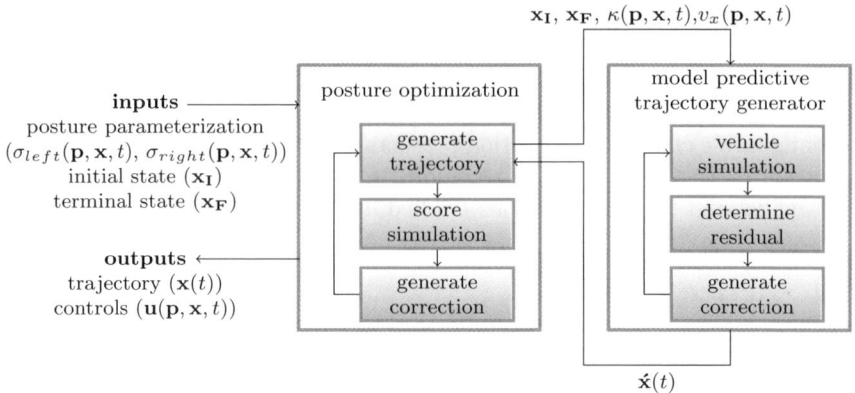

Fig. 2 The above diagram illustrates tiered optimization for trajectory generation with posture optimization. The inner layer generates trajectories that satisfy boundary state constraints with the posture parameters determined by the outer layer. The outer layer optimizes the posture parameters by maximizing the stability of the vehicle over the candidate trajectory $\acute{\mathbf{x}}(t)$

becomes continuum search for the optimal chassis configuration along the trajectory that satisfies the boundary state constraints.

The presented approach is an extension of model-predictive trajectory generation that optimizes vehicle posture in addition to satisfying boundary state constraints. It leverages parameterized profiles to describe continuous inputs in order to reduce the scope of the continuum search. For a mobile robot with an actively reconfigurable chassis as shown in Figure 1, inputs can be defined for the curvature ($\kappa(\mathbf{p}, \mathbf{x}, t)$), linear velocity ($v(\mathbf{p}, \mathbf{x}, t)$), and sidearm angles ($\sigma_{left}(\mathbf{p}, \mathbf{x}, t)$, $\sigma_{right}(\mathbf{p}, \mathbf{x}, t)$), where $\mathbf{p}$ is the parameterization describing the shape of the inputs.

2.2.1 Tiered Optimization

An overview of the tiered optimization technique is shown in Figure 2. The boundary state constraints are satisfied in the inner loop by correcting the curvature and velocity command profile parameters using a model-predictive trajectory generator. The posture optimization is governed by the outer loop, which modifies the commanded sidearm angles based on the accumulated attitude of the generated trajectory.

The outer layer of the optimization determines parameters satisfying the relation:

$$\dot{\mathbf{x}} = f(\mathbf{x}, \kappa, v, \sigma_{left}, \sigma_{right}) \tag{1}$$

$$\mathbf{p} = \arg \min_{\mathbf{p}} G(\sum_t f(x_{t-1}, \kappa, v, \sigma_{left}(\mathbf{p}, \mathbf{x}, t), \sigma_{right}(\mathbf{p}, \mathbf{x}, t)) \Delta t) \tag{2}$$

The cost functional G scores the execution of the controls parameterized by **p**. Gradient descent is used to determine the minimum cost setting of the chassis sidearm control parameters. The inner layer of the optimization is the constrained form of the model-predictive trajectory generator in [4] to produce continuous trajectories which consider the shape of the sidearm control profiles. In this formulation, the inputs associated with steering and velocity along the trajectory are not assumed to be independent of those governing the configuration of the chassis. As the configuration iteratively adjusts, traction can improve, motion over rough terrain can alter, and the inputs required to satisfy the boundary state constraints may change.

2.2.2 Predictive Motion Model

The predictive motion model is used to estimate the vehicle state response to the control inputs. One such example is a predictive motion model based on the Scarab mobile robot used for the experiments described in Section 3. It has a mass of approximately 300kg, has a maximum speed of 0.04 m/s, and is assumed to be quasi-static. The kinematic model is an adaptation of the work for a planetary rover [11] that represents Scarab through a series of Denavitt-Hartenberg Parameters. To represent the configuration of the chassis, the passive major chassis angle (β) and the angles of the chassis side arms ($\sigma_{left}, \sigma_{right}$) are included in the vehicle state (**x**). The passive chassis angle, roll, pitch, and elevation were determined as the vehicle moved through the environment using the technique described in [11].

2.2.3 Cost Function

The outer optimization layer for the chassis sidearm angles maximizes the vehicle stability. A standard tool for kinematically representing vehicle stability is the stability polygon. The stability polygon is the convex hull of the contact points of the vehicle with the ground. The cost function optimized for vehicle stability is the sum over the trajectory of the vehicle of the distance between the center of the stability polygon and the intersection of the projected center of mass with the stability polygon (Figure 3).

As the projection of the center of mass of the vehicle gets closer to the edges of the stability polygon the vehicle comes closer to tipping over. If the centre of mass is outside the polygon then the gravity vector pulls the vehicle away from its contact points inducing tip over. Minimizing the distance from the centre of the support polygon, which maximizes the perpendicular distance to the edges, increases the stability of the vehicle. Maximizing the distance of the centre of mass from the edges of the support polygon is proportional to maximizing the normalized energy stability criterion as described in [3]. Because of the computational simplicity that the distance-to-edge cost function was chosen over energy margin based cost functions.

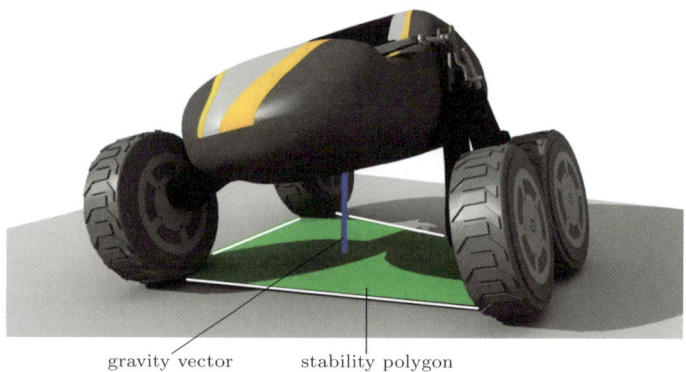

gravity vector   stability polygon

Fig. 3 The stability polygon of the vehicle with the vehicle center of mass projected onto the convex hull of the contact points of the wheels with the ground. The cost function used in the active chassis optimization minimizes the distance between the intersection of the gravity vector and the center of the stability polygon.

3   Experiments and Results

The experiments in this paper were designed to demonstrate the ability for the tiered optimization technique to improve the stability margin for the mobile robot operating in natural environments. The experiments were conducted first in simulation and then on the robotic platform described below.

The target platform for these experiments was Scarab [12], a four-wheeled planetary rover with an actively reconfigurable chassis built at the Robotics Institute's Field Robotics Center (Figure 4). For the simulation and field

Fig. 4 Scarab, a four-wheeled prototype lunar prospector with an independent articulating chassis. In this configuration, the robot has exploited the articulating chassis to minimize roll of the body, thereby limiting exposure to tip over.

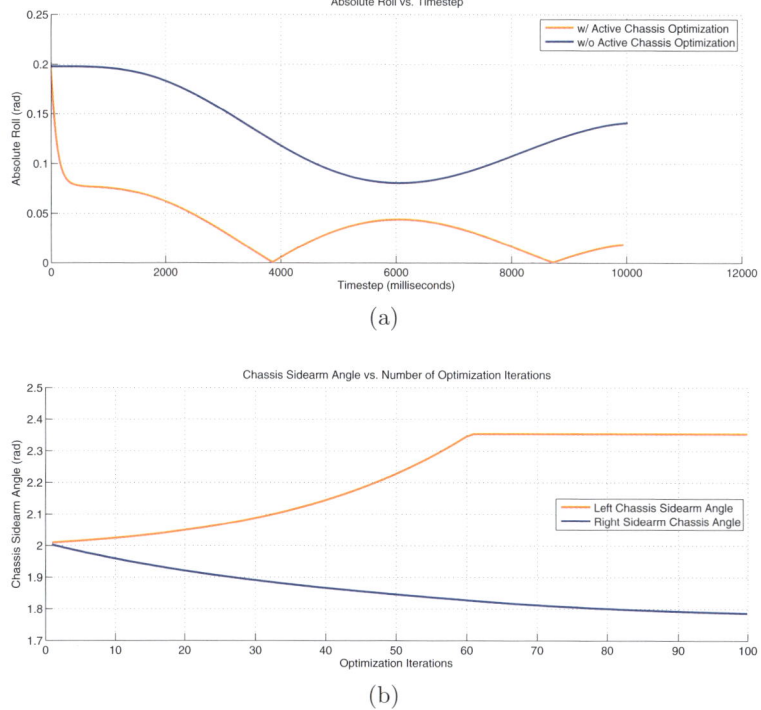

Fig. 5 The cost of the simulated run of the vehicle. The absolute roll experienced by Scarab with and without posture optimization is show in (a). The changes to the sidearm angle as a function of optimization iteration are shown in (b).

experiments, the freedoms of each active chassis elements were constrained such that the wheelbase was between 0.94m and 1.20m.

3.1 Simulation Experiments

In order to test the ability for the posture optimization to maximize stability over the perceived environment a hill was chosen for this experiment as it most resembles the terrain that is expected to be encountered during crater wall navigation. Unlike the field experiment the trajectory generator was given an omniscient view of the terrain.

For consistency with the latter field experiments, the absolute roll of the vehicle is used instead of the stability margin. This is due to the lack of sensing of the major bogie angle on the target platform, which prevents direct computation of the stability criterion.

Figure 5 shows that the reconfigurable chassis provides a trajectory with roll that is consistently lower than that of the fixed chassis. On average the

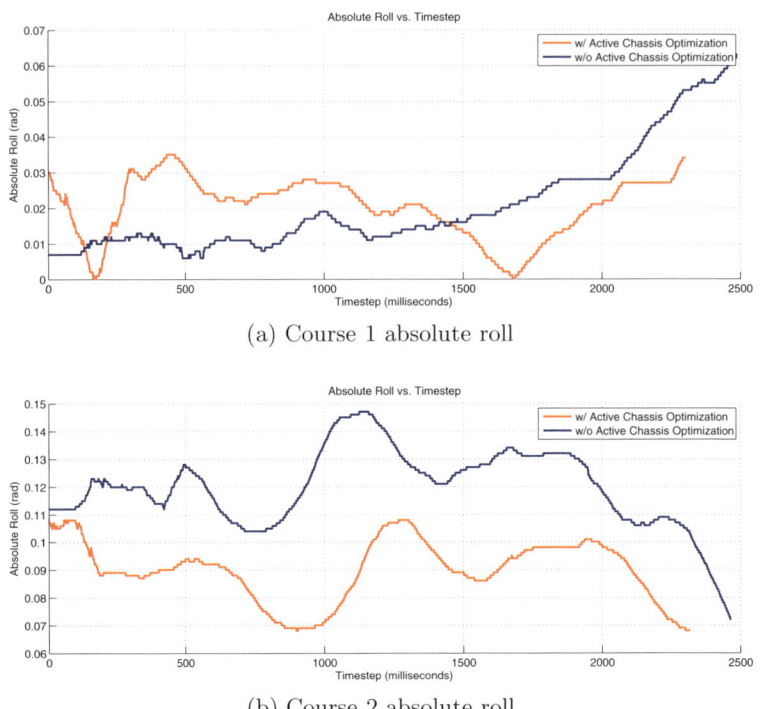

(a) Course 1 absolute roll

(b) Course 2 absolute roll

Fig. 6 The roll of the vehicle while driving up a slope. The orange line represents the roll of the vehicle using the active chassis while the blue line represents the roll of the vehicle using a fixed chassis.

roll of the vehicle with the active chassis is twenty-eight percent of the roll of the vehicle with the fixed chassis. Figure 5 shows the settings of the chassis sidearm angles as the cost of the trajectory is reduced. Over one hundred iterations were required for the sidearm angle parameters to converge, this can be reduced by increasing the step size of the gradient descent algorithm used to optimize the parameters.

3.2 Field Experiments

The field experiments were conducted in a natural terrain similar to the previously described simulation experiments. Instead of assuming perfect knowledge of the terrain shape, an active range sensing perception system was used. A trimesh representing the terrain shape was estimated by processing the observed point cloud. The triangulated meshes were used for planning trajectories and simulating executions of planned actions. For this experiment, the predictive motion model was assumed to be entirely kinematic, although generally any model of dynamics or wheel/terrain interaction could be applied.

As in the previous experiment, vehicle roll was taken as an indicator of stability. In this experiment the robot was position on a slope and two trajectories were planned and executed. The first trajectory assumed a fixed, nominal chassis configuration. The second trajectory optimized the input profiles governing the vehicle posture. Each motion plan execution originated from the same initial state, as verified by the onboard inertial navigation system.

Figure 6 present the vehicle roll for fixed and adaptive chassis while driving over the varied terrain over two different courses. On average, the predictively controlled chassis reduces the roll of the vehicle body while traversing the sloped terrain.

4 Conclusion

The presented technique optimizes wheelbase inputs for a mobile robot with an actively reconfigurable chassis to improve its stability. Increased stability is important on slopes or when traversing rough or uneven terrain. By reasoning deliberatively about actions taken, the reachability of stable configurations and the consequences of these actions can be determined. Separating the trajectory generation and posture optimization with a tiered approach can efficiently determine solutions which satisfy the two-point boundary value problem while optimizing an underconstrained portion of the system. While this technique produces only locally optimal solutions in the continuum the low-order parameterized functions resist problems of numerous local optima.

There are several logical extensions to the work presented in this paper. Uncertainty in the terrain model must be addressed and it is recommended that the trajectory generator be coupled with a online system that adjusts the vehicle's configuration to maximize instantaneous stability. Next is to integrate a higher fidelity predictive motion model that simulates the effect of wheel/terrain interaction and a dynamic (rather than kinematic) vehicle model. The application of actively reconfigurable chassis trajectory optimization to sampling-based navigators is important because it obviates the need to sample the space of chassis sidearm inputs. Lastly, this technique can be used in a motion planning structure to generate edges that consider the shape of the terrain and the configuration of the active chassis.

Acknowledgements. We gratefully acknowledge the contributions of our technical team and the assistance of Dominic Jonak, Scott Moreland, and James Teza. This research was supported by NASA under grants, NNX07AE30G, John Caruso, Project Manager, and NNX08AJ99G, Robert Ambrose, Program Scientist.

References

1. Diaz-Calderon, A., Kelly, A.: Development of a terrain adaptive stability prediction system for mass articulating mobile robots. In: FSR, pp. 343–354 (2003)
2. Farritor, S., Hacot, H., Dubowsky, S.: Physics-based planning for planetary exploration, vol. 1, pp. 278–283 (May 1998)
3. Hirose, S., Tsukagoshi, H., Yoneda, K.: Normalized energy stability margin and its contour of walking vehicles on rough terrain, vol. 1, pp. 181–186 (2001)
4. Howard, T.M., Kelly, A.: Optimal rough terrain trajectory generation for wheeled mobile robots. IJRR 26(2), 141–166 (2007)
5. Iagnemma, K., Rzepniewski, A., Dubowsky, S., Pirjanian, P., Huntsberger, T., Schenker, P.: Mobile robot kinematic reconfigurability for rough-terrain (2000)
6. Ishigami, G., Nagatani, K., Yoshida, K.: Path planning for planetary exploration rovers and its evaluation based on wheel slip dynamics. In: Proceedings of the 2007 IEEE ICRA, April 2007, pp. 2361–2366 (2007)
7. JPL. JPL Robotics: System: SRR: Sample-Return Rover (2009)
8. Messuri, D., Klein, C.: Automatic body regulation for maintaining stability of a legged vehicle during rough-terrain locomotion. Robotics and Automation, IEEE Journal of 1(3), 132–141 (1985)
9. Nakamura, S., Faragalli, M., Mizukami, N., Nakatani, I., Kunii, Y., Kubota, T.: Wheeled robot with movable center of mass for traversing over rough terrain. In: Proceedings of the 2007 IEEE/RSJ IROS (2007)
10. Schenker, P.S., Pirjanian, P., Balaram, B., Ali, K.S., Trebi-ollennu, A., Huntsberger, T.L., Aghazarian, H., Kennedy, B.A., Baumgartner, E.T., Rzepniewski, A., Dubowsky, S.: Reconfigurable robots for all terrain exploration. In: Massachusetts Institute of Technology, pp. 419–6 (2000)
11. Tarokh, M., McDermott, G.: Kinematics modeling and analyses of articulated rovers. Technical Report CS/10/2005, University of California - San Diego (2005)
12. Wettergreen, D., Jonak, D., Kohanbash, D., Moreland, S.J., Spiker, S., Teza, J., Whittaker, W.L.: Design and experimentation of a rover concept for lunar crater resource survey. In: 47th AIAA Aerospace Sciences Meeting Including The New Horizons Forum and Aerospace Exposition (January 2009)

Turning Efficiency Prediction for Skid Steer Robots Using Single Wheel Testing

Daniel Flippo, Richard Heller, and David P. Miller

Abstract. To date, most field robots use wheels as their means of locomotion (especially true of planetary exploration robots). In many cases these robots are required to travel significant distances, with limited power, and over rough terrain. All of which make wheels a major component contributing to their performance. It is through experimentation and iteration that effective wheel design, for a given rover in a given mission, can be achieved. To do this, the SWEET (Suspension and Wheel Evaluation and Experimentation Testbed) simulates the rover environment using a single wheel methodology. The wheels currently being tested belong to the SR2 skid steer Mars rover designed and built at the University of Oklahoma. Simulating a skid steer turn with SWEET is achieved by varying the spinning rate of the platform under the wheel, which is rotating at a certain rate, and recording the forces incurred. These forces interact in such a way that the relevant mobility properties for a rover can be predicted. This experimentation method allows for cheap and timely iterative single wheel design.

1   Introduction

Compared with automotive wheels very little research has been done in the area of interplanetary wheel design. To fill the gap in the understanding of rover wheel design and wheel to soil interaction, testing machines have been designed by various

Daniel Flippo
University of Oklahoma, Norman OK, 73069
e-mail: baldflippo@gmail.com

Richard Heller
University of Oklahoma, Norman OK, 73069
e-mail: richardheller@ou.edu

David P. Miller
University of Oklahoma, Norman OK, 73069
e-mail: dpmiller@ou.edu

A. Howard et al. (Eds.): Field and Service Robotics 7, STAR 62, pp. 479–488.
springerlink.com

institutions. In 1971, NASA tested a single Lunar Rover Vehicle wheel on a testing device called a dynamometer system [6] and now uses devices such as the variable terrain tilt platform (VTTP), at JPL, to gain a better understanding of entire rover systems in a sloped environment. The VTTP is a 16 x 16 ft table that can tilt up to 25 degrees and can be left bare or covered with terrain [4] but is meant to incorporate the total rover assembly and is used in a design validation role rather than an iterative design role. At the Massachusetts Institute of Technology a testing device (FSRL) tests a single driven wheel through different mediums to better understand wheel to soil interaction [3]. A similar device is used at Tohoku University to refine rover steering and other parameters [9]. Other comparable devices test wheels for Earth's terrain are [5, 8, 2]. The University of Oklahoma's testing apparatus named SWEET [1] is unique in that it allows for true turn testing.

All these test beds allow the simulation of aspects of real-life operations. Full assembly test beds are more difficult and expensive to use since they require the full rover, a full compliment of wheels, and much more space. Issues with the wheel design may also be conflated with other aspects of the rover design when a full system is tested, making iterative improvement of the wheel more difficult. Single wheel testing machines, on the other hand, allow a designer to iteratively design a wheel in a much less costly and timelier fashion than full assembly testing. For these single wheel testing machines to be of any use, the data that they give must have some significance in the real world. Their performance in the single wheel testing machine must transfer to predict the behavior exhibited on a multi-wheel rover doing typical maneuvers in field conditions.

Skid steering turn performance is an example of a typical maneuver, beyond the domain of most single wheel test systems. If it can be demonstrated that a model can transform data from a single wheel test to predict the turning efficiency of a rover, then skid steer turning is one more behavior that can be studied and improved upon cheaply and thoroughly using the single wheel testing method. This paper describes a method to test skid steer rover wheels on a single wheel test apparatus and then predict its real world turning performance on a skid steer rover. The predictions are then compared to full assembly tests fitted with four identical wheels to the one tested.

2   Theory of Single Wheel to Full Assembly Correlation

Skid steering is an unintuitive process in that there are multiple forces, due to the lateral sliding, that must take place for a skid steer rover to turn. When a rover initiates a turn its rotation (in the $X - Y$ plane) will accelerate up to a certain spin rate Ω (Fig. 1) at which point it will stabilize and the moment about its center (M_o) will equal zero.

$$\Sigma M_o = 0 .\tag{1}$$

$$\Sigma(F_y R\cos(\Theta)) - \Sigma(F_x R\sin(\Theta)) = 0 .\tag{2}$$

$$F_y = F_x \tan(\Theta) .\tag{3}$$

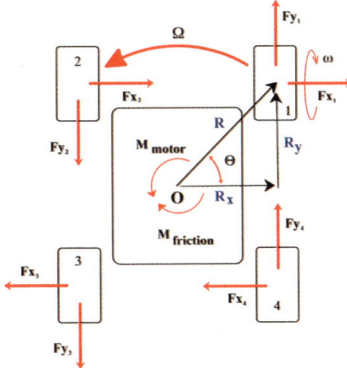

Fig. 1 Skid Steer Force Body Diagram

Equation 3 describes a relationship between F_x and F_y at the turning equilibrium point and is dependent upon the rover geometry (Θ). If the rover were slender (Fig. 2-a) then Θ would be larger than $\frac{\pi}{4}$ and F_x would be much smaller than F_y. If $\Theta = \frac{\pi}{2}$ then $F_y = \infty$. This would mean that no matter how much force a wheel could exert on the ground the rover's spin rate Ω would always be zero. If, on the other hand, Θ were equal to zero, as in Fig. 2-b, then F_y (which is really the net force of power and friction) would be equal to zero. This configuration is better known as Ackerman steering which means that the wheels have no lateral slip and if there is no longitudinal slip then the turning rate can be calculated by eq. 4.

$$\Omega = \frac{\omega r}{R}, F_y = 0 . \tag{4}$$

where ω is the wheel angular velocity in radians per second, r is the wheel radius, and R is the distance from the center of the wheel to the center of rotation of the rover.

Equation 4 refers to the ideal turning rate Ω_{IDEAL} without longitudinal slipping for an Ackerman steering geometry. To calculate Ω_{IDEAL} for a skid steer rover ($\Theta \neq 0$), Θ must be taken into account and is reflected in eq. 5. Ω_{IDEAL} refers to the theoretical maximum a skid steer rover can spin, but F_y, at Ω_{IDEAL}, is still not zero.

$$\Omega_{IDEAL} = \frac{\omega r}{R} \cos(\Theta), F_y \neq 0 . \tag{5}$$

To find the value of $\Omega_{F_y=0}$, which is the spin rate at which there is no longer a net force in the Y direction, the longitudinal velocity (V_y)(Fig. 2b) of the ground under the wheel must be equal to the velocity of the wheel rim (ωr) therefore making $F_y = 0$ (no slip). Equation 10 explains this relationship. Loose soils that cause more viscous friction, such as sand, will alter the slope of the force curves and decreasing the $\Omega_{Predicted}$ and $\Omega_{Fy=0}$ values.

$$V_y = \omega r \, . \tag{6}$$

$$V_y = \cos(\Theta)V_{ground} \, . \tag{7}$$

$$V_{ground} = \Omega R \, . \tag{8}$$

$$\omega r = \Omega R \cos(\Theta) \, . \tag{9}$$

$$\Omega_{Fy=0} = \frac{\omega r}{R \cos(\Theta)} = \frac{\omega r}{R_x} \, . \tag{10}$$

For the right front wheel of a rover pivoting in the counter clockwise direction, the ground must move under it in the opposite direction ($-\Omega \frac{rad}{sec}$) and the relationship of the forces on the wheel, as the spin rate (Ω) of the ground under the wheel

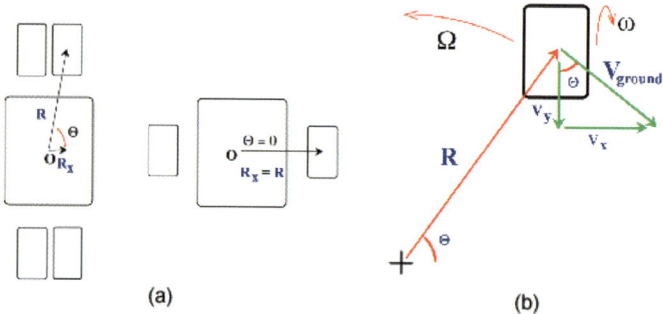

(a)   (b)

Fig. 2 a) Skid Steer Geometry Configurations; b) Skid Steer Kinematics

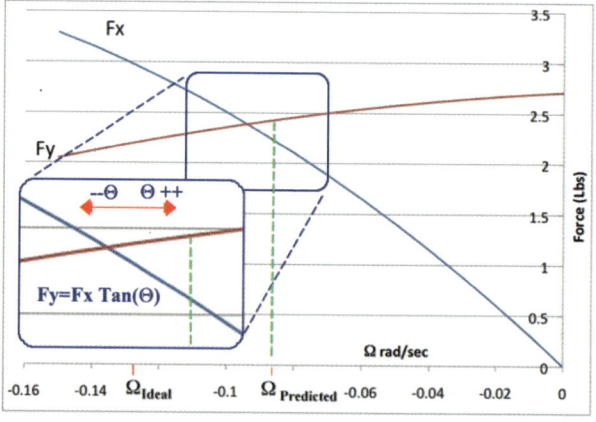

Fig. 3 Force vs Spin Rate Example

increases, can be shown in illustration 3. When the simulated rover's spin rate (Ω) is equal to zero the wheel being tested rotates (ω) but does not move. This causes a force in the Y direction which is just the kinetic friction ($F_y = \mu_k N$) between the wheel and ground. For a blank wheel on smooth ground there is no F_x at $\Omega = 0$, but for a treaded wheel F_x could be non-zero which will be one value to focus on when testing new wheels. As the spin rate of the ground under the wheel, increases F_x increases while F_y decreases until they intersect. This meeting point would represent the equilibrium spin rate ($\Omega_{Predicted}$) of a square rover ($\Theta = \frac{\pi}{4}$). To find the equilibrium point, of a rectangular rover, eq. 3 adds the needed constraint between F_x and F_y. For the SR2 [7] rover $\Theta = .8477\ rad$ when combined with eq. 3 simplifies to eq. 11.

$$F_y = 1.133 F_x . \tag{11}$$

In essence what we are doing is operating the wheel and the ground under the wheel independently, by varying the ground speed (Ω) while keeping the wheel spin rate (ω) constant, and observing the behavior of the forces acting on the wheel. When the forces satisfy eq. 11 the corresponding Ω is the predicted rover spin rate. In Fig. 3 this relationship gives a point just right of the cross point and corresponds to a $\Omega_{Predicted}$ value which is the predicted spin rate of a rover fitted with four wheels with the same orientation, relative to the rover center, and identical tread to the wheel tested.

It should be noted how a rover's geometry affects this relationship. As Θ increases above $\frac{\pi}{4}$ the rover is more slender (Fig. 2) which makes turns less efficient and $\Omega_{Predicted}$ becomes smaller. If, on the other hand, Θ decreases its $\Omega_{Predicted}$ value increases until $\Theta = 0$ and $\Omega_{Predicted} = \frac{\omega r}{R}$ which is an Ackerman steering geometry.

3   Validation Experiments

To do single wheel testing the Suspension and Wheel Experimentation and Evaluation Testbed (SWEET) is used. The testbed (Fig. 4) has a 10 x 10 ft footprint and a weighted drop down test leg, incorporating a driven wheel and a six-axis, force torque sensor which stays stationary in the X and Y directions but allows movement along the Z-axis via a counterbalance system.

SWEET differs from most testbeds in that the table can move in the X and Y directions underneath the test stand, as well as rotate in the X, Y-plane. This added advantage gives the apparatus the unique ability to measure forces and torques in a true turn allowing the study of skid steer turning.

SWEET was programmed to simulate a skid steer turn and fitted with a .109 meter diameter blank wheel on simple carpet (Fig. 4). Parameters were set to mimic our in-house four wheel skid steer rover's (SR2 [7]) geometry and loading. The test variables were wheel spin rates ($\omega = .3, .4,$ and $.5\ \frac{rad}{sec}$) and turn rates ($\Omega = 0, -.01,$ $-.02 \ldots\ldots -.12\ \frac{rad}{sec}$) with 5 trials of each. Post processing, of the data, was done with several C programs that averaged all trials, performed 2nd and 3rd order regression curve fitting, and calculated $\Omega_{SWEET Predicted}$.

 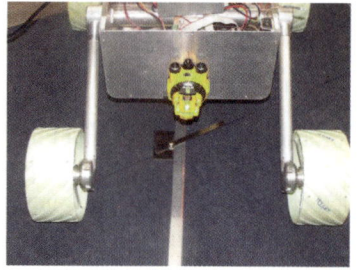

Fig. 4 a) SWEET Testbed; b) SR2 rover spin rate testing

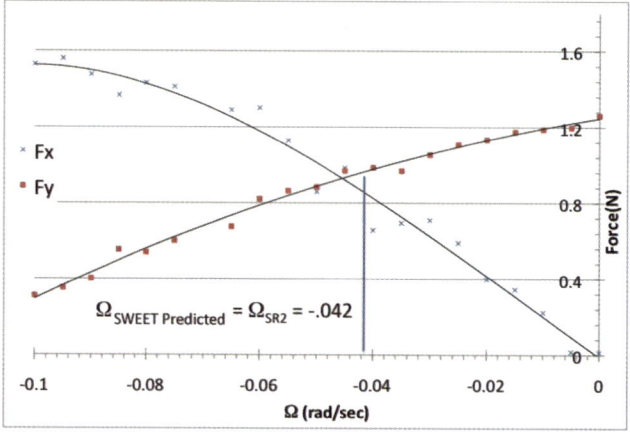

Fig. 5 Results for $\omega = .3\frac{rad}{sec}$

The results, for SWEET's skid steer turn test, are shown in Figs. 5, 6, and 7.

SR2 (Fig. 4) was then fitted with four blank wheels and turned on the same carpet to validate the results. Tests were done for three different wheel speeds ($\omega = .3, .4,$ and $.5\frac{rad}{sec}$) measuring the spin rate of the rover during the test (by measuring the angle between an onboard laser level mark and the initial position and dividing by the elapsed time), which are given in table 1 along with $\Omega_{SWEETPredicted}$ and percentage error. These results show a definite validation of the SWEET single wheel test within 3%.

Table 1 Ω_{SR2} and $\Omega_{SWEETPredicted}$ results in $\frac{rad}{sec}$

ω	Ω_{SR2}	$\Omega_{SWEETPredicted}$	Error
0.3	-.042	-.042	0%
0.4	-.056	-.057	1.8%
0.5	-.066	-.064	3.0%

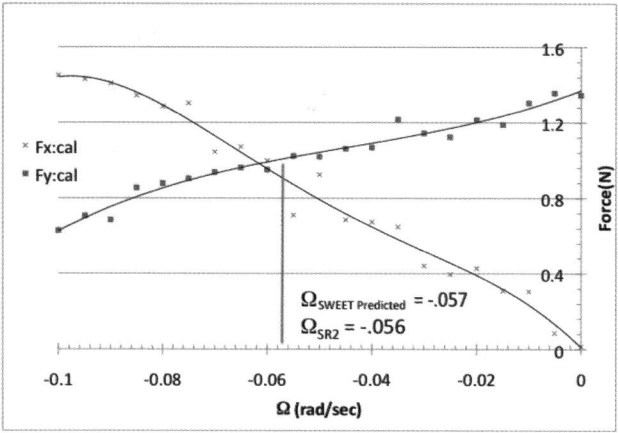

Fig. 6 Results for $\omega = .4\frac{rad}{sec}$

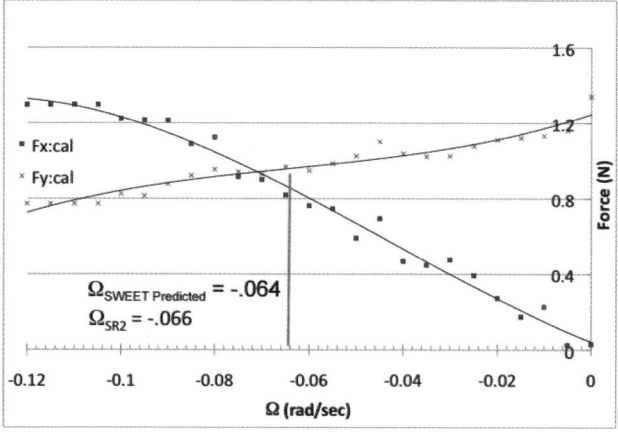

Fig. 7 Results for $\omega = .5\frac{rad}{sec}$

4 Skid Steer Experiments with Non-blank Wheels

In considering a non-blank wheel, particularly a directional patterned wheel such as Fig. 8 there is a possibility of a force along the X axis induced by the tread pattern. If the wheel is mounted on the correct side then the additional force will benefit the turning efficiency by offsetting the frictional force produced by the turn. The theoretical ideal turning rate for a directional treaded wheel has to include any V_x produced by the tread.

$$V_t = V_y \cos(\Theta) + V_x \sin(\Theta) . \tag{12}$$

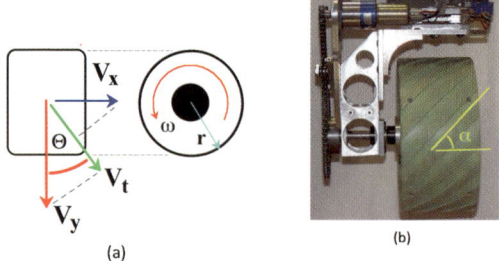

Fig. 8 a) Kinematic explanation of treaded wheel; b)Measuring α on a treaded wheel

$$V_y = \omega r \,. \tag{13}$$

$$V_t = \Omega R \,. \tag{14}$$

$$\Omega_{IDEAL} = \frac{1}{R}(\omega r \cos(\Theta) + V_x \sin(\Theta)) \,. \tag{15}$$

if Ω_{IDEAL} were related to the tread design only (such as a bolt screwing into a nut) and ignored any soil interaction V_x would be a function of ω, α, and r (equation 16 and Fig. 8). Which would give the Ω_{IDEAL} in equation 17.

$$V_x = \frac{\omega r}{\tan(\alpha)} \,. \tag{16}$$

$$\Omega_{IDEAL} = \frac{\omega r}{R}\left(\cos(\Theta) + \frac{\sin(\Theta)}{\tan(\alpha)}\right) \,. \tag{17}$$

Two directional patterned wheels, with diameter of .102 meters (Fig. 8b), were tested in SWEET. Figures 9 and 10 show the performance of the two oppositely patterned wheels dubbed 'left' and 'right' which correspond to their proper

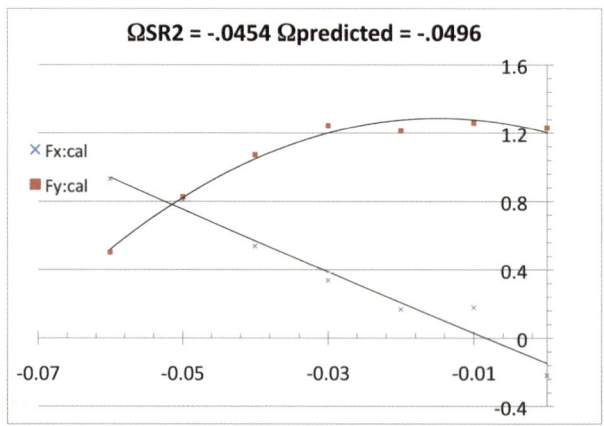

Fig. 9 Results for right treaded wheel rotating at $\omega = .3\frac{rad}{sec}$ in the right front position

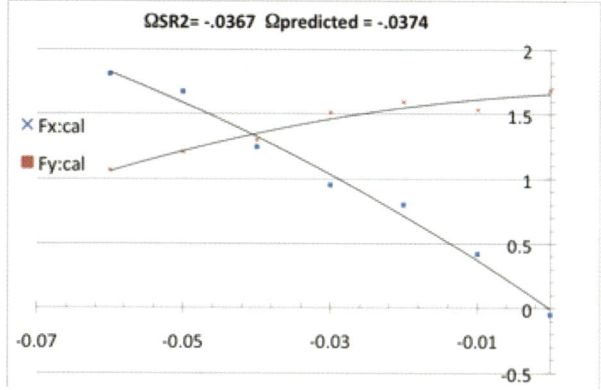

Fig. 10 Results for left treaded wheel rotating at $\omega = .3 \frac{rad}{sec}$ in the right front position

Table 2 Ω_{SR2} and $\Omega_{Predicted}$ results for treaded wheels

Wheel	Ω_{SR2}	$\Omega_{Predicted}$	Error
Left	-.0367	-.0374	1.9%
Right	-.0454	-.0496	9.25%

orientation on the rover. Again the tests were run simulating the right front side of a rover turning in a counter-clock-wise fashion. The tests were run on padded carpet, and not a hard surface, to focus on how the tread itself interacts with the surface and the treads affect on turning performance. Figure 10 shows the results of a left wheel in that position producing a $\Omega_{SWEETPredicted}$ value of -.037 $\frac{rad}{sec}$ while the right wheel gives a $\Omega_{Predicted}$ value of -.050 $\frac{rad}{sec}$ (table 2). The left wheel can be visualized as trying to screw itself to the right fighting against the turn when placed on the right side, the right wheel is trying to screw itself left benefiting the turn.

5 Conclusions

This paper discusses and demonstrates a method that allows the results from a single wheel test to be used to predict turning efficiency for a full assembly skid steer rover. Three different wheels were tested and predicted turn rates were within 10% of full assembly tests which probably can be refined by increasing the sample size. Future work will be to test on sand and other terrain, test and evaluate interesting wheel types, and iterate tread design on conventional wheels to better ascertain a wheel's performance on different media all of which without the cost and time of full assembly tests.

References

1. Flippo, D.: Design of rover wheel testing apparatus. In: ISAIRAS (2008)
2. Fukami, K., Ueno, M., Hashiguchi, K., Okayasu, T.: Mathematical models for soil displacement under a rigid wheel. Journal of Terramechanics 43(3 SPEC ISS), 287–301 (2006); Soil displacement; Soil bin test; Rigid wheel; Displacement increment; Gaussian function; Displacement locus; First derivative; Second derivative
3. Iagnemma, K., Shibly, H., Dubowsky, S.: A laboratory single wheel testbed for studying planetary rover wheel-terrain interaction. Technical report, MIT Field and Space Robotics Laboratory, 77 Massachusetts Avenue, Room 3-435a (January 2005)
4. Lindemann, R.: Platform for testing robotic vehicles on simulated terrain. Technical report, NASA, June 01 (2006)
5. Masami, U., Masaaki, O., Takeshi, S., Koichi, H., Takashi, O.: A precise prediction model of traveling performance for a rigid wheel on sandy ground. Journal of the Japanese Society of Agricultural Machinery 61(2), 101–110 (1999)
6. Melzer, K.J., Green, A.J.: Performance of the boeing lrv wheels in a lunar soil simulant. Technical Report 1, Mobility and Environmental Division U.S. Army Engineer Waterways Experiment Station (December 1971)
7. Roman, M.J.: Design and analysis of a four wheeled planetary rover. Master's thesis, University of Oklahoma (August 2005)
8. Shmulevich, I., Ronai, D., Wolf, D.: A new field single wheel tester. Journal of Terramechanics 33(3), 133–140 (1996)
9. Yoshida, K., Ishigami, G.: Steering characteristics of a rigid wheel for exploration on loose soil. In: 2004 IEEE/RSJ International Conference on Intelligent Robots and Systems (IROS 2004), Proceedings, September 28 - October 2, vol. 4, pp. 3995–4000 (2004)

Field Experiments in Mobility and Navigation with a Lunar Rover Prototype

David Wettergreen, Dominic Jonak, David Kohanbash, Scott Moreland, Spencer Spiker, and James Teza

Abstract. Scarab is a prototype rover for lunar missions to survey resources, particularly water ice, in polar craters. It is designed as a prospector that would use a deep coring drill and apply soil analysis instruments. Its chassis can transform to stabilize its drill in contact with the ground and can also adjust posture to ascend and descent steep slopes. Scarab has undergone field testing at lunar analogue sites in Washington and Hawaii in an effort to quantify and validate its mobility and navigation capabilities. We report on results of experiments in slope ascent and descent and in autonomous kilometer-distance navigation in darkness.

1 Introduction

To discover and measure the resources of the moon, robotic systems will have to survive extremes from blazing sunlight to frigid darkness as well as dust, vacuum, and isolation. Scarab is a prospecting rover developed to perform the necessary science operations to locate volatiles and validate *in situ* resource utilization methods. [5] (Fig. 1) It is a terrestrial concept vehicle designed to deploy a deep coring drill

Fig. 1 Scarab lunar rover prototype on unconsolidated sandy soil in eastern Washington

and to transport soil analysis instruments. The vehicle design employs a passive kinematic suspension with active posture adjustability. Its chassis can lower to stabilize a coring drill in contact with the ground and can also adjust to control

David Wettergreen
Carnegie Mellon University, 5000 Forbes Avenue. Pittsburgh, PA 15213,
e-mail: dsw@ri.cmu.edu

A. Howard et al. (Eds.): Field and Service Robotics 7, STAR 62, pp. 489–498.
springerlink.com   © Springer-Verlag Berlin Heidelberg 2010

roll, meaning rotation about its longitudinal axis, by independently adjusting its side-frames. This allows it to drive cross-slope and turn switchbacks to better ascend and descend unconsolidated soil.

Scarab is intended to operate on and within lunar craters, particularly in polar regions. Because the interior slopes and crater floor are sometimes in shadow, or in some cases in permanent darkness, active sensing methods are needed for terrain modeling and autonomous navigation. Scarab employs laser range scanners with autonomous navigation algorithms to build models of the surrounding terrain to detect obstacles and then determine efficient and safe paths.

In this paper we review results from field experiments at Moses Lake Dunes, Washington and Mauna Kea, Hawaii to measure and verify the prototype rover's ability to meet the demands of a lunar polar crater prospecting mission.

2 Rover Configuration

Scarab was conceived as a work machine with a serialized mission: drive, charge batteries, drill, charge again, analyze soil samples, charge and repeat. The number of repetitions might be 25, leading to 25 kilometers of traverse, 25 cores, and 25 sites surveyed. For some craters, 100 repetitions might be more desirable to characterize the environment and resources.

There are many factors effecting the rover configuration but the drill mechanism and its operation dominate. The requirement to transport and stabilize a deep coring drill literally became central to the design while requirements for ascent and descent in cratered terrain shaped many aspects and fine details.

Drilling requires a stiff platform into which thrust loads, torques and vibrations are transmitted and hole alignment is maintained. Placement of the drill in line with the vehicle's center-of-mass maximizes the mass that can be applied in down force. (Fig. 2) Drilling operations receive three benefits of this feature; first, lowering the chassis allows the full stroke of the drill to be used in the soil resulting in mass savings overall. Secondly, the rover can lean and therefore re-stabilize and place the rover center-of-mass over the drill core. Lastly, under low gravity conditions, the drill torques are counteracted by the increased leverage arm created by spreading the rover wheelbase.

The rationale for the vehicle weight and size is based on the 1 m long, 3 cm diameter drill that is likely to be employed in a lunar mission. Not only does the rover have to support the drill but also it must provide sufficient weight against which the drill can react its downward thrust and torque about the bit. Drill thrusts are expected to reach 250 N and 50 Nm torque. The system weight on lunar surface must react drilling 250 N downforce and maintain 150 N on wheels for stability against uplift and spin, therefore total weight on lunar surface must be greater than 400 N. The weight in lunar gravity (400 N / 1.622 m/s^2 = 250 kg) leads to a minimum 250 kg vehicle mass. [1]

The Scarab rover's chassis and suspension was designed around the drill. This component of the payload is significant in mass (50 kg) and imposes forces on the chassis during transport and while interacting with the ground.

Scarab's chassis allows it to passively conform to the terrain. The suspension has active and passive elements for improved traction on slope terrain. The active element, as previously discussed with respect to drilling, allows the rover to level the body, leading to increased stability and traction efficiency. The passive element, sometimes called an averaging (or differencing) linkage provides a mechanical release allowing the two rover suspension side-frames to pivot independently.

The averaging linkage ensures the body is pitch-averaged between the rocker arms. Scarab's body has three contact points. On either side, the body is connected to the pivot in the rocker arms. On top, the body hangs from the differencing linkage. This linkage runs across the top of the body and also connects to the rocker arms. Scarab actively controls its roll using the rocker arms by changing the height of each side independently, thus controlling

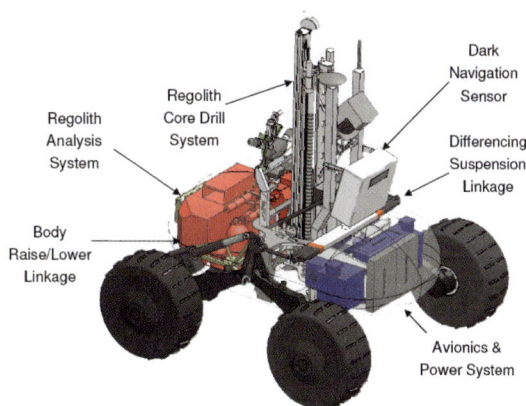

Fig. 2 Scarab rover configuration showing placement of sensors, avionics, and payload. There are drive motors in each of four wheels and two linkages for adjusting sideframe height. An averaging linkage allows all four wheels to maintain ground contact in rough terrain.

Table 1 Scarab Rover Specifications

Mass	280 kg
Weight	2740 N Earth surface
	450 N Lunar surface
Locomotion speed	3 - 6 cm/s
Wheel diameter	65 cm
Track width	1.4 m
Wheelbase	0.8 - 1.4 m
	1.2 m nominal
Aspect ratio (track/wheelbase)	1:1.0 low stance
	1:1.2 nominal stance
	1:1.7 high stance
CG planar location	On geometric center
CG height	0.48 m low stance
	0.64 m nominal stance
	0.74 m high stance
Static pitch-over	56° low stance
	43° nominal stance
	30° high stance
Static roll-over	61° low stance
	53° nominal stance
	49° high stance
Maximum straddle	0.55 m
Minimum straddle	0.00 m (ground contact)

the roll. In contrast the pitch is passive. Scarab's wheels conform to the terrain, which rotates the rocker arms and swivels the differencing linkage. The linkage is constructed such that the body is forced to move up or down by half the angle between the two rocker arms. As the center-of-mass of the rover is located midway between the side frames, equal loading occurs on all four wheels even on drastically uneven terrain.

3 Mobility Experiments

Testing Scarab in the field has been critical in proving the concept for lunar mobility and quantifying performance. Experiments have been conducted in numerous conditions with several findings of importance. However it is understood that continuing experimentation is needed to provide the data for a fully validated performance model and, most important, to enable extrapolation of terrestrial results to the lunar environment.

Moses Lake Sand Dunes in Washington was chosen as a test site for its varied terrain (slopes, pits, etc.), low moisture content, varied soil types (strengths, size distribution) and wide open space. These qualities provided grounds for mobility traction testing and long distance dark navigation traverses. Steep slope ascent/descent in loose soil and tests of new slope climbing techniques and algorithms were the focus of these field tests in June 2008.

Another lunar analogue site, located on Mauna Kea, Hawaii, is at high altitude with dry, deep, basaltic volcanic ash allows repeated mobility and navigation experiments. The soil composition and mechanical properties at this site were ideal for the regolith sampling hardware experiments. The objectives of these tests in November 2008 were to demonstrate roving, drilling, sample acquisition, processing and analysis. The rover was able to autonomously traverse kilometers of rough terrain, inspect a drill site, drill to 1 m depth, process the core samples and analyze the composition of the captured soil and demonstrate extraction of water from soil.

Characterization of Scarab as a system for difficult terrain mobility was first quantified in the laboratory in statics tests and in sandboxes. [1] The independently actuated rocker arms of the Scarab rover allow for actively controlled center-of-mass shifting. The JPL Sample Return Rover has similar capability [2]. Benefits of this feature include decreased slip during cross-slope maneuvers. Scarab was tested normal to the slope and leaning to maintain vertical posture with cross-slope of 10°, 15° and 20°. A surveying total station tracked a prism on the rover to millimeter accuracy and recorded instantaneous slip measurements.

The outcome, expressed as percentage downhill slip with respect to cross-slope distance, appears in Table 2.

The considerable decrease in downhill slip (2.5x at 20° incline) arises from increased traction due to equalization of wheel loading in highly cohesive soils and edging effects of the

Table 2 Downhill Slip

Slope	Normal	Leaning	Change
10°	6%	2%	-4%
15°	22%	8%	-14%
20°	37%	15%	-22%

wheel profile in frictional soils. The significance of this outcome lies in the ability to descend and navigate steeper slopes with while maintaining adequate control authority.

A widely used metric for measuring the total tractive ability of a vehicle is drawbar pull. This is the value the vehicle can pull in a specified material while maintaining forward progress. The maximum drawbar pull value corresponds to the inflection in the load/slip curve where the soil fails and the wheel enters the high slip regime. This value is informative when comparing different designs and also used for determining the slope a vehicle can ascend for the specified material. Drawbar pull experiments were conducted in Washington and Hawaii to evaluate of the effects of rover mass properties, wheel design, and soil properties on tractive performance. (Fig. 3) Both the drawbar pull value and power values derived from this test are used as metrics to determine performance.

The key observations are the range of tractive values that occur with changing soil properties. For high bearing strength materials, the level of looseness and compaction does result in slightly varied traction and power (shear strength and sinkage respectively). The overall mass also has little effect on the normalized drawbar pull value (percentage of vehicle weight) although with extremely low bearing strength materials, this does not hold true as a result of excessive sinkage. The shear strength comes from cohesiveness and internal friction. As a result, the drawbar pull values can be representative of slopes achievable for only highly cohesive material as the normal loading of the surface is constant throughout testing. The most significant effect on traction has resulted in wheel design. Experiments involving different traction surfaces, wheel diameter and ground pressures have shown a large range of drawbar pull values. Differences of 50% have been achievable through traction surface/grouser modifications. Lowering ground pressure and reducing sinkage has moderate effects on traction but results in large differences in driving power (up to 50% during experiments). Drawbar pull tests performed as lab and field experiments have highlighted wheel design as a leading element in tractive and power design requirements. (Table 3.) This is important because wheel design is generally independent of the suspension design and can be optimized for traction and power efficiencies.

Fig. 3 Drawbar pull experimental setup. Weight is added to the sled with the rover in motion while slip is continuously monitored.

Active control methods can also lead to increased tractive performance. Techniques such as "inch-worming" can increase the mobility

of a rover. (Fig. 4) To begin the cycle of inch-
worming, the body lowers
while expanding the
wheelbase and rolling the
front wheels forward while
the rear wheels remains
static. In the second half of
the cycle, the body raises
and the wheel base short-
ens while the rear wheels
rolls forward and the front
wheels remains static.
Non-rotating wheels pro-
vide a fixed point of reaction with no slipping. To
achieve these benefits,
Scarab's inch-worming al-

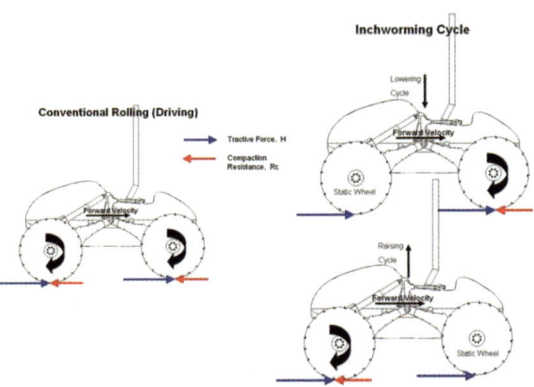

Fig. 4 Conventional rolling versus inch-worming
where one wheel pair is synchronized to the sidearm expansion/contraction and the other reacts
forces into the ground.

gorithm relies on eliminating the compaction resistances on two of the four
wheels, by remaining static with respect to ground, for a resulting net tractive increase. Experimentally we have found that the inch-worming technique is best
suited when wheels become entrenched under high slip. It allows the rover to
move forward (or back up) out of this situation.

Table 3

	Soil Depth	Lunar Wheel	Rubber Tire	Difference
Locomotion Power	7.5 cm	100W, 100W w/grouser	158W	0.58
	5.0 cm	95W	160W	0.68
	2.5 cm	95W	103W	0.08
	1 cm	98W	117W	0.19
Maximum Drawbar Pull	7.5 cm	23%, 32% w/grouser	28%	0.18, 0.28
	5.0 cm	24%	32%	0.25
	2.5 cm	32%	39%	0.18
	1 cm	33%	50%	0.34

Actively-positioned center-of-mass can also increase steepness of slopes traversable: distributing the load amongst the rover's wheels leads to more efficient
traction. Center-of-mass shifting (body leaning) was tested and heading slip, the
slip in the commanded direction with respect to the commanded velocity, showed
increase the steepness of slopes ascendable. The experiments were conducted
with a 25° angle attack from the horizontal. This value was determined

experimentally to have adequate uphill progress and low slip. It was shown that with the transformable suspension of the Scarab rover, slopes of 20° loose, dry, volcanic ash can be ascended with low risk.

4 Navigation Experiments

Scarab navigates autonomously on kilometer scales. A route planning algorithm generates intermediate goals (typically with 100 m spacing) and the operator may specify multiple goals, Scarab will reach each goal in order.

Scarab uses an on-board inertial measurement device (Honeywell HG1700) and optical ground speed sensor to enable it to estimate position and velocity with 1 - 3 % error on distance

Fig. 5 Scarab navigating in darkness. Laser scanner perceives terrain ahead and an underbody optical velocity sensor detects slip.

traveled. (Fig. 5) Laser ranging provides measurements to build models of the surrounding terrain to detect obstacles and then determine efficient and safe paths. (Fig. 6)

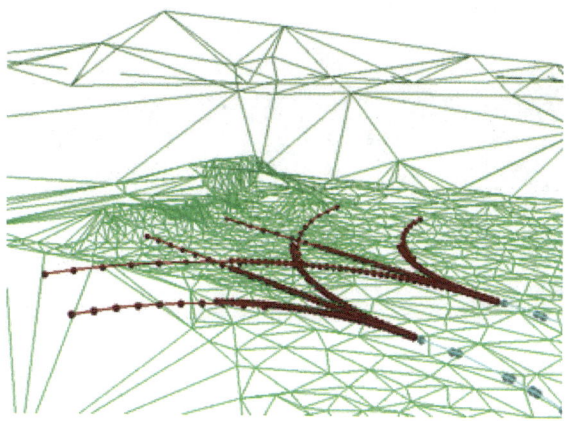

Fig. 6 Mesh terrain model used to represent obstacles and to evaluate and refine the path. Modeling and evaluation iterate to navigate to the goal.

Scarab periodically scans the terrain using a laser rangefinder developed by Neptec Design Group. [3] Previous autonomous rovers have used stereo cameras for high density terrain observations with low power. [7] In the scenario that Scarab addresses, polar lunar craters, there will be significant areas of slope and crater floor in shadow and in some cases, perpetual darkness so active sensing devices are required. Scarab acquires a scan after appreciable driving (more than 3 m) or turning (greater than 10°) or after time has elapsed (more than 100 sec). The sensor produces a dense array of ranges and takes several seconds, so motion must stop to avoid warping the data. The navigation algorithms assume a static world, meaning the terrain does not change between scans. Each 3D cloud of range points is incorporated into the terrain model. The range points are filtered (to remove noise and artifacts) and transformed into the vehicle's coordinate frame. Coarse data reduction on the point cloud is applied and the point cloud is transformed into a mesh. The mesh is then further reduced to eliminate redundant data. Finally the mesh is aligned with prior data to generate the terrain model that is used to identify obstacles and select the best path to the goal. Many candidate vehicle motions are evaluated in the near- and far-field. The near-field analysis involves simulating vehicle motion on the

Fig. 7 Autonomous descent into a large pit. Rendered views are terrain model with path of rover at rim, intermediate, and viewing the floor. Images show rover during descent. Scarab discovered a moderate slope and reached the floor autonomously.

mesh to identify collision and slope hazards and assess their severity. The far-field analysis applies heuristic search to estimate the progress each potential move will make toward the goal. A cost function combines safety in the near-field with progress in the far-field.

Our experimental approach has been to conduct 1 km traverses in a variety of terrains with progressive improvements to algorithms. At both the Moses Lake and Mauna Kea sites, Scarab autonomously completed the following objectives: travel over 3 km, perform 2 night traverses and simulate crater descent. Traverses are kilometer scale and performed after sunset, they account for most of the total distance traveled at each site. Crater descent was conducted with a long (100 m) traverse that included descending a steep (10°) slope.

Scarab completed a total of 3.6 km in 27 traverses in Washington. The first dark traverse was 1.2 km with 4 interventions due to sensor faults and one due to a controller error. These faults are recoverable; they do not jeopardize the rover and are easily resolved by resetting a device. A second dark traverse used an alternative navigation algorithm [4] and completed 1.1 km with two interventions due to localization errors.

Scarab traveled 3.0 km in 20 traverses at Mauna Kea, most of this was accomplished during the two overnight traverses. The first dark traverse was split into two parts; after 199 m the traverse was paused for logistic reasons and later Scarab resumed for an additional 779 m before stopping due to a software error. The second overnight traverse was also split in two; the first part was 312 m and stopped on a software error, the second was 989 m and ended with a CANBus fault. All of these errors are recoverable remotely.

At each site, Scarab autonomously completed a simulated crater descent using available analogue terrain. At Moses Lake, Scarab drove into a 9 m deep pit with 10°-20°sloped sides. (Fig. 7) This was safely accomplished including two undirected switchback maneuvers. At Mauna Kea, Scarab repeatedly drove down a winding drainage channel. The route was over 100 m long and descended over 25 m with an average grade of about 10°.

Traverse termination conditions for both field tests are shown in Chart 1. No interventions were required to stop the vehicle from driving into a hazard (zero emergency stops). At Moses Lake, most traverses (15 of 25) ended with a recoverable fault. On Mauna Kea the navigation method had improved and most traverses ended by reaching the goal (8 of 20) or

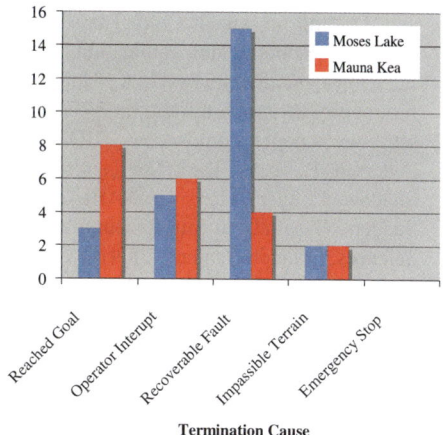

Chart 1 Termination conditions in autonomous navigation experiments.

stopping the traverse for other reasons (6 of 20). Recoverable faults are those that could be remotely corrected and thus would not be mission ending in a lunar scenario.

These results are far from perfect but indications are that reliability is improving and will reach the level of previous planetary rover prototypes.

5 Conclusion

The Scarab rover has been uniquely configured for the transport and stabilization of a coring drill and associated soil analysis instruments. The benefits of central-mounting and active body height and roll control are apparent in deployment of the drill and improved ability to ascend and descend cross-slope.

Field experimentation has quantified drawbar pull and slope climbing ability as well as power required for these activities under a variety of soil and terrain conditions. Field demonstrations have also proven the capability of the laser-based navigation system for kilometer-scale autonomous traverse, including autonomous descent into a crater. In total the mobility and navigation requirements for a lunar surface prospecting mission have been demonstrated in analogue terrain.

Acknowledgments. We gratefully acknowledge the contributions of our technical team and the assistance of Phillipe Ayoub, Paul Bartlett, Deborah Sigel, John Thornton, Chuck Whittaker, and William Whittaker. This research was supported by NASA under grants, NNX07AE30G, John Caruso, Project Manager, and NNX08AJ99G, Robert Ambrose, Program Scientist.

References

1. Bartlett, P., Wettergreen, D., Whittaker, W.: Design of the Scarab Rover for Mobility and Drilling in Lunar Cold Traps. In: International Symposium on Artificial Intelligence, Robotics and Automation in Space (iSAIRAS), Los Angeles (February 2008)
2. Iagnemma, K., Rzepniewski, A., Dubowsky, S., et al.: Mobile robot kine-matic reconfigurability for rough-terrain. In: SPIE 2000 (2000)
3. http://www.neptec.com/Neptec_TriDAR.html
4. Pedersen, L., et al.: Dark Navigation: Sensing and Rover Navigation in Per-manently Shadowed Lunar Craters. In: iSAIRAS, Los Angeles (February 2008)
5. Sanders, G., et al.: In Situ Resource Utilization (ISRU) Program. In: AIAA Aerospace Sciences, Orlando (January 2009)
6. Spudis, P.: Ice on the Moon. The Space Review (November 2006)
7. Wettergreen, D., et al.: Long-distance Autonomous Survey and Mapping in Robotic Investigation of Life in the Atacama. In: iSAIRAS, Los Angeles (February 2008)
8. Wettergreen, D., et al.: Design and Experimentation of a Rover Concept for Lunar Crater Resource Survey. In: AIAA Aerospace Sciences, Orlando (January 2009)

Rover-Based Surface and Subsurface Modeling for Planetary Exploration

Paul Furgale, Tim Barfoot, and Nadeem Ghafoor

Abstract. We develop and test a technique for the creation of coupled surface and subsurface models. Images from a stereo camera are used to estimate the motion of a rover that is collecting ground penetrating radar (GPR) data. The motion estimate and raw sensor data are used to build two novel data products: (1) A three-dimensional, photorealistic surface model coupled with a ribbon of GPR data, and (2) a two-dimensional, topography-corrected GPR radargram with the reconstructed surface topography plotted above. Each result is derived from only the onboard sensors of the rover, as would be required in a planetary exploration setting. These techniques were tested using data collected in a Mars analogue environment on Devon Island in the Canadian High Arctic. GPR transects were gathered over polygonal patterned ground similar to that seen by the Phoenix lander on Mars. Using the techniques developed here, scientists may remotely explore the interaction of the surface topography and subsurface structure as if they were on site.

1 Introduction

The use of ground penetrating radar (GPR) together with a stereo camera on planetary exploration rovers has been proposed several times [1, 7] and is now in development for the European Space Agency's (ESA) ExoMars project (2014) [22]. Used together, surface and subsurface imaging will aid in the search for liquid water and evidence of life. The ESA mission proposes using the stereo camera for site selection and survey, while the GPR will then be used to characterize the subsurface stratigraphy, and to select sites for drilling.

Paul Furgale and Tim Barfoot
University of Toronto Institute for Aerospace Studies
e-mail: {paul.furgale, tim.barfoot}@utoronto.ca

Nadeem Ghafoor
MDA Space Missions, Toronto
e-mail: Nadeem.Ghafoor@mdacorporation.com

A. Howard et al. (Eds.): Field and Service Robotics 7, STAR 62, pp. 499–508.

Despite this interest, there are still several open issues regarding the use of GPR on a rover platform:

1. Rovers must be able to deliver information about the surface (topography, substrate particle size distribution, and/or the presence of any existing outcrops) that enables the operator to give local geologic context to the subsurface data. The location of the GPR traverse must be known with respect to the surface data captured by the rover, so that the scientific interpretation of the data is as close as possible to a direct (human) site survey.
2. For a more complete interpretation of GPR data, the radargram (i.e., the two-dimensional subsurface profile) should be corrected for topography (e.g., [15]). As planetary exploration rovers have no access to a global positioning system (GPS) equivalent, topographic profiles must be generated using other onboard sensors.
3. A flight-ready GPR antenna must satisfy size, mass and power consumption constraints and the integration must minimize interference from the rover's motors and metal chassis.

This paper addresses items 1 and 2 by using stereo imagery to enhance the GPR data. Stereo cameras have been deployed on the Mars Exploration Rovers and are planned for both the Mars Science Laboratory (2011) and the ExoMars Mission (2014) [22]. Visual odometry (VO)[17, 20, 3, 11]—full 6-degree-of-freedom motion estimation using a stereo camera as the primary sensor—is central to the work described in this paper. Our visual odometry algorithm produces motion estimates with accuracy between 0.5% and 5.3% of distance traveled.

On Earth, producing a site survey using GPR on rough terrain involves several manual steps:

1. Place fiduciary markers (e.g., flags) along the intended transect and survey their locations (e.g., using differential global positioning (DGPS)).
2. Drag the antenna along the transect at a constant speed to collect many GPR traces, manually inserting a mark into the data to note the time at which the antenna passes each fiduciary marker.
3. Linearly interpolate these manually-generated markers to correct the horizontal spacing of the GPR traces along the transect.
4. If the surface is not flat, correct the vertical offset of the GPR traces using surface topography manually collected with a DGPS (Step 1).
5. Concatenate the corrected traces into a raster image called a *radargram*.

Our technique uses a VO estimate to fully automate this procedure. Further, we produce two novel data products that may be used to explore the interaction of surface topography and subsurface structure: (1) A three-dimensional, photorealistic surface model coupled with a ribbon of GPR data, and (2) a two-dimensional, topography-corrected GPR radargram with the reconstructed surface topography plotted above.

These techniques have been tested using data gathered at two sites near the Haughton-Mars Project Research Station (HMPRS) on Devon Island, Nunavut, Canada. The sites exhibit polygonally patterned ground, a periglacial landform often indicative of subsurface ice deposits [16]. Stereo images were captured during GPR transects and our integrated surface/subsurface modeling techniques were applied to the resulting data.

The rest of the paper is organized as follows. Our coupled surface and subsurface modeling system is described in Section 2. Sections 3 and 4 outline our field tests on Devon Island and the associated results. Our conclusions are provided in Section 5.

2   Integrated Surface and Subsurface Modeling

This section will describe our integrated surface/subsurface modeling system. Data flow through the main processing blocks of our system can be seen in Figure 1. The images captured from a calibrated stereo camera are first undistorted and rectified. This process accounts for lens distortion and aligns the images as if they came from perfect pinhole cameras with parallel optical axes.

Our algorithm uses Speeded-Up Robust Features (SURF)—an algorithm to detect and describe scale-and-rotation-invariant features [4]—for both *matching* (across stereo pairs) and *tracking* (over time). This is a class of feature pioneered by Lowe [14]. Lowe's Scale-Invariant Feature Transform (SIFT) algorithm has been used previously for object recognition [14], simultaneous localization and mapping [6], and visual odometry [3]. SURF is a similar algorithm that is much faster to compute because it uses integral images to approximate the operations used by SIFT to find and describe features. After two consecutive stereo pairs have been matched, features are tracked between frames. Feature descriptor matches between the consecutive left images are used as candidate tracks. We use a version of RANSAC [5] to simultaneously reject outlier feature tracks and produce a coarse motion estimate that is used to initialize our maximum likelihood solution.

Our maximum likelihood solution is similar to the one developed by Matthies [18] and deployed on the Mars Exploration Rovers [17]. At each timestep, N tracked features pass the outlier rejection stage. For each feature i, we triangulate the three-dimensional location of the point in each of the two consecutive stereo images. This results in a pair of points $\mathbf{p}_1^{i1}$ and $\mathbf{p}_2^{i2}$. As the world is assumed to be rigid, we now

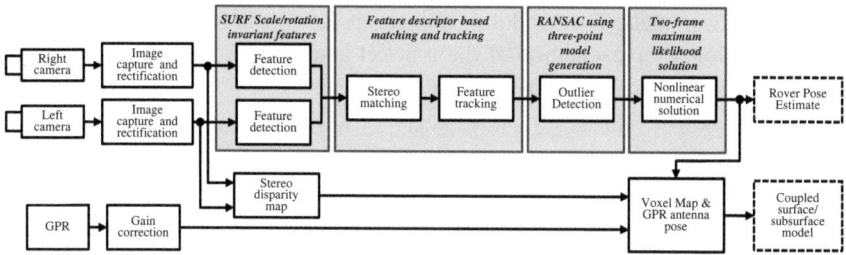

Fig. 1 An overview of the major processing blocks of our system.

seek the rotation $\mathbf{C}_{12}$ and translation ρ_1^{21} that align these two point clouds. This results in the objective function, J, which we seek to minimize:

$$J(\mathbf{C}_{12}, \rho_1^{21}) := \frac{1}{2} \sum_{i=1}^{N} \left(\mathbf{p}_1^{i1} - \left(\mathbf{C}_{12}\mathbf{p}_2^{i2} + \rho_1^{21}\right)\right)^T \mathbf{W}_i \left(\mathbf{p}_1^{i1} - \left(\mathbf{C}_{12}\mathbf{p}_2^{i2} + \rho_1^{21}\right)\right) \quad (1)$$

where $\mathbf{W}_i$ is a weighting matrix. We use the inverse covariance of $\varepsilon_i := \mathbf{p}_1^{i1} - \left(\mathbf{C}_{12}\mathbf{p}_2^{i2} + \rho_1^{21}\right)$ for $\mathbf{W}_i$, and thus J is a Mahalonabis distance, and finding the variables that minimize J also maximizes the joint likelihood of all the data. For further details, please refer to [8].

The motion estimates between each consecutive pair of images are then stacked up to give an estimate of the rover's entire traverse. As the robot is rigid, we obtain the transformation from the camera frame to the GPR frame through calibration, and so the visual odometry estimate also gives us the position of the GPR at each point along the traverse. Knowing the position of the camera and the GPR, we can transform all of the raw data into a common coordinate frame, $\mathscr{F}_0$. This gives us the following intermediate data products, all expressed in $\mathscr{F}_0$:

1. an estimate of the rover's position for each stereo image,
2. the sparse points used to compute the motion estimate,
3. larger point clouds for each stereo image obtained from dense stereo processing,
4. the position of the GPR at each data collection point.

These intermediate results are used to build the higher-level data products described below.

2.1 Three-Dimensional Surface and Subsurface Model

The first data product is a photorealistic, three-dimensional model of the surface, coupled with a model of the subsurface. Point sets derived from dense-stereo processing are aligned using the VO motion estimate [21]. The resulting point cloud is meshed and mapped with texture from the original images [2]. The GPR scan is modeled as a ribbon running under the surface mesh. The known transformation between the stereo camera and the GPR antenna is used to couple the surface and subsurface models. The resulting coupled model allows geologists to inspect a three-dimensional representation of the transect and explore the interaction of the surface morphology and the subsurface scan.

2.2 Two-Dimensional Topography-Corrected Radargram

The second data product is a two-dimensional, topography corrected radargram. The position and attitude of the antenna at each GPR trace is interpolated from the VO estimate. This estimate is used in place of the DGPS survey to perform both the horizontal correction and the vertical correction. The profile of the surface below the antenna is estimated by fitting a spline to feature locations along the transect as

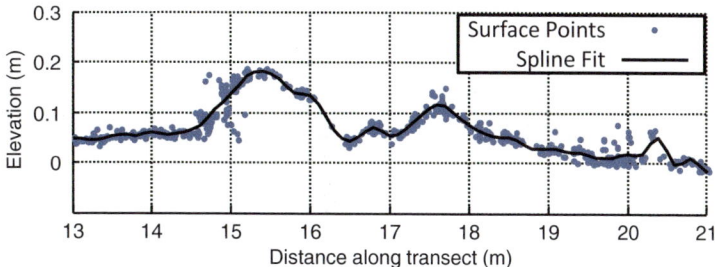

Fig. 2 Sparse surface points and the spline fit along the GPR antenna's path. This section is a polygon trough from transect *poly-2AS-1*.

shown in Figure 2. The spline improves on the topographic correction as it is able to capture narrow features over which the rover may drive.

3 Field Testing

The experiments described in this paper were conducted on Devon Island in the Canadian High Arctic, as part of the Haughton-Mars Project. The Haughton-Mars Project Research Station (HMPRS) is situated just outside the northwest area of the Haughton impact crater, which is located at 75°22′ N latitude and 89°41′ W longitude. Our experiments were conducted approximately 10 kilometers northeast of HMPRS near Lake Orbiter. This site was selected based on ongoing research into the polygonal terrain it hosts. Image sequences from the stereo camera and GPR data were logged at two sites:

1. The Lake Orbiter Transects: Five straight-line transects were taken at the Lake Orbiter site (Figure 3(a)). Each transect is approximately 60 meters long.
2. The Mock Rover Transect: One transect, approximately 357 meters long, at a site that had not been previously studied (Figure 3(b)).

In our experiments, a rover was simulated using a pushcart equipped with rover engineering sensors (i.e., stereo camera, inclinometers, sun sensor, wheel odometers), a ground-penetrating radar, an on-board computer, and two independent GPS systems (one Real-Time Kinematic) used for ground-truth positioning (see Figure 4). Although this was not an actuated rover, our focus in this work is on problems of estimation, and thus it was entirely sufficient as a means to gather data. The GPR (and cart) we used was a Sensors&Software Noggin 250 MHz system [1]. Efforts were made to minimize the effect of the rover body on the GPR data quality (e.g., using plastic parts where possible). The stereo camera was a Point Gray Research Bumblebee XB3 with a 24 cm baseline and 70° field of view, mounted approximately 1 m above the surface pointing downward by approximately 20°. Each image of the stereo pair was captured at 1280×960 pixel resolution.

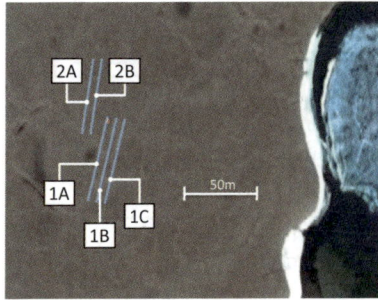

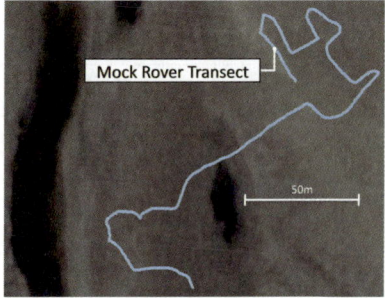

(a) The Lake Orbiter Transects.
$75°29'35''N$, $89°52'57''W$.

(b) The Mock Rover Transect.
$75°28'56''N$, $89°52'11''W$.

Fig. 3 Locations and transects on Devon Island, Nunavut, Canada used for field testing our integrated surface/subsurface modeling technique.

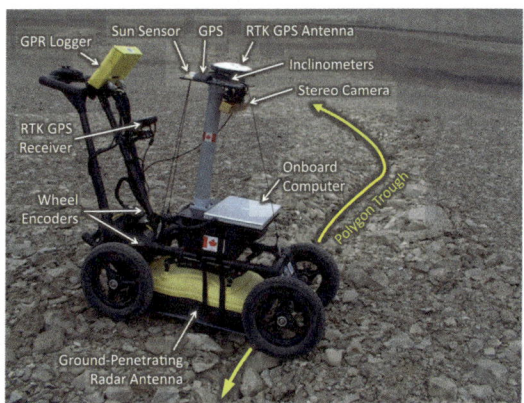

Fig. 4 The rover platform used for field testing.

4   Results

The visual odometry algorithm described in Section 2 was used to process all data collected at the Lake Orbiter site. We used a real-time kinematic GPS unit as ground-truth for our motion estimate. We determine the initial heading through a least-squares fit of the estimated track to the GPS for a small number of poses at the start of the traverse. These poses are then discarded and are not used when evaluating the linear position error. This is similar to the method used by [19].

The results are shown in Table 1, which lists the distance traveled and errors for all datasets collected. On the short Lake Orbiter transects (50 to 60 meters), position errors ranged from 0.5% to 5.3% of distance traveled. The results of the estimation on the Mock Rover Transect are plotted in Figure 4. This estimate accumulated 1.63% position error over this 357.3 meter traverse.

Our results approach those reported by other frame-to-frame VO algorithms [10, 20], and we believe this class of algorithm is suitable for applications such as this, which require a motion estimate over a short distances. To use VO over longer

Fig. 5 Track plots of GPS and the VO estimate for the 357 meter Mock Rover Transect.

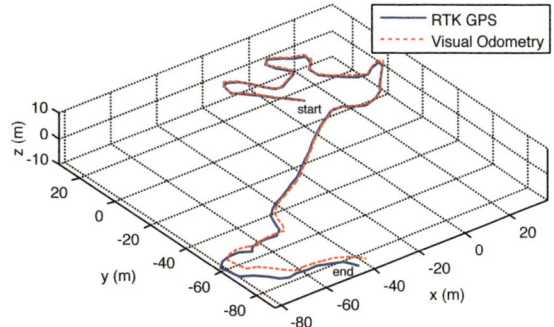

Table 1 Visual Odometry motion estimate results.

Transect	Distance Traveled (m)	Linear Error (m)	Percent Error	Number of Images
Mock Rover	357.30	5.83	1.63%	4818
poly-1AS-1	54.23	2.87	5.29%	333
poly-1BN-1	59.63	0.68	1.14%	316
poly-1BS-1	60.06	1.24	2.06%	317
poly-1CN-1	60.67	1.98	3.26%	327
poly-2AN-1	51.49	1.04	2.01%	270
poly-2AS-1	50.16	0.25	0.51%	263
poly-2BN-1	49.47	1.16	2.34%	260
poly-2BS-1	49.05	0.96	1.96%	258

distances, the work of Konolige et al. offers insights into increasing performance, albeit at the cost of increased computational complexity [11].

4.1 Coupled Surface/Subsurface Models

The complete coupled surface/subsurface model of the Mock Rover Transect is shown in Figure 6. The texture-mapped triangle mesh of the surface is displayed above the ribbon of GPR data. The model may be inspected using a Virtual Reality Modeling Language viewer and rendered from any viewpoint.

Polygonal terrain—a network of interconnected trough-like depressions in the ground—is a landform commonly found throughout the polar regions of both Earth [13] and Mars [12]. In terrestrial environments, these features are often indicative of subsurface ice bodies termed ice wedges [16]. As noted by Hinkel et al. [9], ice wedges "produce exceedingly complex, high amplitude hyperbolic reflections" (p.187) due to the conical shape of the emitted GPR pulse. As a result, while ice wedges themselves are roughly triangular in shape—wider at the top and progressively narrowing with depth—their appearance on a radargram more resembles an inverted hyperbola.

Figure 7 shows the entire corrected GPR radargram produced from data collected at the Mock Rover site. Points A-C illustrate three such examples of hyperbolic

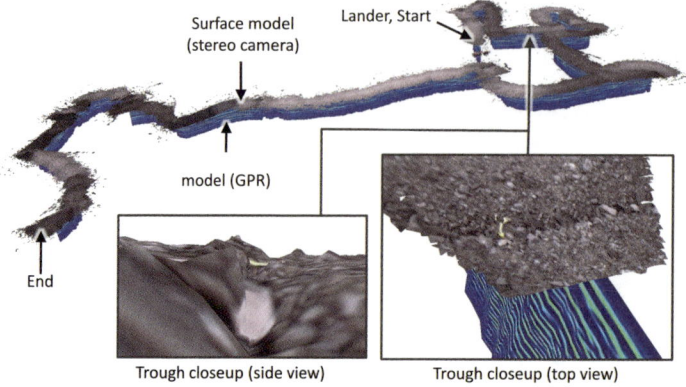

Fig. 6 Screenshots of the coupled surface/subsurface model.

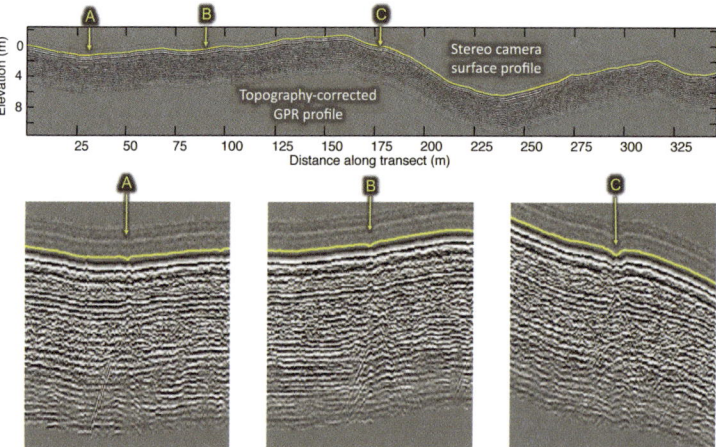

Fig. 7 GPR transect corrected for topography with surface profile plotted above.

subsurface reflections detected within the radargram. At these and other locations along the transect, the hyperbolic reflectors are found immediately beneath the troughs as indicated by small V-shaped depressions in the stereo camera surface profile. Because polygon troughs are the most obvious surface expression of ice wedge locations [16], the successful coupling of our surface/subsurface model is further supported.

5 Conclusion and Future Work

We have presented a coupled surface/subsurface modeling method for planetary exploration. Our method uses only a stereo camera and a ground penetrating radar unit to produce:

1. An estimate of the rover's trajectory over the course of the traverse.
2. A photorealistic three-dimensional surface/subsurface model.
3. Topography-corrected GPR traces plotted with a two-dimensional profile of the surface along the transect.

The models and corrections allow operators to work remotely, surveying the data as if they were on site. Currently, there are several manual steps in terrestrial GPR site surveying. Our approach allows GPR collection to be carried out in an automated manner, on a rover, thereby enabling planetary exploration. Subsurface stratigraphy can be examined in the context of the surface morphology, a key scientific technique used by field geologists to identify sites worthy of further study.

We collected our data in a Mars analogue environment at sites of scientific interest. Visual odometry estimates were produced for approximately 800 meters of traverse. The coupled models were generated using only the types of sensors that are slated to fly on future rover missions, such as ExoMars.

As mentioned in Section 1, a flight rover's metal chassis and interference from the motors may corrupt the GPR signal. Our future work will include a return to Devon Island in July, 2009 to address this issue by direct comparison of GPR data collected with and without an actuated rover platform.

Acknowledgements. Funding for this work was provided by The Canadian Space Agency's Canadian Analogue Research Network (CARN) program and the Natural Sciences and Engineering Research Council of Canada (NSERC). Gordon Osinski, Tim Haltigin, and Kevin Williams offered invaluable scientific guidance in the field and in production of the final data products. The Mars Institute and the Haughton-Mars Project provided infrastructure on Devon Island. Members of the communities of Resolute Bay, Grise Fjord, and Pond Inlet acted as guides, and protected us and our equipment from polar bears. Tom Lamarche from the Canadian Space Agency helped with our field testing. Peter Annan and David Redman from Sensors & Software Inc. helped us with equipment, and offered advice on the use of GPR in the field. Piotr Jasiobedzki and Stephen Se were instrumental in developing MDA Space Mission's Instant Scene Modeler and Ho-Kong Ng converted our motion estimates and images into surface models. Our algorithm used the SURF library developed by Herbert Bay, Luc Van Gool and Tinne Tuytelaars and available at `http://www.vision.ee.ethz.ch/~surf/`.

References

1. Barfoot, T., D'Eleuterio, G., Annan, P.: Subsurface surveying by a rover equipped with ground penetrating radar. In: Proc. IEEE/RSJ Int. Conf. Intell. Robot. Syst (IROS 2003), Las Vegas, NV, pp. 2541–2546 (2003)
2. Barfoot, T., Se, S., Jasobedzki, P.: Visual Motion Estimation and Terrain Modelling for Planetary Rovers. In: Intelligence for Space Robotics, ch. 4. TSI Press, Albuquerquem (2006)
3. Barfoot, T.D.: Online visual motion estimation using fastslam with sift features. In: Proc. IEEE/RSJ Int. Conf. Intell. Robot. Syst (IROS 2005), pp. 579–585 (2005)
4. Bay, H., Ess, A., Tuytelaars, T., Van Gool, L.: Speeded-up robust features (SURF). Comput. Vis. Image Underst. 110(3), 346–359 (2008)

 5. Fischler, M.A., Bolles, R.C.: Random sample consensus: a paradigm for model fitting with applications to image analysis and automated cartography. Commun. ACM 24(6), 381–395 (1981)
 6. Folkesson, J., Christensen, H.: Sift based graphical slam on a packbot. In: Int. Conf. Field Serv. Robot. - FSR 2007 (2007)
 7. Fong, T., Allan, M., Bouyssounouse, X., Bualat, M., Deans, M., Edwards, L., Fluckiger, L., Keely, L., Lee, S., Lees, D., To, V., Utz, H.: Robotics site survey at haughton crater. In: Proc. 9th Int. Symp. Artif. Intell., Robot. Autom. Space (iSAIRAS), Los Angeles, CA (2008)
 8. Furgale, P., Barfoot, T., Ghafoor, N., Haltigin, T., Williams, K., Osinski, G.: Field testing of an integrated surface/subsurface modeling technique for planetary exploration (2008) (in Preparation)
 9. Hinkel, K., Doolittle, J., Bockheim, J., Nelson, F., Paetzold, R., Kimble, J., Travis, R.: Detection of subsurface permafrost features with ground-penetrating radar, barrow, alaska. Permafr. and Periglac. Process. 12, 179–190 (2001)
10. Howard, A.: Real-time stereo visual odometry for autonomous ground vehicles. In: IEEE/RSJ Int. Conf. on Intell. Robot. and Syst., IROS 2008, pp. 3946–3952 (2008), doi:10.1109/IROS.2008.4651147
11. Konolige, K., Agrawal, M., Solà, J.: Large scale visual odometry for rough terrain. In: Proc. Int. Symp. Res. Robot, ISRR (2007)
12. Kuzmin, R., Zabalueva, E.: Polygonal terrains on mars: Preliminary results of global mapping of their spatial distribution. In: Lunar and Planet. Sci. XXXIV, p. Abstract 1912 (2003)
13. Lachenbruch, A.: Mechanics of thermal contraction cracks and ice-wedge polygons in permafrost. Special paper to the Geological Society of America 70, 69 (1962)
14. Lowe, D.G.: Distinctive image features from scale-invariant keypoints. Int. J. Comput. Vis. 60(2), 91–110 (2004)
15. Lunt, I.A., Bridge, J.S.: Evolution and deposits of a gravelly braid bar, Sagavanirktok river, Alaska. Sedimentol 51(3), 415–432 (2004)
16. Mackay, J.: Periglacial features developed on the exposed lake bottoms of seven lakes that drained rapidly after 1950, Tuktoyaktuk Peninsula area, Western Arctic coast, Canada. Permafr. Periglacial Process. 10, 39–63 (1999)
17. Maimone, M., Cheng, Y., Matthies, L.: Two years of visual odometry on the Mars Exploration Rovers. J. Field Robot. 24(3), 169–186 (2007), Maimone, Mark Cheng, Yang Matthies, Larry
18. Matthies, L.: Dynamic stereo vision. Ph.D. thesis, Carnegie Mellon University Computer Science Department (1989); Chairman-Steven A. Shafer
19. Nistér, D., Naroditsky, O., Bergen, J.: Visual odometry. In: Proc. 2004 IEEE Comput. Soc. Conf. Comput. Vis. Pattern Recognit (CVPR 2004), vol. 01, pp. 652–659 (2004)
20. Nistér, D., Naroditsky, O., Bergen, J.: Visual odometry for ground vehicle applications. J. Field Robot. 23(1), 3 (2006)
21. Se, S., Jasiobedzki, P.: Stereo-vision based 3d modeling and localization for unmanned vehicles. Int. J. Intell. Control Syst. 13(1), 47–58 (2008), Special Issue on Field Robotics and Intelligent Systems
22. Vago, J., Gardini, B., Kminek, G., Baglioni, P., Gianfiglio, G., Santovincenzo, A., Bayón, S., van Winnendael, M.: ExoMars - searching for life on the Red Planet. ESA Bull. 126, 16–23 (2006)

Author Index

Springer Tracts in Advanced Robotics

Edited by B. Siciliano, O. Khatib and F. Groen

Further volumes of this series can be found on our homepage: springer.com

Printing: Ten Brink, Meppel, The Netherlands
Binding: Stürtz, Würzburg, Germany